AF598455

RODD'S CHEMISTRY OF CARBON COMPOUNDS

ELSEVIER PUBLISHING COMPANY
335 JAN VAN GALENSTRAAT
P.O.BOX 211, AMSTERDAM, THE NETHERLANDS

ELSEVIER PUBLISHING CO. LTD.
BARKING, ESSEX, ENGLAND

AMERICAN ELSEVIER PUBLISHING COMPANY, INC.
52 VANDERBILT AVENUE, NEW YORK, N.Y. 10017

LIBRARY OF CONGRESS CARD NUMBER: 64-4605

ISBN 0-444-40775-8

WITH 2 TABLES

PRINTED IN THE NETHERLANDS

RODD'S CHEMISTRY OF CARBON COMPOUNDS

ADVISORS

Professor Sir ROBERT ROBINSON, O.M., M.A. (Oxon.), D.SC. (Manc.), HON. D.SC. (Lond., Liv., Wales, Dunelm, Sheff., Belfast, Bris., Oxon., Nott., Strath., Delhi, Sydney, Zagreb), HON. SC.D. (Cantab.), HON. LL.D. (Manc., Edin., Birm., St. Andrews, Glas., Liv.), HON. D. PHARM. (Madrid and Paris), HON. F.R.S.E., F.R.S., *London*

Chairman

Professor A. R. BATTERSBY, M.SC. (Manc.), PH.D. (St. Andrews), D.SC. (Bris.), F.R.S., *Cambridge*

Professor R. N. HASZELDINE, M.A., PH.D., SC.D. (Cantab.), PH.D., SC.D. (Birm.), F.R.I.C., *Manchester*

Professor R. D. HAWORTH, D.SC., PH.D. (Manc.), B.SC. (Oxon.), F.R.I.C., F.R.S., *Sheffield*

Professor Sir EDMUND HIRST, M.A., PH.D. (St. Andrews), D.SC. (Birm.), HON. LL.D. (St. Andrews, Aberdeen, Birm., Strath.), HON. D.SC. (Dublin), F.R.I.C., F.R.S., *Edinburgh*

(The late) Dr. E. H. Rodd, D.SC. (Lond.), F.C.G.I., F.R.I.C., *Newton Abbot, Devon*

Professor Lord TODD, M.A. (Cantab.), D.SC. (Glas.), D. PHIL. (Oxon.), DR. PHIL. NAT. (Frankfurt), HON. LL.D. (Glas., Edin., Melb., Calif.), HON. DR. RER. NAT. (Kiel), HON. D. MET. (Sheff.), HON. D.SC. (Oxon., Dunelm, Lond., Exe., Leic., Liv., Adel., Alig., Madrid, Stras., Wales, Strath.), F.R.I.C., F.R.S., *Cambridge*

RODD'S CHEMISTRY OF CARBON COMPOUNDS

VOLUME I

GENERAL INTRODUCTION

ALIPHATIC COMPOUNDS

★

VOLUME II

ALICYCLIC COMPOUNDS

★

VOLUME III

AROMATIC COMPOUNDS

★

VOLUME IV

HETEROCYCLIC COMPOUNDS

★

VOLUME V

MISCELLANEOUS

GENERAL INDEX

★

RODD'S CHEMISTRY OF CARBON COMPOUNDS

A modern comprehensive treatise

SECOND EDITION

Edited by

S. COFFEY

M.Sc. (London), D.Sc. (Leyden), F.R.I.C.

formerly of

I.C.I. Dyestuffs Division, Blackley, Manchester

VOLUME II PART E

STEROIDS

(continued from Vol. II Part D)

CUMULATIVE INDEX VOL. II PARTS A–E

ELSEVIER PUBLISHING COMPANY

AMSTERDAM LONDON NEW YORK

1971

CONTRIBUTORS TO THIS VOLUME

J. ELKS, D.SC., F.R.I.C.
Glaxo Research Ltd., Greenford, Middlesex

T. W. GOODWIN, D.SC., F.R.I.C., F.R.S.,
Department of Biochemistry, The University, Liverpool, L69 3BX.

R. E. FAIRBAIRN, B.SC., PH.D., F.R.I.C.,
formerly of Research Department, Dyestuffs Division,
I.C.I. Ltd., Manchester 9 (Index).

PREFACE TO VOLUMES IID AND IIE

The "Isoprene Story" referred to in the prefaces of Volumes IIB and IIC is continued to its climax in these last two sub-volumes of Volume II, which deal systematically with the chemistry of the steroids, the extensive range of widely occurring, natural substances, possessing a tetracyclic carbon skeleton, which play such a profound role in biological processes. The mode of presentation adopted in the first Edition, by Professor C. W. Shoppee and Mrs. Eileen Shoppee, has again been followed and the new authors of Chapters 15, 16, 17 and 18, who have freely drawn on the Shoppee text, desire that their indebtedness to these original authors be duly acknowledged. The new ideas and techniques now being applied to the examination of natural products and the elucidation of their structure, have, in this series also provided an enormous amount of new knowledge, which has been assessed and fitted into the original framework. In spite of the huge amount of work expended on this field in the last two or three decades, research on steroids, particularly the synthetic, biochemical, biological and pharmacological aspects, is still very topical and currently of great interest.

Before discussing the actual contents of these volumes something should be said about steroid nomenclature. In the earlier period, as various steroids were isolated and characterised and allotted trivial names, their chemistry and inter-relationships being imperfectly understood, there gradually evolved a rather unsystematic nomenclature. This still persists to some extent in current literature but is gradually being superseded by the more rational system based on a limited number of parent or fundamental compounds laid down in the IUPAC Rules, the 1951 and 1957 versions of which form the basis of the present text. These rules are under constant supervision and certain modifications in nomenclature are proposed in IUPAC/IUB Revised Tentative Rules for Steroid Nomenclature (1967), which were published in 1968 as the present account was going to press. These Tentative Rules are likely to be approved officially and they are printed in full, for reference, as an appendix to Volume IID.

There are three chapters in Volume IID: Chapter 15 deals with the Sterols and with Bile Acids and related compounds, Chapter 16 with Sex Hormones and Adrenocortical Hormones, and Chapter 17 with Cardiotonic Glycosides and Aglycons and with Toad Poisons.

Volume IIE completes the story and comprises two chapters, Chapter 18

dealing with Steroid Saponins and Sapogenins, and Chapter 19 with an account of the present state of work on the biogenesis of terpenes and steroids, currently the most topical aspect of steroid chemistry.

Volume II E concludes with the Cumulative Index for Volume II (Parts A–E).

This preface to the last part of Volume II must end on a sad note and record with profound regret the death of Dr. E. H. Rodd on July 22nd, 1970. All those who were involved in the gigantic task of completing the first edition of the Chemistry of Carbon Compounds remember him with both great respect and lasting affection.

September 1970 S. Coffey

CONTENTS

PREFACE TO VOLUMES IID AND IIE IX
OFFICIAL PUBLICATIONS; SCIENTIFIC JOURNALS AND PERIODICALS XIII
LIST OF COMMON ABBREVIATIONS AND SYMBOLS USED XIV

Chapter 18: Steroid Saponins and Sapogenins

J. ELKS

1. The saponins 1
2. The sapogenins 6
   a. The nucleus 7
   b. The side-chain 7

   (i) Pseudosapogenins and cyclopseudosapogenins, 9 – (ii) Normal and iso-sapogenins: configuration at $C_{(25)}$, 11 – (iii) Configuration of the sapogenins and cyclopseudosapogenins at $C_{(20)}$ and $C_{(22)}$, 14 – (iv) Nomenclature, 18 – (v) Reactions of the sapogenin side-chain, 19 – (vi) Synthesis of the sapogenin side-chain, 21 – (vii) Biogenesis of the steroid sapogenins, 23

   c. Individual sapogenins 29

   (i) Sapogenins with normal side-chains, 29 – (ii) Sapogenins with modified side-chains; kryptogenin and its congeners, 43 – (iii) Sapogenins bearing hydroxyl groups in Ring F, 46 – (iv) Sapogenins with Ring F unsaturated, 51 – (v) Aminospirostans, 52

Chapter 19: The Biogenesis of Terpenes and Steroids

T. W. GOODWIN

1. Introduction 54
2. Biosynthesis of isopentenyl pyrophosphate and dimethylallyl pyrophosphate 55
3. Hemiterpenes 61
   a. Classification 61

   (i) Free hemiterpenes, 61 – (ii) Phenolic hemiterpenes, 62 – (iii) Alkaloids with hemiterpene residues, 64 – (iv) Isopentenyl adenosine and related compounds, 64

   b. Biosynthesis of hemiterpenoids 65
4. Monoterpenes 66
   a. The basic precursor and classification 66

   (i) Acyclic monoterpenes, 67 – (ii) Monocyclic monoterpenes, 68 [1. Cyclohexanes, 68 – 2. Cyclopropanes, cyclopentanes and cycloheptanes, 68] –

(iii) Bicyclic monoterpenes, 69 – (iv) Tricyclic monoterpenes, 69 – (v) Phenolic monoterpenoids, 70 – (vi) Alkaloids with monoterpene residues, 70 – (vii) Atypical monoterpenes, 70

b. Biochemical studies on monoterpene biosynthesis 71
c. Biogenetic considerations 76

(i) General, 76 – (ii) Oxidations, 76 – (iii) Aromatization, 77 – (iv) Cycloheptanes, 78 – (v) Furan rings, 78

5. Sesquiterpenes 79
a. The basic unit 79
b. Classification 80
c. Biosynthetic studies 81
d. Biogenetic relationships 84
6. Diterpenes 85
a. The basic unit and classification 85

(i) Monocyclic diterpenes, 86 – (ii) Bicyclic diterpenes, 86 – (iii) Tricyclic diterpenes, 86, – (iv) Tetracyclic diterpenes, 87 – (v) Diterpene derivatives, 87

b. Biosynthetic studies 88
7. Sesterterpenes 92
8. Triterpenes 93
a. The basic unit 93
b. Biosynthesis of tetracyclic triterpenoids 94

(i) In animals, 94 – (ii) Synthesis in higher plants and micro-organisms, 103 – (iii) Mechanism of alkylation, 108 – (iv) Formation of $C_{(22)}$ double bond in plant sterols, 109 – (v) $C_{(21)}$ Sterols in plants, 110 [1. Pregnane derivatives, 110 – 2. Cardenolides, 111] – (vi) $C_{(18)}$ Sterols in plants, 112 – (vii) Spirostanols $C_{(27)}$ in plants, 112 – (viii) Bufadienolides, 113 – (ix) Ecdysones, 113 – (x) Compounds other than sterols with the cyclopentenophenanthrene nucleus, 115

c. Biosynthesis of pentacyclic triterpenes 116
d. Degraded triterpenes 120
9. Tetraterpenoids 120
a. Biosynthesis of tetraterpenoids 120

(i) Formation of cyclic carotenoids, 126 – (ii) Insertion of oxygen, 126 – (iii) Degradation products of β-carotene, 132

10. Polyterpenoids 133
11. Terpenoid quinones 134
12. Regulation of terpenoid synthesis in seedlings 135

Cumulative Index Vols. IIA–E 139

VOLUMES IIA, B, C AND D

Vol. IIA Monocarbocyclic compounds C_3–C_5

Vol. IIB Six- and higher-membered monocyclic compounds

Vol. IIC Polycarbocyclic compounds, excluding steroids

Vol. IID Steroids (continued in Vol. IIE)

OFFICIAL PUBLICATIONS

B.P.	British (United Kingdom) Patent
F.P.	French Patent
G.P.	German Patent
Sw.P.	Swiss Patent
U.S.P.	United States Patent
B.I.O.S.	British Intelligence Objectives Sub-Committee Reports, H.M. Stationery Office, London.
C.I.O.S.	Combined Intelligence Objectives Sub-Committee Reports
F.I.A.T.	Field Information Agency, Technical Reports of U.S. Group Control Council for Germany
B.S.	British Standards Specification
A.S.T.M.	American Society for Testing and Materials

SCIENTIFIC JOURNALS AND PERIODICALS

With few obvious and self-explanatory modifications the abbreviations used in references to journals and periodicals comprising the extensive literature on organic chemistry, are those used in the World List of Scientific Periodicals.

LIST OF COMMON ABBREVIATIONS AND SYMBOLS USED

A	acid
Å	Ångström units
Ac	acetyl
a	axial
as	asymmetrical
at.	atmosphere
B	base
Bu	butyl
b.p.	boiling point
C, mC and μC	curie, millicurie and microcurie
c, *C*	concentration
conc.	concentrated
crit.	critical
D	Debye unit, 1×10^{-18} e.s.u.
D	dissociation energy
D	dextro-rotatory; dextro configuration
DL	optically inactive (externally compensated)
d	density
dec. or decomp.	with decomposition
deriv.	derivative
E	energy; extinction; electromeric effect
E1, E2	uni- and bi-molecular elimination mechanisms
E1cB	unimolecular elimination in conjugate base
E.S.R.	electron spin resonance
Et	ethyl
e	nuclear charge; equatorial
f	oscillator strength
f.p.	freezing point
G	free energy
G.L.C.	gas liquid chromatography
g	spectroscopic splitting factor, 2.0023
H	applied magnetic field; heat content
h	Planck's constant
Hz	hertz
I	spin quantum number; intensity; inductive effect
I.R.	infrared
J	coupling constant in N.M.R. spectra
K	dissociation constant
k	Boltzmann constant; velocity constant
kcal.	kilocalories
L	laevorotatory; laevo configuration
M	molecular weight; molar; mesomeric effect
Me	methyl
m	mass; mole; molecule; *meta*-
ml	millilitre
m.p.	melting point

Ms	mesyl (methanesulphonyl)
[M]	molecular rotation
N	Avogadro number; normal
N.M.R.	nuclear magnetic resonance
n	normal; refractive index; principal quantum number
o	*ortho-*
O.R.D.	optical rotatory dispersion
P	polarisation; probability; orbital state
Pr	propyl
Ph	phenyl
p	*para-*; orbital
P.M.R.	proton magnetic resonance
R	clockwise configuration
S	counterclockwise config.; entropy; net spin of incompleted electronic shells; orbital state
S_N1, S_N2	uni- and bi-molecular nucleophilic substitution mechanisms
S_Ni	internal nucleophilic substitution mechanisms
s	symmetrical; orbital
sec	secondary
soln.	solution
T	absolute temperature
Tosyl	*p*-toluenesulphonyl
Trityl	triphenylmethyl
t	time
temp.	temperature (in degrees centigrade)
tert	tertiary
U	potential energy
U.V.	ultraviolet
v	velocity
α	optical rotation (in water unless otherwise stated)
[α]	specific optical rotation
α_A	atomic susceptibility
α_E	electronic susceptibility
ε	dielectric constant; extinction coefficient
μ	microns (10^{-4} cm); dipole moment; magnetic moment
μ_B	Bohr magneton
μg	microgram (10^{-6} g.)
λ	wavelength
ν	frequency; wave number
χ, χ_d, χ_μ	magnetic, diamagnetic and paramagnetic susceptibilities
$\sim$	about
(+)	dextrorotatory
(—)	laevorotatory
$\ominus$	negative charge
$\oplus$	positive charge

Chapter 18

Steroid Saponins and Sapogenins

J. ELKS

The steroid sapogenins are a group of C_{27} compounds having the carbon skeleton of cholesterol, but oxygenated at carbon atoms 16, 22, and 26; in all but a few examples, this oxygenation takes the form of 16, 22- and 22, 26-oxide bridges in the characteristic spiroacetal grouping:

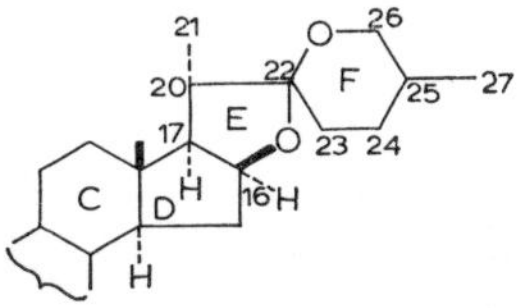

Almost without exception, they carry a hydroxyl group at $C_{(3)}$ and most of them are also substituted at other positions in the steroid nucleus or side-chain by hydroxyl or, less commonly, by oxo-groups.

The sapogenins are present as glycosides, the saponins, in a variety of plants. They achieved economic importance with the discovery, by Russell E. Marker, of a simple method for their conversion into 20-oxopregnanes and, in particular, of an efficient preparation of progesterone from diosgenin. A few sapogenins were later found to be valuable starting materials for the manufacture of cortisone and related compounds, and this discovery stimulated the search for new sapogenins and new sources of existing ones. Diosgenin and hecogenin are probably the only two that are used on a considerable scale, but they must largely have displaced the starting materials used in the earlier commercial preparations of the cortical hormones and their analogues.

1. The saponins

As their name implies, the saponins give soapy solutions, and crude plant extracts containing them have found use as detergents and for the production

TABLE 1.

SAPONINS

Name	Source*	M.P.	$[\alpha]_D$†	Sapogenin	Sugars	Refs.
Digitonin	*Digitalis purpurea* (seeds) *D. lanata* (seeds)	244–285°	—40°	Digitogenin	D-Glucose (2) ‡ D-Galactose (2) D-Xylose (1)	1, 2, 3
Desglucodigitonin	*Digitalis purpurea* (seeds)	—	—	Digitogenin	D-Glucose (1) D-Galactose (2) D-Xylose (1)	3
Gitonin	*Digitalis purpurea* (seeds) *D. germanicum*	240–300°	—51°	Gitogenin	D-Glucose (1) D-Galactose (2) D-Xylose (1)	3, 4
F-Gitonin	*Digitalis purpurea* (leaves)	251–255°	—66°	Gitogenin	D-Glucose (2) ‡ D-Galactose (1) D-Xylose (1)	5
Tigonin	*Digitalis lanata* (leaves) *D. purpurea* (seeds)	—	—	Tigogenin	D-Glucose (2) D-Galactose (2) D-Xylose (1)	3, 6
Lanatigonin I	*Digitalis lanata* (seeds)	275–285°	+42·5°	Tigogenin	D-Glucose (2) ‡ D-Galactose (2) D-Xylose (1)	7
Amolonin	*Chlorogalum pomeridianum*	—	—75·5°	Tigogenin	D-Glucose (3) D-Galactose (1) L-Rhamnose (2)	8
Digalonin	*Digitalis purpurea* (seeds)	250–295°	—	Digalogenin	D-Glucose (2) D-Galactose (2) D-Xylose (1)	3, 9
Dioscin	*Dioscorea tokoro* Makino	275–277°	—115° (C_2H_5OH)	Diosgenin	D-Glucose (1) ‡ L-Rhamnose (2)	10, 11
Trillarin	*Trillium erectum*	211°	—116° (C_2H_5OH)	Diosgenin	D-Glucose (2)	12

(*continued*)

Table I (*continued*)

Name	Source*	M.P.	$[\alpha]_D$†	Sapogenin	Sugars	Refs.
Trillin	*Trillium erectum* *Dioscorea sativa*	275°	—103° (Dioxan)	Diosgenin	D-Glucose (1)‡	13, 14
Trillioside	*Trillium grandiflorum*	315–320°	—	Diosgenin	D-Glucose (2) L-Rhamnose (1)	15
Gracillin	*Dioscorea gracillima*	290–293°	—88°	Diosgenin	D-Glucose (2)‡ L-Rhamnose (1)	11, 16
Kikuba-saponin	*Dioscorea septemloba*	249–251°	—77° (CH_3OH)	Diosgenin	D-Glucose (3) L-Rhamnose (1)	11, 17
Sarsasaponin	*Radix sarsaparillae* *Yucca schottii* (fruit)	245°	—	Sarsasapogenin	D-Glucose (2) L-Rhamnose (1)	18, 19
Parillin	*Radix sarsaparillae*	220–223°	—64° (C_2H_5OH)	Sarsasapogenin	D-Glucose (3)‡ L-Rhamnose (1)	20
Sarsaparilloside	*Radix sarsaparillae*	—	—44°	Sarsasapogenin	D-Glucose (4)‡ L-Rhamnose (1)	21
Yononin	*Dioscorea Tokoro* Makino	238–240°	—14·5°	Yonogenin	L-Arabinose (1)‡	22, 23
Tokoronin	*Dioscorea Tokoro* Makino	270–274°	—10·6°	Tokorogenin	L-Arabinose (1)	22
Hispidin	*Solanum hispidum* Pers.	312–315°	—64°	Hispidogenin	L-Rhamnose (2)	24
Convallasaponin-A	*Convallaria Keisukei*	238–240°	—40° ($CHCl_3$)	Convallagenin-A	L-Arabinose (1)	25
Convallasaponin-B	*Convallaria Keisukei*	273–274°	—56° ($CHCl_3$-CH_3OH)	Convallagenin-B	L-Arabinose (1)‡	25, 26
Glucoconvalla-saponin-B	*Convallaria Keisukei*	221–222°	—35°	Convallagenin-B	L-Arabinose (1)‡ D-Glucose (1)	26
Convallasaponin-C	*Convallaria Keisukei*	218–221°	—90° ($CHCl_3$-CH_3OH)	Isorhodeasapogenin	L-Arabinose (1)‡ L-Rhamnose (2)	25, 27
Convallasaponin-D	*Convallaria Keisukei*	264–265°	—66°	Rhodeasapogenin	D-Glucose (1)‡ D-Xylose (1) L-Rhamnose (2)	26

* The list of sources is not comprehensive.
† In pyridine, unless otherwise stated.
‡ Complete structure has been elucidated.

References

1 *O. Schmiedeberg*, Arch. exp. Path. Pharmak., 1875, **3**, 16; *H. Kiliani*, Ber., 1890, **23**, 1555; 1891, **24**, 339, 3951; 1901, **34**, 3561; 1910, **43**, 3562; 1916, **49**, 701; 1918, **51**, 1613.
2 *K. Száhlender*, Arch. Pharm. Berl., 1936, **274**, 446.
3 *R. Tschesche* and *G. Wulff*, Tetrahedron, 1963, **19**, 621.
4 *A. Windaus* and *A. Schneckenburger*, Ber., 1913, **46**, 2628; *Kiliani, ibid.*, 1916, **49** 701; 1918, **51**, 1613.
5 *T. Kawasaki et al.*, Chem. pharm. Bull. Tokyo, 1964, **12**, 1311; Tetrahedron, 1965, **21**, 299.
6 *Tschesche*, Ber., 1936, **69**, 1665.
7 *Tschesche* and *G. Balle*, Tetrahedron, 1963, **19**, 2323.
8 *P. C. Jurs* and *C. R. Noller*, J. Amer. chem. Soc., 1936, **58**, 1251.
9 *Tschesche* and *Wulff*, Ber., 1961, **94**, 2019.
10 *T. Tsukamoto et al.*, J. pharm. Soc. Japan, 1936, **56**, 802; 1957, **77**, 1225; Pharm. Bull. Tokyo, 1956, **4**, 35.
11 *Kawasaki* and *T. Yamauchi*, Chem. pharm. Bull. Tokyo, 1962, **10**, 703.
12 *D. C. Grove, G. L. Jenkins* and *M. R. Thompson*, J. Amer. pharm. Ass., 1938, **27**, 457; *R. E. Marker* and *J. Krueger*, J. Amer. chem. Soc., 1940, **62**, 2548.
13 *Marker* and *Krueger, ibid.*, 1940, **62**, 2548, 3349; *S. Lieberman et al., ibid.*, 1942, **64**, 2581.
14 *Wei-Yuan Huang, Yuh-Cheng Chen* and *Jen-Hung Chu*, Acta chim. sin., 1956, **22** 409 (C.A., 1958, **52**, 11098).
15 *J. A. Mockle*, Canad. pharm. J., 1957, **90**, 162.
16 *Tsukamoto* and *Kawasaki*, J. pharm. Soc. Japan, 1954, **74**, 1127; Pharm. Bull. Tokyo, 1956, **4**, 104.
17 *Kawasaki et al.*, J. pharm. Soc. Japan, 1957, **77**, 1221; Chem. pharm. Bull. Tokyo, 1962, **10**, 698, 703.
18 *A. W. van der Haar*, Rec. Trav. chim. Pays-Bas, 1929, **48**, 726.
19 *Marker* and *J. Lopez*, J. Amer. chem. Soc., 1947, **69**, 2389.
20 *Tschesche, R. Kottler* and *Wulff*, Ann., 1966, **699**, 212.
21 *Tschesche, G. Lüdke* and *Wulff*, Tetrahedron Letters, 1967, 2785.
22 *Kawasaki* and *Yamauchi*, J. pharm. Soc. Japan, 1963, **83**, 757.
23 *Kawasaki* and *K. Miyahara*, Tetrahedron, 1965, **21**, 3633.
24 *P. C. Maiti* and *S. Mookherjea*, Chem. and Ind., 1965, 1653.
25 *M. Kimura, M. Tohma* and *I. Yoshizawa*, Chem. pharm. Bull., Tokyo, 1966, **14**, 50.
26 *Yoshizawa, Tohma* and *Kimura, ibid.*, 1967, **15**, 129.
27 *Kimura, Tohma* and *Yoshizawa, ibid.*, 1966, **14**, 55.

of stable foams. They cause haemolysis, even in very dilute solution, a property that has been used for their detection in plant extracts. However; they are not toxic to mammals when taken orally, probably because they are not absorbed from the gut, and they are used as fish-poisons.

The saponins form molecular complexes with several classes of compounds; notable among these are the very insoluble complexes formed between digitonin, on the one hand, and cholesterol and related steroid alcohols, on the other (see Vol. II D, p. 39).

The first source of steroid saponins was *Digitalis purpurea*, more important as a source of the cardiac glycosides. *O. Schmiedeberg* (Arch. exp. Path. Pharmak., 1875, **3**, 16) isolated from commercial digitalis extracts a material that he named digitonin; this was subsequently shown to contain 10–20% of a second glycoside, named gitonin (*A. Windaus* and *A. Schneckenburger*, Ber., 1913, **46**, 2628). A third saponin, tigonin, was isolated much later by *R. Tschesche* (*ibid.*, 1936, **69**, 1665) from a related plant, *Digitalis lanata*. (Still more recent work has shown the presence, in *Digitalis* sp., of several other saponins, which are listed in Table 1.) Richer sources of saponins were subsequently found, many of them by Marker and his colleagues, in a number of other plants, notably among Mexican and U.S. members of the *Liliaceae*, *Dioscoreaceae* and *Scrophulariaceae* families.

Table 1 lists the better characterised saponins and indicates the nature of both the sapogenin and the sugar moieties. It will be seen that the sugars are common ones, in contrast to those found in the cardiac glycosides (see Vol. II D, p. 361).

The saponins are not easy to isolate in a pure state and they had been relatively little studied until recently, most workers having hydrolysed the crude plant extract before isolation of the sugar-free sapogenins. The hydrolysis may be done enzymatically (by enzymes that occur along with the saponin in the plant, or by fungal enzymes), by mineral acid, or by a combination of both methods (*M. E. Wall et al.*, J. Amer. chem. Soc., 1954, **76**, 2938, 3515; 1955, **77**, 1238; *P. C. Spensley*, Chem. and Ind., 1952, 426; 1956, 229). Enzymatic and chemical hydrolysis usually yield the same product, though artefacts may be formed by acid hydrolysis (*W. J. Peal*, Chem. and Ind., 1957, 1451; *Wall, S. Serota* and *L. P. Witnauer*, J. Amer. chem. Soc., 1955, **77**, 3086; see also pp. 42 and 44–49).

The complete structures of several saponins have now been established by conventional degradative methods (see Table 1). With few exceptions, the sugar is linked to the steroid through the 3-hydroxyl group. Among the exceptions are yononin, the 2-α-L-arabinoside of a 2β,3α-diol, convallasaponin-B, the 5-α-L-arabinoside of a 1β,3β,4β,5β-tetra-ol, glucoconvallasaponin-B, the corresponding 3-β-D-glucoside 5α-L-arabinoside and convallasaponin-D with sugar residues attached at both the 1- and 3-hydroxyl groups of rhodeasapogenin (for references, see Table 1). A more interesting case is that of sarsaparilloside, in which ring F is opened and the resulting hydroxyl group at $C_{(26)}$, as well as that at $C_{(3)}$ has formed a glycoside;

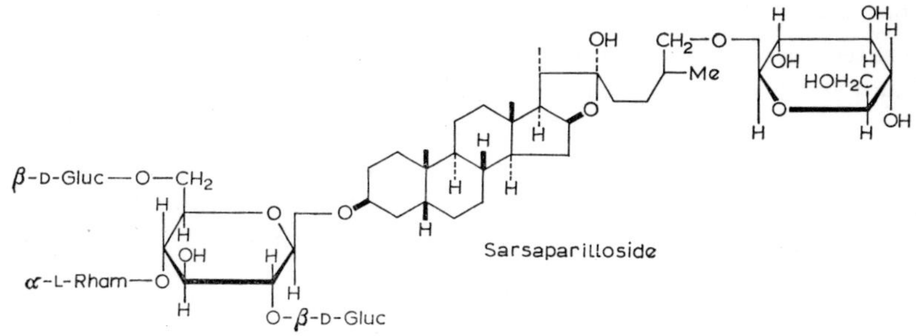

Sarsaparilloside

after removal of the sugar from the 26-hydroxyl group, recyclisation of ring F occurs very readily (*Tschesche, G. Lüdke* and *G. Wulff*, Tetrahedron Letters 1967, 2785). *R. E. Marker* and *J. Lopez* (J. Amer. chem. Soc., 1947, **69**, 2389) had suggested a similar structure for certain saponins and regarded the derived sapogenins with their closed ring F as artefacts, but this has been disproved for most saponins (*inter al., E. S. Rothman, Wall* and *C. R. Eddy, ibid.*, 1952, **74**, 4013); sarsaparilloside provides the first proven example of such an open-chain saponin.

The polysaccharide chains of the saponins are usually branched. resembling in this sense, those associated with some steroidal alkaloids of the *Solanum* class. As an example, the complete structure of dioscin is illustrated below:

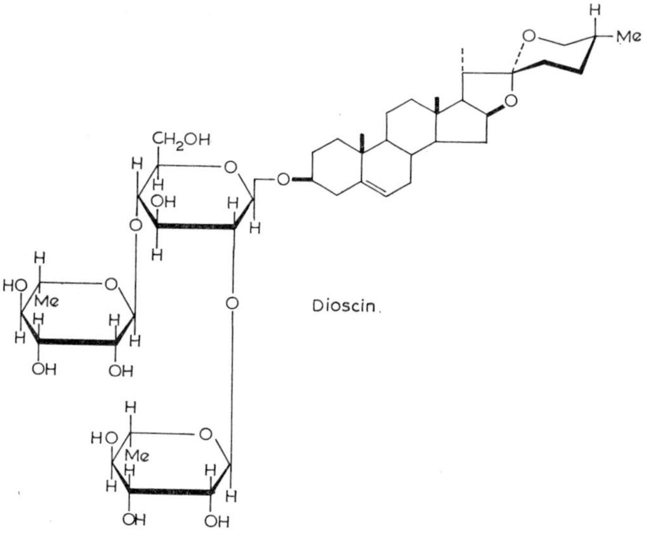

Dioscin.

2. The sapogenins

The early work, with which H. Kiliani and A. Windaus are particularly associated, provided a number of degradation products, but was confused by the persistent belief that the sapogenins were C_{26} compounds (see *L. F. Fieser* and *M. Fieser*, "Steroids", Reinhold Publishing Corporation, New York, 1959, pp. 814–816). The correct C_{27} formulae were firmly established as late as 1936 by application of special analytical techniques (*L. F. Fieser* and *R. P. Jacobsen*, J. Amer. chem. Soc., 1936, **58**, 943; *J. C. E. Simpson* and *W. A. Jacobs*, J. biol. Chem., 1935, **109**, 573; cf. *R. Tschesche* and *A. Hagedorn*, Ber., 1935, **68**, 1412).

(a) The nucleus

The first indication of the presence of the steroid nucleus came from the isolation of Diels' hydrocarbon (Vol. IID, p. 370) from the products of dehydrogenation of sarsasapogenin or gitogenin with selenium (*Jacobs* and *Simpson*, J. biol. Chem., 1934, **105**, 501; J. Amer. chem. Soc., 1934, **56**, 1424). The single hydroxyl group of sarsasapogenin was shown to be at $C_{(3)}$ by its conversion into an oxo group, reaction of this with methylmagnesium iodide to give the methyl carbinol and, finally, dehydrogenation with selenium to 7-methyl-1,2-cyclopentenophenanthrene (*G. A. R. Kon et al.*, J. chem. Soc., 1937, 414; 1939, 794).

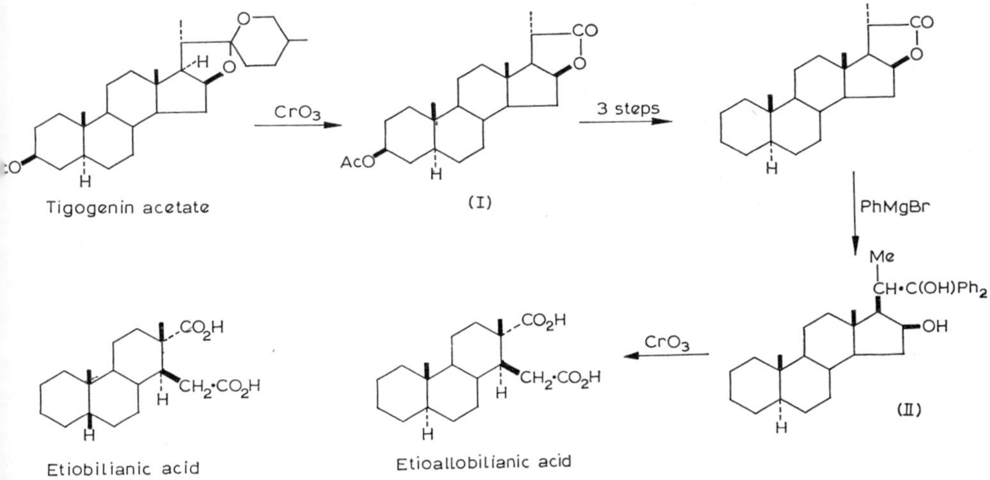

Scheme 1.

Further evidence came from conversion of tigogenin acetate into the known etioallobilianic acid by the method shown in Scheme 1 (*Tschesche* and *Hagedorn*, *loc. cit.*). In addition to confirming the presence of the steroid ring system, this degradation proved that the configurations at the ring junctions in tigogenin were identical with those of cholestane. A similar degradation, applied to sarsasapogenin acetate or smilagenin acetate, led to etiobilianic acid, and so showed these sapogenins to belong to the coprostane (5β) series (*S. N. Farmer* and *Kon*, J. chem. Soc., 1937, 414).

(b) The side-chain

The formation of carbonyl compounds with eight carbon atoms by degradation of sarsasapogenin or tigogenin suggested the presence of a C_8-side-chain, in conformity with the C_{27} formulation for the intact sapogenin

molecule (*L. Ruzicka* and *A. G. van Veen*, Hoppe-Seyl. Z., 1929, **184**, 69; *Jacobs* and *Simpson*, J. biol. Chem., 1934, **105**, 501; J. Amer. chem. Soc., 1934, **56**, 1424). The formation of the etiobilianic acids (see Scheme 1) showed the side-chain to be attached to the nucleus at carbon atoms 16 and 17.

The first sapogenins to be investigated contained two oxygen atoms besides those in the hydroxyl groups. They appeared to be oxidic and it seemed probable that one of them was in a tetrahydrofuran ring attached to ring D, both because of the stability of the lactone I and because the diol II was readily dehydrated to a cyclic ether (*H. P. Kaufmann* and *C. Fuchs*, Ber., 1923, **56**, 2527; *Tschesche* and *Hagedorn, loc. cit.*).

Some early formulations of the structure of the side-chain proved to be incorrect and will not be considered here. The correct structure was put forward by *R. E. Marker*, who considered that various reactions of the sapogenins in acid media (including hydrogenation, bromination and Clemmensen reduction) were better explained as those of a ketone group, protected as a spiroacetal, than as those of a true tetrahydrofuran (*Marker* and *E. Rohrmann*, J. Amer. chem. Soc., 1939, **61**, 846). The following degradations accord with Marker's view.

Oxidation of tigogenin acetate with chromic acid gives, together with the lactone I, a dioxo-acid as major product (*Tschesche* and *Hagedorn, loc. cit.*).

Sarsasapogenin acetate yields an analogous product known as sarsasapogenoic acid, which contains two relatively unreactive carbonyl groups and which is dehydrated by base to a carboxylic acid (anhydrosarsasapogenoic acid; III), whose ultra-violet spectrum suggests the presence of an α,β-unsaturated ketone grouping (Scheme 2). Reduction of this compound gives a saturated hydroxy-acid that can readily be converted into a lactone.

Oxidation of the anhydro-acid (III) with permanganate gives a hydroxy-oxo-dicarboxylic acid, IV, shown by Marker to have the formula $C_{27}H_{42}O_7$; further oxidation leads to the oxo-acid, V, which, like compound IV, is degraded by hypoiodite to 3β-hydroxyetiobilianic acid (VI) and iodoform. Again, ozonolysis of the anhydro-acid III yields the oxo-acid V, which is reduced by sodium and alcohol to the lactone VII (*Fieser* and *Jacobsen*, J. Amer. chem. Soc., 1938, **60**, 28, 2753; *Marker et al., ibid.*, 1939, **61**, 2072; 1941, **63**, 2274; 1942, **64**, 147, 180, 721).

Catalytic reduction of sarsasapogenoic acid leads to the tetrahydrofuran VIII; the reverse reaction is accomplished by oxidation with chromic acid, which leads also to compounds V, VI and VII (*Fieser et al., ibid.*, 1938, **60**, 2753; 1939, **61**, 1849; *Marker* and *Rohrmann, ibid.*, 1939, **61**, 3477). Compound VIII can also be prepared by catalytic reduction of the original sapogenin and oxidation of the "dihydrosapogenin" IX, so produced, with chromic acid (*Marker et al., ibid.*, 1939, **61**, 846, 2072, 3477; 1942, **64**, 721).

Confirmatory evidence for the structure of the side-chain comes from the reactions to be described and from the synthesis of sapogenins from simpler steroids (p. 21).

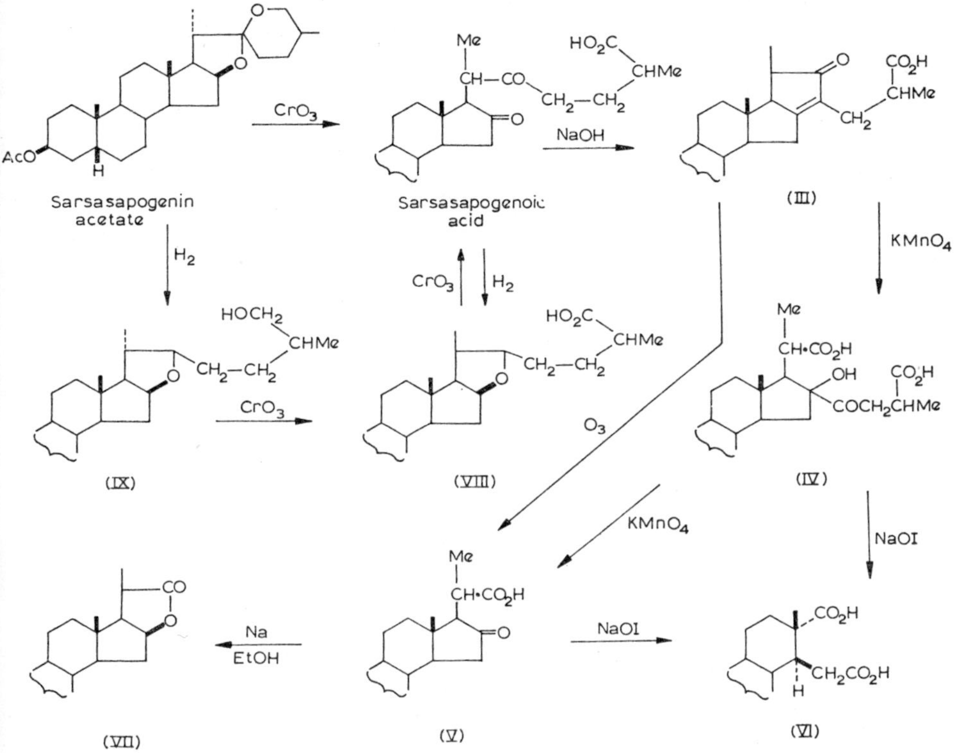

Scheme 2.

(*i*) *Pseudosapogenins and cyclopseudosapogenins*

Marker and his colleagues made the important observation that sapogenins, on being heated to ~200° with acetic anhydride, undergo fission of the tetrahydropyran ring, with introduction of a 20,22-double-bond (J. Amer. chem. Soc., 1939, **61**, 3592; 1940, **62**, 518, 648, 898, 2525, 2532; 1941, **63**, 774); this is illustrated for diosgenin (as acetate) in Scheme 3. Oxidation of the resulting "pseudodiosgenin diacetate" (X) with chromic acid results in cleavage of the new double bond to give the 16β-acyloxy-20-ketone XI. This compound readily undergoes a β-elimination of the acyloxy-group*

* The eliminated acid, namely 5-acetoxy-4-methylpentanoic acid (which, as explained below, may be in either of the enantiomorphic forms, depending upon the starting sapogenin) has been synthesised in racemic form (*R. Brettle* and *F. S. Holland*, J. chem. Soc., 1962, 4836).

under alkaline or acid conditions, to give 3β-hydroxy- (or 3β-acetoxy)-pregna-5,16-dien-20-one, from which progesterone is readily prepared (*Marker et al., ibid.*, 1940, **62,** 518, 898, 2525; 1941, **63,** 774).

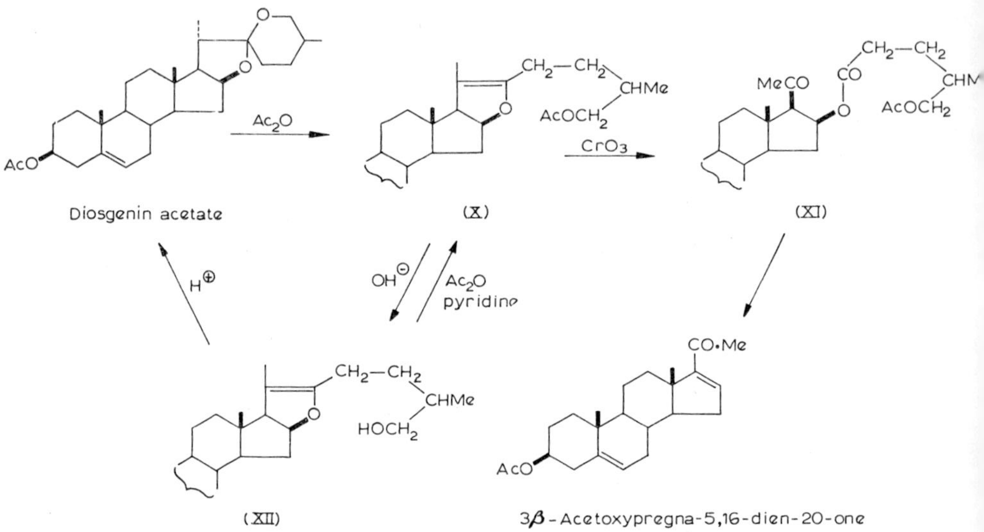

Scheme 3.

This degradation is generally applicable to sapogenins and it provided the key to their use for the large-scale production not only of progesterone, but also of the sex hormones and, more important, of the cortical hormones and their analogues. It has, therefore, been studied in some detail.

Because there is some disadvantage in the use of acetic anhydride above its normal boiling-point, considerable effort has gone into the search for alternative methods of preparing the pseudosapogenins (as X). These include (a) the use of high-boiling anhydrides at ordinary pressure (which give the corresponding 26-acyloxy derivatives of XII) (*Marker* and *Rohrmann, ibid.*, 1940, **62,** 518; *A. F. B. Cameron et al.*, J. chem. Soc., 1955, 2807) and (b) the use of boiling acetic anhydride in presence of Lewis acids or, better still, pyridine hydrochloride or methylamine hydrochloride and pyridine (*Cameron et al., loc. cit., D. H. Gould, H. Staeudle* and *E. B. Hershberg*, J. Amer. chem. Soc., 1952, **74,** 3685; *W. G. Dauben* and *G. J. Fonken, ibid.*, 1954, **76,** 4618; *M. E. Wall, H. E. Kenney* and *E. S. Rothman, ibid.*, 1955, **77,** 5665; *Organon Labs. Ltd.*, B.P., 749,697/1954). Chromic acid is probably the most generally useful reagent for cleavage of the 20,22-double-bond of the pseudosapogenins, but hydrogen peroxide in acetic

acid and permanganate-periodate have also been used (*G. P. Mueller, R. E. Stobaugh* and *R. S. Winniford*, J. Amer. chem. Soc., 1953, **75**, 4888; *Marker, E. M. Jones* and *E. L. Wittbecker, ibid.*, 1942, **64**, 468; *Wall* and *S. Serota*, J. org. Chem., 1959, **24**, 741). A variety of reagents has been used for the elimination reaction; methods involving use of a base seem prone to side-reactions, including addition of solvent to give a 16α-substituted-20-ketone – though this latter reaction can be avoided by use of such solvents as *tert*-butanol or tetrahydrofuran (*Marker*, J. Amer. chem. Soc., 1949, **71**, 4149; cf. *D. K. Fukushima* and *T. F. Gallagher, ibid.*, 1951, 73, 196; *E. M. Chamberlin et al., ibid.*, 1953, **75**, 3477; *Wall, Kenney* and *Rothman, loc. cit., Mueller, Stobaugh* and *Winniford, loc. cit.*). The elimination has also been carried out on adsorbents such as alumina and silica and by simple warming with acetic acid (*Cameron et al., loc. cit., Mueller, Stobaugh* and *Winniford, loc. cit.*).

The pseudosapogenin 3,26-di-esters (as X) can be hydrolysed by alkali to the free pseudosapogenins (XII), which are re-cyclised by hot mineral acid to the starting sapogenin (see below). However, much gentler treatment with mineral acid or, better, acetic acid, produces stereo-isomers of the sapogenins, known as the cyclopseudosapogenins. These compounds differ from the naturally-occurring parent isomers in undergoing much readier cleavage of ring F. Boiling acetic anhydride, without a catalyst, serves to convert them into the pseudosapogenin acetates (X), and the corresponding 26-ols XII, result from mere heating of the cyclopseudo compounds. Again, oxidation of the cyclopseudosapogenins with chromic acid-acetic acid, and subsequent treatment with base, gives the Δ^{16}-20-ketones (as 3β-acetoxypregna-5,16-dien-20-one) (Scheme 3). As might be expected, the cyclopseudosapogenins are isomerised by mineral acid to the more stable parent sapogenin, the pseudosapogenin presumably being an intermediate (*R. K. Callow et al.*, J. chem. Soc., 1955, 1966; *Wall et al.*, J. Amer. chem. Soc., 1955, **77**, 1230, 5661; *J. B. Ziegler, W. E. Rosen* and *A. C. Shabica, ibid.*, p. 1223).

The stereochemistry of these compounds is discussed below.

(*ii*) *Normal and iso-sapogenins: configuration at* $C_{(25)}$

Stereo-isomerism of a different kind is found among the naturally-occurring steroid sapogenins, several pairs being known whose members differ only in the configuration of their side-chains. Further, as first observed by *Marker* and *Rohrmann* (J. Amer. chem. Soc., 1939, **61**, 846), one member of such a pair is converted into the other by the action of boiling alcoholic hydrochloric acid. [The reaction is reversible, the equilibrium lying well to one side (*Wall, Serota* and *Witnauer, ibid.*, 1955, **77**, 3086)]. The following are examples of such pairs, the less stable isomer being named first: sarsasapogenin and smilagenin; yamogenin and diosgenin; neotigogenin and tigogenin. The less stable compounds were called "normal "or "neo-" sapogenins, the more stable were named "isosapogenins"; these names can now, with advantage, be replaced by more systematic ones (see p. 18).

The isomers can readily be distinguished from each other and from the corresponding cyclopseudosapogenins by their infrared spectra. The spiroacetal system is responsible for a complex pattern of bands in the 800–1100 cm.$^{-1}$ region (absent from the spectra of the pseudo- and dihydro-sapogenins), whose details are characteristic of the particular isomeric type (*Wall et al., ibid.*, 1955, **77**, 5661; Analyt. Chem., 1952, **24**, 1337; 1953, **25**, 266; *C. R. Eddy, M-A. Barnes* and *C. S. Fenske, ibid.*, 1955, **27**, 1067; *R. N. Jones, E. Katzenellenbogen* and *K. Dobriner*, J. Amer. chem. Soc., 1953, **75**, 158; *Callow et al., loc. cit.*). The proton magnetic resonance spectra of the normal and isosapogenins also show characteristic differences, associated with the resonances due to the hydrogen atoms attached to $C_{(26)}$ and $C_{(27)}$ (*J. P. Kutney*, Steroids, 1963, **2**, 225).

The isomerisation of normal to iso-sapogenins with acid suggested that the spiroacetal group was involved in the change and, hence, that the two series were epimeric at $C_{(22)}$. This appeared to receive support from the mistaken beliefs (a) that sarsasapogenin and smilagenin, on treatment with acetic anhydride, gave the same pseudosapogenin diacetate which reverted to sarsasapogenin on treatment of the free 3,26-diol with mineral acid, and (b) that both sapogenins gave the same dihydro-compound (as XVI)(Scheme 4) on catalytic hydrogenation (*Marker et al.*, J. Amer. chem. Soc., 1939, **61**, 846; 1940, **62**, 648). These experiments were repeated, much later, by *I. Scheer, R. B. Kostic* and *E. Mosettig* (*ibid.*, 1953, **75**, 4871; 1955, **77**, 641, cf. *Wall et al., ibid.*, 1953, **75**, 4437) who showed (a) that pseudosarsasapogenin (XIVa) and pseudosmilagenin (XIVb) are distinct compounds, each reverting to its own parent (XIIIa and XIIIb, respectively) on treatment with acid and (b) that both pseudosapogenins yield the same Δ^{16}-20-ketone on oxidative degradation, but that, in addition, pseudosarsasapogenin gives (+)-α-methylglutaric acid (XVa), whereas smilagenin gives the (—)-isomer (XVb). This provided unequivocal evidence that the two sapogenins are of opposite absolute configuration at $C_{(25)}$, the "normal" and "iso"-sapogenin having the L- and D-configuration respectively (*S*- and *R*-, respectively, in the Cahn-Ingold-Prelog convention). [Still more certain evidence for the absolute configurations came from an alternative degradation of an isosapogenin to (+)-methylsuccinic acid, of well-established absolute configuration (*V. H. T. James*, J. chem. Soc., 1955, 637)]. Dihydrosarsasapogenin (XVIa) and dihydrosmilagenin (XVIb) were, again, shown to be distinct compounds; that they were isomeric *only* at $C_{(25)}$ was shown by their conversion into the same 26-desoxy compound XVII, in which asymmetry at $C_{(25)}$ has been destroyed. This also proved that the original sapogenins had identical configurations at $C_{(20)}$ but, in view of the possibility of inversion at $C_{(22)}$ during the hydrogenation to give XVI, it did not decide with certainty whether the sapogenins had the same or different configurations at this centre.

Evidence that the two series do, indeed, differ only at $C_{(25)}$ was provided by the conversion of neotigogenin and tigogenin into the same diosphenol (XVIII), in which the asymmetry at this centre has been lost, by a series of reactions (see Scheme 4) that could be shown not to affect the other asymmetric centres (*R. K. Callow* and *P. N. Massy-Beresford, ibid.*, 1957, 4482).

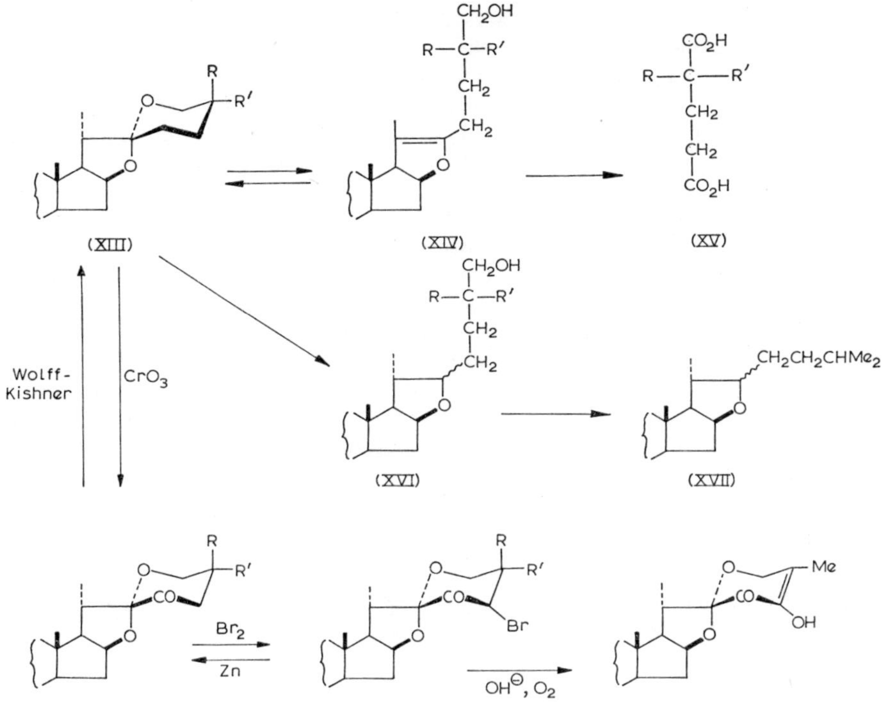

(XIII) (XIV) (XV)

Wolff-Kishner ; CrO_3

(XVI) (XVII)

Br_2 ; Zn ; $OH^{\ominus}$, O_2

(XVIII)

(a) R = Me, R′ = H; normal sapogenins
(b) R = H, R′ = Me; isosapogenins

Scheme 4.

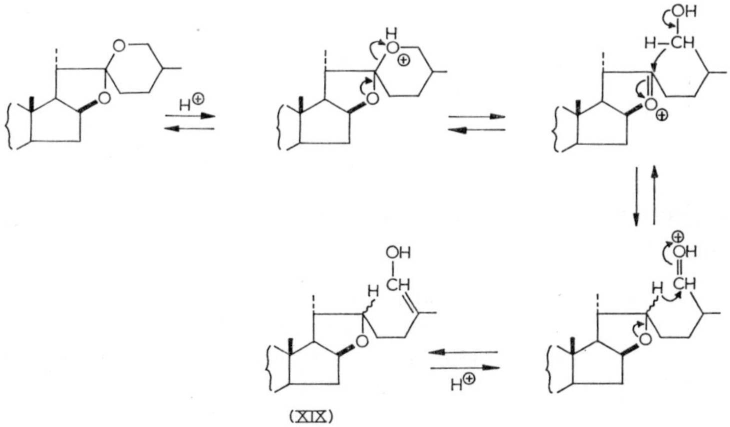

$H^{\oplus}$

(XIX)

Scheme 5.

The mechanism of the acid-catalysed interconversion of the normal and iso-sapogenins became more puzzling when it was thus established that it involved epimerisation at $C_{(25)}$ rather than at $C_{(22)}$. *R. B. Woodward, F. Sondheimer* and *Y. Mazur* (J. Amer. chem. Soc., 1958, **80**, 6693) suggested the mechanism illustrated in Scheme 5: asymmetry at $C_{(25)}$ is lost in the enolised aldehyde XIX, and reversal of the reactions can thus lead to equilibration of the $C_{(25)}$-epimers.

The same authors isolated an aldehyde (XX≡XIX) (see Scheme 5a) by oxidation of dihydrotigogenin 3-acetate with dichromate; in accordance with the suggested mechanism, treatment of this compound with alcoholic hydrochloric acid gave a mixture of tigogenin and neotigogenin as major and minor products. Still further evidence for formation of the aldehyde XX during the normal-iso interconversion came from the production of its ethylene dithioacetal XXI when tigogenin acetate was treated with ethanedithiol in presence of boron trifluoride; the structure of compound XXI was proved by its reduction, with Raney nickel, to the known 16β,22-epoxycholestan-3β-ol (as XVII) (*C. Djerassi et al.*, J. org. Chem., 1959, **24**, 1) (Scheme 5a).

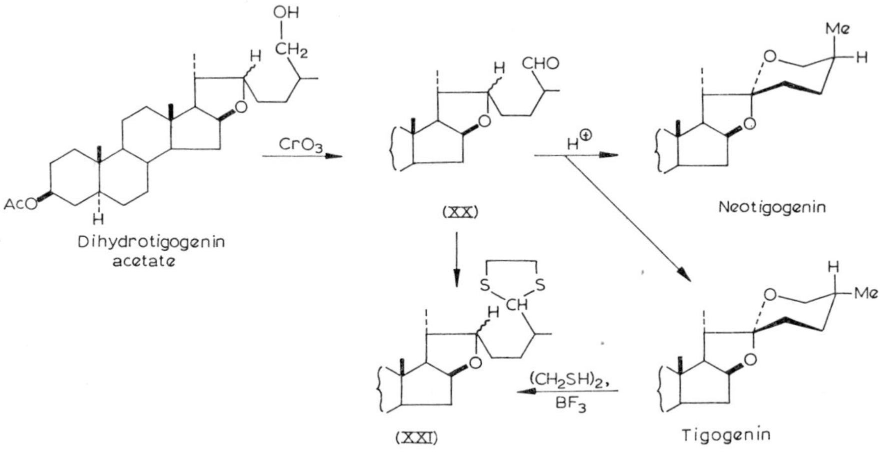

OH
CH_2
H
AcO
H
Dihydrotigogenin acetate
CrO_3
H CHO
(XX)
$H^{\oplus}$
Me
H
Neotigogenin
S S
CH
H
(XXI)
$(CH_2SH)_2$, BF_3
H
Me
Tigogenin

Scheme 5a.

(iii) Configuration of the sapogenins and cyclopseudosapogenins at $C_{(20)}$ and $C_{(22)}$

Biogenetic considerations would suggest that the naturally-occurring sapogenins have the same configuration at position 20 as the natural C_{24}, C_{27}, C_{28}, and C_{29} steroids; *i.e.* that the 20-methyl group is in the rearward (α) position. Evidence for this supposition came from the observation that the lactone XXII, derived from tigogenin, is more stable than its $C_{(20)}$-epimer XXIII, the latter being converted completely into the former on treatment with base (*J. W. Corcoran*

and *H. Hirschmann*, J. Amer. chem. Soc., 1956, **78**, 2325; *Mazur, N. Danieli* and *F. Sondheimer, ibid.*, 1960, **82**, 5889). This is easily explained by the proposed structures, examination of models showing severe interaction between the 20- and 13-methyl groups in XXIII, which is absent from XXII:

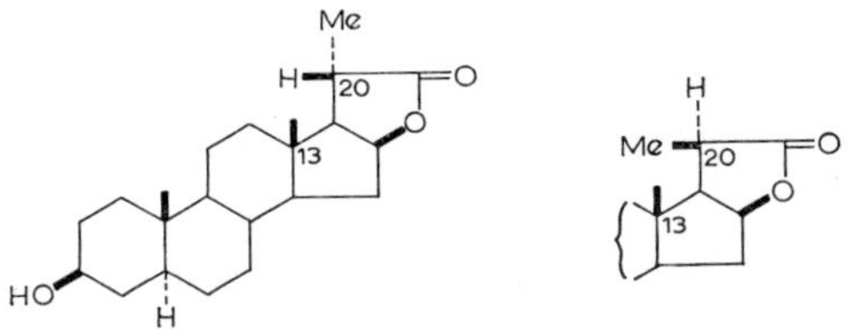

(XXII) (XXIII)

Wall and his colleagues (*ibid.*, 1955, **77**, 1230) suggested that the cyclopseudosapogenins are stereoisomeric at $C_{(20)}$ with the sapogenins proper and that they owe their instability to the methyl-methyl interaction mentioned above. The following evidence, due to the same group (*Wall* and *Walens, ibid.*, 1955, **77**, 5661; 1958, **80**, 1984) lends support to the suggestion (see Scheme 6). Oxidation of cyclopseudotigogenin acetate with chromic acid gives a 20-hydroxy-derivative XXIV, which can be dehydrated to the $\Delta^{20(21)}$-olefin XXV. This compound can be reconverted (a) to the starting cyclopseudo compound by catalytic reduction and (b) to the 20-hydroxy-compound XXIV by treatment with osmium tetroxide to give the 20,21-diol XXVI, conversion into the 21-tosylate and reduction of this compound by lithium aluminium hydride. On the reasonable assumption that both osmylation and catalytic reduction involve attack on the

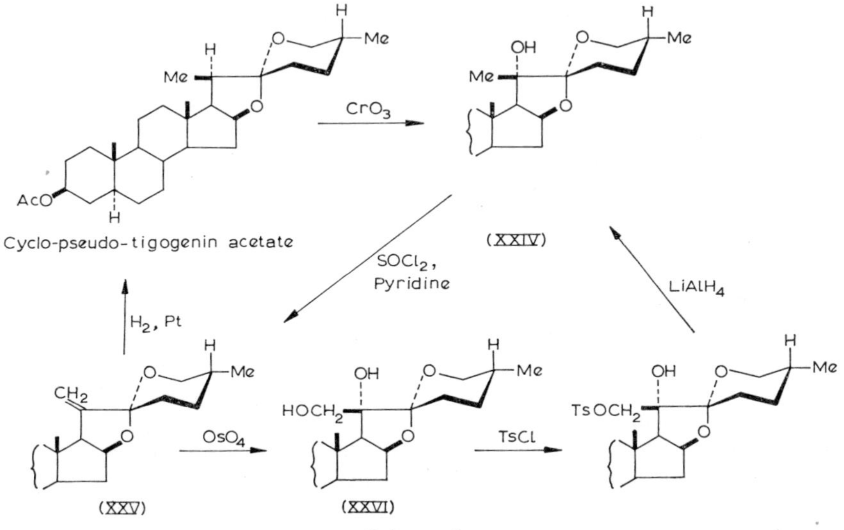

Cyclo-pseudo-tigogenin acetate (XXIV) (XXV) (XXVI)

Scheme 6.

less hindered rear-side of the 20,21-double-bond, the configurations at $C_{(20)}$ must be as shown in Scheme 6. However, *Fieser* and *Fieser* (*op. cit.*, p. 828) show that a different, though less probable conclusion can be drawn.

Rather less direct are the arguments for the configurations of the various isomers at $C_{(22)}$. They rest partly on the fact that a normal sapogenin is unstable with respect to the corresponding iso-sapogenin (as indicated by the position of the equilibrium between them under acid conditions and by the faster conversion of the normal sapogenin into the pseudo-compound), whereas the reverse is true in the cyclopseudo-series (as indicated by the relative rates of conversion to the pseudo-compounds) (*Wall* and *Serota*, J. Amer. chem. Soc., 1957, **79**, 6481). Conformational analysis of the various possible forms shows that these facts are best explained by assigning to the four isomeric series the structures XXVII–XXX shown in Scheme 7 (the cyclopseudo-compounds can be taken to have the same absolute configuration at $C_{(25)}$ as the parent sapogenin, in the light of their interconvertibility *via* the corresponding pseudosapogenin) (cf. *Fieser* and *Fieser*, *op. cit.*, pp. 824–830).

(XXVII)
Normal sapogenin

(XXVIII)
Isosapogenin

(XXIX)
Cyclopseudo-normal sapogenin

(XXX)
Cyclopseudo-iso-sapogenin

Scheme 7.

In agreement with the proposed structure for the cyclopseudo-iso compounds (XXX) is the strong hydrogen-bonding between the 20α-hydroxy-group and the ring F oxygen atom in compound XXIV (*Wall* and *Walens*, J. Amer. chem. Soc., 1958, **80**, 1984). Further evidence for the structure of the cyclopseudo-normal-compounds (XXIX) came from a study of the 20α-hydroxy-compound XXXI (Scheme 8) which results from the action of per-acid on pseudosarsasapogenin (the same reaction can be used in the isosapogenin series to prepare compounds of type XXIV). Dehydration of the 20-hydroxy-compound XXXI gives the

20-methylene compound XXXIII. Osmylation of the latter, tosylation of the resulting 20,21-diol and, finally, reduction with lithium aluminium hydride gives compound XXXI again, suggesting that the latter has a 20α-hydroxyl group (see above); its infrared spectrum shows the expected strong hydrogen-bonding between the hydroxyl group and the ring F oxygen atom. The combination of configurations at positions 20, 22, and 25 in compound XXXI had been predicted to be unfavourable; in agreement with this, it is isomerised under very mildly acidic conditions to a compound whose hydroxyl group is unbonded and which must have structure XXXII, *i.e.* of 20α-hydroxycyclopseudosarsasapogenin. An attempt to prepare the 22-epimer (XXXV) of cyclopseudosarsasapogenin was unsuccessful, because the 20(21)-dehydro-compound XXXIII, on hydrogenation under ostensibly neutral conditions, yielded cyclopseudosarsasapogenin(XXXIV) itself. Presumably compound XXXV is formed first but is so unstable that it epimerizes spontaneously to XXXIV (*Wall, Walens* and *Tyson*, J. org. Chem., 1961, **26**, 5054).

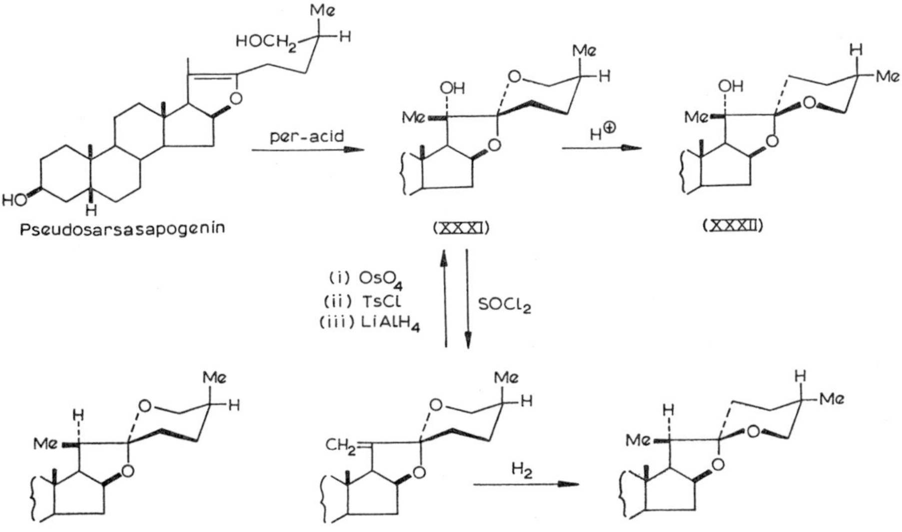

Scheme 8.

In any particular set of isomers, the normal (XXVII), iso- (XXVIII) and cyclopseudo-iso compounds (XXX) have similar rotations, whereas the cyclopseudo-normal compound (XXIX) is much more dextrorotatory; this was the first indication that the last isomer differed from the other three in configuration at $C_{(22)}$ (*Wall*, Experientia, 1955, **11**, 340). A rather similar situation is disclosed by measurements of optical rotatory dispersion of the isomers (*Djerassi* and *R. Ehrlich*, J. Amer. chem. Soc., 1956, **78**, 440). Studies of proton magnetic resonance also support these structures (*W. E. Rosen et al., ibid.*, 1959, **81**, 1687; *Kutney*,

Steroids, 1963, **2,** 225; *G. F. H. Green, J. E. Page* and *S. Staniforth,* J. chem. Soc. B, 1966, 807; *D. H. Williams* and *N. S. Bhacca,* Tetrahedron, 1965, **21,** 1641; *P. M. Boll* and *W. von Philipsborn,* Acta Chem. Scand., 1965, **19,** 1365). In particular, the $C_{(27)}$-methyl resonance in compounds of type XXVII appears significantly downfield from that in compounds of types XXVIII, XXIX and XXX; this accords with the difference in conformation of the methyl group – axial in the first, equatorial in the other three structures. Again, the $C_{(18)}$- and $C_{(21)}$-methyl resonances in the 20β-methyl compounds XXIX and XXX appear downfield from those in the 20α-methyl-compounds XXVII and XXVIII; this can be attributed to steric interaction between $C_{(18)}$ and $C_{(21)}$ in the former pair.

(iv) Nomenclature

For the purpose of systematic nomenclature, the name "spirostan" is given to the skeleton XXXVI which specifies configuration at all asymmetric centres other than $C_{(5)}$ and $C_{(25)}$. The first of these is defined by the usual α,β-convention, but this cannot be applied at $C_{(25)}$. Several proposals have been put forward

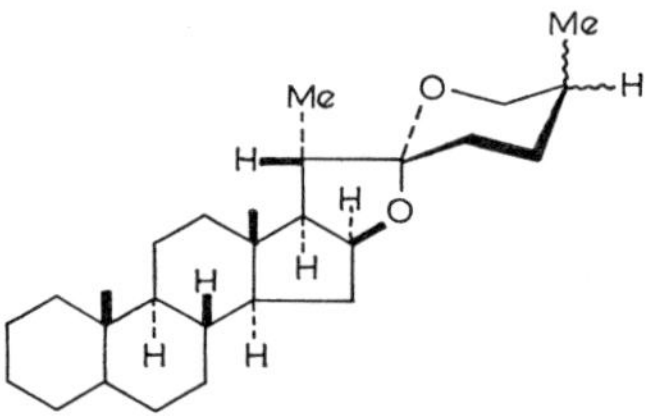

(XXXVI) 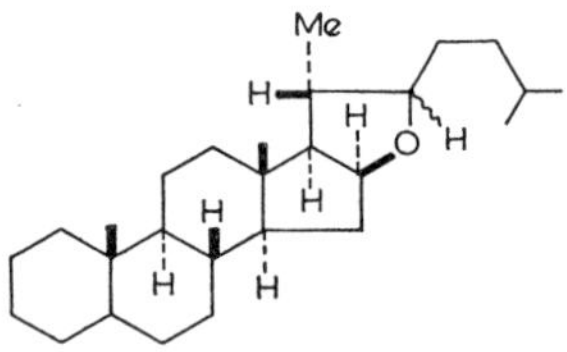 (XXXVII)

for specifying the configuration systematically at all centres in rings E and F (*Fieser* and *Fieser,* Tetrahedron, 1960, **8,** 360; *G. P. Mueller* and *G. R. Pettit,* Experientia, 1962, **18,** 404; *Pettit, ibid.,* 1963, **19,** 124). The author would recommend the use of the sequence rule procedure (*R. S. Cahn, C. Ingold* and *V. Prelog,* Angew. Chem. internat. Edn., 1966, **5,** 385) since this seems likely to win official acceptance. Thus, a "normal sapogenin" (XXVII) is named as (25*S*)... spirost..., an "isosapogenin", (XXVIII) as (25*R*)... spirost..., substituents and configurations in the nucleus being specified in the usual way. If the configurations at positions 20 and 22 are different from those specified in formula XXXVI, they, too, should be specified by the sequence rule, as should the configurations at carbon atoms 23, 24 and 26 if these bear substituents; thus, compounds of types XXIX and XXX would bear the prefixes (20*R*, 22*S*, 25*S*)- and (20*R*, 22*R*, 25*R*)-, respectively. It should be noted that substitution on or near the asymmetric centre may necessitate a change of prefix. Examples of the use of the system will be found later in this chapter.

In the same way, the name "furostan" is given to the skeleton XXXVII, without implication as to configuration at $C_{(5)}$ or $C_{(22)}$. The former is specified in the usual steroid convention, the latter (if known) by the sequence rule: again, the same rule is used to indicate configuration in the side-chain and at $C_{(20)}$, if this last differs from that shown in formula XXXVII.

(v) Reactions of the sapogenin side chain

Clemmensen reduction of the sapogenin side-chain gives a cholestane-16β,26-diol derivative XXXVIII (*Marker* and *Rohrmann*, J. Amer. chem. Soc., 1939, **61,** 846); the product from diosgenin has been converted into cholesterol (*Marker* and *D. L. Turner, ibid.*, 1941, **63,** 767; cf. *I. Scheer* and *E. Mosettig, ibid.*, 1955, **77,** 1820).

Treatment of sapogenins with persulphate, peracetic acid or performic acid leads to a Baeyer-Villiger type of oxidation, the product, after basic hydrolysis, being the 16β,20α-diol XLI (*Marker et al., ibid.*, 1940, **62,** 525, 2532; *K. Morita et al.*, Chem. pharm. Bull. Tokyo, 1963, **11,** 90, 95, 103, 139). Somewhat akin to this is the microbiological degradation, by *Fusarium solani*, of some sapogenins to 16-oxoandrostanes XXXIX; the pathway may involve the 16-oxopregnan-20-ol which could undergo retro-aldol reaction to give the 16-oxo compound (*E. Kondo* and *T. Mitsugi*, J. Amer. chem. Soc., 1966, **88,** 4737) (Scheme 9).

Treatment of diosgenin acetate with hydrogen chloride in acetic anhydride gives, in low yield, the 16β-acetoxy-26-chloro-22-ketone XL (*F. C. Uhle*, J. org. Chem., 1962, **27,** 656; cf. *R. S. Miner* and *E. S. Wallis, ibid.*, 1956, **21,** 715).

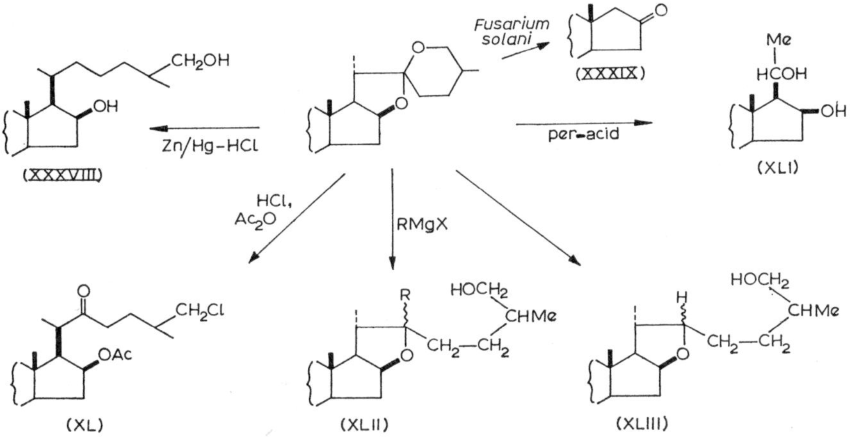

Scheme 9.

Reference has been made on p. 8 to the opening of ring F by platinum-catalysed hydrogenation, the product being a "dihydrosapogenin" XLIII (of uncertain configuration at position 22). The same reduction can be done with lithium aluminium hydride in presence of hydrogen chloride, aluminium chloride or boron trifluoride (*H. M. Doukas* and *T. D. Fontaine*, J. Amer. chem. Soc., 1953, **75,** 5355; *Pettit et al.*, J. org. Chem., 1960, **25,** 84; 1961, **26,** 4553). Catalytic reduction of pseudosapogenins (or of cyclopseudosapogenins) also gives dihydrosapogenins that differ, however, from those obtained by reducing the sapogenins;

in them, for example, ring E is much less stable to oxidation, and this has been used as evidence for their having a 20β-methyl-group (cf. p. 15) (*Marker et al.*, J. Amer. chem. Soc., 1940, **62**, 521, 2532; 1942, **64**, 1655; *Wall, Serota* and *Eddy*, *ibid.*, 1955, **77**, 1230; *D. A. H. Taylor*, Chem. and Ind., 1954, 1066).

A similar opening of ring F occurs when sapogenins are treated with Grignard reagents, the product in this instance being the 22-alkylfurostan-26-ol XLII (*Marker* and co-workers, J. Amer. chem. Soc., 1940, **62**, 900; 1941, **63**, 772; *R. D. Youssefyeh, Y. Mazur* and *F. Sondheimer*, Chem. and Ind., 1964, 421).

Among reactions of the intact side-chain, the best studied is bromination. *Marker* and *Rohrmann* (J. Amer. chem. Soc., 1939, **61**, 846, 1921) observed that sarsasapogenin acetate gives a mono-bromo derivative that is reduced to its parent by sodium and alcohol or zinc and acetic acid. Since chromic oxidation of the bromo compound gives 3β-hydroxy-16-oxobisnorcholanoic acid, the bromine atom must be in the side-chain, and it was taken to be at $C_{(23)}$, α- to the potential carbonyl group at $C_{(22)}$ (*Marker et al.*, *ibid.*, 1941, **63**, 1032). This conclusion has been confirmed by P.M.R. spectroscopy (*J. P. Kutney et al.*, Tetrahedron, 1964, **20**, 1999).

Subsequent study has shown that both (25*S*)- and (25*R*)- sapogenins form mono- and di-bromo derivatives (*Marker* and *Rohrmann*, J. Amer. chem. Soc., 1939, **61**, 1516; *Djerassi, H. Martinez* and *G. Rosenkranz*, J. org. Chem., 1951, **16**, 303; *Ziegler, Rosen* and *Shabica*, J. Amer. chem. Soc., 1955, **77**, 1223; *Wall* and *Jones*, *ibid.*, 1957, **79**, 3222) and that, at least in the 25*R*-series, monobromination gives two isomers (*G. P. Mueller* and *L. L. Norton*, *ibid.*, 1954, **76**, 749; *Wall* and *Jones*, *loc. cit.*). The configurations of these compounds have been elucidated by infrared, P.M.R. and X-ray studies (*D. H. R. Barton, J. E. Page* and *C. W. Shoppee*, J. chem. Soc., 1956, 331; *R. K. Callow et al.*, *ibid.* C, 1966, 288; *Kutney et al.*, Tetrahedron, 1964, **20**, 1999). The 23-bromine atom is very stable to alkali (*Marker et al.*, J. Amer. chem. Soc., 1947, **69**, 2167; *Djerassi*,

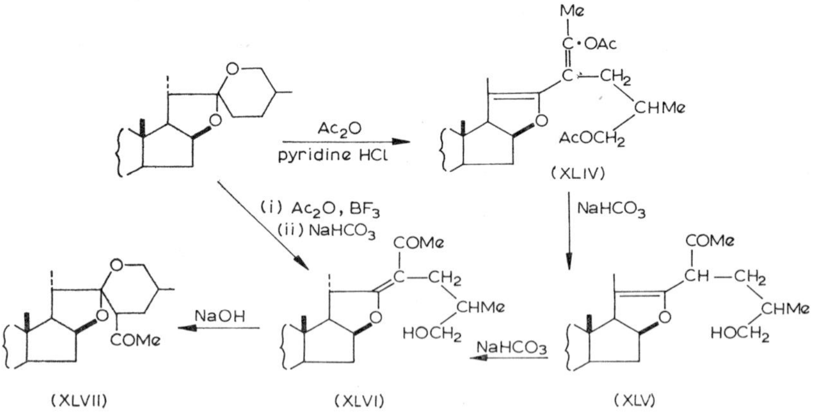

Scheme 10.

Martinez and *Rosenkranz, loc. cit.*); in addition, it appears to stabilise the sapogenin ring-system, for example against attack by selenium dioxide and by acid halides (*Wall* and *Jones*, J. Amer. chem. Soc., 1957, **79**, 3222). The bromine is readily removed by reduction, for which purpose zinc is commonly used; sodium iodide-acetic acid has proved useful when zinc would have caused reduction elsewhere (*D. N. Kirk, D. K. Patel* and *V. Petrow*, J. chem. Soc., 1957, 1046).

Acylation at $C_{(23)}$ occurs to a minor extent during acid-catalysed conversion of sapogenins into pseudosapogenin esters. *Uhle* (J. org. Chem., 1965, **30**, 3915) isolated the product of prolonged reaction of diosgenin with acetic anhydride and pyridine hydrochloride and showed that it was the enol acetate XLIV of the 23-acetylfurost-20(22)-en (Scheme 10). Treatment with bicarbonate gave, successively, the free 23-acetyl compound XLV and its isomer XLVI, in which the double-bond has moved into conjugation with the carbonyl group. Compounds of type XLVI had earlier been obtained from sapogenins and acetic anhydride in presence of perchloric acid or boron trifluoride; on treatment with strong alkali they are cyclised to the 23-acetylspirostans XLVII (*Uhle, loc. cit.*; *J. A. Zderic, L. Cervantes* and *M. T. Galvan*, J. Amer. chem. Soc., 1962, **84**, 102; *A. F. B. Cameron et al.*, J. chem. Soc., 1955, 2807).

(*vi*) *Synthesis of the sapogenin side-chain*

Y. Mazur, N. Danieli and *F. Sondheimer*, by the series of reactions outlined in Scheme 11, converted 3β-hydroxy-5α-androstan-17-one into a mixture of tigogenin and neotigogenin (J. Amer. chem. Soc., 1960, **82**, 5889).

The 17-ketone XLVIII was converted by enol acetylation, epoxidation of the 16,17-double-bond, and acid-catalysed rearrangement of the resulting 16α, 17α-epoxy-17β-acetoxy compound into 3β,16α-diacetoxy-5α-androstan-17-one (XLIX). A Reformatsky reaction with ethyl α-bromopropionate, reacetylation of the 3- and 16-hydroxyl groups and elimination of the tertiary 17-hydroxyl group gave a mixture of the unsaturated acetoxy-ester L and the corresponding lactone LI. Catalytic reduction of the former gave ethyl 3β,16α-diacetoxy-bisnor-5α,20-isocholanoate (LII) by rear-side addition; alkaline hydrolysis, re-esterification of the resulting dihydroxy-acid with diazomethane and oxidation with chromic acid then gave methyl 3,16-dioxobisnor-5α,20-isocholanoate (LIII).

Catalytic reduction of the unsaturated lactone LI proceeded from the front side to give the saturated lactone LV, which was not opened by base; however, reduction with lithium aluminium hydride gave the 3,16,22-triol, which was converted, by oxidation with chromic acid and esterification, into methyl 3,16-dioxobisnor-5α,17α-cholanoate (LVI). Treatment of isomers LIII and LVI with potassium hydroxide resulted in epimerization at $C_{(20)}$ and $C_{(17)}$, respectively, the product from both, after re-esterification, being methyl 3,16-dioxobisnor-5α-cholanoate (LIV), with the "natural" configuration at both centres. [In bisnorcholanic acids lacking an oxo-group at $C_{(16)}$ the 20-iso-configuration is the more stable; the influence of the 16-oxo-group has been attributed to a field

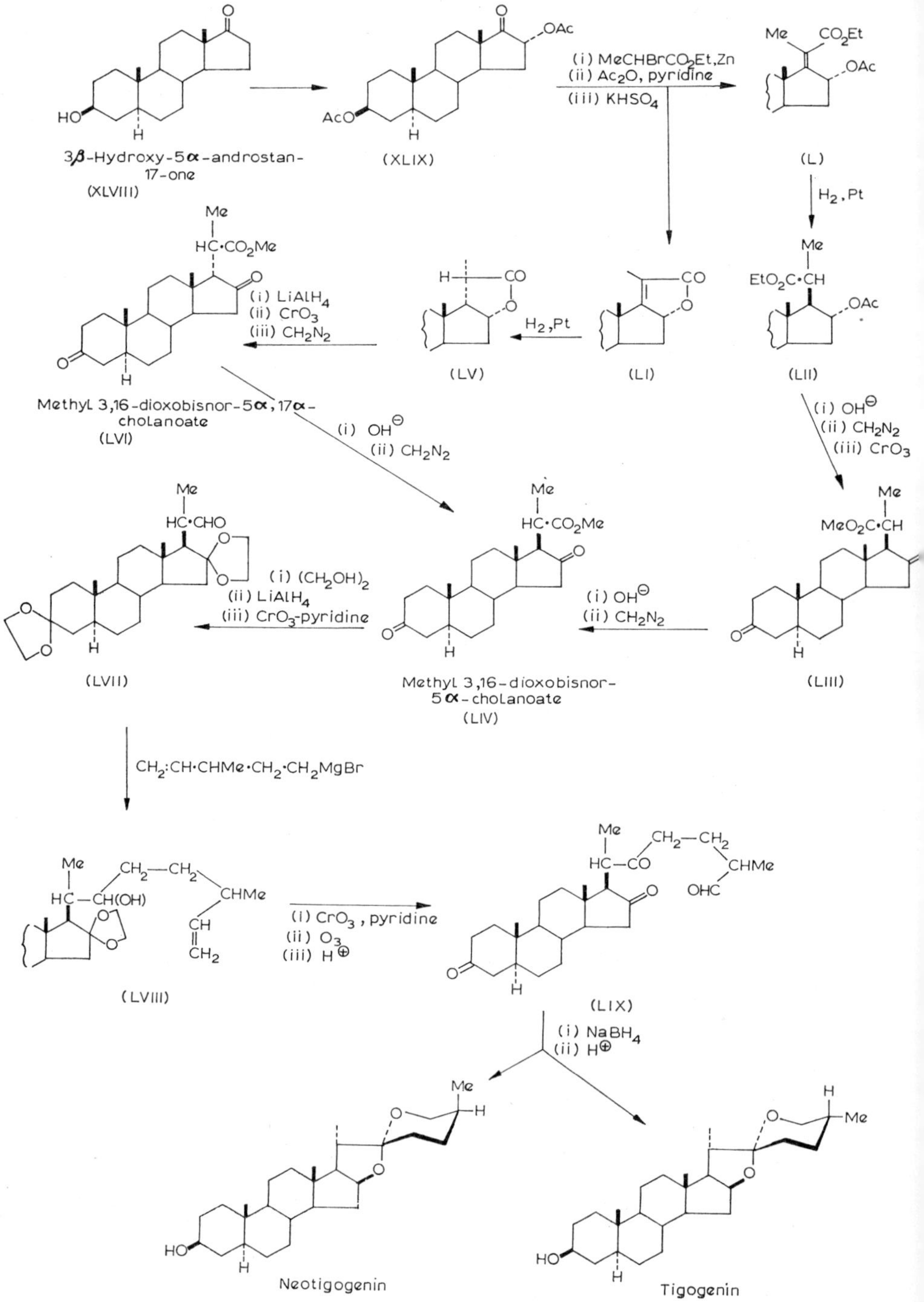

3β-Hydroxy-5α-androstan-17-one (XLVIII)

(XLIX)

(i) $MeCHBrCO_2Et$, Zn
(ii) Ac_2O, pyridine
(iii) $KHSO_4$

(L)

H_2, Pt

(LII)

(i) $OH^{\ominus}$
(ii) CH_2N_2
(iii) CrO_3

(LIII)

(LI)

H_2, Pt

(LV)

(i) $LiAlH_4$
(ii) CrO_3
(iii) CH_2N_2

Methyl 3,16-dioxobisnor-5α,17α-cholanoate (LVI)

(i) $OH^{\ominus}$
(ii) CH_2N_2

(i) $OH^{\ominus}$
(ii) CH_2N_2

Methyl 3,16-dioxobisnor-5α-cholanoate (LIV)

(i) $(CH_2OH)_2$
(ii) $LiAlH_4$
(iii) CrO_3-pyridine

(LVII)

CH_2:CH·CHMe·CH_2·CH_2MgBr

(LVIII)

(i) CrO_3, pyridine
(ii) O_3
(iii) $H^{\oplus}$

(LIX)

(i) $NaBH_4$
(ii) $H^{\oplus}$

Neotigogenin

Tigogenin

Scheme 11.

effect between it and the carboxy-carbonyl group (*Mazur* and *Sondheimer*, Experientia, 1960, **16**, 181)].

The oxo groups of compound LIV were then protected as their ethylene acetal derivatives; reduction of the ester group with lithium aluminium hydride and subsequent oxidation of the 22-ol with chromic anhydride-pyridine gave the bisnorcholanaldehyde LVII. The required side-chain was then built up by reaction of this aldehyde with 3-methylpent-4-enylmagnesium bromide. Oxidation of the resulting carbinol LVIII, again with chromic anhydride-pyridine, to the corresponding 22-ketone, cleavage of the terminal double-bond by ozonolysis and acid-catalysed removal of the acetal groups then gave 3,16,22-trioxo-5α-cholestan-26-al (LIX). Sodium borohydride served to reduce the aldehyde and the 3- and 16-oxo groups, and treatment of the product with acid gave a 1:1 mixture of tigogenin and neotigogenin.

Since the starting material, 3β-hydroxy-5α-androstan-17-one (XLVIII) has been prepared by total synthesis, this constitutes a formal total synthesis of tigogenin, neotigogenin and of a number of other naturally-occurring sapogenins, into which they can be converted.

A second synthesis of the sapogenin side-chain, reported by *S. V. Kessar* and his co-workers (Tetrahedron Letters, 1966, 4319; Tetrahedron, 1968, **24**, 887, 893, 899) is shorter and more stereoselective (see Scheme 12).

The readily accessible 16α,17α-epoxy-20-ketone LX is converted by hydrazine in presence of alkali into a mixture of isomeric 16α-hydroxypregn-17(20)-enes (LXI) and (LXII). Only the latter is easily oxidised by manganese dioxide and the resulting 16-oxo-compound LXIII is easily separated from unchanged alcohol LXI. 2-Methyl-5-nitropentyl acetate (LXIV) is synthesised from 2-methylpent-4-enoic acid as shown; this acid can first be resolved, so determining the configuration at the atom destined to become $C_{(25)}$. Base-catalysed Michael addition of the (*R*)-isomer of nitro-compound LXIV to the α,β-unsaturated ketone LXIII gave, in rather poor yield, the nitro-ketone LXV, resulting from "rearward attack" on the steroid olefin. A Nef reaction then gave, in "modest yield", the open-chain sapogenin, kryptogenin (see p. 43). Kryptogenin can be reduced by borohydride to diosgenin (LXVI; R=H,R′=Me), but this latter compound was obtained more conveniently and more efficiently by direct treatment of the nitroketone LXV with sodium borohydride in boiling ethanol. Use of the (*S*)-isomer of LXIV gave, in the same way, yamogenin LXVI; R=Me, R′=H); the overall yield from dehydropregnenolone was 20%.

The flexibility of this route has been demonstrated by its use, illustrated in Scheme 12, for the synthesis of the 27-hydroxysapogenin, isonarthogenin (*Kessar, A. L. Rampal* and *Y. P. Gupta, ibid.*, 1968, **24**, 905).

(*vii*) *Biogenesis of the steroid sapogenins*

This subject has not been much studied, but it is known that labelled mevalonate is incorporated into the steroid moiety of saponins by the growing plant and

[*text continued p. 29*]

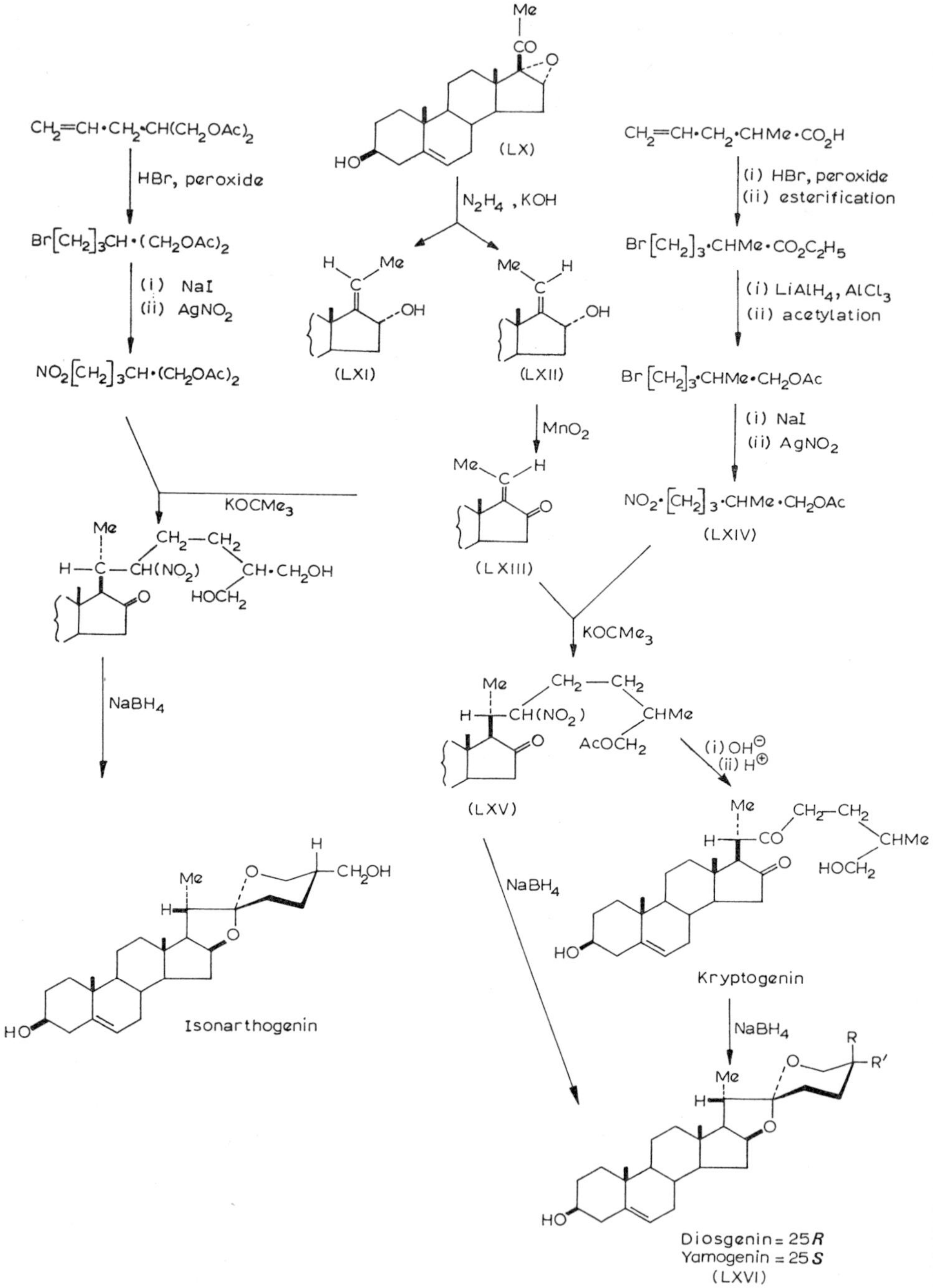
CH2=CH•CH2•CH(CH2OAc)2
HBr, peroxide
Br[CH2]3CH•(CH2OAc)2
(i) NaI
(ii) AgNO2
NO2[CH2]3CH•(CH2OAc)2
KOCMe3
NaBH4
Isonarthogenin
(LX)
N2H4 , KOH
(LXI)
(LXII)
MnO2
(LXIII)
KOCMe3
(LXV)
NaBH4
CH2=CH•CH2•CHMe•CO2H
(i) HBr, peroxide
(ii) esterification
Br[CH2]3•CHMe•CO2C2H5
(i) LiAlH4, AlCl3
(ii) acetylation
Br[CH2]3•CHMe•CH2OAc
(i) NaI
(ii) AgNO2
NO2•[CH2]3•CHMe•CH2OAc
(LXIV)
(i) OH⊖
(ii) H⊕
Kryptogenin
NaBH4
Diosgenin = 25R
Yamogenin = 25S
(LXVI)

Scheme 12.

NATURALLY-OCCURRING SAPOGENINS[a]

Name[b]	Source[c]	Configuration $C_{(25)}$	Configuration $C_{(5)}$	Structure Substituents OH	Structure Substituents :O	Double bond	Sapogenin M.P.	Sapogenin $[\alpha]_D$[d]	Acetate[e] M.P.	Acetate[e] $[\alpha]_D$[d]
Sarsasapogenin[1,2]	*Smilax* sp. *Radix sarsaparilla* *Yucca schottii* (fruit)	*S*	β	3β	—	—	198–199°	—75°	143–144°	—70°
Smilagenin[1,2]	*Radix sarsaparilla* *Yucca schottii* (flowers)	*R*	β	3β	—	—	187–188°	—66°	150°	—60°
Neotigogenin[3]	*Chlorogalum pomeridianum* *Digitalis* sp.	*S*	α	3β	—	—	203°	—75°	179°	—75°
Tigogenin[1,2,4,5]	*Chlorogalum pomeridianum* *Yucca* sp.; *Agave* sp.	*R*	α	3β	—	—	208°	—67°	206–208°	—74°
Yamogenin[6]	*Dioscorea* sp.	*S*	—	3β	—	5	201°	—129°	182°	—119°
Diosgenin[1,2,7]	*Dioscorea* sp. *Trillium* sp.	*R*	—	3β	—	5	208°	—129°	199–202°	—121°
Rhodeasapogenin[8]	*Rhodea japonica* Roth.	*S*	β	$1\beta,3\beta$	—	—	293–295°	—72°	185–187°	—70·5°
Isorhodeasapogenin[8]	*Rhodea japonica* Roth.	*R*	β	$1\beta,3\beta$	—	—	245–248°	—71°	205°	—73°
Ruscogenin[9]	*Ruscus aculeatus* L.	*R*	—	$1\beta,3\beta$	—	5	205–210°	—127°	192–194°	—85°
Markogenin[f,10]	*Yucca* sp.	*S*	β	$2\beta,3\beta$	—	—	257°	—70°	185–186°	—84°
Samogenin[6]	*Yucca schottii* *Samuela carnerosana* Trel.	*R*	β	$2\beta,3\beta$	—	—	203–205°	—74°	198°	—84°
Yonogenin[11]	*Dioscorea tokoro* Makino	*R*	β	$2\beta,3\alpha$	—	—	240–243°	—53°	212°	—29°
Neogitogenin[12]	*Digitalis purpurea*; *D. lanata*	*S*	α	$2\alpha,3\beta$	—	—	248°	—	218–220°	—
Gitogenin[1,13]	*Digitalis* sp.; *Yucca schottii*; *Agave* sp.	*R*	α	$2\alpha,3\beta$	—	—	264–267°	—78°	242–243°	—97°
Lilagenin[6,14]	*Lilium* sp.	*S*	—	$2\alpha,3\beta$	—	5	246°	—	155°	—
Yuccagenin[6]	*Yucca* sp.	*R*	—	$2\alpha,3\beta$	—	5	248°	—122°	178°	—139°

(*continued*)

Table 2 (*continued*)

Name[b]	Source[c]	Configuration $C_{(25)}$	Configuration $C_{(5)}$	Structure Substituents OH	Structure Substituents :O	Double bond	Sapogenin M.P.	Sapogenin $[\alpha]_D$[d]	Acetate[e] M.P.	Acetate[e] $[\alpha]_D$[d]
Neochlorogenin[35]	*Solanum paniculatum* L.	*S*	α	3β,6α	—	—	261–265°	—65°	200–202°	—51°
Chlorogenin[4]	*Chlorogalum pomeridianum*	*R*	α	3β,6α	—	—	276°	—45°	155°	—38°
Laxogenin[15]	*Smilax Sieboldi* Miq.	*R*	α	3β	6	—	210–212°	—86°	219–222°	—89°
Neonogiragenin[16]	*Metanarthecium luteo-viride* Maxim.	*S*	β	3β,11α	—	—	128–131°	—75°	142–144°	—78°
Nogiragenin[17]	*Metanarthecium luteo-viride* Maxim.	*R*	β	3β,11α	—	—	200–201°	—71°	208–209°	—74°
Tamusgenin[36]	*Tamus edulis* Lowe	*R*	—	3β	11	5	180–182°	—75°	209–213°	—81°
Rockogenin[6]	*Agave gracilipes* Trel.	*R*	α	3β,12β	—	—	216–220°	—63°	206–209°	—65°[h]
Chiapagenin[18, 19]	*Dioscorea chiapasensis* Matuda	*S*	—	3β,12β	—	5	249–251°	—126°	191–193°	—127°
Isochiapagenin[19]	*Dioscorea chiapasensis* Matuda	*R*	—	3β,12β	—	5	236–237°	—121°	206–207°	—120°
Heloniogenin[20]	*Heloniopsis orientalis* (Thunb.) C., Tanaka	*R*	—	3β,12α	—	5	212–213°	—91°[g]	184–185°	—58°[g]
Willagenin[21]	*Yucca filifera*	*S*	β	3β	12	—	166–168°	+5°	183–185°	—1°
Sisalagenin[22]	*Agave sisalana* Perrine	*S*	α	3β	12	—	244–246°	—4.5°	228–232°	—12°
Hecogenin[6, 23]	*Agave* sp. *Hechtia texensis*	*R*	α	3β	12	—	268°	+10°	245°	—1°
Hispidogenin[37]	*Solanum hispidum* Pers.; *Asparagus umbellatus* Link	*R*	α	12β	3	—	208°	—46°	212°	—51°
Neobotogenin[24] (= correllogenin)	*Dioscorea spiculiflora*	*S*	—	3β	12	5	209–211°	—69°	213–215°	—67°
Botogenin[24] (= gentrogenin)	*Dioscorea spiculiflora*	*R*	—	3β	12	5	215–216°	—57°	227°	—56°
9-Dehydrohecogenin[25]	*Agave* sp.	*R*	α	3β	12	9	223–226°	—10°	215–217°	—9°

(*continued*)

Name[b]	Source[c]	Configuration $C_{(25)}$	Configuration $C_{(5)}$	Structure Substituents OH	Structure Substituents :O	Structure Substituents Double bond	Sapogenin M.P.	Sapogenin $[\alpha]_D$[d]	Acetate[e] M.P.	Acetate[e] $[\alpha]_D$[d]
Neodigalogenin[12]	*Digitalis purpurea*; *D. lanata*	*S*	α	3β,15β	—	—	223–226°	—	211–212°[j]	—88°
Digalogenin[12, 26]	*Digitalis purpurea*; *D. lanata*	*R*	α	3β,15β	—	—	218–219°	—75·5°	238–239°[j]	—70°
Tokorogenin[27]	*Dioscorea tokoro*	*R*	β	1β,2β.3α	—	—	266–268°	—50°[k]	255°	—8°[g]
Convallagenin-A[38, 39]	*Convallaria keisukei*	*S*	β	1β,3β,5β	—	—	268–269°	—28°[m]	208–210°[s]	—78°
Diotigenin[40]	*Dioscorea tenuipes*	*S*	β	2β,3α,4β	—	—	—	—	—	—
Agapanthagenin[28]	*Agapanthus* sp.	*R*	α	2α,3β,5α	—	—	285°	—	298–299°[l]	—101°
Magogenin[29]	*Maguey cacaya*	*R*	α	2ξ,3ξ,6ξ	—	—	284°	—	214°	—
Metagenin[30]	*Metanarthecium luteo-viride*-Maxim.	*R*	β	2β,3β,11α	—	—	273–274°	—82°[m]	250–252°	—77·5°
Neomexogenin[31]	*Agave roezliana*	*S*	β	2β,3β	12	—	221–222°	—	162–164°	—
Mexogenin[6]	*Yucca schottii*; *Samuela carnerosana* Trel.	*R*	β	2β,3β	12	—	237–238°	—6°	208°	—
Agavogenin[6]	*Agave huachucensis* Baker	*R*	α	2α,3β,12ξ	—	—	242°	—62°[n]	230°	—98°[n]
Neomanogenin[32]	*Yucca schottii*; *Agave striata*	*S*	α	2α,3β	12	—	242°	—	222°	—
Manogenin[6]	*Agave* sp.; *Manfreda maculosa* Hook	*R*	α	2α,3β	12	—	242°	—5°	259°	—45°
Neokammogenin[32]	*Dioscorea mexicana*	*S*	—	2α,3β	12	5	230°	—	203–205°	—
Kammogenin[6]	*Yucca* sp.; *Samuela carnerosana* Trel.	*R*	—	2α,3β	12	5	245°	—53°	259°	—83°
9-Dehydromanogenin[25]	*Agave* sp.	*R*	α	2α,3β	12	9	240°	—16°[n]	263°	—62°[n]
Neodigitogenin[12, 33]	*Digitalis purpurea*	*S*	α	2α,3β,15β	—	—	277–279°	—82°[o]	229–232°[l]	—114°
Digitogenin[1]	*Digitalis purpurea*; *D. lanata*	*R*	α	2α,3β,15β	—	—	296°	—80°	235–236°[l]	—103°
Kogagenin[11]	*Dioscorea tokoro* Makino	*R*	β	1β,2β,3α,5β	—	—	318–322°	—27°[o]	249–252°[p]	—26°
Convallagenin-B[38, 41]	*Convallaria keisukei*	*S*	β	1β,3β,4β,5β	—	—	277–278°	—43°[m]	228–230°[q]	—46·5°
Kitigenin[34]	*Reineckia carnea* Kunth.	*R*	β	1β,3β,4β,5β	—	—	298°	—35°[o]	219–220°[q]	—54°
Cacogenin[29]	*Maguey cacaya*	*R*	α	2ξ,3ξ,6ξ	12	—	278°	—	248°	—
Pentologenin[34]	*Reineckia carnea* Kunth.	*R*	ξ	1ξ,2ξ,3ξ,4ξ,5	—	—	320°	—54·5°[m]	165–168°[r]	—

Footnotes to Table 2

a This list does not include compounds in which the usual spirostan side-chain is modified; these are considered individually on pp. 43–52.
b The compounds are listed in order of increasing number of oxygen atoms: ketonic compounds follow the corresponding alcohols: unsaturated compounds follow the corresponding saturated compounds.
c List of sources is not comprehensive.
d In chloroform, unless otherwise noted.
e Fully acetylated compound, unless otherwise noted.
f *Marker et al.* (J. Amer. chem. Soc., 1947, **69**, 2167) attributed this structure to a different compound, which they called texogenin.
g Solvent unspecified.
h In acetone.
j 3-Monoacetate.
k In ethanol.
l 2,3-Diacetate.
m In chloroform-methanol.
n In dioxane.
o In pyridine.
p 1,2,3-Triacetate.
q 1,3,4-Triacetate.
r 1,2,3,4-Tetra-acetate.
s 1,3-Diacetate.

References

1 See references to Table 1.
2 *R. E. Marker et al.*, J. Amer. chem. Soc., 1943, **65**, 1199.
3 *L. H. Goodson* and *C. R. Noller, ibid.*, 1939, **61**, 2420; *Marker, E. Rohrmann* and *E. M. Jones, ibid.*, 1940, **62**, 1162.
4 *Poe Liang* and *Noller, ibid.*, 1935, **57**, 525.
5 *Marker et al., ibid.*, 1947, **69**, 2375, 2403.
6 *Idem, ibid.*, p. 2167.
7 *T. Tsukamoto* and *Y. Ueno*, J. pharm. Soc. Japan, 1936, **56**, 802; *Marker et al.*, J. Amer. chem. Soc., 1940, **62**, 2542; 1947, **69**, 2242.
8 *H. Nawa*, Proc. imp. Acad. Japan, 1957, **33**, 570; Chem. pharm. Bull. Tokyo, 1958, **6**, 255; J. pharm. Soc. Japan, 1953, **73**, 1192, 1195, 1197.
9 *H. Lapin* and *C. Sannié*, Bull. Soc. chim. Fr., 1955, 1552, 1556.
10 *M. E. Wall et al.*, J. Amer. chem. Soc., 1953, **75**, 4437.
11 *K. Takeda, T. Okanishi* and *A. Shimaoka*, Chem. pharm. Bull. Tokyo, 1958, **6**, 532.
12 *R. Tschesche, G. Wulff* and *G. Balle*, Tetrahedron, 1962, **18**, 959.
13 *Marker* and *J. Lopez*, J. Amer. chem. Soc., 1947, **69**, 2375, 2403.
14 *Marker et al., ibid.*, 1940, **62**, 2620.
15 *A. Akahori* and *F. Yasuda*, J. pharm. Soc. Japan, 1963, **83**, 557; *Okanishi, Akahori* and *Yasuda*, Chem. Pharm. Bull. Tokyo, 1965, **13**, 545.
16 *H. Minato* and *Shimaoka*, Chem. pharm. Bull. Tokyo, 1963, **11**, 876.
17 *Takeda et al., ibid.*, 1961, **9**, 388.
18 *S. Serota, R. P. Koob* and *Wall*, J. org. Chem., 1960, **25**, 1768.
19 *I. T. Harrison, M. Velasco* and *C. Djerassi, ibid.*, 1961, **26**, 155.
20 *Okanishi, Akahori* and *Yasuda*, Chem. pharm. Bull. Tokyo, 1962, **10**, 1195.

21 *H. E. Kenney* and *Wall*, J. org. Chem., 1957, **22,** 468.
22 *R. K. Callow* and *V. H. T. James*, J. chem. Soc., 1955, 1671.
23 *Callow, J. W. Cornforth* and *P. C. Spensley*, Chem. and Ind., 1951, 699; *Spensley*, *ibid.*, 1952, 426.
24 *H. A. Walens, Serota* and *Wall*, J. org. Chem., 1957, **22,** 182.
25 *R. B. Wagner, R. F. Forker* and *P. F. Spitzer*, J. Amer. chem. Soc., 1951, **73,** 2494.
26 *Tschesche* and *Wulff*, Ber., 1961, **94,** 2019.
27 *M. Nishikawa et al.*, J. pharm. Soc. Japan, 1954, **74,** 1165.
28 *T. Stephen*, J. chem. Soc., 1956, 1167.
29 *Marker*, J. Amer. chem. Soc., 1947, **69,** 2399.
30 *Takeda et al.*, J. pharm. Soc. Japan, 1957, **77,** 175.
31 *Marker* and *J. Lopez*, J. Amer. chem. Soc., 1947, **69,** 2373, 2383.
32 *Idem, ibid.*, p. 2375.
33 *D. L. Klass, M. Fieser* and *L. F. Fieser, ibid.*, 1955, **77,** 3829.
34 *Takeda et al.*, Chem. pharm. Bull. Tokyo, 1961, **9,** 631.
35 *K. Schreiber, H. Ripperger* and *H. Budzikiewicz*, Tetrahedron Letters, 1965, 3999, Ber., 1967, **100,** 1741.
36 *A. G. González, R. Freire* and *E. S. López*, An. R. Soc. esp. Fís. Quím., 1967, **63B,** 966.
37 *P. C. Maiti* and *S. Mookherjea*, Chem. and Ind., 1965, 1653; *R. F. Díaz, R. F. Barreira* and *González*, An. R. Soc. esp. Fís. Quím, 1967, **63B,** 927.
38 *M. Kimura, M. Tohma* and *I. Yoshizawa*, Chem. pharm. Bull. Tokyo, 1966, **14,** 50
39 *Idem, ibid.*, 1967, **15,** 1204.
40 *M. Ogata, Yasuda* and *Takeda*, J. chem. Soc. (C), 1967, 2397; *Takeda, G. Lukacs* and *Yasuda, ibid.*, 1968, 1041.
41 *M. Kimura, M. Tohma* and *I. Yoshizawa*, Chem. pharm. Bull., Tokyo, 1967, **15,** 1713.

it is thought that the biogenetic pathway is much the same as that to cholesterol. However, cholesterol is not, itself, an intermediate on the major pathway; the carbon atoms that ultimately become $C_{(26)}$ and $C_{(27)}$ remain distinct, whereas they would become equivalent in cholesterol (*R. Joly* and *C. Tamm*, Tetrahedron Letters, 1967, 3535; *R. D. Bennett et al.*, Archs. Biochem. Biophys., 1963, **103,** 74, cf. *Tschesche* and *H. Hulpke*, Z. Naturf., 1966, **21б,** 494; *Bennett* and *E. Heftmann*, Phytochemistry, 1965, **4,** 577).

(c) Individual sapogenins

Table 2 lists the naturally-occurring sapogenins with the usual spirostan side-chain; some compounds in which the side-chain is modified are described later in this section (see pp. 43–52).

Inter-relationship of members of pairs epimeric at $C_{(25)}$ has already been discussed; in the following account of individual sapogenins, only one member of any such pair is usually considered.

(i) Sapogenins with normal side-chains

Evidence that **sarsasapogenin** is a 3β-hydroxy-5β-compound comes from its degradation to 3β-hydroxyetiobilianic acid (p. 8). Sarsasapogenin can be

converted, *via* the 5β-3-ketone, into the Δ^4-3-ketone which, on reduction with sodium and alcohol, gives the 3β-hydroxy-5α-compound, **neotigogenin**; **smilagenin,** the (25*R*)-epimer of sarsasapogenin, can similarly be converted into **tigogenin** (*R. E. Marker et al.*, J. Amer. chem. Soc., 1940, **62,** 647, 1162). Further evidence for the β-configuration of the 3-hydroxyl group in tigogenin and neotigogenin comes from their formation of insoluble digitonides (*Marker* and *E. Rohrmann, ibid.*, p. 647; *C. R. Noller, ibid.*, 1939, **61,** 2717), and their degradation to simpler compounds of known configuration at $C_{(3)}$ provides more certain proof.

The structure of **diosgenin** as the 5,6-dehydro derivative of tigogenin follows from its conversion into the latter by catalytic hydrogenation (*T. Tsukamoto et al.*, J. pharm. Soc. Japan, 1936, **56,** 802; 1937, **57,** 9, 283), from its conversion into cholesterol (see p. 19) and from its oxidation, under Oppenauer conditions, to the Δ^4-3-ketone (diosgenone) (*Marker, Tsukamoto* and *D. L. Turner,* J. Amer. chem. Soc., 1940, **62,** 2525). Diosgenin is probably the most abundant of the sapogenins and is now widely used as a cheap starting material for steroid synthesis – notably for the manufacture of steroid hormones and their analogues *via* 3β-hydroxypregna-5,16-dien-20-one (see Vol. II D, pp. 232, 233, 254, 326, 328).

Isorhodeasapogenin was tentatively assigned the $1\beta,3\beta$-dihydroxy-5β-structure as a result of work on the structure of tokorogenin (see below). This was confirmed by its degradation to the known methyl $1\beta,3\beta$-diacetoxy-5β-etianate (*H. Nawa,* Chem. pharm. Bull. Tokyo, 1958, **6,** 255).

Ruscogenin, a mono-unsaturated diol, was characterised as a hydroxy derivative of diosgenin, since it could be selectively oxidised with chromic acid to a hydroxy-ketone LXVII that gave diosgenin (LXVIII) on Wolff-Kishner reduction (Scheme 13) (*H. Lapin* and *C. Sannié,* Bull. Soc. chim. Fr., 1955, 1552). Oppenauer oxidation of ruscogenin was accompanied by elimination of the second

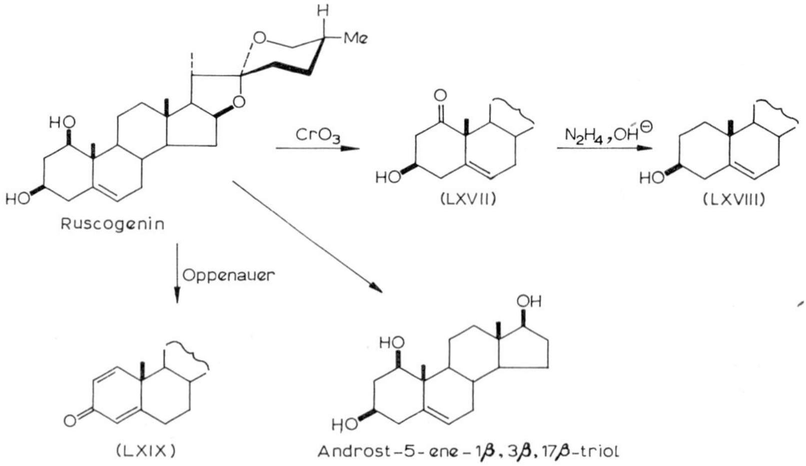

Scheme 13.

hydroxyl group, to give the $\Delta^{1,4}$-3-ketone LXIX, thus showing this hydroxy-group to be at $C_{(1)}$ (*D. Burn, B. Ellis* and *V. Petrow,* J. chem. Soc., 1958, 795; cf. *A. L. Nussbaum et al.*, J. Amer. chem. Soc., 1959, **81**, 5230; *Lapin,* Bull. Soc. chim. Fr., 1957, 1501); proof of its β-configuration was provided by the conversion of ruscogenin into authentic androst-5-ene-1β,3β,17β-triol (*W. R. Benn, F. Colton* and *R. Pappo,* J. Amer. chem. Soc., 1957, **79**, 3920).

Samogenin (LXX; R=H) (Scheme 14) was shown by *Marker et al.* to be a vicinal diol by its oxidation to a dicarboxylic acid LXXI containing the same number of carbon atoms (*ibid.*, 1947, **69**, 2167). It can be converted, *via* the dimesylate (LXX; R=Ms) and Δ^2-olefin LXXII, into 3-deoxysmilagenin (LXXIII) and is, therefore, a (25*R*)-5β-spirostandiol. The *cis*-relationship of the hydroxyl groups is shown by formation of an acetonide and by regeneration of samogenin through the action of osmium tetroxide on the Δ^2-olefin LXXII. The latter compound also gives an epoxide, which is reduced by lithium aluminium hydride to smilagenin (LXXIV), and must therefore be the 2β,3β-epoxide. [The alternative possibility, that it was the 3β,4β-epoxide, was ruled out by

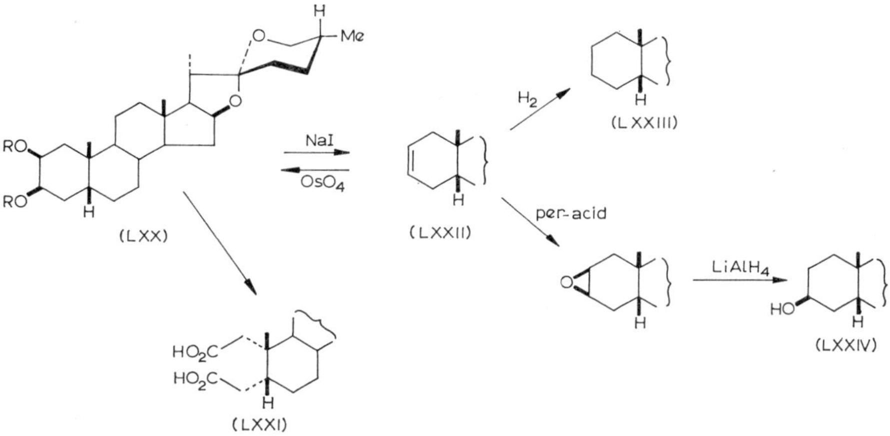

Scheme 14.

synthesis of (25*R*)-5β-spirost-3-en ,which differed from LXXII]. Assuming that osmylation and epoxidation of the olefin involve attack from the same side of the molecule, samogenin must have the 2β,3β-dihydroxy-structure (LXX; R=H) (*C. Djerassi* and *J. Fishman, ibid.*, 1955, **77**, 4291).

Yonogenin, a diol of the (25*R*)-series, is oxidised to the same dicarboxylic acid LXXI as is samogenin; it is, therefore, an isomeric 2,3-dihydroxy-5β compound. It does not yield an acetonide, excluding a *cis*-relationship of the hydroxyl groups, and its formation of a dicathylate suggests that both hydroxyl groups are equatorial (*i.e.* 2β,3α). In agreement with this, it proved identical with the 3-episamogenin prepared by treatment of samogenin with sodium

ethoxide (*K. Takeda, T. Okanishi* and *A. Shimaoka*, Chem. pharm. Bull. Tokyo, 1958, **6**, 532).

Gitogenin (LXXVI) is oxidised by chromic acid to the same dicarboxylic acid, gitogenic acid (LXXVII), as is tigogenin, indicating the (25*R*)-2,3-dihydroxy-5α-spirostan structure (*R. Tschesche*, Ber., 1935, **68**, 1090; *A. Windaus* and *A. Schneckenburger*, *ibid.*, 1913, **46**, 2628; *W. A. Jacobs* and *J. C. E. Simpson*, J. biol. Chem., 1935, **110**, 429). The failure of gitogenin to yield an acetonide suggests that it is a *trans*-glycol; since it proved to be different from (25*R*)-5α-spirostan-2β,3α-diol, the 2α,3β-dihydroxy-structure LXXVI was suggested (*J. Pataki, G. Rosenkranz* and *Djerassi*, J. Amer. chem. Soc., 1951, **73**, 5375). This was confirmed by synthesis of gitogenin from diosgenin as shown in Scheme 15; the α-configuration of the 2-acetoxy-group in LXXV followed from the known course of the acetolysis reaction in related series. Proof of the β-configuration of the 3-hydroxy-group then followed from the non-identity of gitogenin with (25*R*)-5α-spirostan-2α,3α-diol (*J. Herran, Rosenkranz* and *F. Sondheimer*, *ibid.*, 1954 **76**, 5531).

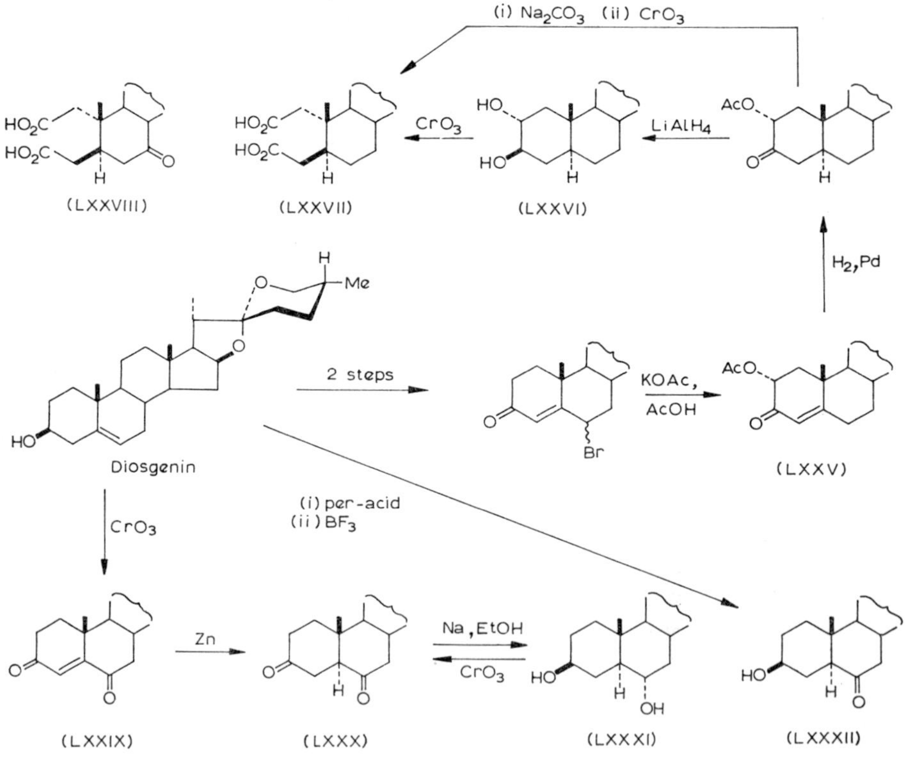

(i) Na_2CO_3 (ii) CrO_3

CrO_3 LiAlH₄

(LXXVIII) (LXXVII) (LXXVI)

H_2,Pd

Diosgenin

2 steps

KOAc, AcOH

(LXXV)

CrO_3

(i) per-acid (ii) BF_3

Zn

Na, EtOH

CrO_3

(LXXIX) (LXXX) (LXXXI) (LXXXII)

Scheme 15.

Yuccagenin is a dehydro-derivative of gitogenin, into which it is converted by hydrogenation. The double-bond was shown to be at 5,6 by a series of transformations that led to 7-oxogitogenic acid (LXXVIII), which could also be obtained in a similar way from diosgenin (*Marker et al., ibid.*, 1947, **69**, 2167, 2401).

Chlorogenin (LXXXI), a diol, is oxidised by chromic acid to the corresponding diketone, chlorogenone (LXXX), which is stable to acid and yields a pyridazone derivative on treatment with hydrazine; *Marker* and *Rohrmann* (*ibid.*, 1939, **61**, 946) correctly deduced that the sapogenin is a 3,6-diol of the 5α-series. Further, diosgenin, on chromic oxidation, gives the Δ^4-3,6-dione LXXIX, which is reduced by zinc to chlorogenone (LXXX). Reduction of the latter with sodium and ethanol yields chlorogenin (LXXXI), whereas reduction with hydrogen gives the isomeric "β-chlorogenin". From the known course of such reductions in related compounds, *Marker* assigned the 3β,6α-configurations to the hydroxyl groups of chlorogenin, "β-chlorogenin" being the 3β,6β-diol (*Marker et al., ibid.*, 1940, **62**, 2537, 3006, 3009).

Laxogenin, a hydroxy-ketone, yields tigogenin on Wolff-Kishner reduction and a mixture of chlorogenin and β-chlorogenin on reduction with sodium borohydride; it is, therefore, (25*R*)-3β-hydroxy-5α-spirostan-6-one (LXXXII). This was confirmed by its synthesis from diosgenin, whose 5α,6α-epoxide yielded laxogenin on treatment with boron trifluoride. It has been shown that laxogenin exists in the saponin in the 5α-configuration, rather than arising by epimerisation of a 5β-compound under the acid conditions of hydrolysis (*A. Akahori* and *F. Yasuda*, J. pharm. Soc. Japan, 1963, **83**, 557; *T. Okanishi, Akahori* and *Yasuda*, Chem. pharm. Bull. Tokyo, 1965, **13**, 545).

Nogiragenin, characterised as a (25*R*)-spirostandiol, is oxidised by *N*-bromoacetamide or under Oppenauer conditions to the known (25*R*)-11α-hydroxy-5β-spirostan-3-one and by chromic acid to the corresponding 3,11-dione; it is, therefore, a (25*R*)-5β-spirostan-3,11α-diol. The β-configuration of the 3-hydroxyl group was shown by the formation of nogiragenin on reduction of (25*R*)-11α-acetoxy-2β,3β-epoxy-5β-spirostan with lithium aluminium hydride (*K. Takeda et al., ibid.*, 1961, **9**, 388; Steroids, 1963, **2**, 27).

Metagenin, a congener of nogiragenin, is a (25*R*)-spirostantriol. It forms an acetonide, whose free hydroxyl group yields an oxo group on oxidation; removal of the protecting group and Wolff-Kishner reduction then gives samogenin, showing metagenin to be (25*R*)-5β-spirostan-2β,3β,x-triol (*Takeda* and *K. Hamamoto*, Chem. pharm. Bull. Tokyo, 1960, **8**, 1004). The reactions of the third hydroxyl group suggested that it was in the 11α-position, and this was confirmed by the conversion of metagenin into (25*R*)-11α-acetoxy-5β-spirostan, identical with an authentic sample (*Hamamoto, ibid.*, 1960, **8**, 1099; 1961, **9**, 32).

Hecogenin has already been mentioned as one of the economically important steroid sapogenins. It owes this importance to its possession of an oxo group at $C_{(12)}$ (which facilitates its conversion into 11-oxygenated compounds, such as cortisone) and to its presence in the waste-liquor from the manufacture of sisal. That hecogenin is an oxo derivative of tigogenin follows from its conversion into

the latter by Wolff-Kishner reduction. The ketone group is rather (but not completely) unreactive to carbonyl reagents and *Marker*, having eliminated some less likely positions, suggested that it was at $C_{(12)}$ (J. Amer. chem. Soc., 1947, **69**, 2167). This was confirmed by a conversion of hecogenin into 5α-pregnane-3,12,20-trione, identical with material obtained from desoxycholic acid (*R. B. Wagner, J. A. Moore* and *R. F. Forker, ibid.*, 1950, **72**, 1856).

Botogenin (renamed **gentrogenin** by *Wall et al.*) was shown to be the 5,6-dehydro-derivative of hecogenin by its conversion into the latter on catalytic reduction, and by its conversion into diosgenin on Wolff-Kishner reduction. In the same way **neobotogenin** (**correllogenin**) has been related to sisalagenin and to yamogenin (*H. A. Walens, S. Serota* and *M. E. Wall*, J. org. Chem., 1957, **22**, 182).

Manogenin (LXXXIII; R=H) (Scheme 16) is reduced under Wolff-Kishner conditions to gitogenin and is, therefore, an oxo derivative of the latter. That the oxo group is at $C_{(12)}$ was shown (a) by the oxidation of manogenin to the same oxo-dicarboxylic acid, hecogenic acid, as is given by hecogenin and (b)

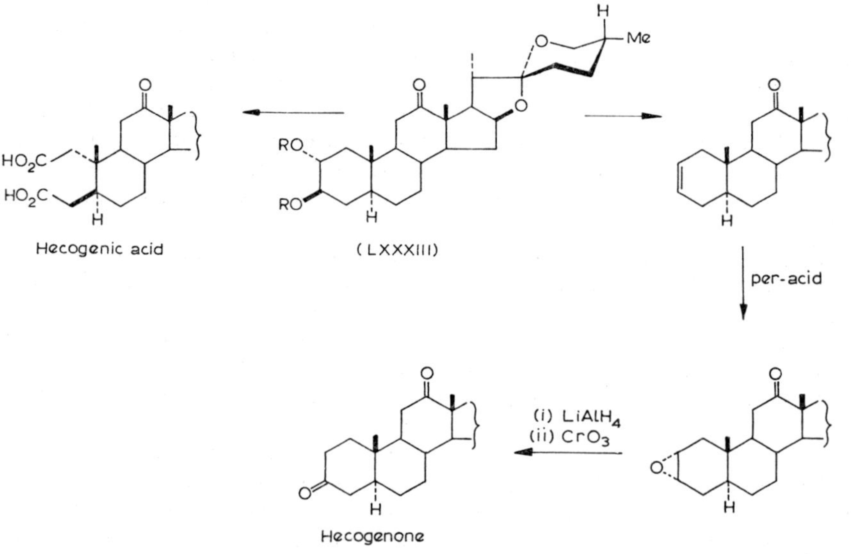

Scheme 16.

by conversion of manogenin *via* the 2,3-dimesylate (LXXXIII; R=Ms), Δ^2-olefin and 2α,3α-epoxide into a 3,12-diol, which was oxidised by chromic acid to hecogenone, (25*R*)-5α-spirostan-3,12-dione (*Marker et al.*, J. Amer. chem. Soc., 1947, **69**, 2167; *N. L. Wendler, H. L. Slates* and *M. Tishler, ibid.*, 1952, **74**, 4894; cf. *Slates* and *Wendler, ibid.*, 1956, **78**, 3749). **Kammogenin,** the 5,6-dehydro derivative of manogenin, yields the latter on catalytic hydrogenation and is

converted into yuccagenin by Wolff-Kishner reduction (*Marker et al.*, *ibid.*, 1947, **69**, 2167).

Of the remaining 12-oxospirostans, **willagenin**, a hydroxy-ketone of the (25*S*)-series, is converted by Wolff-Kishner reduction into sarsasapogenin, of which it is, therefore an oxo derivative. The position of the oxo group has not been proven rigorously, but has been assigned to $C_{(12)}$, partly by elimination, partly on the basis of infrared and optical rotatory characteristics (*H. E. Kenney* and *Wall*, J. org. Chem., 1957, **22**, 468).

Mexogenin has been characterised, by Wolff-Kishner reduction, as an oxo derivative of samogenin. Again, the position of the oxo group is not certain, but it is thought to be at position 12, partly on the basis of its reactivity, partly from spectroscopic evidence (*Marker et al.*, J. Amer. chem. Soc., 1947, **69**, 2167; *Djerassi et al.*, *ibid.*, 1955, **77**, 4291; 1956, **78**, 440).

9-Dehydrohecogenin and **9-dehydromanogenin** occur together with the saturated 12-ketones. Their ultraviolet spectra show the presence of an α,β-unsaturated ketone grouping, and they have been related to hecogenin and manogenin, respectively, by reduction (*Wagner*, *Forker* and *Spitzer*, *ibid.*, 1951, **73**, 2494). 9-Dehydrohecogenin acetate has been synthesised from hecogenin acetate (a) by 11,23-dibromination, dehydrobromination to give 23-bromo-9-dehydrohecogenin acetate and removal of the side-chain bromine with zinc (*C. Djerassi*, *H. Martinez* and *G. Rosenkranz*, J. org. Chem., 1951, **16**, 303) and (b) by direct dehydrogenation with selenium dioxide in presence of pyridine (*Djerassi et al.*, J. chem. Soc., 1961, 1859).

As well as these 12-oxosapogenins, several 12-hydroxy derivatives have been characterised. Thus, **rockogenin** was recognised as a (25*R*)-5α-spirostan-3β,12-diol by its conversion into hecogenone (see Scheme 16) on oxidation with chromic acid and by its formation from the corresponding 3β-hydroxy-12-ketone, hecogenin, by reduction with hydrogen in presence of platinum, or with sodium and ethanol (*Marker et al.*, J. Amer. chem. Soc., 1947, **69**, 2167). Reduction of hecogenin with lithium aluminium hydride gives both rockogenin and its 12-epimer. The $C_{(12)}$ configurations of these two 3β,12-diols were assigned on the basis of their rotations and of the relative rates of hydrolysis of their acetates and were confirmed by the fact that solvolysis of the mesylate of the axial 12α-hydroxy compound resulted in simple elimination to the Δ^{11}-olefin, whereas that of the equatorial 12β-hydroxy-compound (rockogenin) underwent rearrangement to C-nor-D-homo-compounds (*Wendler et al.*, *ibid.*, 1954, **76**, 4013).

Chiapagenin is converted into rockogenin by catalytic hydrogenation and subsequent acid-catalysed epimerisation at $C_{(25)}$; it is, therefore, a (25*S*)-spirosten-3β,12β-diol. The 5,6-position of the double bond is shown by formation of correllogenin acetate on oxidation of chiapagenin 3-mono-acetate with chromic acid and, conversely, by reduction of correllogenin with lithium in liquid ammonia to give chiapagenin (*Serota*, *R. P. Koob* and *Wall*, J. org. Chem., 1960, **25**, 1768; *I. T. Harrison*, *M. Velasco* and *Djerassi*, *ibid.*, 1961, **26**, 155.

Heloniogenin is a (25*R*)-spirost-5-en-3β,12-diol, its 3-mono-acetate being oxidised by chromic acid to gentrogenin acetate; since it differs from isochiapa-

genin, it must be the 12α-epimer. Reduction of gentrogenin with sodium borohydride or lithium aluminium hydride gives a mixture of isochiapagenin and heloniogenin (*T. Okanishi, A. Akahori* and *F. Yasuda,* Chem. pharm. Bull. Tokyo, 1962, **10,** 1195).

Agavogenin was characterised as a (25*R*)-5α-spirostan-2,3,12-triol, since it gave hecogenic acid on oxidation with chromic acid. It can be obtained from manogenin by reduction with hydrogen or with sodium and ethanol and it is, therefore, assigned the 2α,3β,12-trihydroxy-structure, the 12-hydroxyl group probably having the β-configuration (*Marker et al.,* J. Amer. chem. Soc., 1947, **69,** 2167).

Magogenin is characterised as a (25*R*)-2,3,6-trihydroxy-5α- compound by its oxidation to the same 6-oxo-dicarboxylic acid as is given by chlorogenin. The configurations of the hydroxyl groups are not known. **Cacogenin** gives magogenin on Wolff-Kishner reduction and is, therefore an oxo derivative of it. Since oxidation to a dioxo-dicarboxylic acid and partial Clemmensen reduction yields hecogenic acid (Scheme 16) the oxo-group must be at position 12 (*Marker, ibid.,* 1947, **69,** 2399).

Digitogenin, although the first sapogenin to be isolated, was late to be completely characterised. Since oxidation of this triol with chromic acid gives an oxo-dicarboxylic acid, digitogenic acid, that is reduced under Wolff-Kishner conditions to gitogenic acid, two of the hydroxyl groups must be at positions 2 and 3 (*Tschesche,* Ber., 1935, **68,** 1090). Digitogenin was shown, in fact, to be a hydroxy derivative of gitogenin, into which it could be converted by formation of the 2,3-dicathylate, oxidation of the remaining hydroxyl-group to ketone and reduction of the resulting compound LXXXIV *via* the dithioacetal (*D. L. Klass, M. Fieser* and *L. F. Fieser,* J. Amer. chem. Soc., 1955, **77,** 3829; cf. *Djerassi T. T. Grossnickle* and *L. B. High, ibid.,* 1956, **78,** 3166) (Scheme 17.) Since digitogenic acid is epimerised by base, it was assumed that the ketone group (and hence the third hydroxyl group in digitogenin) is α- to a bridge-head CH-group, and *Marker* correctly suggested that it was at $C_{(15)}$ (*Tschesche* and *A. Hagedorn,* Ber., 1936, **69,** 797; *Marker, D. L. Turner* and *P. R. Ulshafer,* J. Amer. chem. Soc., 1942, **64,** 1843). Evidence for this came later from the infrared spectra of the oxo compounds, which showed the presence of a 5-membered ring ketone. Further, digitogenin can be converted, *via* its 2,3-dimesylate and the Δ^2-olefin LXXXV, into the 2α,3α-epoxide LXXXVI; this, on reduction with lithium aluminium hydride, gives the 3α,15β-diol LXXXVII, which can be degraded, by standard methods, to 5α-pregnane-3,15,20-trione, identical with authentic material prepared from 15β-hydroxyprogesterone (*Djerassi, Grossnickle* and *High, loc. cit.*). This established the position of the third hydroxyl group of digitogenin and provided final proof of the 14α-configuration. Assignment of the β-configuration to the 15-hydroxyl group is based upon the difficulty of acylating it (consistent with the hindered 15β-position) and by its rotation contribution (*Klass, Fieser* and *Fieser, loc. cit.*; *Djerassi et al.,* J. Amer. chem. Soc., 1955, **77,** 3673; cf. *Djerassi, Grossnickle* and *High, loc. cit.*).

The more recently isolated **digalogenin** yields, on oxidation with chromic

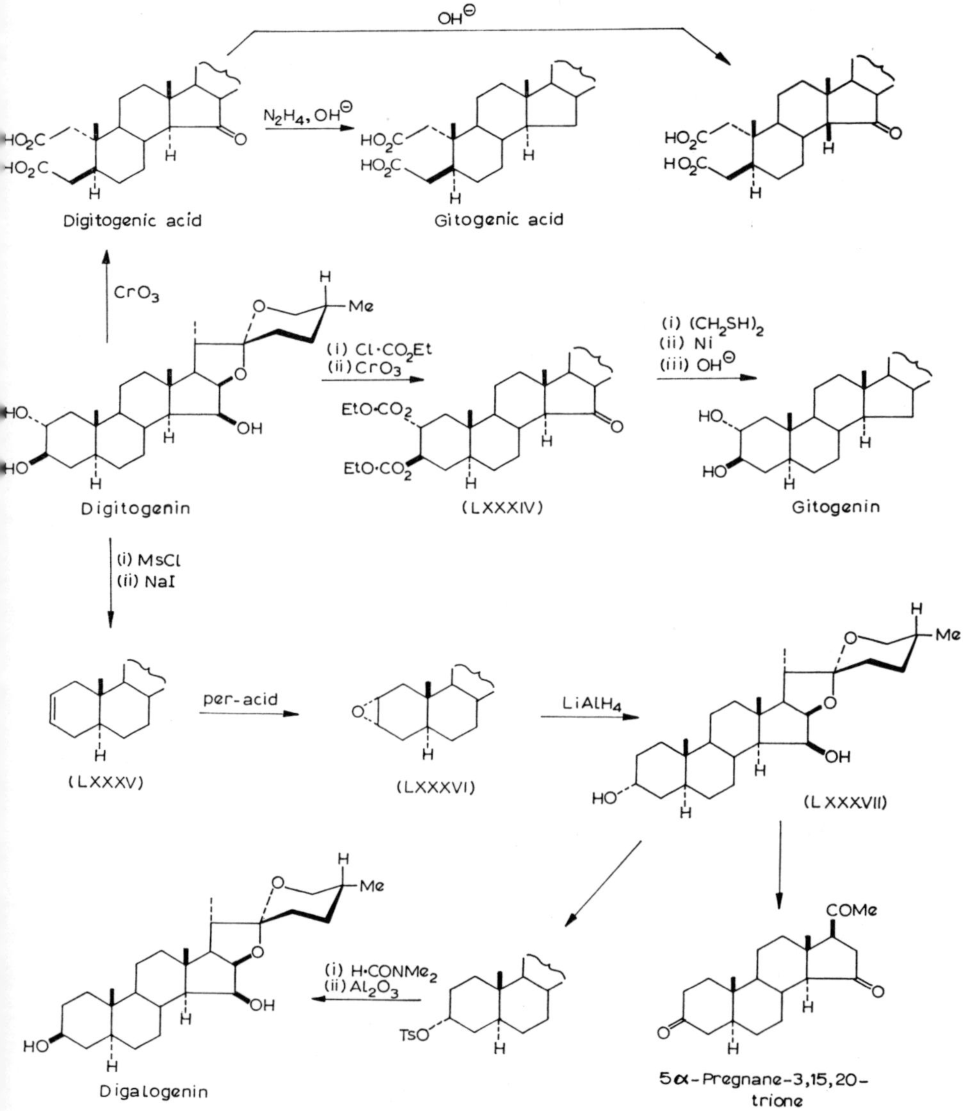

Scheme 17.

acid, a diketone, shown by infrared spectroscopy to have one carbonyl group in a six-membered, the other in a five-membered ring. This suggested that it is a 3,15-diol and proof that it is (25*R*)-5α-spirostan-3β,15β-diol came from its

synthesis from the $3\alpha,15\beta$-diol LXXXVII by epimerisation of the 3-hydroxy-group, as shown (*Tschesche* and *Wulff*, Ber., 1961, **94**, 2019).

An interesting series of sapogenins, containing 3, 4 or 5 hydroxyl groups in Ring A, has been isolated, mostly by Japanese workers. **Tokorogenin** is a (25*R*)-spirostantriol, which is oxidised by chromic acid to a dibasic acid, tokorogenic acid, (Scheme 18) containing one less carbon atom than the sapogenin and in which one carboxyl group is attached to a fully substituted carbon atom; this is consistent with the 1,2,3-triol structure. Tokorogenin forms an acetonide whose free hydroxyl group can be tosylated; removal of the protecting group and treatment of the resulting triol monotosylate LXXXVIII with base gives a hydroxy-epoxide LXXXIX. Hence tokorogenin contains two vicinal hydroxyl groups in the *cis*-relationship, the third hydroxyl group being vicinal and *trans* to one of these. Reduction of the hydroxy-epoxide LXXXIX with lithium aluminium hydride yields isorhodeasapogenin (XC); this is not a vicinal diol and, therefore, if the usual axial opening of the epoxide is assumed, it must be the 1,3-diaxial diol. Tokorogenin was shown to have the 5β-configuration by con-

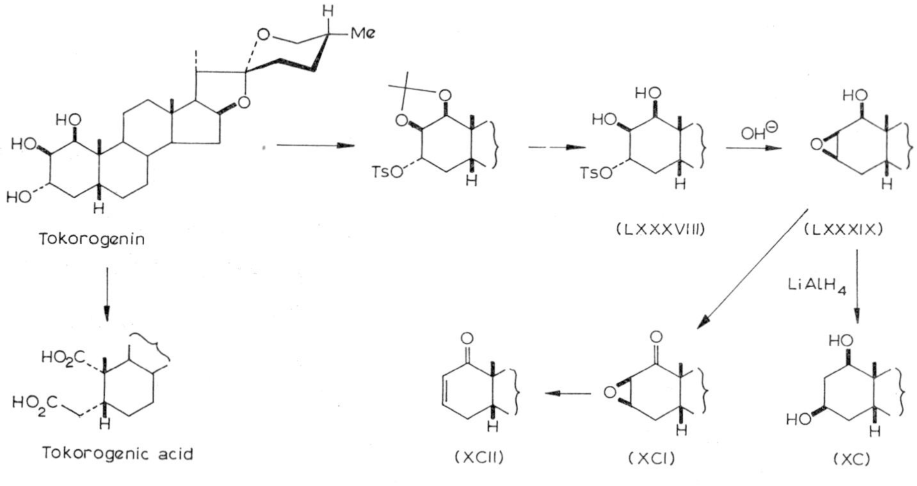

Scheme 18.

versions into samogenic acid and smilagenone. Isorhodeasapogenin must, therefore, be (25*R*)-5β-spirostan-$1\beta,3\beta$-diol (XC) (for independent proof of structure, see p. 30), and tokorogenin must be either the $1\beta,2\beta,3\alpha$- or the $1\alpha,2\beta,3\beta$-triol. A decision can be reached between these two in the following way: oxidation of the hydroxy-epoxide LXXXIX to the corresponding oxo-epoxide XCI and reduction with chromous chloride yields an α,β-unsaturated ketone XCII with the properties of a Δ^2-1-ketone, rather than those of a Δ^1-3-ketone. This is consistent only with the $1\beta,2\beta,3\alpha$-triol formulation for tokorogenin, since the

1α,2β,3β-triol would have given rise to the 3β-hydroxy-1β,2β-epoxide and, thence, the Δ^1-3-ketone (*K. Morita*, Pharm. Bull. Tokyo, 1957, **5**, 494; Bull. chem. Soc. Japan, 1959, **32**, 476, 791, 796).

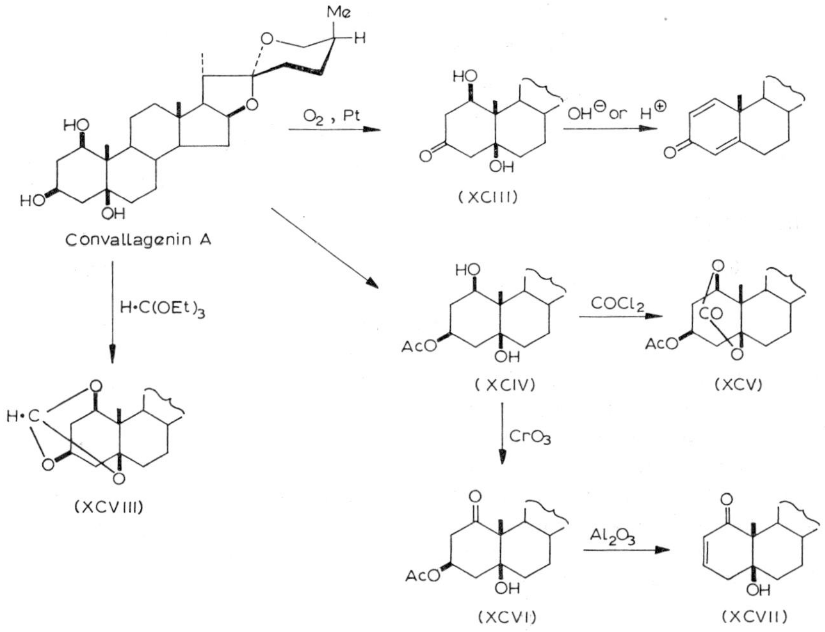

Scheme 19.

Convallagenin-A, a (25*S*)-spirostantriol, yields a mono- and a di-acetate; the third hydroxyl group is inert to chromic acid and is, therefore, tertiary. Oxidation of the triol with oxygen in presence of platinum gives a dihydroxy-ketone XCIII (Scheme 19), which is converted, by acid or base, into the $\Delta^{1,4}$-3-ketone; hence convallagenin-A is a 1,3,5-triol. The ORD spectrum of the dihydroxyketone XCIII suggests that rings A and B are *cis*-fused; *i.e.* that the 5-hydroxyl group is in the β-configuration. Since the 3-mono-acetate XCIV yields a 1,5-cyclic carbonate XCV, the 1-hydroxy-group must also have the β-configuration. Oxidation of the mono-acetate with chromic acid gives the 1-oxo compound XCVI, from which the 3-acetoxy group is eliminated, with formation of compound XCVII under very mildly basic conditions; this suggests that the 3-hydroxyl group in convallagenin-A is axial, *i.e.* in the β-configuration. Further evidence for this structure includes the formation of the cyclic orthoformate XCVIII when convallagenin-A is treated with ethyl orthoformate (*M. Kimura, M. Tohma* and *I. Yoshizawa*, Chem. pharm. Bull. Tokyo, 1967, **15**, 1204).

Agapanthagenin (XCIX; R=H) (Scheme 20), is a (25*R*)-triol that yields only

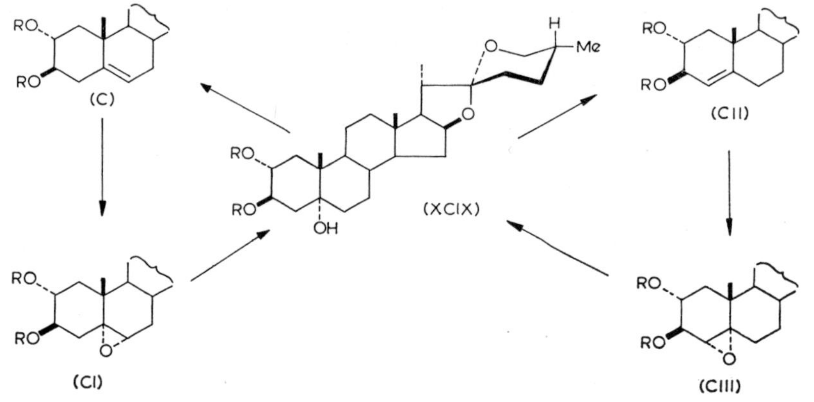

Scheme 20.

a diacetate under normal conditions. Dehydration gives yuccagenin (C; R=H) and its Δ^4-isomer CII; R=H), showing it to be a $2\alpha,3\beta,5$-triol. Yuccagenin diacetate (C; R=Ac) and its isomer (CII; R=Ac) yield epoxides (CI and CIII; R=Ac), formulated as $5\alpha,6\alpha$- and $4\alpha,5\alpha$-, respectively; these are both reduced to agapanthagenin by lithium aluminium hydride, showing the sapogenin to be (25*R*)-5α-spirostan-$2\alpha,3\beta,5$-triol (XCIX; R=H) (*T. Stephen*, J. chem. Soc., 1956, 1167; *G. E. A. Mathew* and *Stephen, ibid.*, 1957, 262).

Kogagenin, a (25*R*)-spirostan-tetra-ol, gives a triacetate CIV, which is stable

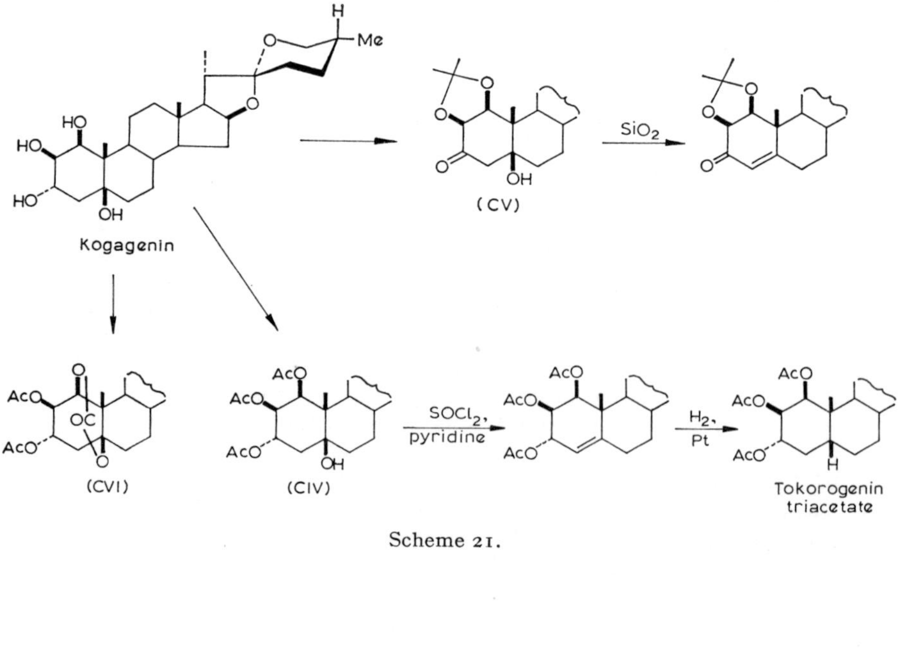

Scheme 21.

to chromic acid and which, on dehydration and hydrogenation of the resulting olefin, yields tokorogenin triacetate (Scheme 21). It is, therefore, a hydroxy derivative of tokorogenin, the fourth hydroxyl group being tertiary. The hydroxyl group is readily eliminated from the 3-oxo-1,2-acetonide CV, showing it to be at position 5; its β-configuration was suggested by the O.R.D. spectrum of the 3-ketone CV and confirmed by formation of a cyclic 1,5-carbonate CVI from kogagenin 2,3-diacetate (*T. Kubota et al.*, Tetrahedron, 1959, **7**, 62; 1960, **10**, 1. Chem. pharm. Bull. Tokyo, 1959, **7**, 898).

Kitigenin (Scheme 22), another (25*R*)-spirostantetraol, can be oxidised to (25*R*)-des-A-spirostan-5-one (CVII), showing that all four hydroxyl groups are in ring A. It gives a di- and a tri-acetate: the former, CVIII, on oxidation, yields a diacetoxy-hydroxy-ketone CIX, in which the hydroxyl group must be tertiary and therefore at $C_{(5)}$. Kitigenin tri-acetate CX, can be dehydrated to an enol acetate, which must be a 4-acetoxy-Δ^4 compound. This, in turn, is converted by lithium aluminium hydride, under mild conditions, into an oxo-diol CXI, or the 1β,4-dihydroxy-3-ketone, which is readily oxidised to the 4-hydroxy-$\Delta^{1,4}$-3-ketone CXII, establishing that one hydroxyl group in kitigenin is at $C_{(3)}$ and suggesting, in view of its ready elimination from a 3-ketone, that another is at $C_{(1)}$. Confirmation for the 3-, 4- and 5-hydroxyl groups came from reaction of the

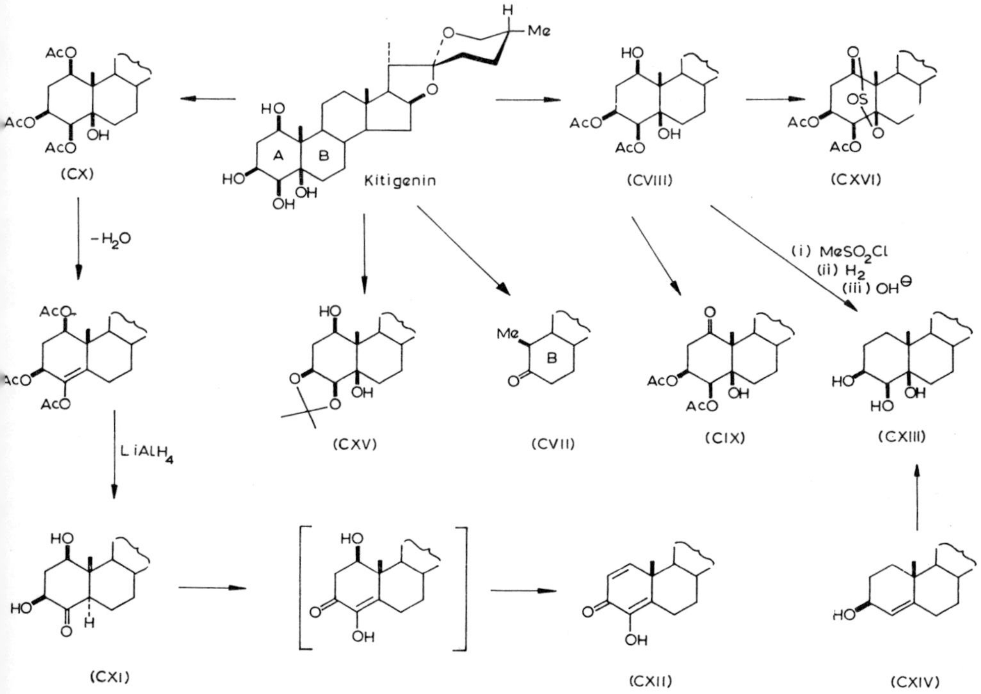

Kitigenin (CX) (CVIII) (CXVI) $-H_2O$ (i) $MeSO_2Cl$ (ii) H_2 (iii) $OH^\ominus$ (CXV) (CVII) (CIX) (CXIII) $LiAlH_4$ (CXI) (CXII) (CXIV)

Scheme 22.

original diacetate CVIII with methanesulphonyl chloride, which resulted in elimination of the 1-hydroxyl group. Reduction of the resulting olefin and hydrolysis gave a product, CXIII, identical with a triol obtained by the action of osmium tetroxide on (25*R*)-3β-hydroxyspirost-4-en (CXIV); kitigenin, therefore, contains a 3β-hydroxyl group as well as hydroxyl groups at 4 and 5 that are in the *cis*-relationship. Kitigenin yields an acetonide CXV, which can be shown to be formed by the 3- and 4-hydroxyl groups; these must, therefore, be *cis*, and the 3-, 4- and 5-hydroxyl groups must all be in the β-configuration. Confirmation of the presence of a 1-hydroxyl group came from a conversion of kitigenin into (25*R*)-5α-spirostan-1-one. Finally, kitigenin 3,4-diacetate (CVIII) yields a 1,5-cyclic sulphite CXVI, showing that the 1-hydroxyl group also has the β-configuration (*K. Sasaki et al.*, Chem. pharm. Bull. Tokyo, 1961, **9**, 631, 684, 693).

Pentologenin is characterised as a (25*R*)-spirostan-penta-ol. All of the hydroxyl groups must be in ring A, because oxidation with lead tetra-acetate yields the known (25*R*)-10-formyl-des-A-spirostan-5-one (*K. Takeda et al., ibid.*, p. 631) (see Scheme 23).

Both **luvigenin** and **meteogenin** are isolated from *Metanarthecium luteo-viride* Maxim. *Luvigenin*, m.p. 183–184°, $[\alpha]_D$ —35° ($CHCl_3$), was characterised as a (25*R*)-spirostan containing no further oxygen and having ring A aromatic. It was shown to be (25*R*)-4-methyl-19-norspirosta-1,3,5(10)-trien by its synthesis from (25*R*)-spirosta-1,4-dien-3-one (CXVII), by reduction with lithium aluminium hydride and treatment of the resulting dienol with acid (this type of re-arrangement has been observed with other $\Delta^{1,4}$-3-ols under similar conditions) (*Takeda et al.*, Tetrahedron, 1961, **15**, 183).

Scheme 23.

Meteogenin, m.p. 157–158°; $[\alpha]_D$ —174° ($CHCl_3$); 11-*acetate*, m.p. 147–148°, 162–163°; $[\alpha]_D$ —163° ($CHCl_3$), also has ring A aromatic, but it contains a hydroxyl group that, as shown by the infrared spectrum of the ketone obtained from it by oxidation, is not adjacent to the benzene ring. Meteogenin yields an acetate and can be dehydrated with formation of a compound CXVIII, containing a double-bond in conjugation with the aromatic ring. Reduction of this anhydro-compound gives an isomer of luvigenin, shown to be (25*R*)-1-methyl-19-nor-spirosta-1,3,5(10)-trien (CXIX) by its synthesis from diosgenin *via* (25*R*)-spirosta-1,4,6-trien-3-one. Meteogenin is found in association with 11α-hydroxy-spirostans, suggesting that the hydroxyl group is in this position. Finally, hecogenin was converted into (25*R*)-1-methyl-19-norspirosta-1,3,5(10)-trien-11α-ol, which proved to be identical with meteogenin (*K. Igarashi*, Chem. pharm. Bull. Tokyo, 1961, **9**, 722).

Both of these compounds are assumed to be artefacts that arise, perhaps from glycosides of $\Delta^{1,4}$- or $\Delta^{1,5}$-3-ols, during the acid-catalysed hydrolysis of the crude saponin mixture.

(*ii*) *Sapogenins with modified side-chains; kryptogenin and its congeners*

Kryptogenin, m.p. 192–193°, $[\alpha]_D$ —200° ($CHCl_3$); *diacetate*, m.p. 152–153°, $[\alpha]_D$ —182° (dioxan), was first isolated by *Marker* from Beth root (*Trillium erectum*) and was later found to accompany diosgenin in several Mexican *Dioscorea* species (*Marker et al.*, J. Amer. chem. Soc., 1943, **65**, 739; 1947, **69**,

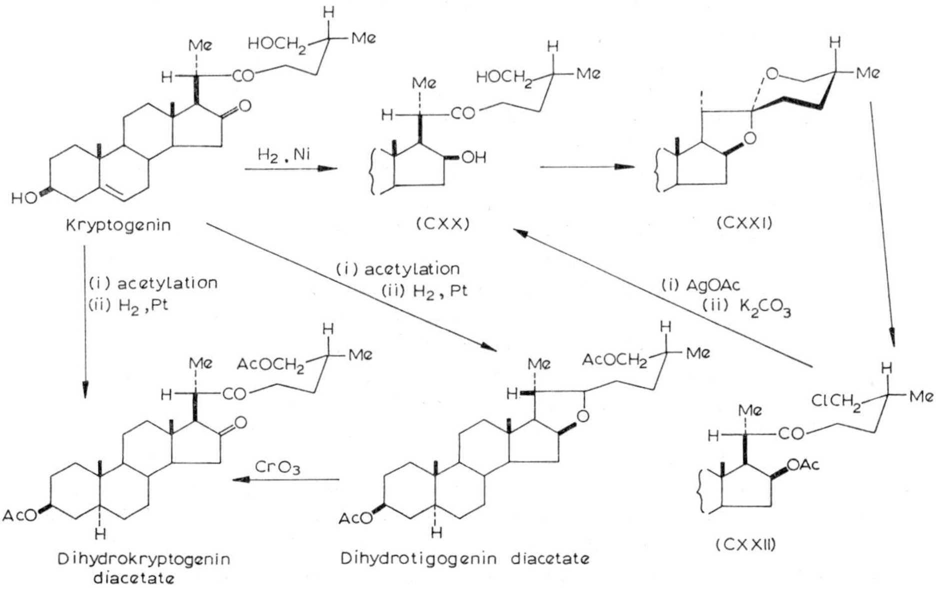

Kryptogenin
H_2, Ni
(CXX)
(CXXI)
(i) acetylation (ii) H_2, Pt
(i) acetylation (ii) H_2, Pt
(i) AgOAc (ii) K_2CO_3
CrO_3
Dihydrokryptogenin diacetate
Dihydrotigogenin diacetate
(CXXII)

Scheme 24.

2167) and in *Chionographis japonica* Maxim. (*Takeda et al.*, Chem. pharm. Bull. Tokyo, 1965, **13**, 691). It differs from the more usual sapogenins in that both of the spiroacetal rings are opened and the 16-oxygen function is present as a ketone; thus, it yields a diacetate and a dioxime. Hydrogenation of the diacetate under vigorous conditions results in reduction both of the double-bond and of the oxo groups and in ring-closure, to give dihydrotigogenin diacetate. Conversely, oxidation of dihydrotigogenin diacetate with chromic acid yields 5α,6-dihydrokryptogenin diacetate, also obtained by gentle hydrogenation of kryptogenin diacetate (Scheme 24). Kryptogenin is itself reduced under similar conditions to tigogenin or dihydrotigogenin (*Marker et al.*, J. Amer. chem. Soc., 1947, **69**, 2167; *I. Scheer, M. J. Thompson* and *E. Mosettig, ibid.*, 1957, **79**, 3218). Kryptogenin has also been related to diosgenin; thus, selective reduction of the 16-oxo-group gives the 16β,26-dihydroxy-22-ketone CXX, which, in acetic acid, cyclises to diosgenin (CXXI) (*S. Kaufmann* and *G. Rosenkranz, ibid.*, 1948, **70**, 3502). Further, the 16β-acetoxy-26-chloro compound CXXII, obtained from diosgenin acetate by means of hydrogen chloride in acetic anhydride (see p. 19) can be converted by the successive action of silver acetate and potassium carbonate into the same 16β,26-dihydroxy-22-ketone CXX (*R. S. Miner* and *E. S. Wallis*, J. org. Chem., 1956, **21**, 715; cf. *F. C. Uhle, ibid.*, 1962, **27**, 656). These reactions suggest that the configuration of kryptogenin at $C_{(20)}$ is the same as that in cholesterol (see Scheme 24); in view of the acid conditions of the isolation, this configuration is presumably here the stable one. The stability of this form appears to be associated with the presence of the 16-oxo group, since the 20-iso-configuration is favoured in its absence (cf. p. 21: *N. Danieli, Y. Mazur* and *F. Sondheimer*, Chem. and Ind., 1958, 1725).

The 1,4-diketone system of kryptogenin lends itself to a series of cyclisation reactions, some of which are shown in Scheme 25.

The 16α-methoxysapogenin (CXXIII) was first isolated from Beth root extracts that had been subjected to treatment with methanol in presence of hydrogen chloride, and it was named **bethogenin** (*C. R. Noller et al.*, J. Amer. chem. Soc., 1942, **64**, 2581; 1943, **65**, 1435): it is, however, an artefact, being interconvertible with kryptogenin as shown (*Marker et al, ibid.*, 1947, **69**, 2167). Kryptogenin, like the related sarsasapogenoic acid (see p. 8), is dehydrated by base to a cyclopentenone derivative; this compound has been isolated from Beth root and named **fesogenin**, but it too appears to be an artefact (*idem, ibid.*, 1943, **65**, 1248; 1947, **69**, 2167, 2386). The action of acetic anhydride on kryptogenin yields pseudokryptogenin diacetate (CXXIV), a furan derivative; this yields an adduct with maleic anhydride and gives, on hydrogenation, pseudodiosgenin diacetate (CXXVa) (*Kaufmann* and *Rosenkranz, loc. cit.*: *A. L. Nussbaum et al.*, J. org. Chem., 1952, **17**, 426).

Nologenin, m.p. 265–267°; 3,26-*diacetate*, m.p. 179–180°, is probably the form in which the sapogenin occurs in the saponin (nolonin), since hydrolysis of the saponin under slightly alkaline conditions gives only this compound. Mild treatment with acid converts it into the 16α-hydroxysapogenin, **pennogenin,** which in turn, is converted by more vigorous treatment with acid into kryptogenin.

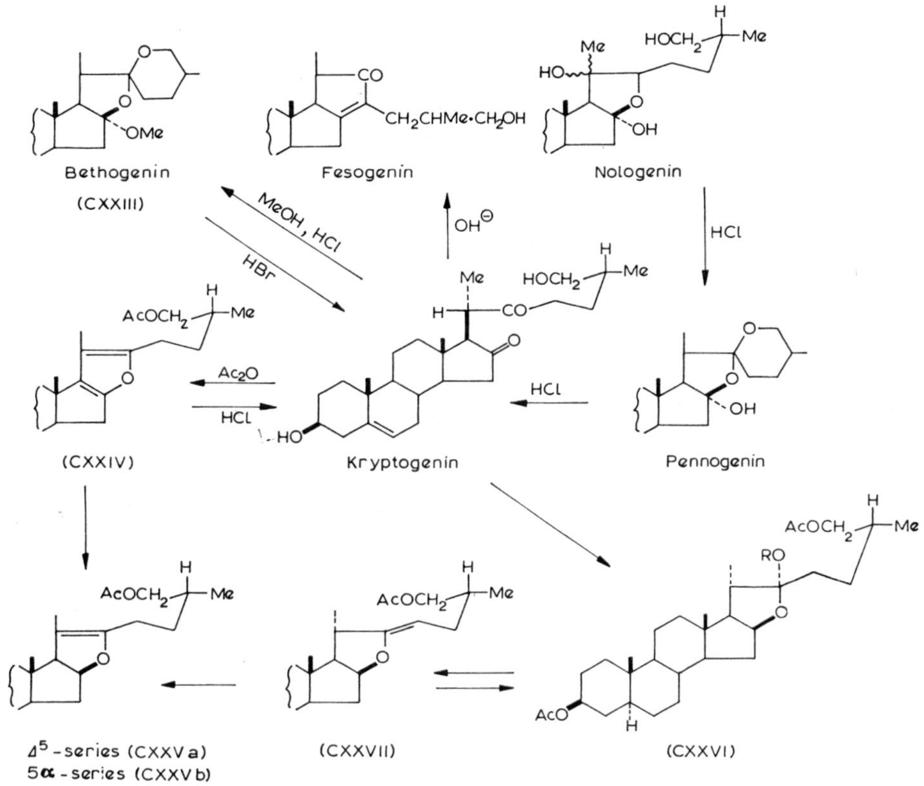

Scheme 25.

Since the two last-named occur only in the acid-hydrolysed saponin, they are probably both artefacts (*Marker et al.*, J. Amer. chem. Soc., 1943, **65**, 1248; 1947, **69**, 2167, 2386). *Marker* (*ibid.*, 1943, **65**, 1248; 1947, **69**, 2167, 2395) regarded both nologenin and pennogenin as 17-hydroxy compounds, but *K. Heusler* and *A. Wettstein* (Ber., 1954, **87**, 1301) cast doubt on some of Marker's evidence and put forward the 16α-hydroxy structures shown in Scheme 25. The infrared spectrum of pennogenin shows the bands characteristic of (25*R*)-spirostans, as does that of bethogenin (*R. N. Jones, E. Katzenellenbogen* and *K. Dobriner*, J. Amer. chem. Soc., 1953, **75**, 158).

Hydrogenation of 5α,6-dihydrokryptogenin diacetate with Raney nickel in ethanol gives the hemiacetal (CXXVI; R=H) together with the corresponding mono-ethyl acetal (CXXVI; R=Et); these compounds readily undergo elimination to the furost-22(23)-en CXXVII, which is reconverted to the acetal (XCXVI; R=Et) by ethanol in presence of acetic acid and is isomerised by phosphoric acid-acetic acid to the $\Delta^{20(22)}$- isomer, *i.e.* pseudotigogenin diacetate (CXXVb).

The 3,26-diol corresponding to CXXVII is cyclised to tigogenin under the mildly acidic conditions that convert the $\Delta^{20(22)}$-isomer into cyclopseudotigogenin (*H. Hirschmann* and *F. B. Hirschmann*, Tetrahedron, 1958, **3**, 243).

Ricogenin, m.p. 225–227°; *diacetate*, m.p. 195–197°, isolated from *Dioscorea macrostachya*, was regarded by *Marker* (J. Amer. chem. Soc., 1949, **71**, 3856) as the 12-oxo derivative of kryptogenin, on the basis of its conversion *via* pseudoricogenin (cf. CXXIV) into pseudobotogenin diacetate and, thence, into 3β-hydroxypregna-5,16-diene-12,20-dione.

(iii) *Sapogenins bearing hydroxyl groups in Ring F*

In the only report of the occurrence of sapogenins in animal material, *F. S. Spring* and his co-workers (J. chem. Soc., 1954, 1218) described the isolation from ox-bile of two compounds, which they named **cholegenin,** m.p. 193°; $[\alpha]_D$ —25° ($CHCl_3$): *diacetate*, m.p. 174–175°, $[\alpha]_D$ —19° ($CHCl_3$), and **isocholegenin,** m.p. 256–257°, $[\alpha]_D$—65° ($CHCl_3$): 3-*monoacetate*, m.p. 187°, $[\alpha]_D$—42° ($CHCl_3$). Both compounds had the properties of dihydroxyspirostans. Thus, standard side-chain degradation gave 3α-hydroxy-5β-pregn-16-en-20-one, and oxidation of cholegenin diacetate with chromic acid gave a lactone identical with that obtained from 3-epismilagenin under similar conditions. They appeared, therefore, to be 5β-spirostan-3α,x-diols, the second hydroxyl group being in ring F. Cholegenin could be oxidised to an oxo-acid, which was reduced by lithium aluminium hydride to yield cholegenin again; the second hydroxyl was therefore primary and it was assigned to $C_{(27)}$. Further, cholegenin was converted by acid into isocholegenin, and the two compounds were regarded as the $C_{(25)}$-epimers of 5β-spirostan-3α,27-diol (CXXVIII) (Scheme 26) (*Y. Mazur* and *Spring*, *ibid.*, p. 1223).

In re-examining the problem, *E. Mosettig* and his co-workers (J. Amer. chem. Soc., 1959, **81**, 5222, 5225) used Spring's material, for they failed to repeat the isolation from ox-bile. They found that, in accordance with the above structures, both compounds yielded the same furostan-triol on catalytic reduction; however, this proved not to be identical with the expected compound CXXIX, which was prepared from 3-episarsasapogenin by a multi-stage process. The furostan-triol yielded only a diacetate and it could be cleaved by periodate, in agreement with structure CXXX. The correctness of this structure was proved by synthesis, again from 3-episarsasapogenin. Cholegenin, in turn, was assigned the unusual structure shown with a 5-membered ring F; in agreement with this, cholegenin yields a diacetate under mild conditions, and its infrared spectrum is different from that of the usual sapogenins. Isocholegenin, on the other hand, yields only a mono-acetate under mild conditions and has a typical isosapogenin spectrum; it is, therefore, regarded as the (25*S*)-25-hydroxyspirostan CXXXI, its formation from cholegenin under mildly acidic conditions being the result of ring opening of the 5-membered ring F, by cleavage of the $C_{(22)}$-O-bond, and re-closure with the primary 27-hydroxyl group to give the 6-membered ring F of CXXXI. The configuration of cholegenin at $C_{(25)}$ follows from that of isocholegenin, since they yield the same triol CXXX.

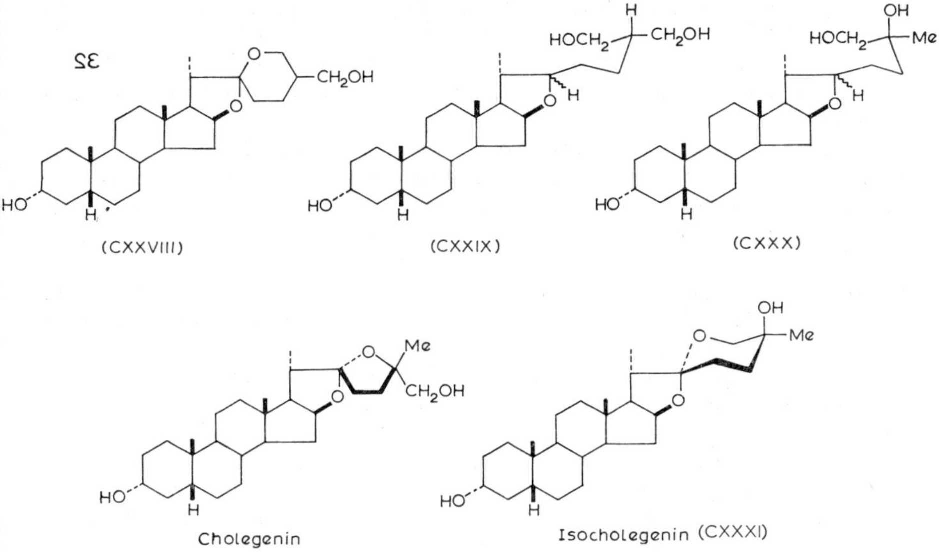
CH2OH
HO
H
(CXXVIII)
H
HOCH2
CH2OH
H
HO
H
(CXXIX)
OH
HOCH2
Me
H
HO
H
(CXXX)
Me
CH2OH
HO
H
Cholegenin
OH
Me
HO
H
Isocholegenin (CXXXI)

Scheme 26.

Sapogenins with ring F hydroxylated have also been isolated from plant sources. Thus, *R. Tschesche* and *K. H. Richert* (Tetrahedron, 1964, **20**, 387) have isolated, by acid hydrolysis of the glycoside fraction from the roots of *Solanum sisymbriifolium*, **nuatigenin**, m.p. 210–214^{0}, $[\alpha]_D$ —93^{0} ($CHCl_3$); *diacetate*, m.p.

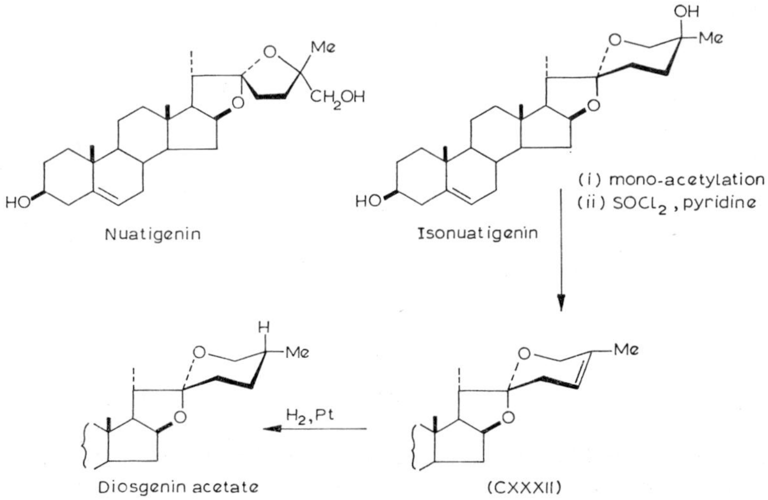
Me
CH2OH
HO
Nuatigenin
OH
Me
HO
Isonuatigenin
(i) mono-acetylation
(ii) SOCl2, pyridine
H
Me
H2, Pt
Me
Diosgenin acetate
(CXXXII)

Scheme 27.

156–159°, $[\alpha]_D$ —95° ($CHCl_3$), and **isonuatigenin,** m.p. 215·5–218·5°, $[\alpha]_D$ —140° ($CHCl_3$); *diacetate,* m.p. 211·5–215·5°, $[\alpha]_D$ —109° ($CHCl_3$). By methods similar to those described above they have been shown to have, respectively, structures shown in Scheme 27, analogous to those of cholegenin and iso-cholegenin.

Confirmation of the configuration of nuatigenin and isonuatigenin at $C_{(25)}$ came from their I.R. spectra, which showed that the side-chain hydroxyl was bonded, in the former to the ring E oxygen atom, in the latter to the ring F oxygen atom. Further, dehydration of the 3-mono-acetate of isonuatigenin with thionyl chloride gave the endocyclic olefin CXXXII, in accordance with the axial conformation of the 25-hydroxyl group; catalytic hydrogenation then yielded diosgenin acetate. Isonuatigenin is probably an artefact, arising from nuatigenin under the acidic conditions of the isolation.

Again, *Reineckia carnea* Kunth. has yielded **reineckiagenin,** m.p. 278–280°, $[\alpha]_D$ —71° (dioxan); 1,3-*diacetate,* m.p. 198–200°, $[\alpha]_D$ —82° ($CHCl_3$): **iso-reineckiagenin,** m.p. 240–242°, $[\alpha]_D$ —66° (dioxan); 1,3-*diacetate,* m.p. 202–204°, $[\alpha]_D$ —87° ($CHCl_3$), and **isocarneagenin,** m.p. 242–244°, $[\alpha]_D$ —63° (1:1-methanol – $CHCl_3$); *triacetate,* m.p. 215–218°, $[\alpha]_D$ —75° ($CHCl_3$) (*Takeda et al.,* Tetrahedron Letters, 1962, 1107; Tetrahedron, 1963, **19,** 759). The first two have been shown to be (25*R*)- and (25*S*)-5β-spirostan-1β,3β,25-triol, respectively; both give diacetates that can be dehydrated to a mixture of Δ^{24}- and $\Delta^{25(27)}$-

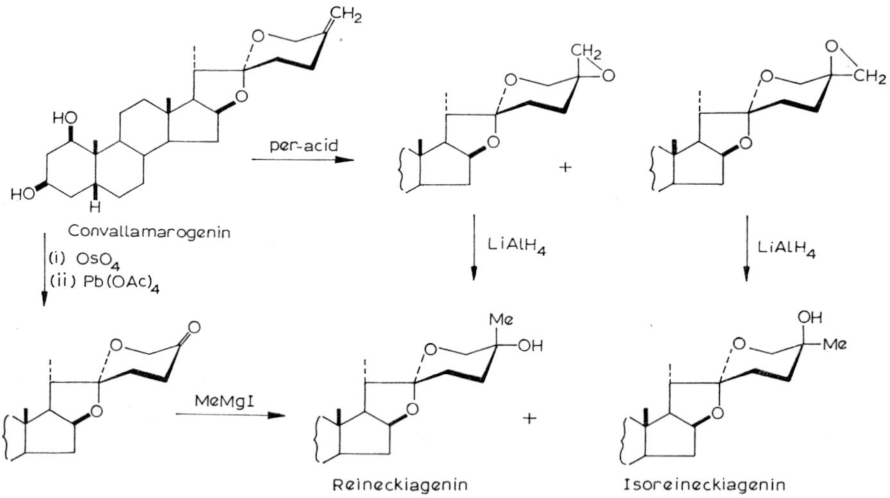

Scheme 28.

olefins. Further, they can be prepared from convallamarogenin (see p. 51) by the methods shown in Scheme 28. The configurations of the two epimers at $C_{(25)}$ are shown by the infrared and proton magnetic resonance spectra.

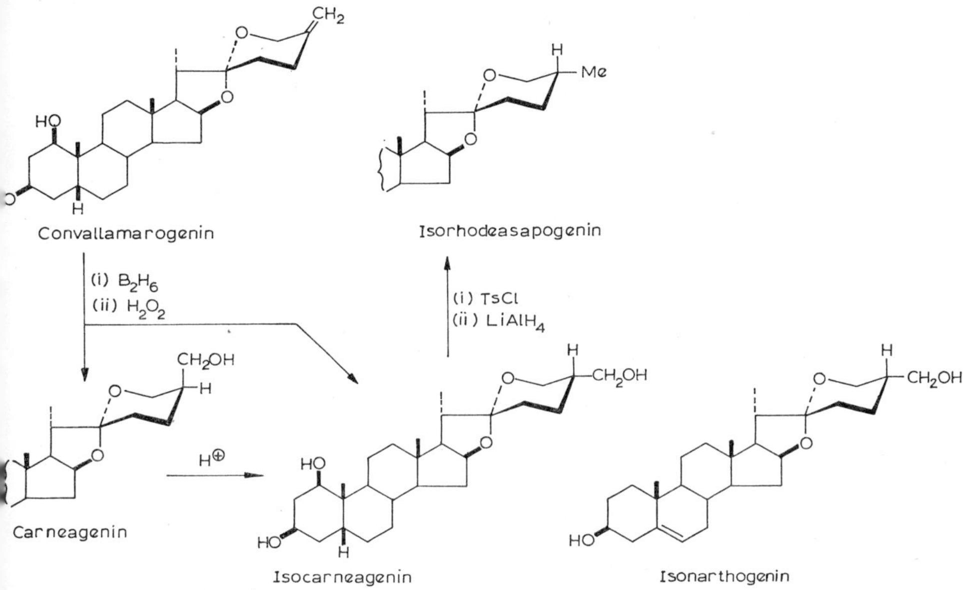

Scheme 29.

Isocarneagenin, which readily yields a triacetate, was shown to be (25*S*)-5β-spirostan-1β,3β,27-triol (see Scheme 29); it yielded a monotosylate that was reduced by lithium aluminium hydride to isorhodeasapogenin, showing it to be a hydroxy derivative of the latter compound. Hydration of the 25,27-double bond of convallamarogenin (p. 51) by successive treatment with diborane and hydrogen peroxide yielded isocarneagenin, together with an isomer, presumably the 25-epimer, carneagenin. Since the latter is readily converted into isocarneagenin by mild treatment with acid, it is possible that the sapogenin occurs in the plant as a glycoside of carneagenin. Again, in view of the experience with cholegenin and isocholegenin, it is possible that reineckiagenin and isoreineckiagenin are artefacts, arising from a compound with a 5-membered ring F, allied to cholegenin.

Narthogenin, m.p. 214–216°, $[\alpha]_D$ —112° ($CHCl_3$); *diacetate,* m.p. 144–146°, $[\alpha]_D$ —98° ($CHCl_3$), and **isonarthogenin,** m.p. 240–242°, $[\alpha]_D$ —110° ($CHCl_3$); *diacetate,* m.p. 160–162°, $[\alpha]_D$ —94° ($CHCl_3$), from *Metanarthecium luteo-viride* Maxim. are spirostandiols, whose infrared spectra suggested a relationship to carneagenin and isocarneagenin. Isonarthogenin was proved to be (25*S*)-spirost-5-en-3β,27-diol by its conversion, *via* the 27-monotosylate, into diosgenin and into 25(27)-dehydrodiosgenin. Narthogenin is converted into isonarthogenin under mildly acid conditions; like isonarthogenin, it readily yields a diacetate

and is regarded as its epimer at position 25 (*H. Minato* and *A. Shimaoka,* Chem. pharm. Bull. Tokyo, 1963, **11,** 876).

The infrared spectra of these sapogenins, hydroxylated at positions 25 and 27 have been discussed by *K. Takeda* and his co-workers (J. chem. Soc., 1963, 4815).

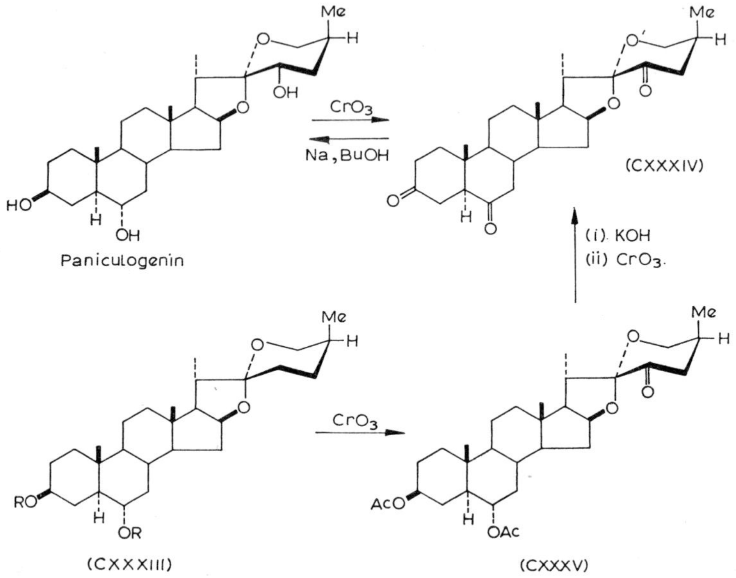

Scheme 30.

Hydroxylation at the 23-position is represented by a single example, **paniculogenin,** isolated from the leaves of *Solanum paniculatum* L. (*H. Ripperger, K. Schreiber* and *H. Budzikiewicz,* Ber., 1967, **100,** 1741). This compound, m.p. 214–216°, $[\alpha]_D$ —45·7° ($CHCl_3$); 3,6-*diacetate,* m.p. 200–202°, $[\alpha]_D$ —51°, a trihydroxyspirostan (see Scheme 30), showed a fragmentation pattern in its mass spectrum that suggested that one of its three hydroxyl groups was attached to $C_{(23)}$ and this was supported by examination of the mass spectrum of a synthetic 23-hydroxyspirostan. From other spectroscopic evidence, paniculogenin was assigned the structure shown, *i.e.* the (23*S*)-hydroxy derivative of its congener, neochlorogenin (CXXXIII; R=H). Confirmation came from the following reactions; oxidation of paniculogenin with chromic acid gave the 3,6,23-trione CXXXIV, identical with material prepared from neochlorogenin diacetate (CXXXIII; R=Ac) by oxidation with chromic acid to the 23-oxo derivative CXXXV, removal of the acetyl groups and further oxidation of the 3- and 6-hydroxyl groups. Reduction of the triketone CXXXIV with sodium and butanol gave back paniculogenin; this suggests that the 23-hydroxyl group, like those at positions 3 and 6, is equatorial, and so leads to the structure shown.

(*iv*) *Sapogenins with ring F unsaturated*

Neoruscogenin, m.p. 198–201°, $[\alpha]_D$ —118° ($CHCl_3$); *diacetate,* m.p. 132–134°, $[\alpha]_D$ —64° ($CHCl_3$), occurs along with ruscogenin (p. 30) and was first thought to be its (25*S*)-epimer (*C. Sannié* and *H. Lapin,* Bull. Soc. chim. Fr., 1957, 1237). The infrared spectrum does not, however, support this, and the compound was found to be the 25(27)-dehydro derivative of ruscogenin from, *inter alia,* the following evidence: (a) its diacetate absorbs two atoms of hydrogen to give a mixture of the diacetates of ruscogenin (CXXXVI) and of the true (25*S*)- epimer CXXXVII of ruscogenin; (b) ozonolysis yields formaldehyde; (c) pseudoneoruscogenin yields, on oxidation with chromic acid and subsequent hydrolysis, α-methyleneglutaric acid (*J. Robert, R. Vaupré* and *G. Poiget,* C.r. Acad. Sci., Paris, 1960, **250,** 3187) (Scheme 31); the proton magnetic resonance spectrum supports this formulation (*L. Mandell, A. L. Nussbaum* and *E. P. Oliveto,* Tetrahedron Letters, 1960, No. 19, p. 25).

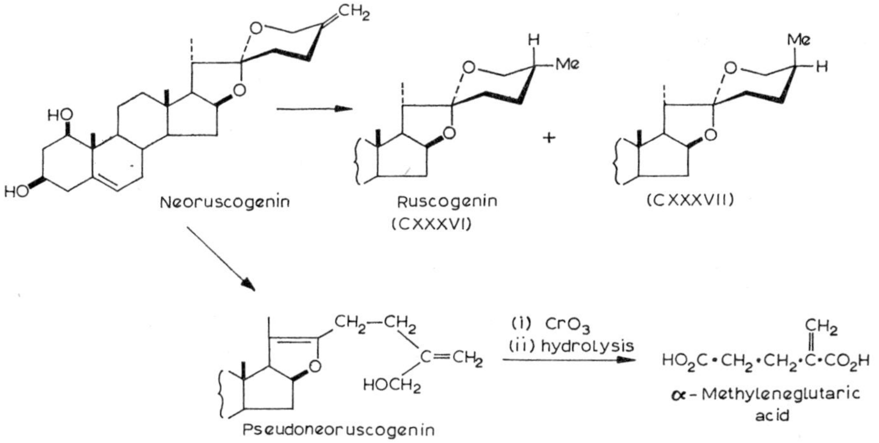

Neoruscogenin → Ruscogenin (CXXXVI) + (CXXXVII)

Neoruscogenin → Pseudoneoruscogenin —(i) CrO_3 (ii) hydrolysis→ $HO_2C \cdot CH_2 \cdot CH_2 \cdot C(=CH_2) \cdot CO_2H$ α-Methyleneglutaric acid

Scheme 31.

Convallamarogenin (see Scheme 29, p. 49), m.p. 259–261°, $[\alpha]_D$ —79° ($CHCl_3$); *diacetate,* m.p. 214–215°, $[\alpha]_D$ —88° ($CHCl_3$), isolated from the roots of *Convallaria majalis* L. (*R. Tschesche, H. Schwarz* and *G. Snatzke,* Ber., 1961, **94,** 1699) and of *Reineckia carnea* Kunth. (*Takeda et al.,* Tetrahedron, 1963, **19,** 759), was characterised as a spirostendiol. The infrared spectrum suggested that the double-bond was in a terminal methylene group; this was confirmed by production of formaldehyde on successive treatment of the compound with osmium tetroxide and periodate. Reduction of the double-bond gave a mixture of $C_{(25)}$-epimers, which was converted by ethanolic hydrochloric acid into isorhodeasapogenin, thus showing convallamarogenin to be 5β-spirost-25(27)-en-1β,3β-diol. These $\Delta^{25(27)}$-compounds are not, as has been suggested, artefacts, formed by

dehydration, during isolation, of 25-hydroxyspirostans, analogous to isocholegenin (*Takeda et al.*, Tetrahedron, 1965, **21,** 2089). They are interconvertible, in the plant, with the saturated compounds of either 25*R* or 25*S*-configuration, but not, apparently, through the 25-hydroxy compounds (*Takeda, Minato* and *Shimaoka*, J. chem. Soc. (C), 1967, 876). Consistent with this finding is the isolation, from *Hosta Kiyosumiensis* F. Maek, of the 25(27)-dehydro derivatives of gitogenin, manogenin, 9-dehydromanogenin and tigogenin, along with the parent 25,27-dihydro-compounds (*Takeda et al., loc. cit.*).

(*v*) *Aminospirostans*

Naturally-occurring compounds in which the ring F oxygen atom of the sapogenins is replaced by NH are found among the *Solanum* alkaloids and will be discussed elsewhere in this work. More closely akin to the steroid sapogenins is the 3β-amino-compound, **jurubidine,** m.p. 182–186°, $[\alpha]_D$ —51° (pyridine), isolated from the roots of *Solanum paniculatum* L. It was characterised spectroscopically, by its deamination to neotigogenin (CXXXVIII) (Scheme 32) and by its synthesis from neotigogenone *via* its oxime (*K. Schreiber, H. Ripperger* and *H. Budzikiewicz*, Tetrahedron Letters, 1965, 3999; Ber., 1967, **100,** 1725).

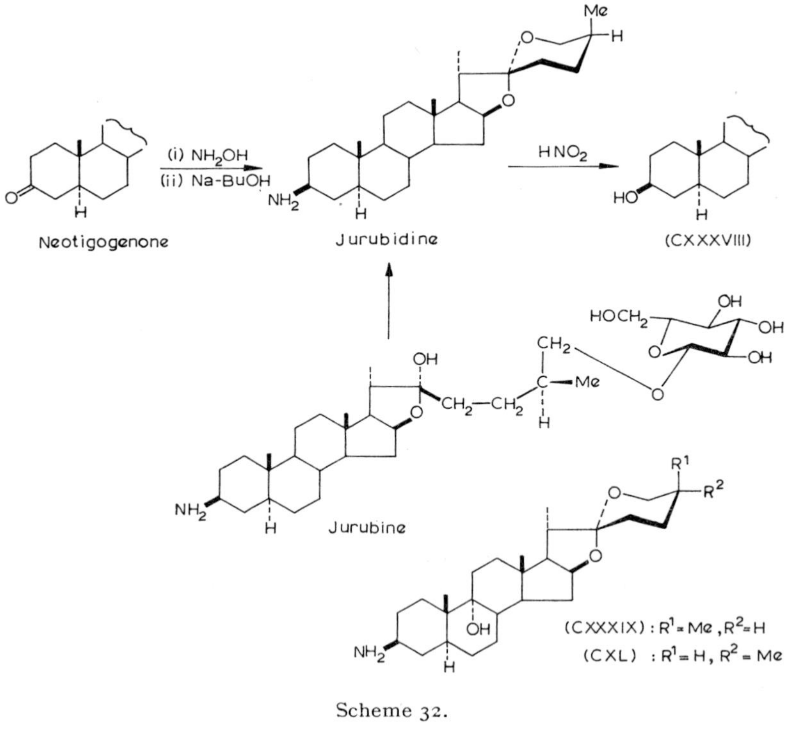

Scheme 32.

Juribidine has been shown to arise by acidic or enzymic hydrolysis of **jurubine,** m.p. 212–214°, $[\alpha]_D$ —31° (pyridine), the glycoside of the open-chain form (*Ripperger, Budzikiewicz* and *Schreiber*, Ber., 1967, **100,** 1725); the true aglycon has not been isolated, presumably cyclising under the reaction conditions to the spirostan jurubidine (cf. pp. 5–6).

Paniculidine, also isolated from *S. paniculatum,* has been identified, largely on spectroscopic evidence, as a mixture of 9α-hydroxyjurubidine (CXXXIX) and its (25*R*)-epimer CXL (*Idem, loc. cit.*).

Chapter 19

The Biogenesis of Terpenes and Steroids

T. W. GOODWIN

1. Introduction

It is clear from the vast amount of structural chemistry carried out on terpenes that they can all be derived from isoprenoid (branched C_5) units. This was first fully appreciated by *Ruzicka* who formulated his well-known "isoprene rule" (see *L. Ruzicka, A. Eschenmoser* and *H. Heusser*, Experientia, 1953, 9, 357). However, when the structure of lanosterol was elucidated, it became apparent that it could not easily be split into isoprene residues;

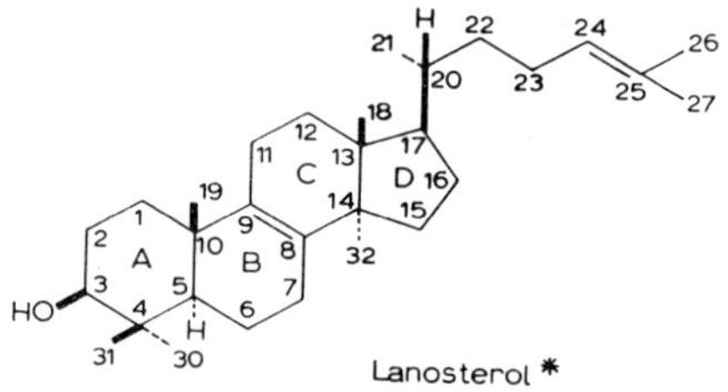

Lanosterol *

this, together with other similar examples led to a revised version of the "isoprene rule" which was termed "the biogenetic isoprene rule" (*Ruzicka*, Proc. chem. Soc., 1959, 341). It states that terpenes are formed by combination of isoprenoid units to form compounds such as geraniol (C_{10}), farnesol (C_{15}), geranylgeraniol (C_{20}), squalene (C_{30}) and others of a similar kind, and that they can be derived from these precursors by accepted cyclization and, in certain cases, rearrangement mechanisms. Biochemical investigations which were going on at this time also strongly pointed to the need for revising

* The method of numbering is that used for steroids (see Vol. IID, Appendix).

the original rule, and such studies culminated in the demonstration that a universal "isoprenoid unit" exists as isopentenyl pyrophosphate (IPP).

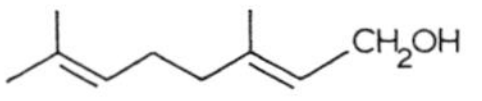

Geraniol

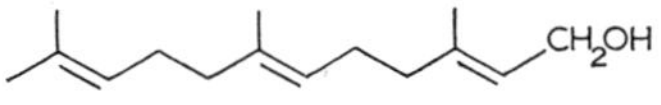

Farnesol

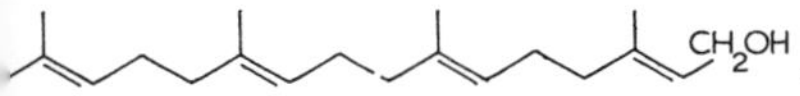

Geranylgeraniol

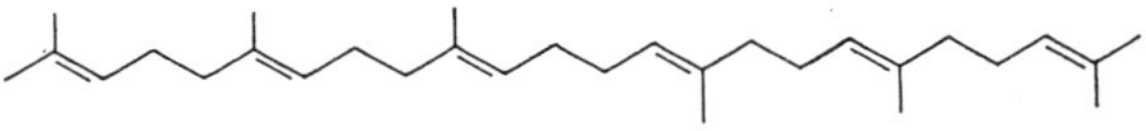

Squalene

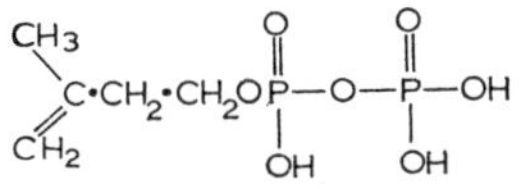

Isopentenyl pyrophosphate

Considerable progress has recently been made in elucidating the mechanisms by which this compound is built up into various terpenes and terpenoids *in vitro*.

2. Biosynthesis of isopentenyl pyrophosphate (IPP) and dimethylallyl pyrophosphate (DMAPP)

The basic building unit is acetyl-coenzyme A (acetyl-CoA), which undergoes the enzymatic transformations indicated in Scheme 1. The reactions leading from acetyl-CoA to β-hydroxy-β-methylglutaryl-CoA (HMG-CoA) (**A, B, C,** Scheme 1) are not exclusive to terpenoid biosynthesis; it is however, the conversion of HMG-CoA into mevalonic acid (MVA) which represents the "point of no return", because as *G. Popják* and *J. W. Cornforth* (Adv. in Enzymol., 1960, **22,** 281) have pointed out, MVA has no metabolic future other than being converted into terpenoids. HMG-CoA, on the other hand, can be irreversibly cleaved to acetyl-CoA and acetoacetate (reaction **E**). MVA was discovered as the acetate-replacing factor for an acetate-requiring mutant of *Lactobacillus acidophilus* (see *K. Folkers et al.*, in "Biosynthesis of terpenes and sterols" Ed. *G. E. W. Wolstenholme* and *M. O'Connor*, Churchill, London, 1959, p. 20); its resemblance to HMG prompted an experiment which revealed that 43% of racemic [2-^{14}C]-MVA was converted into cholesterol by liver homogenates which were not particularly impressive in their ability to convert acetate into cholesterol. As it was later shown that only one enantiomorph was biologically active, the conversion was

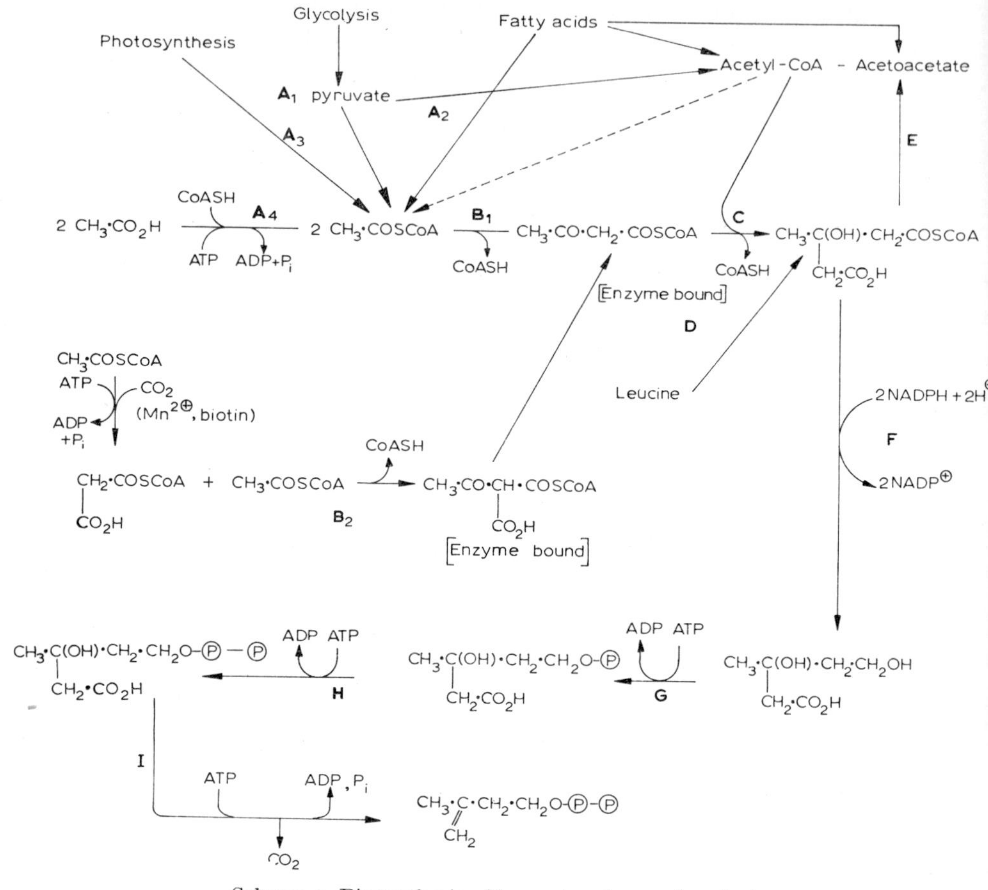

Scheme 1. Biosynthesis of isopentenyl pyrophosphate.

ADP	=	Adenosine diphosphate
ATP	=	Adenosine triphosphate
NAD	=	Nicotinamide adenine dinucleotide
NADPH	=	Nicotinamide adenine dinucleotide phosphate
P_i	=	Phosphate ion
CoA	=	Coenzyme A

almost quantitative. Similar results were obtained with other terpenoids. The discovery of the biosynthetic activity of MVA was a turning point for investigations on terpenoid biosynthesis.

Of the intermediates concerned with MVA formation acetyl-CoA arises mainly from pyruvate produced during glycolysis (reaction $\mathbf{A_1}$ – Scheme 1)

or from the oxidation of fatty acids (**A**$_2$) or from photosynthesis (**A**$_3$); it can also arise from various amino acids, but this generally represents a minor source. Acetate itself can be activated (reaction **A**$_4$) by the enzyme acetothiokinase in the presence of coenzyme A (CoASH) and adenosine triphosphate (ATP). The reaction catalysed by the HMG-CoA condensing enzyme (reaction **C** – Scheme 1) takes place in yeast on a heat-stable acyl carrier protein (ACP) (*H. Rudney et al.*, J. biol. Chem., 1966, **241**, 1226). The HMG-CoA condensing enzyme converts acetoacetyl-ACP into HMG-ACP which then undergoes non-enzymic transacylation with CoASH to form HMG-CoA and releases ACP for further activity. Acetoacetyl-ACP can arise from two molecules of acetyl-CoA (**B**$_1$) or from one of acetyl-CoA and one of malonyl-CoA (**B**$_2$). Which is quantitatively more significant in terpenoid biosynthesis is not yet known. The animal enzyme, which is located in the microsomes, becomes extremely unstable on attempted purification (*Rudney*, in *Wolstenholme* and *O'Connor*, *op. cit.*, p. 75; *F. Lynen et al.*, Biochem. Z., 1958, **330**, 269).

HMG-CoA can also be synthesized from leucine by a pathway which does not involve acetate but which does involve a carbon dioxide-fixing step which is biotin-dependent (*M. J. Coon et al.*, in *Wolstenholme* and *O'Connor*, *op. cit.*, p. 62) (Scheme 2). It is this reaction which is the basis of the fixation of carbon dioxide into sterols (*K. Bloch*, in "Essays in Biochemistry", Ed. *S.*

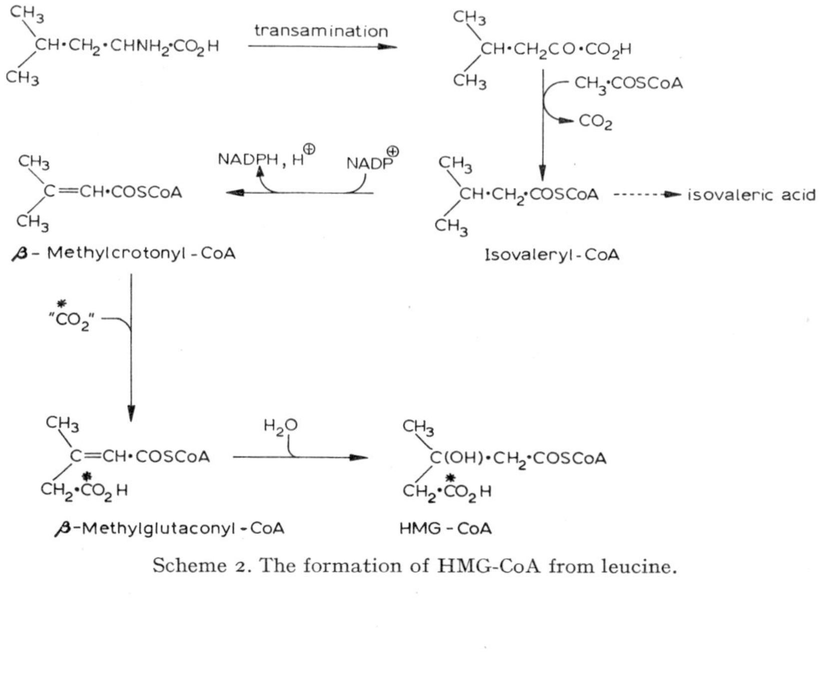

$(CH_3)_2CH{\cdot}CH_2{\cdot}CHNH_2{\cdot}CO_2H$ —transamination→ $(CH_3)_2CH{\cdot}CH_2CO{\cdot}CO_2H$ —($CH_3{\cdot}COSCoA$; CO_2)→ $(CH_3)_2CH{\cdot}CH_2{\cdot}COSCoA$ (Isovaleryl-CoA) ------→ isovaleric acid

Isovaleryl-CoA —($NADP^{\oplus}$ → NADPH, $H^{\oplus}$)→ $(CH_3)_2C{=}CH{\cdot}COSCoA$ (β-Methylcrotonyl-CoA)

β-Methylcrotonyl-CoA —"$\overset{*}{C}O_2$"→ $(CH_3)(CH_2{\cdot}\overset{*}{C}O_2H)C{=}CH{\cdot}COSCoA$ (β-Methylglutaconyl-CoA) —H_2O→ $(CH_3)(CH_2{\cdot}\overset{*}{C}O_2H)C(OH){\cdot}CH_2{\cdot}COSCoA$ (HMG-CoA)

Scheme 2. The formation of HMG-CoA from leucine.

Graff, Chapman and Hall, London 1956) and carotenoids (*T. W. Goodwin*, in *Wolstenholme* and *O'Connor*, *op. cit.*, p. 279).

Reaction **F** (Scheme 1) is the first reaction specific to terpenoid biosynthesis and is essentially irreversible. The enzyme, HMG-CoA reductase, catalyses the reaction without the appearance of free mevaldic acid (*J. J. Ferguson*, *I. F. Durr* and *Rudney*, Proc. nat. Scad. Sci., 1959, 45, 499; *Lynen et al.*, Angew. Chem., 1959, 71, 657). Experiments with [1-^{14}C] and [2-^{14}C] HMG-CoA show that it is the carboxyl group carrying the coenzyme A residue which is reduced to the primary alcohol. *Popják* and *Cornforth* (*loc. cit.*) proposed the following mechanism for the reaction, which explains why free mevaldic aid is not formed:

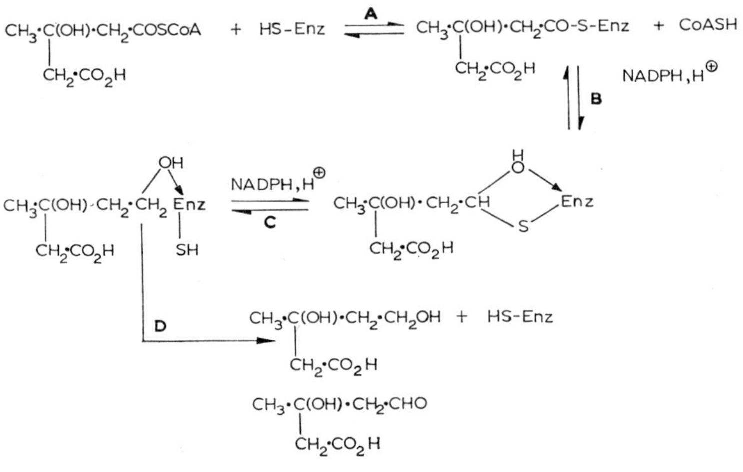

$CH_3{\cdot}C(OH)(CH_2{\cdot}CO_2H){\cdot}CH_2{\cdot}COSCoA$ + HS-Enz ⇌ (**A**) $CH_3{\cdot}C(OH)(CH_2{\cdot}CO_2H){\cdot}CH_2{\cdot}CO{\cdot}S{\cdot}Enz$ + CoASH

⇅ (**B**) NADPH,H$^{\oplus}$

$CH_3{\cdot}C(OH)(CH_2{\cdot}CO_2H){\cdot}CH_2{\cdot}CH(OH){\cdot}S{\cdot}Enz$ (hemithioacetal, O–H → Enz)

⇅ (**C**) NADPH,H$^{\oplus}$

$CH_3{\cdot}C(OH)(CH_2{\cdot}CO_2H){\cdot}CH_2{\cdot}CH_2(OH \to Enz{\cdot}SH)$

→ (**D**) $CH_3{\cdot}C(OH)(CH_2{\cdot}CO_2H){\cdot}CH_2{\cdot}CH_2OH$ + HS-Enz

$CH_3{\cdot}C(OH)(CH_2{\cdot}CO_2H){\cdot}CH_2{\cdot}CHO$

Mevaldic acid

The first step (**A**) is the binding of the substrate as a thiol ester with the liberation of coenzyme A. Reduction with one mole of NADPH (**B**) could yield a hemithioacetal which on further reduction would yield an enzyme-MVA complex (**C**) which, in turn, would irreversibly dissociate to liberate free MVA and regenerate the enzyme for further reaction. If the hemithioacetal did not readily dissociate, and according to *Popják* and *Cornforth* this is likely, especially if the oxygen atoms were further bound by co-ordination with a metallic cation, then free mevaldic acid would not appear during the reaction, but added mevaldic acid would be rapidly taken up and reduced to MVA. The overall irreversibility of the reaction would be achieved if the last step were not readily reversible. However, in addition to HMG-CoA reductase, two additional enzymes exist in yeast; one, partic-

ulate (mevaldic dehydrogenase) converts HMG-CoA into mevaldic acid, and the second, soluble, converts mevaldic acid into MVA (*Lynen et al.*, Fed. Proc., 1959, **18**, 278).

The enzymic reduction of HMG-CoA is stereospecific (*Ferguson et al.*, *loc. cit.*) as is the utilisation of MVA (*T. T. Tchen*, J. biol. Chem., 1958, **233**, 1100). *M. Eberle* and *D. Arigoni* (Helv., 1960, **43**, 1508) have established the absolute configuration (3*R*) of biosynthetic MVA and that of biologically synthesized HMG-CoA has been deduced from this.

$HOOC-CH_2-C(CH_3)(OH)-CH_2-CH_2OH$

Mevalonic acid (MVA)

$HOOC-CH_2-C(CH_3)(OH)-CH_2-COSCoA$

β-Hydroxy-β-methylglutaryl-coenzyme A (HMG-CoA)

In yeast and liver free HMG is not incorporated into terpenoids because of the absence of an activating enzyme which would convert the free acid into its CoA derivative (*P. A. Tavormina, M. H. Gibbs* and *J. W. Huff*, J. Amer. chem. Soc., 1956, **78**, 4498; *F. Dituri et al.*, J. biol. Chem., 1957, **229**, 826). However, HMG is incorporated into sterols and carotenoids in the mould *Phycomyces blakesleeanus* (*C. O. Chichester et al.*, *ibid.*, 1959, **234**, 598; 1962, **237**, 681) and into terpenoids in certain *Lactobacilli* (*K. J. I. Thorne* and *E. Kodicek*, Biochim. Biophys. Acta, 1962, **59**, 273).

Up to MVA the stages in terpene biosynthesis are electrophilic reactions in which coenzyme A acts by increasing the electron deficit at the carbonyl carbon (Scheme 1). After MVA the type of reaction changes and coenzyme A is no longer involved. MVA is "activated" by stepwise conversion into 5-pyrophospho-MVA (reactions **G** and **H**). **G** is catalysed by mevalonic kinase which has been purified from yeast (*Tchen*, J. Amer. chem. Soc., 1957, **79**, 6345), pig liver (*L. A. Witting, H. J. Knaus* and *J. W. Porter*, Fed. Proc., 1959, **18**, 353; *H. R. Levy* and *Popják*, Biochem. J., 1960, **75**, 417), and plant tissues (*J. Battaile* and *W. D. Loomis*, Biochim. Biophys. Acta, 1961, **51**, 545; *I. P. Williamson* and *R. G. O. Kekwick*, Biochem. J., 1965, **96**, 862). Phosphomevalonic kinase (reaction **H**) has been partly purified from yeast autolysates (*Bloch et al.*, J. biol. Chem., 1959, **234**, 2595; *U. Henning, E. M. Moslein* and *Lynen*, Arch. Biochem. Biophys., 1959, **83**, 259), and from liver (*Witting* and *Porter*, J. biol. Chem., 1959, **234**, 2841; *H. Hellig* and *Popják*, J. Lipid Res., 1961, **2**, 235). The final step (**I**, Scheme 1) is catalysed by 5-pyrophosphomevalonic anhydrodecarboxylase which has been purified from yeast (*A. de Waard, A. H. Phillips* and *Bloch*, J. Amer. chem. Soc., 1959,

81, 2913) and pig liver (*Hellig* and *Popják*, Biochem. J., 1961, **80**, 41P), and the following mechanism is proposed (*Popják* and *Cornforth*, *ibid.*, 1966, **101**, 553 :

$$\mathrm{^{\ominus}O_2C{-}CH_2{-}C(CH_3)(OH){-}CH_2{\cdot}CH_2O{-}\text{Ⓟ}{-}\text{Ⓟ}} + \mathrm{ATP} \longrightarrow \mathrm{CH_2{=}C(CH_3){-}CH_2{\cdot}CH_2O{-}\text{Ⓟ}{-}\text{Ⓟ}} + \mathrm{CO_2} + \mathrm{(HO)_2P(O)OH} + \mathrm{^{\ominus}O{-}P(O)(O^{\ominus}){-}O{-}P(O)(O^{\ominus}){-}O{-}Ad}$$

This mechanism is consistent with the following observations: (i) when the reaction is carried out in deuterium oxide, deuterium does not appear in the methylene group of IPP (*Bloch et al.*, J. biol. Chem., 1959, **234**, 2594); (ii) the oxygen of the tertiary hydroxyl of MVA becomes one of the oxygen atoms of the inorganic phosphate formed (*M. Lindberg et al.*, Biochemistry, 1962, **1**, 182); (iii) ATP is required (*Bloch et al.*, *loc. cit.*); (iv) no 3-phosphorylated MVA has been observed as an intermediate, and (v) the elimination of the OH and the $CO_2^{\ominus}$ groupings is *trans* (*Popjak* and *Cornforth*, *loc. cit.*).

$$\mathrm{(CH_3)_2C{=}CH{\cdot}CH_2O{-}\text{Ⓟ}{-}\text{Ⓟ}}$$

Dimethylallyl pyrophosphate (DMAPP)

Before the biosynthesis of higher terpenes can begin, one molecule of dimethylallyl pyrophosphate (DMAPP) must be formed in order to act as starter for chain elongation (see p. 66). IPP is isomerized to DMAPP by isopentenyl pyrophosphate isomerase (prenyl isomerase), which has been purified from yeast (*Lynen et al.*, Angew. Chem., 1959, **71**, 657) and liver (*D. H. Shah*, *W. W. Cleland* and *Porter*, J. biol. Chem., 1965, **240**, 1946; *P. W. Holloway* and *Popják*, Biochem. J., 1968, **106**, 835). The reaction involves prototropic attack and can be formulated thus:

$$\mathrm{CH_2{=}C(CH_3){-}CH_2{\cdot}CH_2O{-}\text{Ⓟ}{-}\text{Ⓟ}} + \mathrm{H^{\oplus}} \longrightarrow \mathrm{(CH_3)_2\overset{\oplus}{C}{-}CH_2{-}CH_2O{-}\text{Ⓟ}{-}\text{Ⓟ}} \longrightarrow \mathrm{(CH_3)_2C{=}CH{\cdot}CH_2O{-}\text{Ⓟ}{-}\text{Ⓟ}} + \mathrm{H^{\oplus}}$$

The hydrogen removal from $C_{(2)}$ of IPP in the formation of *trans* polyprenols is stereospecific in that the pro-*R* hydrogen is eliminated from $C_{(2)}$ of IPP

and the pro-*S* hydrogen at $C_{(2)}$ is retained (*Cornforth et al.*, Proc. roy. Soc. B, 1966, **163**, 492). This followed from experiments with two stereospecifically labelled species of MVA, [2-^{14}C-4*R*-4-$^{3}H_1$]MVA and [2-^{14}C-4*S*-4-$^{3}H_1$]MVA (note: the 4*R* tritium of MVA becomes the 2*S* tritium of IPP). It is not yet known at which side of the double bond of IPP a proton is attached in the first step but it is clear that the newly formed methyl group is *trans* to the $-CH_2OPP$ group (*Arigoni*, Biochem. Soc. Symp., 1960, **19**, 32; *Popják* and *Cornforth*, *loc. cit.*). The isomerization can thus be rewritten more exactly thus:

$$H^{\oplus} + CH_2{=}C(CH_3){-}CH(H)(H^*){-}CH_2O\text{Ⓟ–Ⓟ} \longrightarrow (CH_3)_2C^{\oplus}{-}CH(H)(H^*){-}CH_2O\text{Ⓟ–Ⓟ} \xrightarrow{-H^+} (CH_3)_2C{=}C(H^*){-}CH_2O\text{Ⓟ–Ⓟ}$$

3. Hemiterpenes

(a) Classification

(i) Free hemiterpenes

Hemiterpenes (C_5) as such are not common in nature, but are widely distributed as phenolic derivatives and as constituents of other structures containing non-terpene residues.

Apart from IPP and DMAPP (see previous section) which, as far as one knows, are present only in minute amounts as metabolic intermediates, the main hemiterpenes found naturally are isovaleraldehyde (I), isoamyl alcohol (II), isovaleric acid (III), 2-methylbutyric acid (IV), isopropylideneacetic acid (*β*-methylcrotonic acid) (V), tiglic acid (VI) and angelic acid (VII):

$(CH_3)_2CH{\cdot}CH_2{\cdot}CHO$ (I)

$CH_3{\cdot}CH(CH_3){\cdot}CH_2{\cdot}CH_2OH$ (II)

$(CH_3)_2CH{\cdot}CH_2{\cdot}CO_2H$ (III)

$CH_3{\cdot}CH_2{\cdot}CH(CH_3){\cdot}CO_2H$ (IV)

$(CH_3)_2C{=}CH{\cdot}CO_2H$ (V)

$CH_3{\cdot}CH{=}C(CH_3){\cdot}CO_2H$ (VI)

$CH_3{\cdot}CH{=}C(CH_3){\cdot}CO_2H$ (VII)

Neither isovaleric acid (nor isovaleraldehyde), *β*-methylcrotonic acid nor tiglic acid appears normally to be formed from IPP; they are formed during the catabolism of leucine (Scheme 2), valine, and isoleucine, respectively.

It is possible that all "true" hemiterpenes, in the sense that they are biosynthesized from IPP, occur bound to other molecules.

(ii) Phenolic hemiterpenes

The four possible products which could arise by the condensation of DMAPP with a phenol are indicated in Scheme 3 (see *W. D. Ollis* and *I. O. Sutherland*, "Recent Developments in the Chemistry of Natural Phenolic

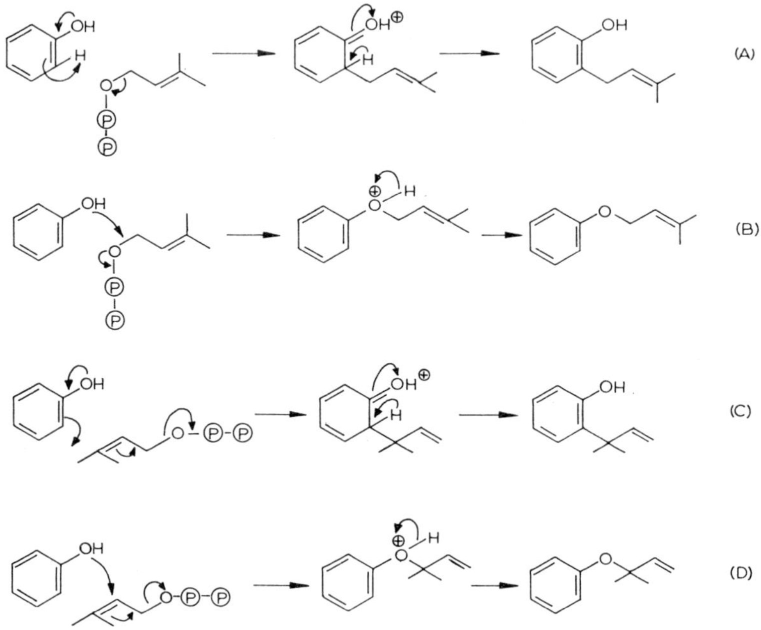

Scheme 3. Mechanism for formation of four types of hemiterpene phenols.

Compounds", Pergamon, London, 1961). These reaction mechanisms could also apply to the formation of phenols with longer terpenoid side chains (see *e.g.* p. 81).

Compounds of types A and B are more common than those of type C and D. Typical examples are (A) flavoglaucin (*A. Quilico* and *C. Cardani*, Gazz., 1953, 83, 1088); (B), foeniculin (*T. A. Geissman* and *E. Hinreiner*, Bot. Review, 1952, 18, 82); (C) echinulin, which also is an example of group A (*C. Cardani et al.*, Tetrahedron Letters, 1959, 1, 10), and (D), the anisoxide isomer (VIII) (*E. Taken*, Reichstoff, 1929, 4, 8), which may be converted into anisoxide during the isolation of the latter (*D. H. R. Barton et al.*, J. chem. Soc., 1958, 4393).

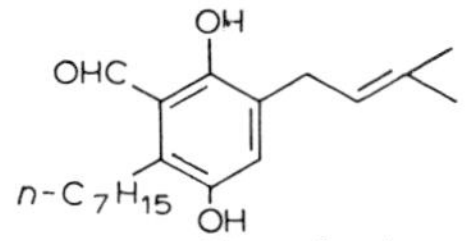

Flavoglaucin 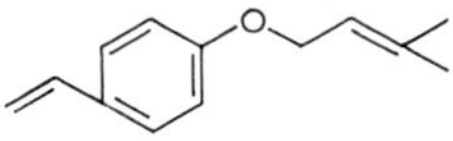 Foeniculin

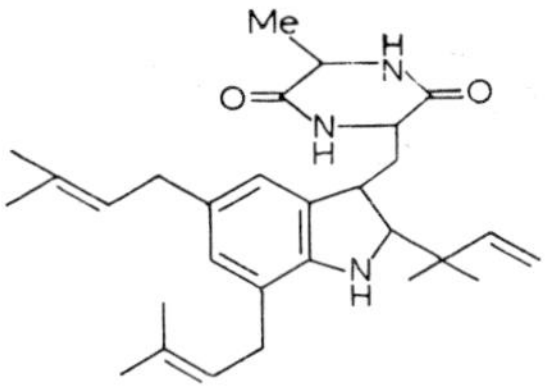

Echinulin 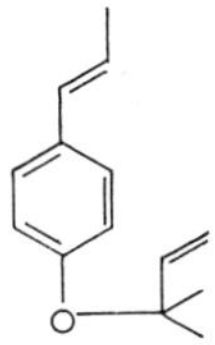 (VIII) 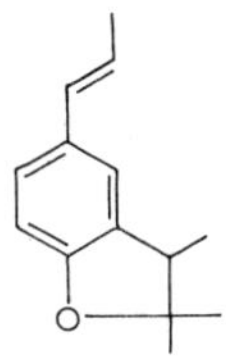 Anisoxide

Structural variations to the dimethylallyl (DMA) side-chain which have been observed include the oxidation to the corresponding isopentenyl epoxide (*e.g.* auraptene; *H. Böhme* and *G. Peitsch*, Ber., 1939, **72**, 773) or isopentane diols [*e.g.* B type-evoxine (*F. W. Eastwood, G. K. Hughes* and *E. Ritchie*, Aust. J. Chem., 1954, **8**, 87) and A type-toddalolactone (*B. B. Dey* and *P. P. Pillay*, Arch. Pharm., 1935, **273**, 223; *E. Spath, Dey* and *E. Tyray*, Ber., 1938, **71**, 1825; 1939, **72**, 53)]:

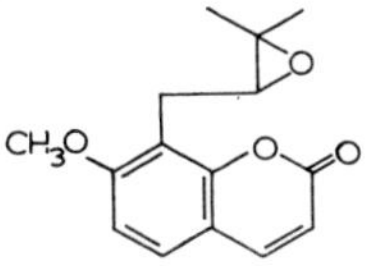

Auraptene 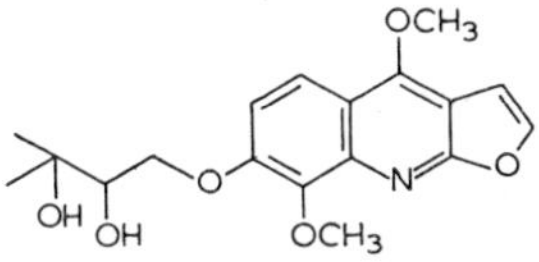 Evoxine 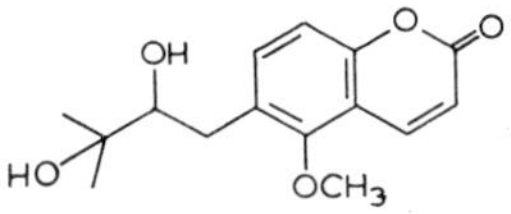 Toddalolactone

Cyclization products of isoprenylated phenols are probably the precursors of 2,2-dimethylchromenes such as flindersine (*R. F. C. Brown et al.*, Aust. J. Chem., 1954, **7**, 348). *Ollis* and *Sutherland* (*loc. cit.*) have proposed the following mechanism for cyclization:

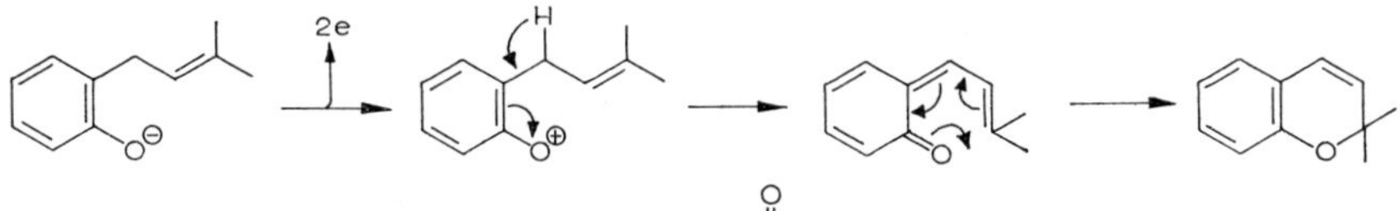

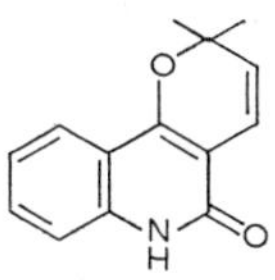

Flindersine 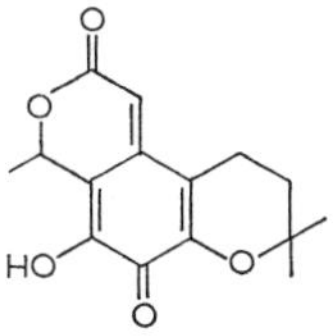 Fuscin

One well authenticated naturally occurring 2,2-dimethylchromane is fuscin (*Barton* and *J. B. Hendrickson*, J. chem. Soc., 1956, 1028).

Ring closure of $\gamma\gamma$-dimethylallyl epoxides could give rise to hydroxyisopropyldihydrobenzofurane, thus (*Ollis* and *Sutherland*, *loc. cit.*):

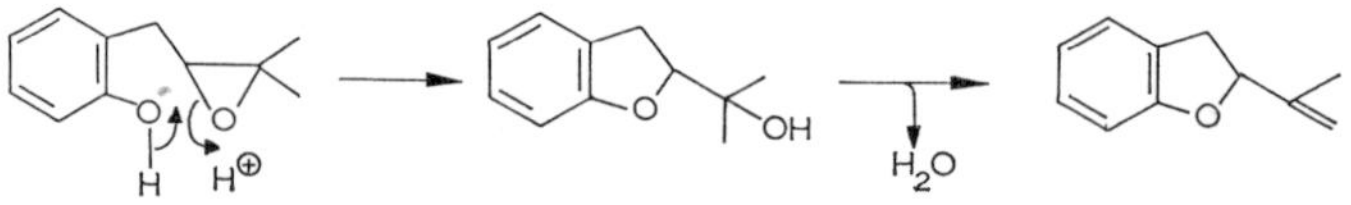

A typical example is visamminol (*W. Bencze* and *H. Schmidt*, Experientia, 1954, **10**, 12). Dehydration of such a compound could yield known natural products such as oreoselone (*Späth, K. Klager* and *C. Schlosser*, Ber., 1931, **64**, 2203); double bond migration could then give rise to derivatives such as munetone (*N. L. Dutta*, J. Ind. chem. Soc., 1959, **36**, 165).

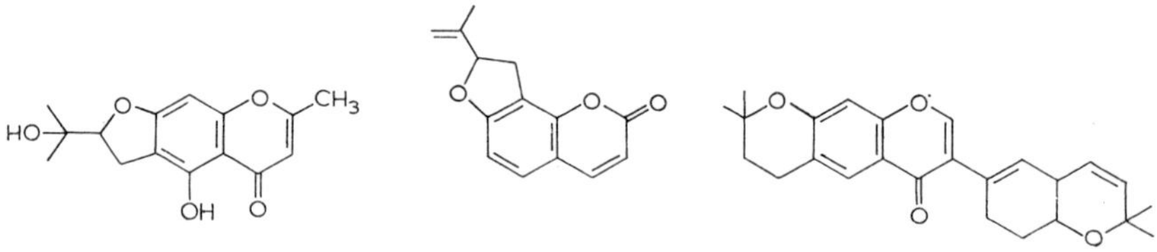

Visamminol Oreoselone Munetone

(iii) Alkaloids with hemiterpene residues

A number of these are known but will be discussed only in relation to biosynthetic issues (see p. 65).

(iv) Isopentenyl adenosine and related compound

An important observation is the occurrence of $N_{(6)}$-(Δ^2-isopentenyl) adenosine (IX) as a component base of transfer ribonucleic acid (RNA) in yeast (*K. Biemann et al.*, Angew. Chem., 1966, **78**, 600; *R. H. Hall et al.*, J. Amer. chem. Soc., 1966, **88**, 2614) and of mammalian *t*-RNA (*M. J. Robins et al.*, Biochemistry, 1967, **6**, 1837). $N_{(6)}$-(*cis*-4-hydroxy-3-methylbut-2-en)adenosine (X) is also present in *t*-RNA in higher plants (*Hall et al.*, Science, 1967, **156**, 69).

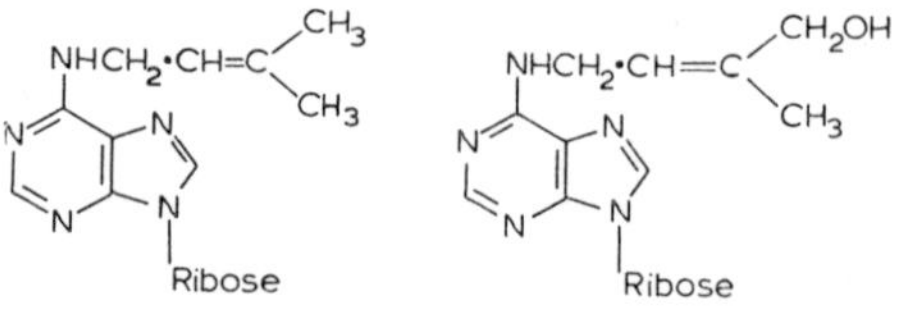

(IX) (X)

(b) Biosynthesis of hemiterpenoids

An overall picture of the biogenesis of hemiterpenoids was discussed by *W. B. Whalley* ("Recent Developments in the Chemistry of Natural Phenolic Compounds", London, 1961, p. 20); although the proposals have a compelling logic, most lack direct biochemical demonstrations. The incorporation of [2-^{14}C]mevalonate into the isoprenoid side chains of the following compounds has been demonstrated: auroglaucin (*A. J. Birch, J. Schofield* and *H. Smith,* Chem. and Ind., 1958, 1321); echinulin (*Birch et al.*, J. chem. Soc., 1961, 3128); the alkaloids elymoclavine and agroclavine (*D. Groger et al.*, Z. Naturforsch., 1960, **15b,** 141; *B. J. McLoughlin* and *Smith,* Tetrahedron Letters, 1960, 7, 1; *S. Battacharji et al.*, J. chem. Soc., 1962, 421), sphaerophysin (*G. Reuter*, Abh. Deut. Akad. Wiss. Berlin, Kl. Chem. Geol. Biol., 1966, 617; *R. J. Suhadolnik* quoted by *G. R. Waller,* Progress in Chemistry of Fats and other Lipids, 1968, **10,** 151), and humulone (*W. D. Loomis et al.*, Abst. 4th Int. Symp. on Chemistry of Natural Products 1966, p. 141) and into the lysergic residue (XI) of the alkaloids ergotamine, ergosine and ergometrine (*E. H. Taylor* and *E. Ramstad,* Nature, 1960, **188,** 494).

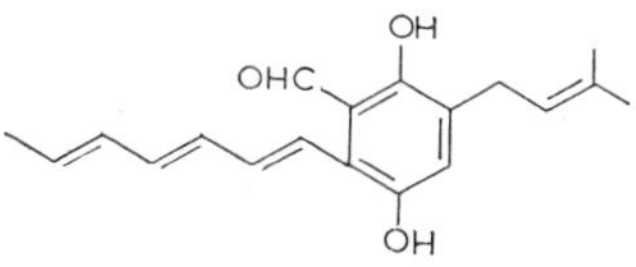

Auroglaucin 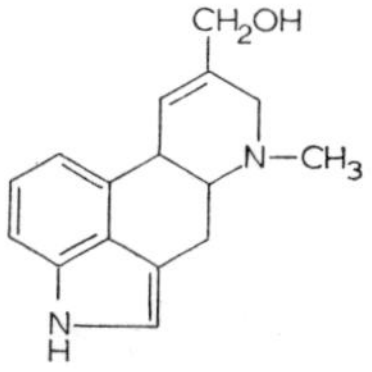 Elymoclavine 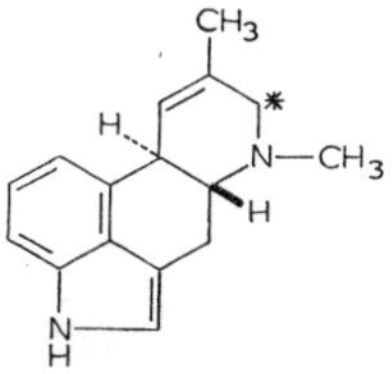 Agroclavine

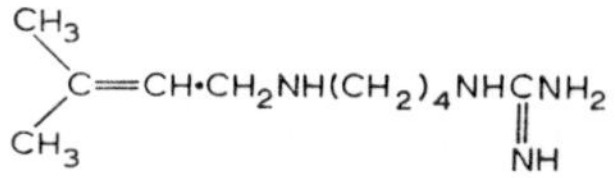

Sphaerophysin 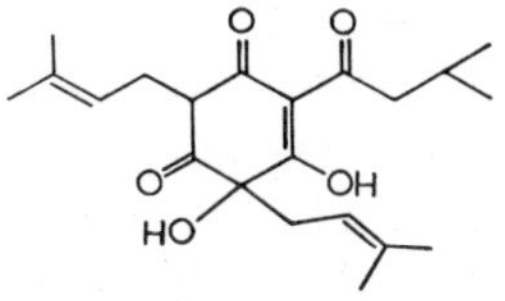 Humulone

The appearance of label from [2-^{14}C]MVA mainly in the starred atom of agroclavine indicates the existence of the IPP → DMAPP pathway (*Birch, McLoughlin* and *Smith, loc. cit.*); similarly [2-^{14}C]MVA led to chanoclavine I labelled almost exclusively at $C_{(7)}$ (starred) (*T. Fehr, W. Acklin* and *D. Arigoni,* Chem. Comm., 1966, 801). [2-^{14}C-$^{3}H_2$]MVA yielded pyroclavine

[and its $C_{(7)}$ epimer, festuclavine] with the same $^{14}C:{}^{3}H$ ratio indicating that the $C_{(2)}$ methylene group in MVA was incorporated intact in the alkaloid (*R. M. Baxter et al.*, J. Amer. chem. Soc., 1962, **84**, 4350). DL-[4-^{14}C]Valine was incorporated almost exclusively into the *gem*-dimethyl groups (starred in formula) of echimidinic acid (*D. H. G. Gnout*, J. chem. Soc. C., 1966, 1968).

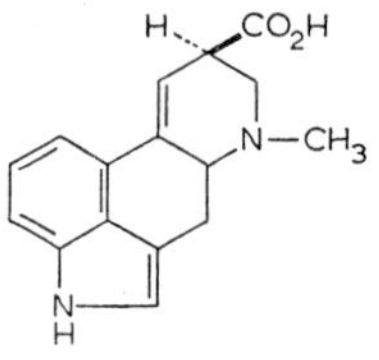

(XI) 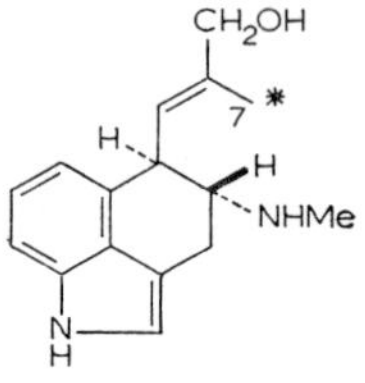 Chanoclavine I 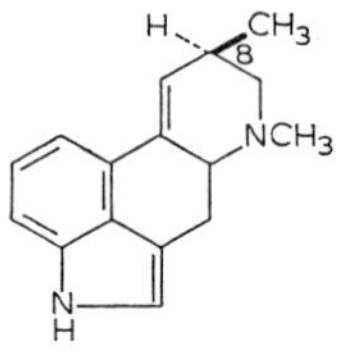 Pyroclavine 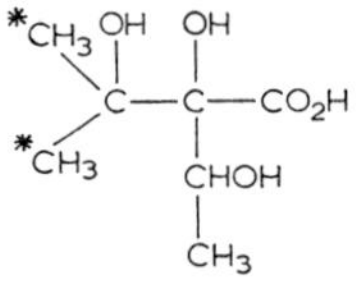 Echimidinic acid

The isopentenyl residue of $N_{(6)}$-(Δ^2-isopentenyl)adenosine (IX) arises from MVA in the bacteria *Lactobacillus acidophilus* and *L. plantarum* (*A. Peterkofsky*, Biochemistry, 1968, **7**, 472; *F. Fitter, L. K. Kline* and *Hall, ibid.*, 1968, **7**, 940).

4. Monoterpenes

(a) The basic precursor and classification

Geranyl pyrophosphate is almost certainly the fundamental C_{10}-precursor of all monoterpenes. It is formed from the condensation of one molecule of DMAPP and IPP with the elimination of an allylic pyrophosphate group in the presence of the enzyme geranyl transferase (farnesyl pyrophosphate synthetase) isolated from yeast (*B. W. Agranoff et al.*, J. Amer. chem. Soc., 1959, **81**, 1255), from *Micrococcus lysodeicticus* (*A. A. Kandutsch et al.*, J. biol. Chem., 1964, **239**, 2507) and from pig liver (*C. R. Benedict, J. Kett* and *J. W. Porter*, Arch. Biochem. Biophys., 1965, **110**, 611; *P. W. Holloway* and *G. Popják*, Biochem. J., 1967, **104**, 57). The *trans* isomer is the only product of the reaction (*Popják et al.*, J. Lipid Res., 1959, **1**, 29). The stereochemistry of this reaction and the analogous reactions leading to *trans*-farnesyl pyrophosphate (C_{15}) and *trans*-geranylgeranyl pyrophosphate (C_{20}) has been elegantly elucidated (*Popják* and *J. W. Cornforth*, Biochem. J., 1966, **101**, 553). They formulate a mechanism involving two steps, thus:

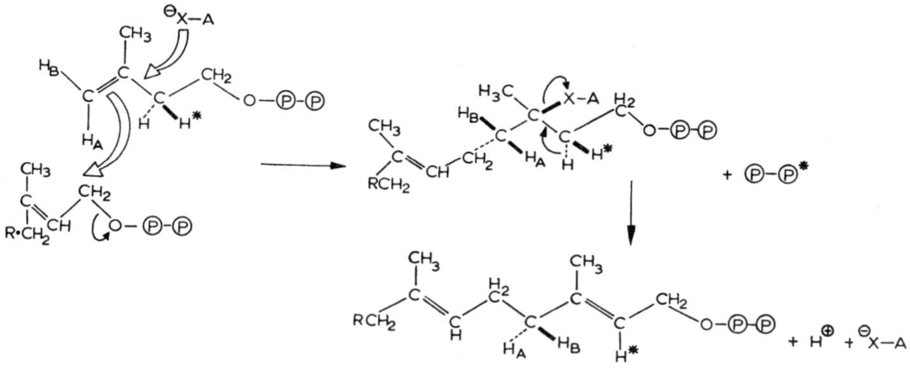

Geranyl pyrophosphate

* The encircled P's are an approved abbreviation for inorganic phosphate.

The first is *trans* addition to the existing double bond followed by *trans* elimination to form the new one. This is in agreement with the stereochemistry of the reaction and with chemical analogies but it requires the intervention of an electron-donating group X which is temporarily attached to $C_{(3)}$ of DMAPP. The nature of X is not yet known but it is possibly an amino acid residue of an enzyme.

(i) Acyclic monoterpenes

Probably all naturally occurring monoterpenes are formed by dehydration, hydrogenation, introduction of oxygen in various forms, rearrangements or isomerization of geraniol; whether any of the changes occurs at the pyrophosphate level or with free geraniol is not known. Typical examples of compounds in which these changes have occurred are (i) dehydration, myrcene; (ii) hydrogenation, (+)-β-citronellol; (iii) introduction of oxygen, linalool oxide and elsholtzione (carbonyl); (iv) rearrangement, (+)-linalool

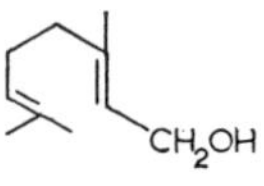

Geraniol  Myrcene 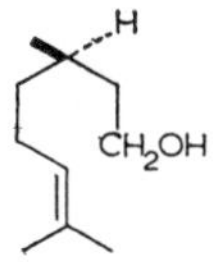 (+)-β-Citronellol 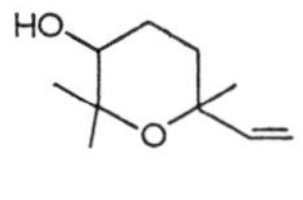 Linalool oxide

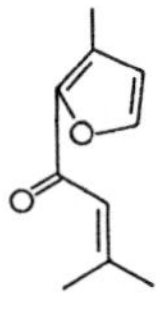

Elsholtzione 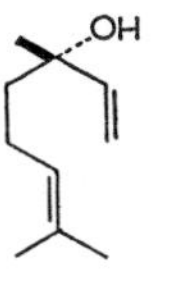 (+)-Linalool 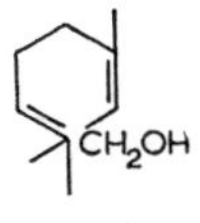 Nerol

and (v) isomerization, nerol (*cis*-geraniol) (see *W. Sandermann* in "Comparative Biochemistry", Vol. IIIA, p. 503, Academic Press, New York 1962; *A. Haagen-Smit* and *C. C. Nimmo*, in "Comprehensive Biochemistry", Vol. 9, Eds. *M. Florkin* and *E. H. Stotz*, Elsevier, Amsterdam 1963, for comprehensive lists of examples).

It is possible that linalool can be formed at the pyrophosphate level because traces have been reported in enzyme experiments (*Popják* and *Cornforth*, Advances in Enzymology, 1960, **22**, 281).

However, allylic rearrangement of geraniol takes place so readily at acid pH that non-enzymic formation of linalool may occur in plant cells. Nerol may be formed from free geraniol, because the existence of neryl pyrophosphate has not yet been clearly demonstrated.

(ii) Monocyclic monoterpenes

(1) *Cyclohexanes.* Two main types are known based on the cyclogeraniol and *p*-menthane nuclei, respectively, both of which can arise from geraniol as its pyrophosphate; on purely chemical grounds, however, neryl pyrophosphate is more likely to be the precursor than geranyl pyrophosphate. On acid hydrolysis the latter gives no cyclic products whilst neryl pyrophosphate gives α-terpineol in 60 % yield.

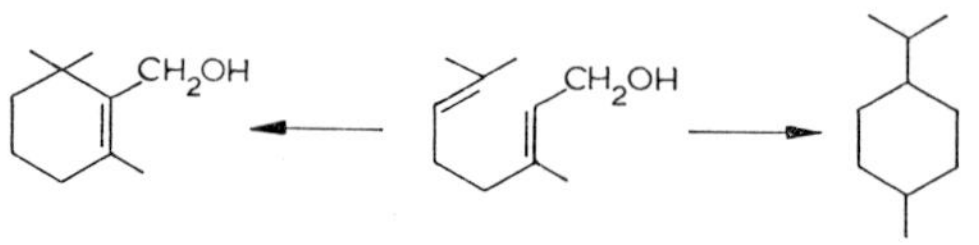

β-Cyclogeraniol Geraniol *p*-Menthane

Variants on the *p*-menthane structure are numerous. Just one example is α-terpineol. Only a few cyclogeraniol derivatives are known, *e.g.* safranal. They may possibly be degradation products of carotenoids (see p. 132) rather than primary biosynthetic products:

α-Terpineol Safranal

(2) *Cyclopropanes, cyclopentanes and cycloheptanes.* Typical examples of these groups, which can be regarded as menthane derivatives, are the monomethyl ester of chrysanthemumdicarboxylic acid (I) (cyclopropane) from

Chrysanthemum cinaerarifolium; 1-acetyl-4-isopropylidenecyclopent-1-ene (II) from *Eucalyptus globulus*, and eucarvone (III) (cycloheptane) from *Asarum sieboldi*. It has been suggested (*Haagen-Smith* and *Nimmo*, *loc. cit.*) that I and II are probably derived from a car-2-ene-type structure IV by oxidation.

CO_2CH_3 CO_2H

(I) (II) (III) (IV)

(iii) Bicyclic monoterpenes

These structures always contain in addition to a six-membered ring, a three, four- or five-membered ring. The basic unit is the menthane structure and the various groups which can be derived therefrom by cyclization are indicated below. Typical examples are car-2-ene (IV), umbellulone, α-pinene, borneol and camphene, which has a norbornane structure.

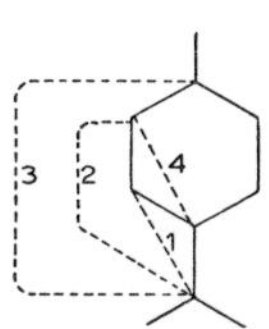

1. Carane 3. Bornane
2. Pinane 4. Thujane

Cyclizations which can occur in the menthane structure (*Haagen-Smit* and *Nimmo*, 1963, *loc. cit.*).

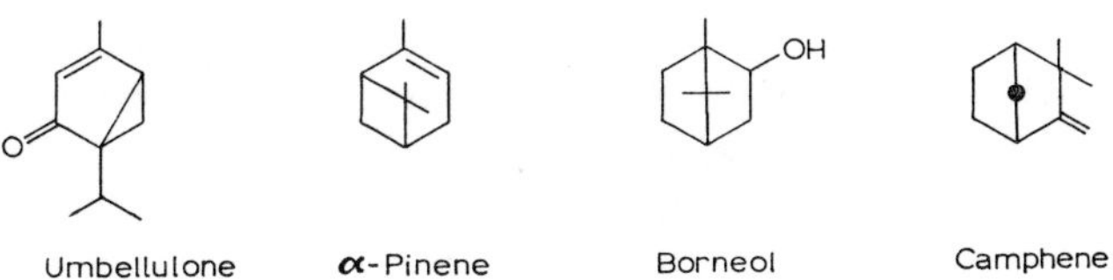

(iv) Tricyclic monoterpenes

Sandalwood (*Santalum album*) oil contains tricyclic compounds such as teresantalol which can formally be derived from a bornane skeleton by linking $C_{(2)}$ and $C_{(6)}$.

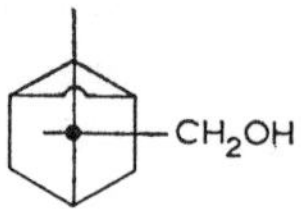

Teresantalol

(v) Phenolic monoterpenoids

Types analogous to **A** and **B** (Scheme 3, p. 62) for hemiterpenes are known for monoterpenes; typical examples are ostruthin (*E. Spath* and *K. Klager*, Ber., 1934, **67**, 859) and collinin (*O. Isler et al.*, Helv., 1958, **41**, 786). Hydration of double bonds can occur as in marmin (*A. Chatterjee* and *A. Chondhury*, Naturwiss., 1955, **42**, 512). No monoterpenes of types **C** and **D** (Scheme 3) appear to be known:

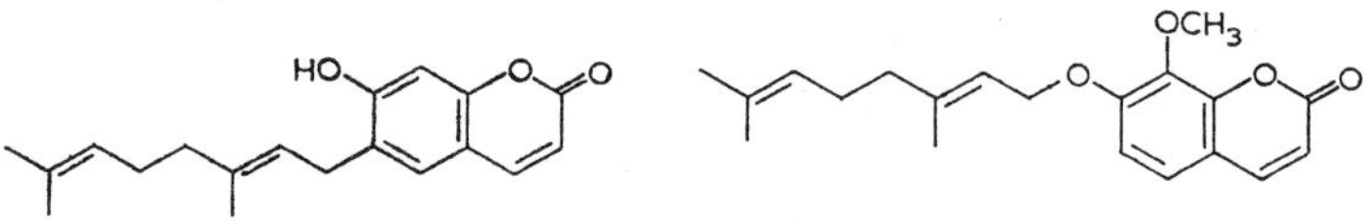

Ostruthin

Collinin

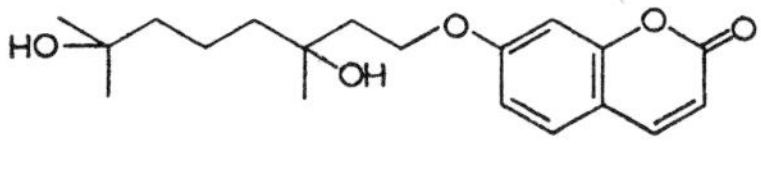

Marmin

(vi) Alkaloids with monoterpene residues

The non-aromatic C_{10} or C_9 residues of a number of indole alkaloids have been shown to be derived from mevalonate (see p. 75).

(vii) Atypical monoterpenes

Phenols which contain cyclic monoterpene residues are also known. A typical example is cannabidiol (*R. Adams, R. B. Baker* and *B. R. Wearn*, J. Amer. chem. Soc., 1940, **62**, 2204), which according to *Whalley* (*loc. cit.*) could arise from the corresponding geranyl resorcinol (V):

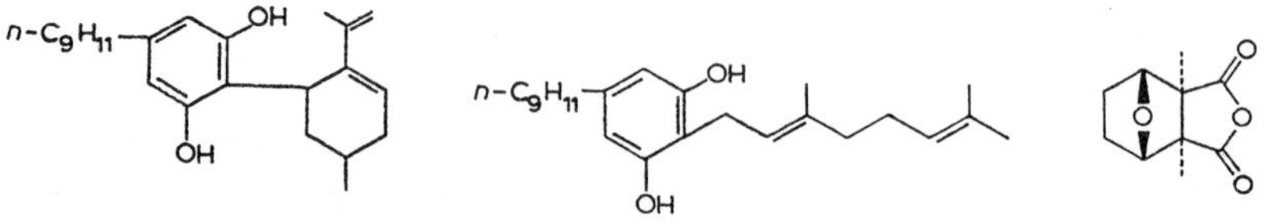

Cannabidiol

(V)

Cantharidin

Cantharidin from dried *Cantharis vesicatoria* appears to be derived from two isoprenoid residues joined tail to tail. Other atypical monoterpenes are artemisia ketone, and chrysanthemic acid. The label from [2-^{14}C]MVA is apparently incorporated into the latter (*M. P. Crowley et al.*, Biochim. Biophys. Acta, 1962, 60, 312) but not into the former (*Waller et al.*, Phytochemistry, 1968, 7, 213). The reason may be that chrysanthemic acid is derived from a "regular" monoterpene such as car-2-ene (see Vol. II C, pp. 152, 252).

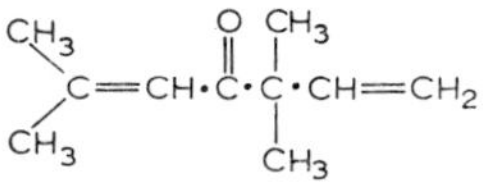

Artemisia ketone

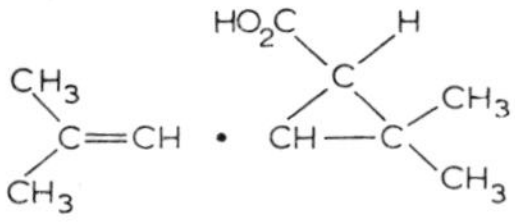

Chrysanthemic acid

(b) Biochemical studies on monoterpene biosynthesis

A major obstacle to the study of the detail of monoterpene biosynthesis is however the difficulty of obtaining significant incorporation of labelled MVA into the compounds (*W. D. Loomis*, in "Terpenes in Plants", Ed. *J. B. Pridham*, Academic Press, New York, 1967, p. 59). For example no incorporation was observed in myrcene (p. 67) in *Humulus*, although incorporation was observed with *Santolina chaemcyparisus* (*Waller et al., loc. cit.*), and in the cases where small incorporations have been reported it is difficult to decide whether it is specific or whether it is due to incorporation of $^{14}CO_2$ liberated during the metabolism of MVA; *e.g.*, $^{14}CO_2$ is well incorporated into terpenes in peppermint and geranium (*Loomis, loc. cit.*), and although one group of workers found [2-^{14}C]MVA to be incorporated specifically into cineole (labelled atoms starred) (*Birch et al.*, Tetrahedron Letters, 1959, No. 3, 1), others

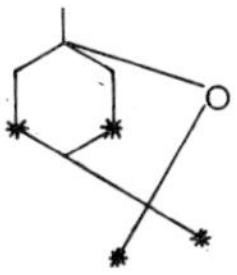

Cineole

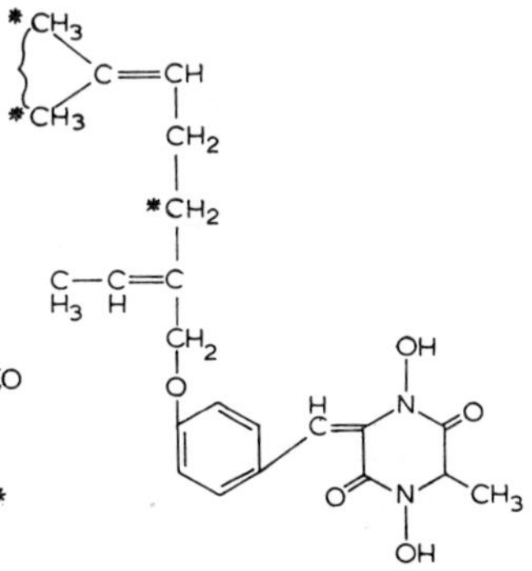

Mycelianamide

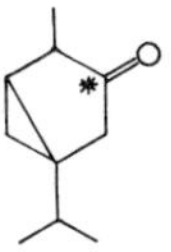

Thujone

found that all the label was randomized indicating that the small amount of label incorporated (0·01%) had been converted into $^*CO_2$ before incorporation (*Arigoni*, quoted in *Loomis, loc. cit.*; see also *R. G. Battee* and *H. W. Youngken*, Lloydia, 1966, **29**, 360). The label from [1-^{14}C]geranyl pyrophosphate is, however, specifically found in the two positions indicated in cineole (*B. Achilladelis* and *J. R. Hanson*, Phytochemistry, 1968, **7**, 313). High incorporation of [2-^{14}C]MVA into the monoterpenes of rose petals has been recorded (*M. J. O. Francis* and *M. O'Connell*, *ibid.*, 1969, **8**, 1705). In the side-chain of mycelianamide (*Birch et al.*, J. chem. Soc., 1962, 1502) the atoms starred arise from $C_{(2)}$ of MVA. In an ingenious experiment, it was shown that only one of the *gem*-dimethyl groups was labelled. Mycelianamide was cleaved by sodium in liquid ammonia and the resultant hydrocarbon fed to a rabbit which metabolized it to a dicarboxylic acid which, on ozonolysis, yielded acetaldehyde free from radio-activity. The reaction scheme is:

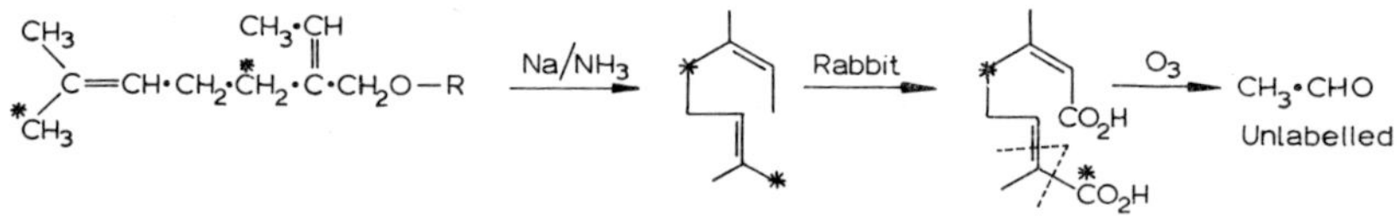

CH_3—C(=CH·CH_2·CH_2·C(=CH·CH_3)·CH_2O—R)—CH_3 → Na/NH$_3$ → → Rabbit → → O_3 → CH_3·CHO Unlabelled

However, the isopropylidene carbons (indicated by a star) in plumieride which arise from [2-^{14}C]MVA *via* citronellal and iridodial have lost their individuality (*D. A. Yeowell* and *H. Schmidt*, Experientia, 1964, **20**, 250):

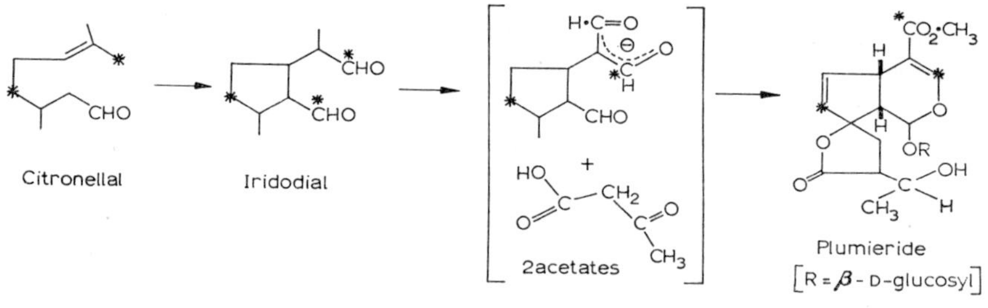

Citronellal → Iridodial → [+ 2acetates] → Plumieride [R = β-D-glucosyl]

Somewhat similar results were obtained with verbenalin synthesized by *Verbena officinalis* in which little randomization occurred between $C_{(6)}$ and $C_{(9)}$ whatever the age of the plant used. However, in young (1 month old) plants the randomization between $C_{(3)}$ and $C_{(8)}$ is essentially complete, whilst in older (four month plants) the majority of the label is at $C_{(3)}$ (*J. E. S. Hüni et al.*, Experientia, 1966, **22**, 656; *A. G. Horodysky* quoted by *Waller*, 1966 *loc. cit.*). With nepetalactone limited randomization occurs at both sites of [2-^{14}C]MVA incorporation (*F. E. Reigner et al.*, Phytochemistry, 1968, **7**, 22).

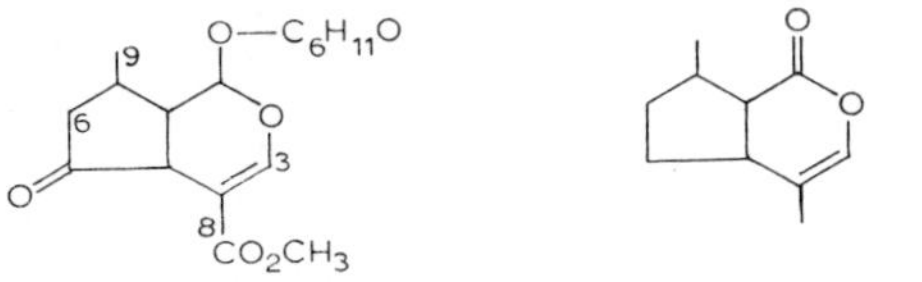

Verbenalin

Nepetalactone

The labelling in the C_{10} residues of some alkaloids indicates similar loss of individuality (*J. E. S. Hüni et al.*, Experientia, 1966, **22**, 656).

It is claimed that label from [2-^{14}C]MVA is incorporated only into the carbonyl group of thujone. That is, one C_5 unit appears to come from a source other than exogenous MVA. The suggestion made is that a large endogenous pool of dimethylallyl pyrophosphate would account for this observation (*D. V. Banthorpe* and *K. W. Turnbull*, Chem. Comm., 1966, 147). Similar results have been obtained with the incorporation of [2-^{14}C]MVA into camphor when all the label was found in $C_{(6)}$ (starred) (*D. V. Banthorpe* and *D. Baxendale*, Chem. Comm., 1968, 1553):

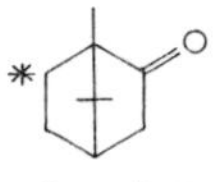

Camphor

Biosynthetic studies on *Mentha* spp. have indicated the pathway in Scheme 4 for the synthesis of (—)-menthol and (+)-isomenthol; the reactions indicated by dotted lines have not been clearly demonstrated (*R. H.*

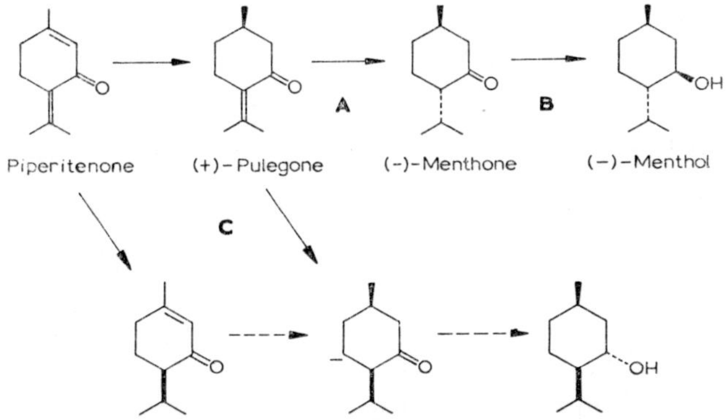

Scheme 4. Metabolic interconversion of monoterpenes in peppermint. (Dotted arrows: postulated reactions.)

Reitsema, J. Amer. pharm. Ass. Sci. Ed. 1958, **47**, 267; *J. Battaile* and *Loomis*, 1961, Biochim. Biophys. Acta, 1961, **51**, 545; *R. H. Reitsema et al.*, J. Pharm. Sci., 1961, **50**, 18; *F. W. Hefendehl*, Planta Med., 1962, **10**, 241).

Reactions marked **A, B** and **C** in the scheme have been demonstrated in cell-free extracts (*Battaile et al.*, Phytochemistry, 1968, **7**, 1159). Genetic studies have also indicated a probable pathway to piperitenone and carvone. A pure line of *Mentha crispa* produces mainly carvone and dihydrocarvone. Self-pollinated first generation offspring produce plants which synthesize carvone plus dihydrocarvone or the pugelone/menthol group in a ratio of 3:1. This is a typical Mendelian segregation of a single gene pair with the dominant allele controlling carvone synthesis (*M. J. Murray*, Genetics, 1960, **45**, 931). The biosynthetic scheme indicated in Scheme 5 is a slight variant on Loomis's original ideas.

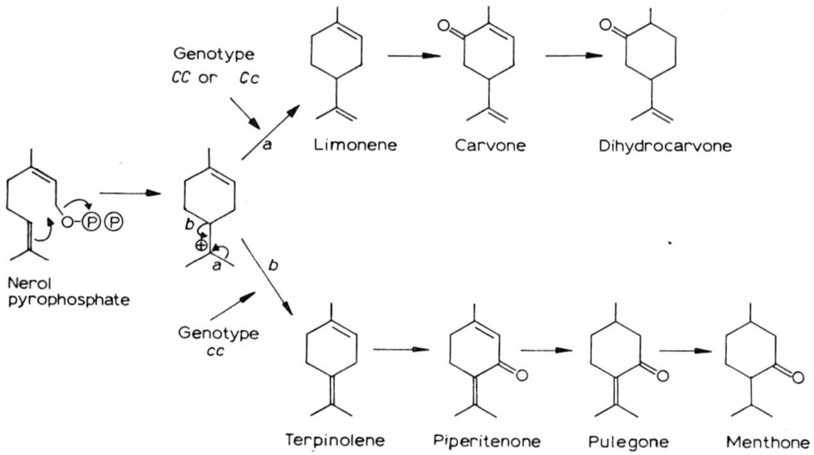

Genotype *CC* or *Cc*
Limonene
Carvone
Dihydrocarvone
Nerol pyrophosphate
Genotype *cc*
Terpinolene
Piperitenone
Pulegone
Menthone

Scheme 5. Possible biosynthetic pathway of monoterpene formation in mints based on genetic studies.

Loss of a proton occurs from one position of the postulated carbonium ion intermediate according to the location of the proton-accepting site on the dehydrogenase, the structure of which is controlled by the gene pairs *CC* and *cc*. This results in the formation of either limonene or terpinolene, both of which could be oxidized by a non-specific α-oxidase to yield carvone and piperitenone. In the latter case the two double bonds working together direct the oxygen into the desired position.

Pulegone is converted into menthol in the rabbit (*R. Teppate*, Arch. Inst.

Pharmacodyn., 1937, **57**, 440); however, it is claimed that when piperitone is metabolised by sheep it retains its 1,2 double bond and is converted, *inter alia,* into thymol (*J. M. Harvey,* University of Queensland Papers, 1942, **1**, No. 23). Harvey also found that the methyl group of (—)-α-phellandrene

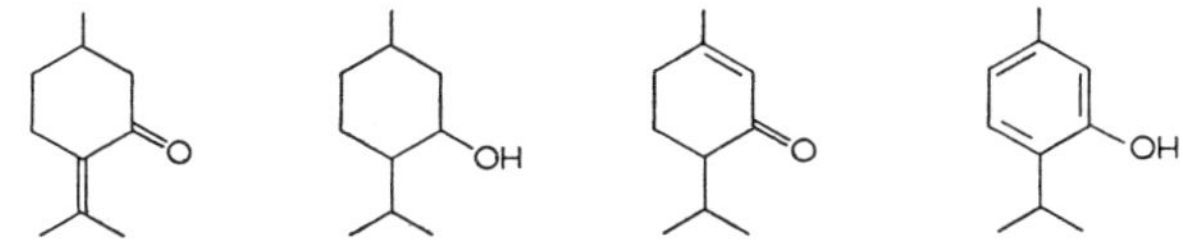

Pulegone Menthol Piperitone Thymol

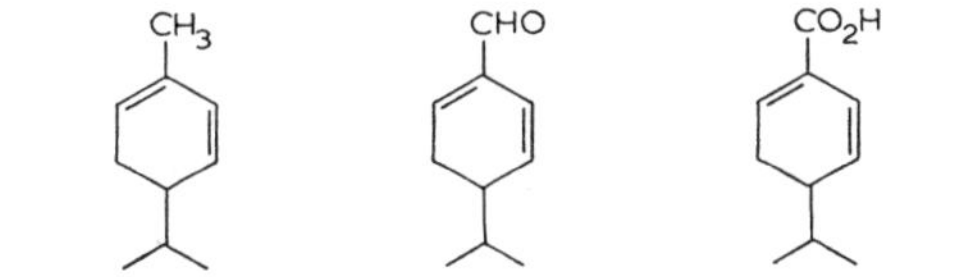

CH_3 CHO CO_2H

(−)-α-Phellandrene Phellandral Phellandric acid

is oxidized by sheep so that phellandral and phellandric acid are excreted in the urine. [For a full list of terpenoid transformations which can occur in animals, see *R. T. Williams,* "Detoxication Mechanisms", 2nd Edn., Chapman and Hall, London, 1959].

More success has been obtained with indole alkaloids which contain a monoterpene residue. [2-^{14}C]MVA has been incorporated specifically into vindoline in *Catharanthus roseus* (*T. Money et al.,* Proc. nat. Acad. Sci. U.S.A., 1965, **53**, 901; *H. Goeggel* and *Arigoni,* Chem. Comm., 1965, 538), reserpine (*Goeggel* and *Arigoni, loc. cit.*) and dehydroaspidospermine (*A. R. Battersby et al.,* Chem. Comm., 1966, 46). Labelled geraniol is also actively incorporated into these alkaloids in the expected position (*G. F. Waller et al.,* Analyt. Biochem., 1966, **6**, 277; *Battersby et al.,* Chem. Comm., 1966, 346, 888; *E. Leete* and *S. Ueda,* Tetrahedron letters, 1966, 4915; *Leete* and *J. N. Wemple,* J. Amer. chem. Soc., 1966, **88**, 4743). Nerol, the *cis* isomer of geraniol, was also incorporated into *Vinca* alkaloids (*Battersby et al., loc. cit.*). Loganin is the intermediate in vindoline synthesis (*idem,* Chem. Comm., 1968, 133). The pattern of incorporation of [2-^{14}C]MVA into β-skytanthine by *Skytanthus acutus* is very similar to that observed with verbenalin (p. 73). Whatever the age of the plant, there is little randomization between $C_{(6)}$ and $C_{(10)}$; in young plants there is considerable randomization between $C_{(3)}$ and $C_{(9)}$ and this almost disappears when experiments are done on older plants (*H. Suda et al.,* J. Amer. chem. Soc., 1967, **89**, 2476):

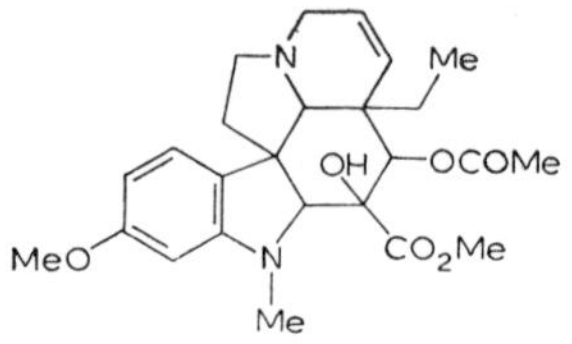

Vindoline

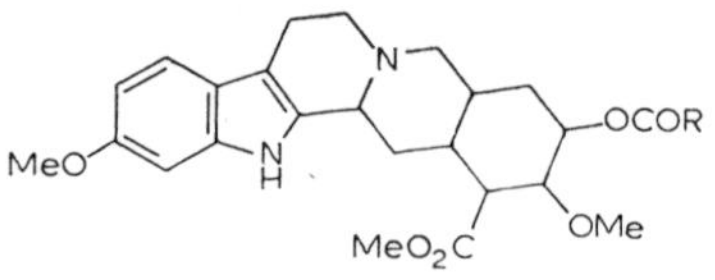

Reserpine (R = 3,4,5-trimethoxyphenyl)

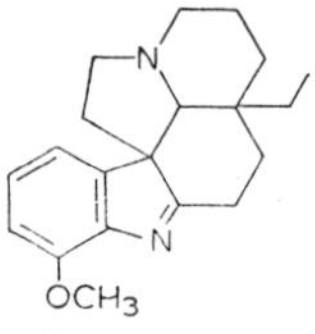

Dehydroaspidospermine

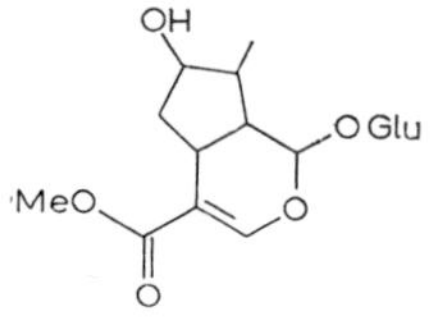

Loganin

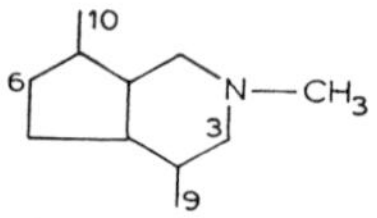

Skytanthine

(c) *Biogenetic considerations*

(i) General

The suggestion that β-pinene is derived from myrcene (*L. Ruzicka et al.*, 1953, *loc. cit.*) is supported by the observation that myrcene is photochemically converted into β-pinene in 9% yield (*K. J. Crowley*, Proc. chem. Soc., 1962, 245):

Myrcene → β-Pinene

(ii) Oxidations

Direct demonstration of oxidations of monoterpenes has only been obtained in mint (Scheme 5, p. 74) and in animals (*Williams, loc. cit.*); however, the co-existence in plants of various compounds with a basic structure at various oxidation states has prompted obvious suggestions for biogenetic interrelationships. For example, α-pinene, verbenol and verbenone exist together in oil of olibanum (*Boswellia carteria*) and the following biogenetic sequence has been suggested (*E. Guenther*, "The Essential Oils", Vol. 4, Van Nostrand, New York, 1950, p. 355):

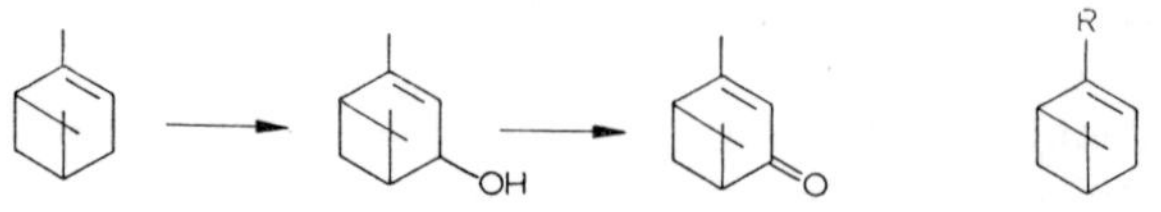

α-Pinene → Verbenol → Verbenone

(VI)

A number of monoterpenes also occur together which suggests that methyl groups can undergo stepwise oxidation. For example, (+)-α-pinene (VI; $R=CH_3$), (+)-myrtenol (VI; $R=CH_2OH$) and (+)-myrtenal (VI, $R=CHO$) occur together in *Eucalyptus globosus*, whilst the corresponding carboxylic acid (VI; $R=CO_2H$) is found in *Chamaecyparis formosensis* (*Guenther, loc. cit.*, p. 484). Direct biochemical evidence for these oxidations has not been obtained in plants, but they do occur in animals. For example, dogs fed camphor (VII) excrete in their urine 3-hydroxycamphor (VIII) and 5-hydroxycamphor (IX) (*Y. Asahina* and *M. Ishidate*, Ber., 1928, **61**, 533); similarly dimethylcamphor (X) is metabolized to the corresponding hydroxy derivative (cf. IX). Uroterpenol (XI) isolated from human urine as a glucuronide (*A. P. Wade et al.*, Biochem. J., 1966, **101**, 727) is probably derived from dietary limonene (XII) (see *Williams, loc. cit.* for further examples). Similar oxidations occur in the higher terpenes (see *Sandermann, loc. cit.*) but they will not be discussed in detail in this article.

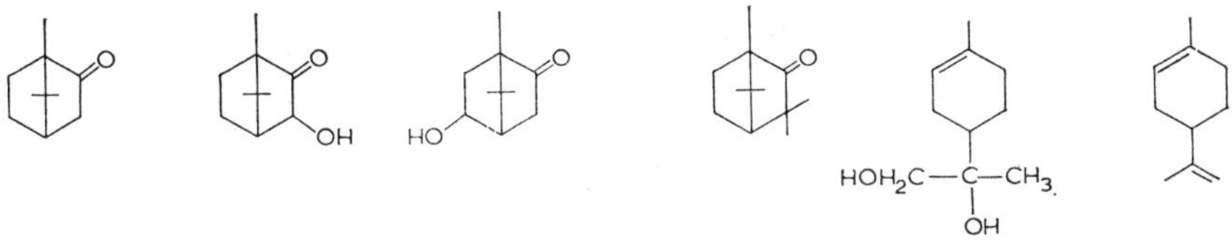

(VII) (VIII) (IX) (X) (XI) (XII)

(iii) Aromatization

Distribution patterns have led to a number of suggestions for the biosynthetic sequence leading to the formation of aromatic terpenoids; for example, the constituents of the oil of *Monarda fistulosa* (wild bergamot) indicate the following pathway (*W. Sandermann*, In "Comparative Biochemistry"; Vol. IIIA, Academic Press New York, 1963, p. 591):

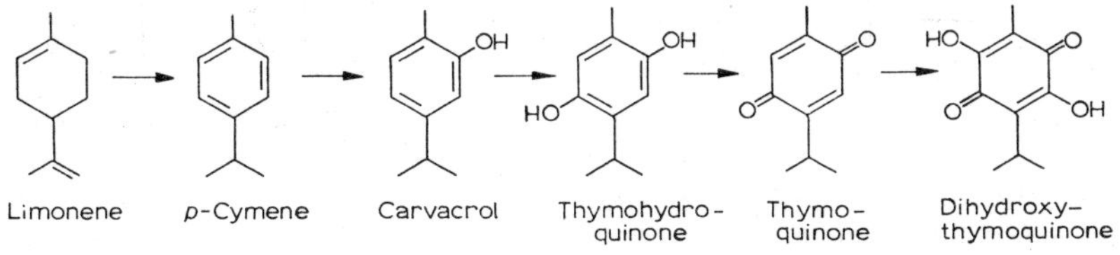

Limonene → p-Cymene → Carvacrol → Thymohydroquinone → Thymoquinone → Dihydroxythymoquinone

Another possibility is that phenols are produced by enolization of an oxo-group; this is suggested by the co-occurrence of cryptone and *p*-isopropylphenol in *Eucalyptus cneroifolia* (*Sandermann, loc. cit.*).

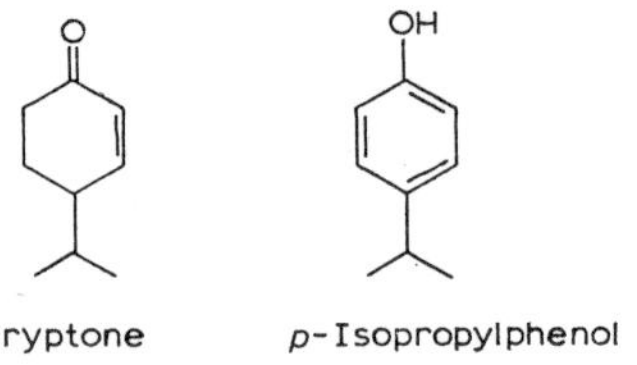

Cryptone p-Isopropylphenol

(*iv*) *Cycloheptanes*

The class of cycloheptanes which are exemplified by γ-thujaplicin, which occurs in the heartwood of conifers (*T. Nozoe,* Prog. Chem. org. nat. Prod., 1956, 13, 232) are considered to be derived from carone (see *H. Erdtman,* "Perspectives in Organic Chemistry", Interscience, New York, 1956, p. 485):

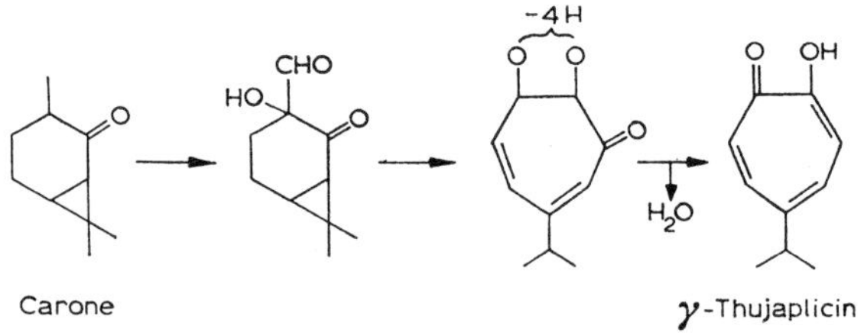

Carone γ-Thujaplicin

No direct evidence is yet available to support these views, however.

(*v*) *Furan rings*

A number of monoterpenes exist with furan ring systems. The biosynthetic mechanisms involved in their formation are not known but the existence of perillene and perilla ketone in *Perilla citriodora* and *P. frutescens*, respectively, has prompted the suggestion that they are formed from geraniol, *via* a dialdehyde thus (*Sandermann, loc. cit.*):

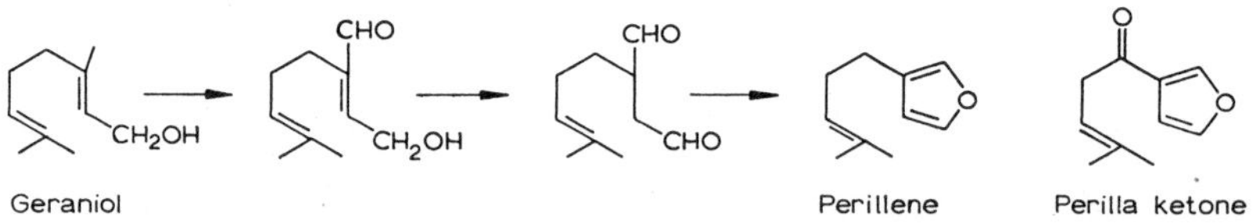

Geraniol Perillene Perilla ketone

Menthofuran, which accumulates in peppermint plants kept under warm night conditions (*Loomis, loc. cit.*), is however, formed from (+)-pulegone (Scheme 5). Probable mechanisms for formation of furan rings in mono- and sesqui-terpenes (next section) have been discussed by *M. D. Sutherland* and *R. J. Park* (In "Terpenes in Plants", Ed. *J. B. Pridham,* Academic Press, New York, 1967, p. 147).

(vi) Bicyclic monoterpenes

No biochemical evidence has yet been provided to explain the formation of bicyclic monoterpenes, but a rational mechanism is shown in Scheme 6.

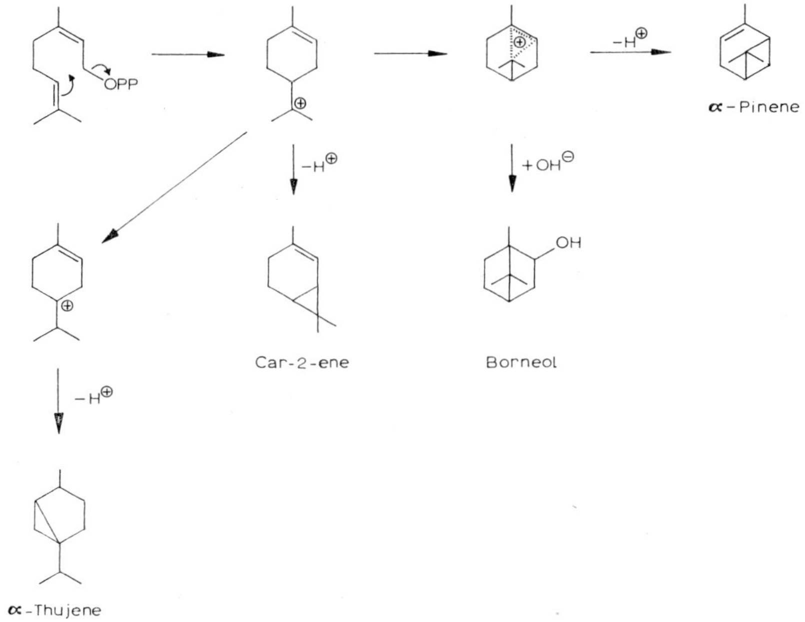

OPP

$-H^{\oplus}$

α-Pinene

$-H^{\oplus}$

$+OH^{\ominus}$

OH

Car-2-ene

Borneol

$-H^{\oplus}$

α-Thujene

Scheme 6.

5. Sesquiterpenes

(a) The basic unit

The enzymic reaction leading to the monoterpene basic unit, geranyl pyrophosphate has been described on p. 66. The same enzyme, farnesyl pyrophosphate synthetase (prenyl transferase), adds on a further molecule of IPP to geranyl pyrophosphate to form *trans*-farnesyl pyrophosphate, the basic unit of sesquiterpene biosynthesis (*F. Lynen et al.*, Angew. Chem., 1958, **70**, 739; *G. Popják* and *J. W. Cornforth*, Adv. in Enzymol., 1960, **22**, 281). The stereochemistry of the reaction is also the same (*Popják* and *Cornforth*, Biochem. J., 1966, **101**, 553).

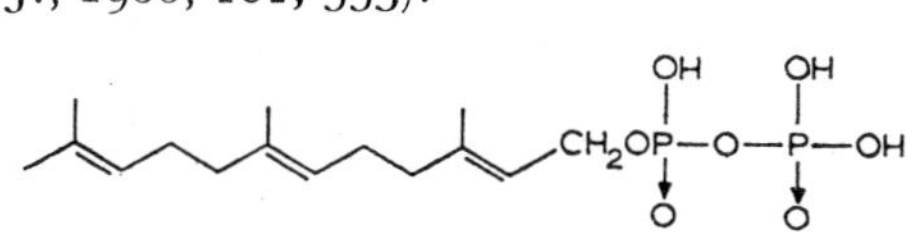

OH OH

$CH_2OP—O—P—OH$

O O

Farnesyl pyrophosphate

Farnesol occurs naturally and all sesquiterpenes can be formally derived from it, but whether free farnesol or the pyrophosphate is involved is not known.

(b) Classification

The main groups of sesquiterpenes are (a) acyclic compounds such as farnesol; (b) monocyclic systems such as bisabolol; (c) bicyclic systems which are sub-divided into four main groups, the cadalene type [*e.g.*, (+)-γ-cadinene], the eudesmol type (*e.g.* α-eudesmol) (*B. Riniker et al.*, J. Amer. chem. Soc., 1954, **76**, 313), the iresin type (*e.g.*, I) (*C. Djerassi* and *S. Burstein, ibid.*, 1958, **80**, 2593), the spiro-4,5-decane type (*e.g.* β-vetivone) (*J. A. Marshall et al., ibid.*, 1967, **89**, 2748, 2750):

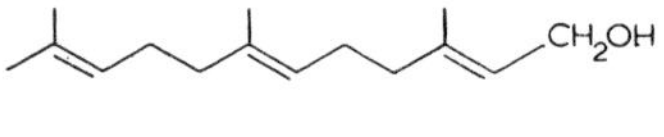

Farnesol 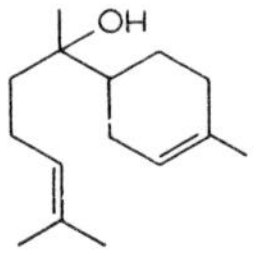　Bisabolol　(+)-γ-Cadinene　α-Eudesmol

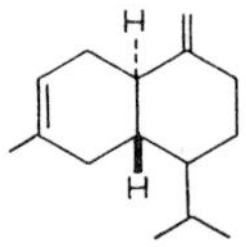

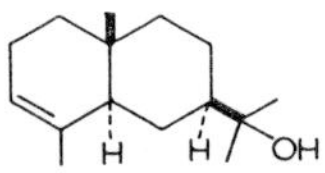

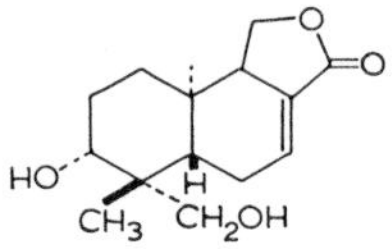

(I)　β-Vetivone

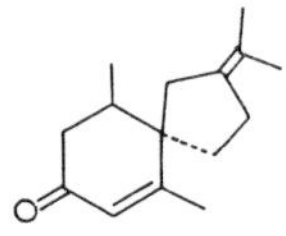

Many other variants exist but only three are important in the present context: the 11 membered ring system of α-humulene (α-caryophyllene) (*F. Šorm et al.*, Coll. Czech. chem. Comm., 1954, **19**, 570), the bicyclic (+)-longifolene, which has a number of biosynthetic peculiarities, and gossypol which is a phenolic bis-sesquiterpene (*R. Adams et al.*, J. Amer. chem. Soc., 1941, **63**, 528, 535, 2439):

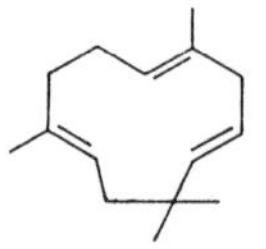

α-Humulene 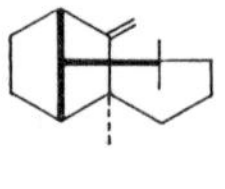　(+)-Longifolene 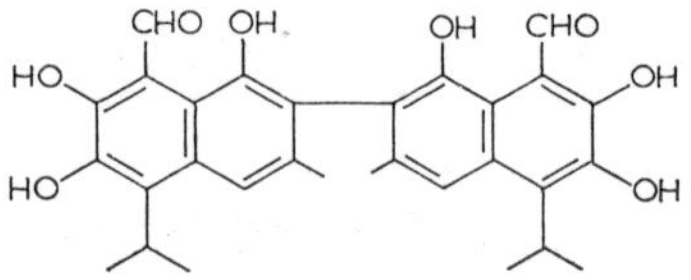　Gossypol

Sesquiterpenoid phenols with C_{15} terpenoid side-chains corresponding to types **A** and **B** (Scheme 3, p. 62) are known, for example, ammoresinol (**A**) (*E. Späth, A. F. J. Simon* and *J. Lintner*, Ber., 1936, **69**, 1656) and umbelliprenin (**B**) (*Späth* and *F. Vierhapper*, *ibid.*, 1938, **71**, 1667).

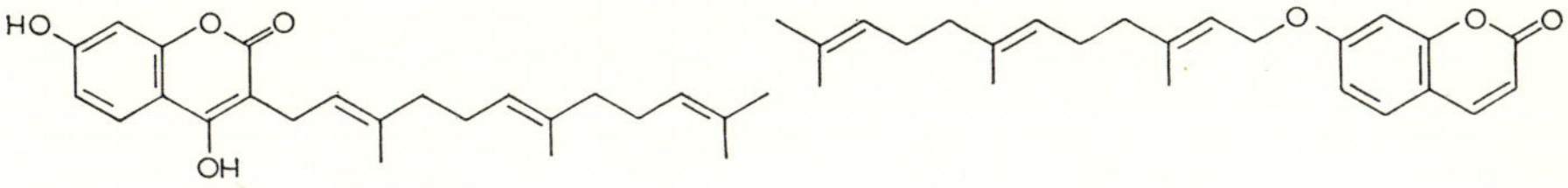

Ammoresinol

Umbelliprenin

(c) Biosynthetic studies

The few tracer experiments so far reported all indicate labelling patterns consistent with the view that the sesquiterpene under examination arises from farnesol pyrophosphate (FPP). The labelling of helminthosporal, the toxin from *Helminthosporium sativium*, synthesised in the presence of [1-^{14}C]-acetate supports the view that it arises *via* an intermediate such as II (*P. de Mayo, E. Y. Spencer* and *R. W. White*, J. Amer. chem. Soc., 1962, **84**, 492; *de Mayo et al.*, Experientia, 1962, **18**, 359). The proposed intermediate sativene(II) has been isolated (*de Mayo* and *R. E. Williams*, J. Amer. chem. Soc., 1965, **87**, 3275).

Helminthosporal

(II)

Similar experiments indicate that in the formation of carotol in *Daucus carota* farnesol pyrophosphate cyclises thus:

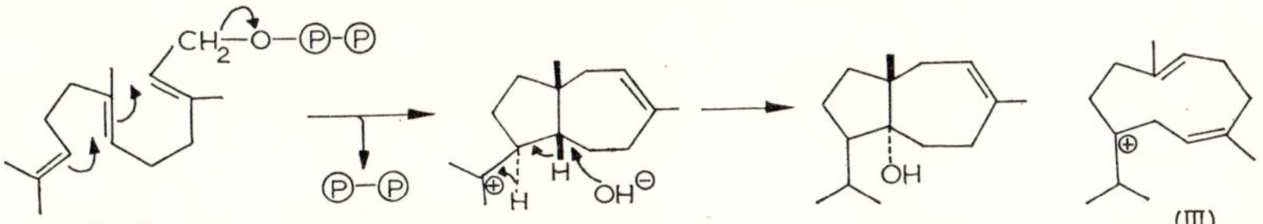

FPP

Carotol

(III)

and that a ten-membered monocyclic intermediate III is not concerned (*M. Soucek*, Coll. chem. Comm., 1962, **27**, 2929).

Degradation of longifolene labelled with [1-^{14}C]acetate (*W. Sandermann* and *K. Bruns*, Ber., 1962, **95**, 1863) indicated a folding of *cis*-farnesol [pyrophosphate] to give a hypothetical ion IV which rearranges to V, which is equivalent to VI; this is then transformed into longifolene by a humulene-type intermediate (*J. H. Richards* and *J. B. Hendrickson*, "Biosynthesis of terpenes, Steroids and Acetogenins", W. A. Benjamin, New York, 1964):

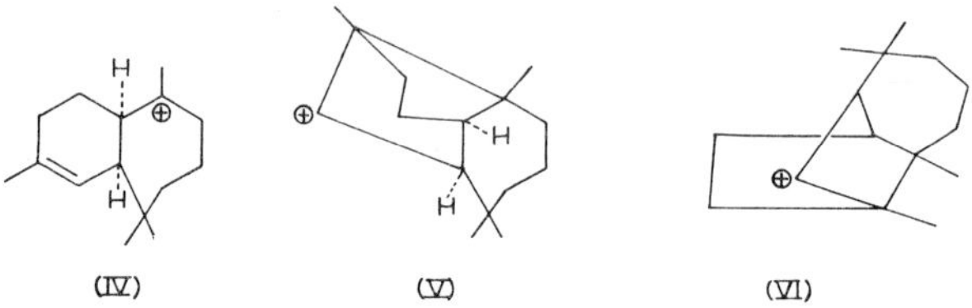

(IV) (V) (VI)

A simple folding of the following type:

is ruled out because ozonolysis would yield labelled formaldehyde (the atoms expected to be labelled are indicated "*"); this does not occur and the observation is thus consistent with the first proposal. [2-^{14}C]Mevalonate is also incorporated into trichothecin by *Trichothecium roseum* (*J. Fishman et al.*, Proc. chem. Soc., 1959, 127) as is farnesyl pyrophosphate (*B. Achilladelis* and *J. R. Hanson*, Phytochemistry, 1968, **7**, 589).

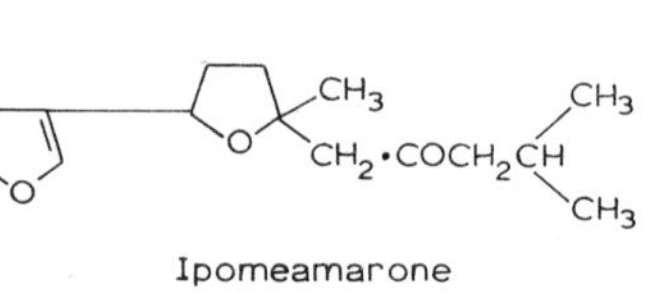

OCO·CH=CH·CH$_3$

Trichothecin

Ipomeamarone

With [2-^{14}C]mevalonate no label could be found in the *cis*-crotonate residue, whilst with [1-^{14}C]acetate 95% of the total activity appeared in this residue.

The expected labelling is observed when [2-^{14}C]mevalonate is incorporated into ipomeamarone synthesized by sweet potatoes infected with blackrot fungus (*T. Akazawa, I. Uritani* and *Y. Akazawa*, Arch. Biochem. Biophys., 1962, **99**, 52), and into gossypol (p. 80) by the roots of various cotton varieties (*P. F. Heinstein, F. H. Smith* and *S. B. Tove*, J. biol. Chem., 1962, **237**, 2643), and into marasmic acid (*J. J. Dugan et al.*, J. Amer. chem. Soc., 1966, **88**, 2838):

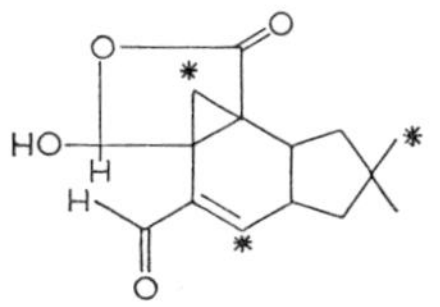

Marasmic acid

Problems of incorporating labelled MVA into sesquiterpenes have been reported for santonin in *Artemesia maritima* (*D. H. R. Barton et al.*, J. chem. Soc. C., 1968, 1813) and coriamyrtin and tutin (*M. Biollaz* and *D. Arigoni*, Chem. Comm., 1969, 633). The latter investigators found low incorporation and uneven distribution [60% of activity at $C_{(12)}$, 10% at $C_{(9)}$ and 10% at

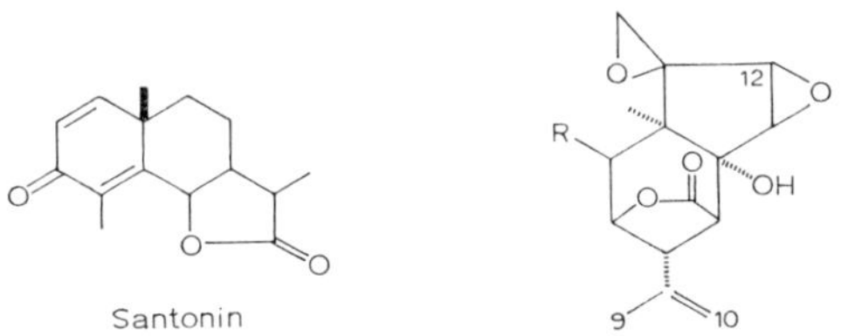

Santonin

Coriamyrtin (R = H)
Tutin (R = OH)

$C_{(10)}$]; however, *A. Corbello et al.* (*ibid.*, 1969, 634) found a relatively even distribution amonst the expected four carbon atoms. In the case of gossypol, neryl pyrophosphate was a much better precursor than geranyl pyrophosphate and *trans,trans*-farnesyl pyrophosphate was ineffective. These observations combined with the isolation of 2-[2-methylbutyl]-3-isopropyl-6-methylphenol as an active intermediate (*P. F. Heinstein*, quoted by *Waller*, *loc. cit.*) suggested that the biosynthetic pathway is:

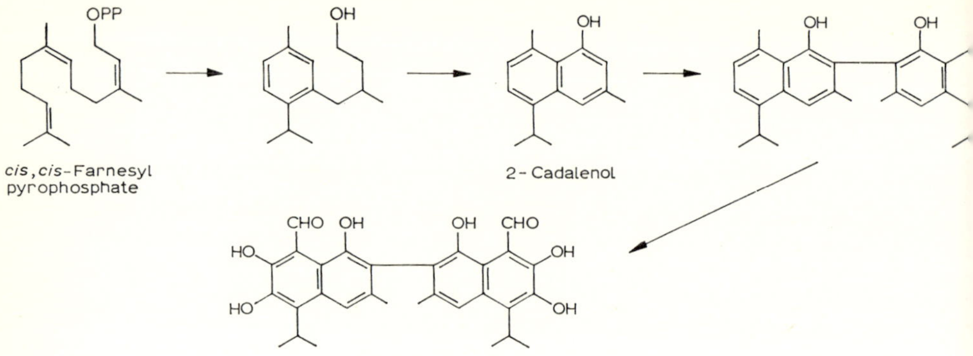

OPP OH OH OH OH

cis,cis-Farnesyl pyrophosphate

2-Cadalenol

CHO OH OH CHO HO OH HO OH

Gossypol

(d) Biogenetic relationships

Scheme 7 indicates how the various sesquiterpene skeletons discussed in this section can arise from farnesol (see *L. Ruzicka*, 1953, *loc. cit.* for full details) *via* certain cyclic ions formed by antiparallel additions.

No direct evidence for their implication in the biosynthesis of a number of these types has yet been proposed, although the work just cited on carotol and longifolene makes at least two pathways highly probable. Iresin (p. 80) is, however, particularly interesting because it could be formed by a cyclization reaction very similar to those observed in the triterpenoid series (*e.g.* sterols, p. 96) *i.e.* one initiated by the formation of an epoxide. A suggestion that iresin and its derivatives are formed by degradation of a triterpene is unlikely because they have the 5β,10α absolute configuration and not the 5α,10β configuration found in most triterpenes.

It has been suggested that farnesiferol a and farnesiferol b could arise by cyclization of the farnesyl side-chain of umbelliprenin (p. 81) (*L. Caglioti et al.*, Helv., 1958, **41**, 2278; 1959, **42**, 2557). In this connection the presence of the hydroxyl group is again particularly significant. The origin of abscisic acid, the regulator of bud dormancy in woody plants (*K. Ohkuma et al.*, Tetrahedron Letters, 1965, **29**, 2529; *Cornforth et al.*, Nature, 1965, **205**, 715, 1269) which is structurally similar to farnesiferol *b*, is mevalonic acid (*D. A. Robinson* and *G. Ryback*, Biochem. J., 1969, **113**, 895). However, the immediate precursor remains to be discovered.

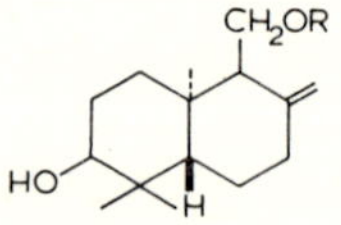

CH_2OR HO H CH_2OR HO O OH CO_2H

Farnesiferol a [R = *p*-coumaroyl]

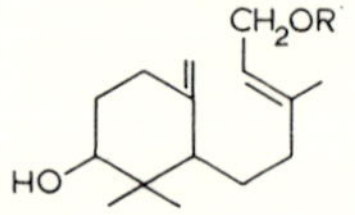

Farnesiferol b [R = *p*-coumaroyl]

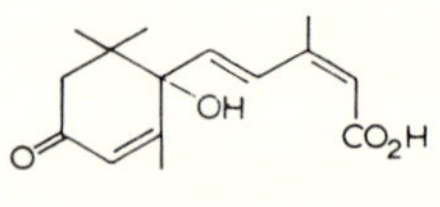

Abscisic acid

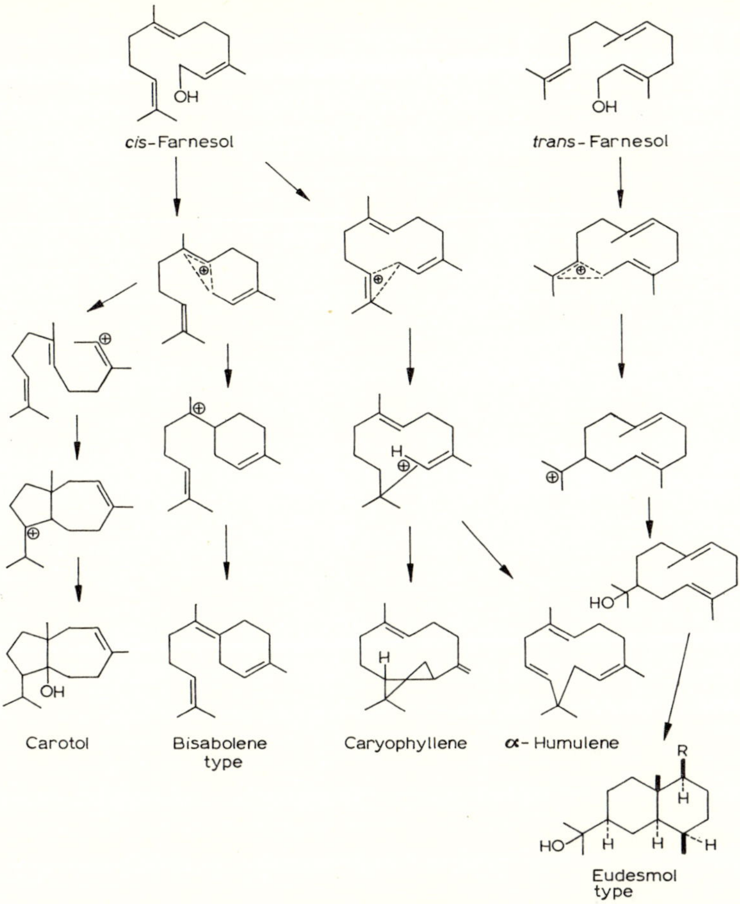

cis-Farnesol

trans-Farnesol

Carotol

Bisabolene type

Caryophyllene

α-Humulene

Eudesmol type

Scheme 7. Proposed biogenesis of various sesquiterpenes from *cis*- and *trans*-farnesol (after *W. Sandermann*, 1963, *loc. cit.*).

6. Diterpenes

(a) The basic unit and classification

The enzyme farnesyl pyrophosphate synthetase (prenyl transferase), which catalyses the formation of geranyl pyrophosphate (C_{10}) (p. 66) and farnesyl pyrophosphate (C_{15}) (p. 79) will add on a further IPP unit to farnesyl pyrophosphate to form geranylgeranyl pyrophosphate. This has been demonstrated in yeast (*E. C. Grob, K. Kirschner* and *F. Lynen,* Chimia,

1962, **15**, 308), *Micrococcus lysodeikticus* (*A. A. Kandutsch et al.*, J. biol. Chem., 1964, **239**, 2507) and pig liver (*C. R. Benedict et al.*, Arch. Biochem. Biophys., 1965, **110**, 611; *P. W. Holloway* and *G. Popják*, Biochem. J., 1967, **104**, 57). Evidence is also accumulating that this compound is a metabolic intermediate in diterpene biosynthesis.

(i) Monocyclic diterpenes

These are comparative rare; one, shown in formula I, corresponding to the menthane derivatives in the monoterpene series (p. 68) occurs in wormwood oil (*F. Šorm, M. Suchy* and *V. Herout*, Coll. Czech. chem. Comm., 1951, **16**, 278). Vitamin A is strictly a diterpene, but biologically it is a degradation product of the animal metabolism of certain of the tetra-

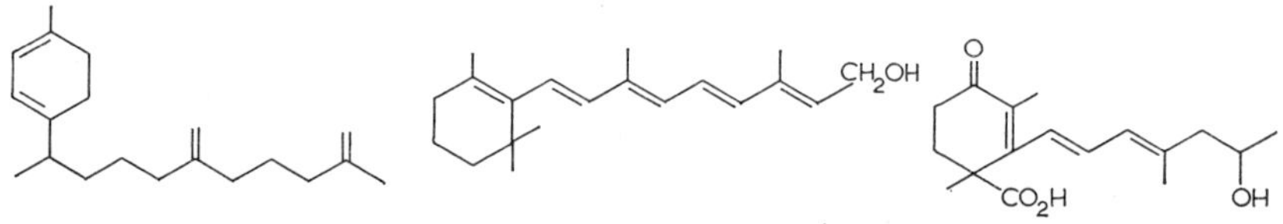

(I) Vitamin A Trisporic acid

terpenes (the carotenoids) (see p. 120 and *T. W. Goodwin*, Biosynthesis of Vitamins and Related Compounds, Academic Press, London, 1963). Trisporic acid, which is the carotene-stimulating factor isolated from mated strains of *Blakeslea trispora* (*L. Caglioti et al.*, Chim. Ind., Milan 1964, **46**, 1) is possibly a degraded triterpene or tetraterpene. Cembrene is discussed on p. 91.

(ii) Bicyclic diterpenes

The two main groups are represented by manool and eperuic acid, which differs from manool in the stereoisomerism of the ring fusion (*F. E. King* and *G. Jones*, J. chem. Soc., 1955, 658):

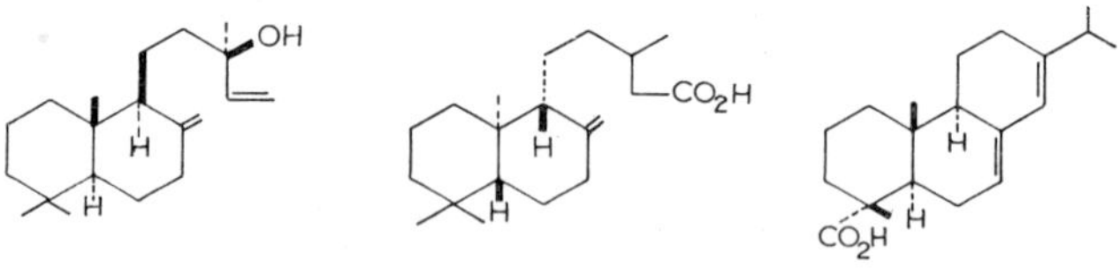

Manool Eperuic acid Abietic acid

(iii) Tricyclic diterpenes

The classical main group is represented by the resinic acids of which abietic acid is a well known example.

(iv) Tetracyclic diterpenes

The main groups are exemplified by phyllocladene and the gibberellins, *e.g.*, gibberellic acid (GA_3) but new and important groups which have particular biosynthetic significance include the taxines (*J. W. Harrison et al.*, J. chem. Soc. C, 1966, 1933) and fusicoccin, a phytotoxic metabolite from *Fusicossum amygdali* (*K. D. Barrow et al.*, Chem. Comm., 1968, 1195; *A. Ballio et al.*, Experientia, 1968, **24**, 631) (see p. 92):

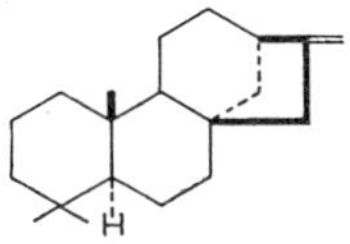

Phyllocladene

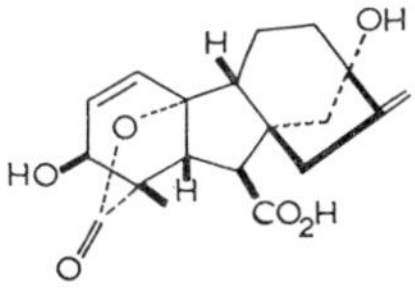

Gibberellic acid (GA_3)

(v) Diterpene derivatives

Phytol (hexahydrogeranylgeraniol) is present as a phytyl side-chain in chlorophyll *a* and *b*, phylloquinone (vitamin K_1) and in a number of well-known chromanes, of which α-tocopherol is probably the best known. The oxidized forms of the tocopherols (*e.g.* α-tocopherolquinone) also occur naturally (see *e.g. J. F. Pennock*, in *J. B. Pridham*, Ed. "Terpenoids in Plants", Academic Press, London, 1967). The tocotrienols which have

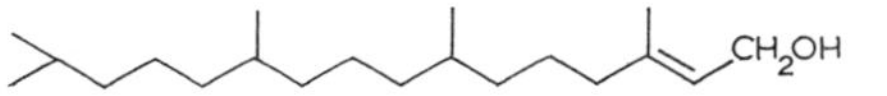

Phytol

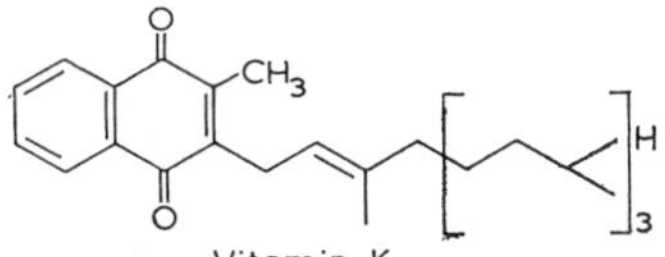

Vitamin K_1

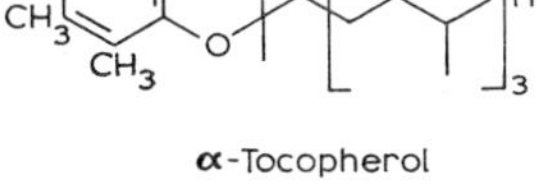

α-Tocopherol

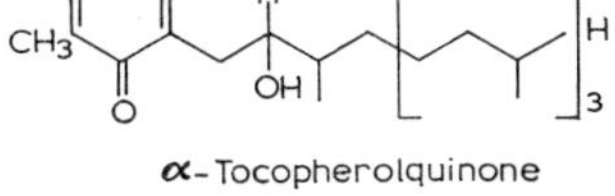

α-Tocopherolquinone

geranylgeraniol instead of phytol as side-chains are present in considerable amounts in *Hevea* (rubber) latex (*P. J. Dunphy et al.*, Nature, 1965, **207**, 521).

(b) Biosynthetic studies

A. J. Birch and his colleagues (Proc. chem. Soc., 1958, p. 223) and *J. J. Britt* and *D. Arigoni* (*ibid.*, 1958, 224) (see also "Biosynthesis of terpenes and sterols", Eds. *Wolstenholme* and *O'Connor*, Churchill, London, 1959) showed that [1-^{14}C]acetate and [2-^{14}C]mevalonate are incorporated into rosenonolactone by *Tricothecium roseum* in the expected fashion (carbons from mevalonate indicated "*") and that a methyl shift from $C_{(10)}$ to $C_{(9)}$ which is presumably initiated by $H^{\oplus}$ attack, occurs during ring closure of the acyclic precursor (all *trans*-geranylgeranyl pyrophosphate) (*Achilladelis* and *Hanson*, Phytochemistry, 1968, 7, 589; Tetrahedron Letters, 1968, 4397).

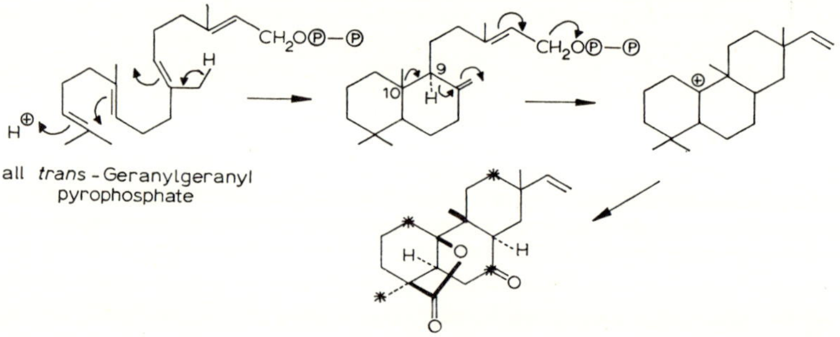

all *trans*-Geranylgeranyl pyrophosphate

Rosenonolactone

It is interesting that the *gem*-dimethyl groups have maintained this individuality; the equatorial group arises from $C_{(2)}$ of mevalonate and the axial methyl group is oxidized. The same maintenance of individuality is encountered in gibberellic acid (p. 90) and soyasapogenol A (p. 119). *T. roseum* also makes a sesquiterpenoid (triothecin, p. 82).

A methyl shift similar to that observed in rosenonolactone probably occurs also during the formation of columbin (*D. H. R. Barton* and *D. Elad*, J. chem. Soc., 1956, 2085, 2090).

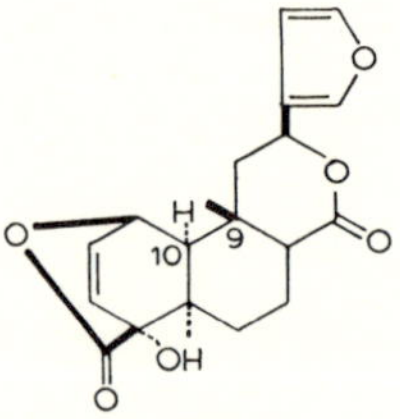

Columbin

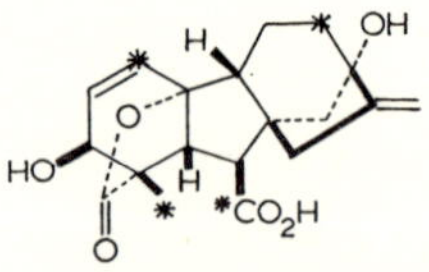

Gibberellic acid (II)

Birch et al. (Proc. chem. Soc., 1958, p. 192; "Biosynthesis of terpenes and

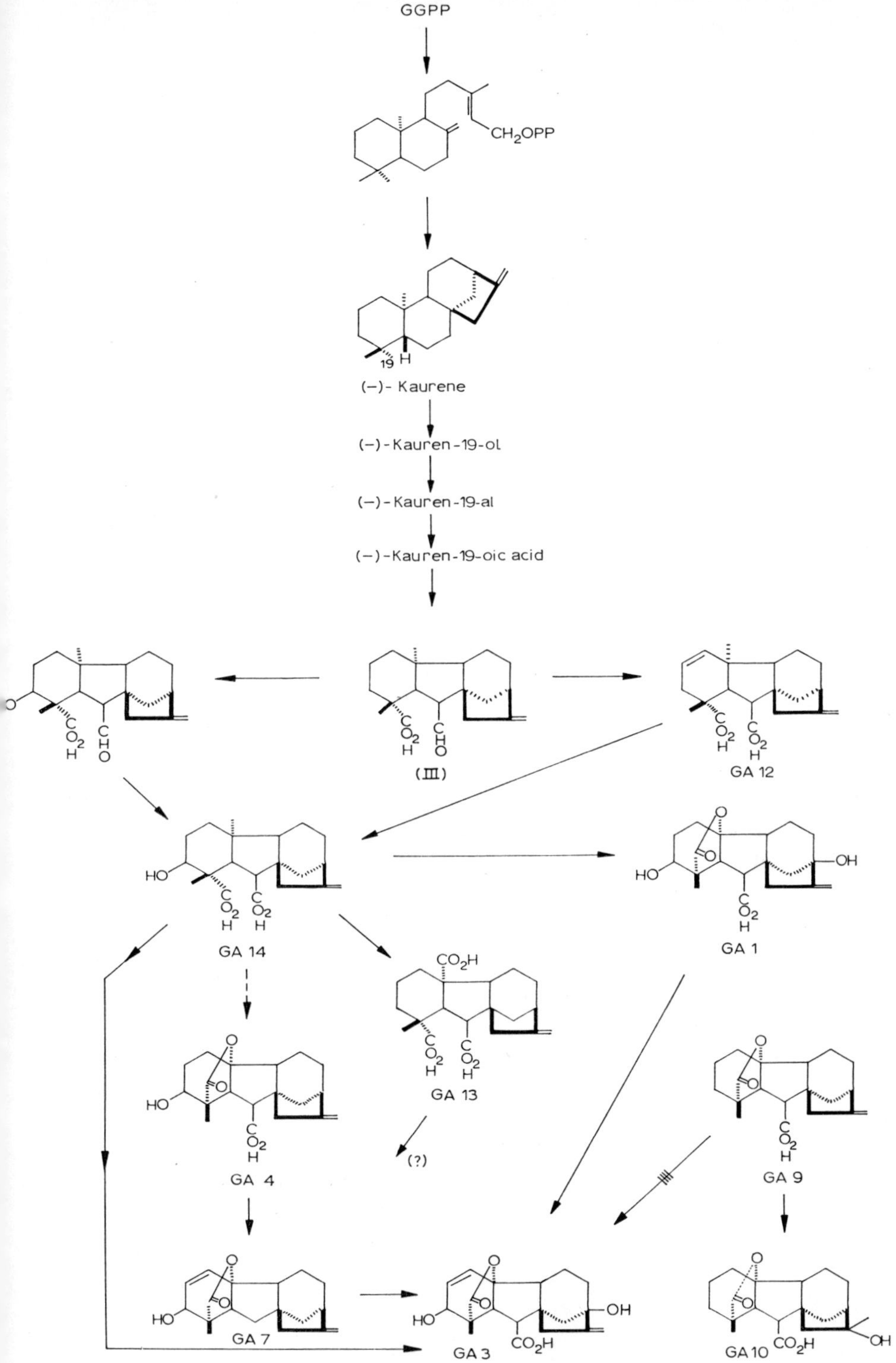

GGPP

CH_2OPP

19 H

(−)- Kaurene

(−)-Kauren-19-ol

(−)-Kauren-19-al

(−)-Kauren-19-oic acid

CO_2H CHO

(III)

CO_2H CO_2H

GA 12

HO

CO_2H CO_2H

GA 14

HO OH

CO_2H

GA 1

CO_2H

CO_2H CO_2H

GA 13

(?)

HO

CO_2H

GA 4

CO_2H

GA 9

HO

GA 7

HO OH

GA 3 CO_2H

GA 10 CO_2H OH

Scheme 8. Pathway of biosynthesis of gibberellins.

sterols", London, 1959, p. 245) first demonstrated that the incorporation of mevalonate into gibberellic acid (GA_3) indicated that it was a degraded diterpene, and an elegant series of degradations demonstrated the positions of the labelling (* in II). This led them to propose a biosynthetic pathway which is basically correct but which is now known to be out of sequence. The currently accepted pathway is given in Scheme 8. Geranylgeranyl pyrophosphate is converted into (—)-kaurene (*D. T. Dennis et al.*, Plant Physiol., 1965, **40**, 948) *via* (—)-labda-8(17),13-dien-15-ol (copalol) pyrophosphate (*J. R. Hanson* and *A. F. White*, J. chem. Soc., 1969, 981; *I. Shechter* and *C. A. West*, J. biol. Chem., 1969, 244, 3200). (—)-Kaurene, first shown to be an intermediate in G.A. synthesis by *B. E. Cross* and *K. Norton* (Tetrahedron Letters, 1962, 145; J. chem. Soc., 1963, 2937; 1964, 295) and later by *J. Graebe et al.*, (J. biol. Chem., 1965, **240**, 1847), undergoes oxidation to (—)-kauren-19-oic acid (*T. A. Geissman et al.*, Phytochemistry, 1966, **5**, 933; *Dennis* and *C. A. West*, J. biol. Chem., 1967, **242**, 3293; *Hanson* and *White*, Phytochemistry, 1968, **7**, 595). The next intermediate is 7β-hydroxy-(—)-kauren-19-oic acid (*Hanson* and *White*, Chem. Comm., 1969, 410), which is first converted into the aldehyde III and then into GA_{14}; GA_{12} is also probably another intermediate between the aldehyde and GA_{14} (*Cross* and *Norton*, Chem. Comm., 1965, 1570). The various ways by which GA_3 can subsequently be formed are indicated in Scheme 8. GA_{13} appears to be a dead end (*Cross et al.*, Tetrahedron, 1968, **24**, 231). There is no indication to show that hydroxylation at $C_{(3)}$ which takes place with retention of configuration (*Hanson* and *White*, J. chem. Soc., 1969, 981) can occur after the introduction of the double bond into ring A, which is formed by *cis* dehydrogenation (*idem*, Chem. Comm., 1969, 1071). Gibberellins can also be formed from kaurene in isolated chloroplasts (*J. L. Stoddart*, Phytochemistry, 1969, **8**, 831).

Hydroxylation of ring A seems to precede lactonization because GA_9 is not converted into GA_3, but it is hydrated to GA_{10}. The labelling of GA_3 with MVA and with [2-^{14}C]acetate indicates that ring D is formed by a phyllocladene-type rearrangement. Phyllocladene (p. 87) itself, an enantiomer of (—)-kaurene, is not a gibberellin precursor (*Birch* and *J. Winter*, J. chem. Soc., 1963, 5547).

(—)-Kaurene is converted into steviol in *Stevia rebaudiana* (*R. D. Bennett et al.*, Phytochem, 1967, 6, 1107; *Hanson* and *White*, Phytochemistry, 1968, **7**, 595), but this is not converted into known gibberellins by *Fusarium moniliforme* although compounds with gibberellin-like activity are formed (*M. Ruddat et al.*, Arch. Biochem. Biophys., 1965, **111**, 187).

The diterpene alkaloids of the lycoctonine-type from *Delphinium* spp. (Vol. II C, p. 381) are MVA (and glycine)-derived (*G. M. Frost et al.*, Chem. and Ind., 1967, 320).

Pleuromutilin, a diterpene even more modified than the gibberellins, has been obtained from *Pleurotus mutilis*; isotope experiments with [2-^{14}C]MVA (radio-active atoms starred in formula) indicate that it is formed by re-arrangement of a normal manool (p. 86) type ion (IV) (*Birch et al.*, Chem. and Ind., London, 1963, 374; Tetrahedron, 1966, Suppl. 8, Part II, 359; *Arigoni*, Gazz. chim. Ital., 1962, **92**, 884; Plenary Lecture (UPAC Symposium, 1968, p. 331).

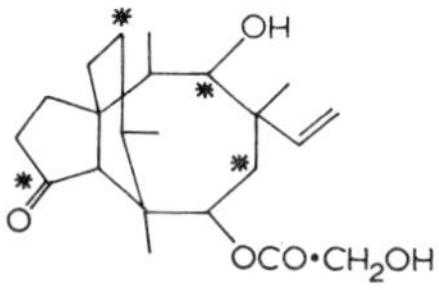

Pleuromutilin 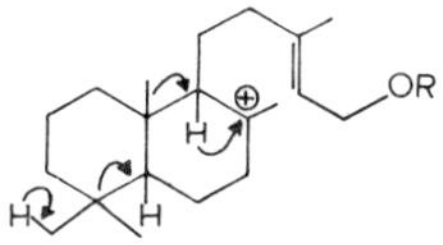 (IV)

Nothing is known of the biosynthesis of the first C_{14} ring compound found in nature, the diterpene cembrene (thundergene) (*W. G. Dauben, W. E. Thiessen* and *P. R. Resnick*, J. Amer. chem. Soc., 1962, **84**, 2015):

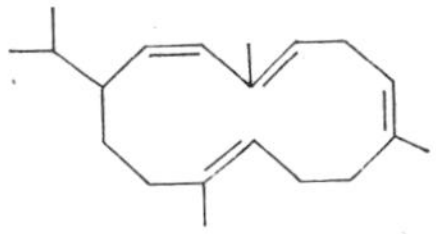

Cembrene

Experiments with stereospecifically labelled MVA have shown that phytol (attached to chlorophyll) is derived from an all *trans* precursor, presumably geranylgeranyl pyrophosphate (*A. R. Wellburn et al.*, Biochem. J., 1966, **100**, 23C).

Cyclization of geranylgeranyl pyrophosphate to a macrocycle diterpene intermediate has been postulated for the formation of the new group of diterpenes the taxines (*J. W. Harrison et al.*, J. chem. Soc. C, 1966, 1933), thus:

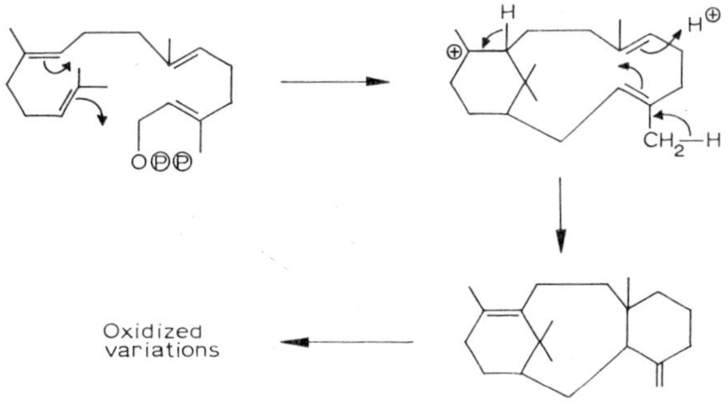

Oxidized variations

The rceently observed fusicoccin, a phytotoxic metabolite of *Fusicossum amygdali*, must be formed by an ophiobolin (see next section) type transformation of geranylgeranyl pyrophosphate, instead of, in the case of ophiobolin geranylfarnesyl pyrophosphate (*K. D. Barrow et al.*, Chem. Comm., 1968, 1195; *A. Ballio et al.*, Experientia, 1968, **24**, 631).

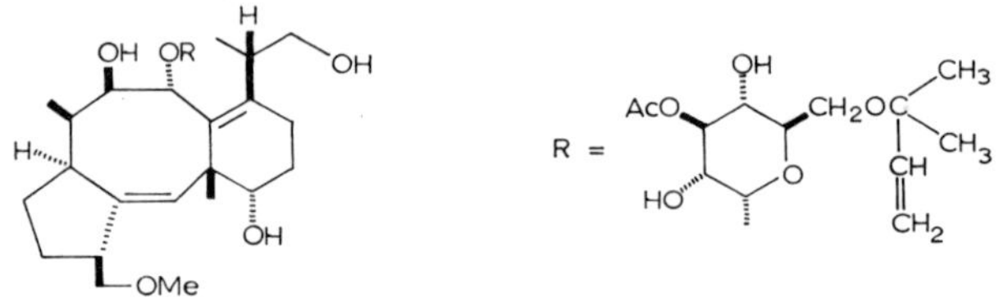

7. Sesterterpenes

This new group of terpenes now has at least three members, the first reported being ophiobolin A from *Cochliobolus miyabeanus* (*S. Nozoe*, J. Amer. chem. Soc., 1965, **87**, 4968) (≡ cochliobolin A) (*L. Canonica et al.*, Tetrahedron Letters, 1966, 1211), zizanin A (I, R=H) and ophiobolin B (≡ zizanin B) (I, R=OH) (*Nozoe, H. Hirai* and *K. Tsuda, ibid.*, 1966, 2211) (≡ cochliobolin B) (*Canonica et al., ibid.*, 1966, 1329). The probable mechanism for the formation of the ophiobolane ring system from geranyl farnesyl pyrophosphate (*idem, ibid.*, 1967, 4657) is:

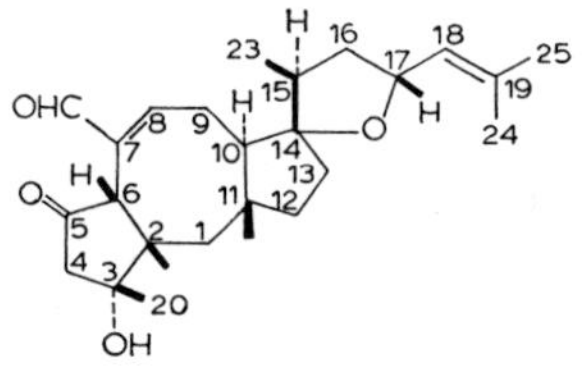

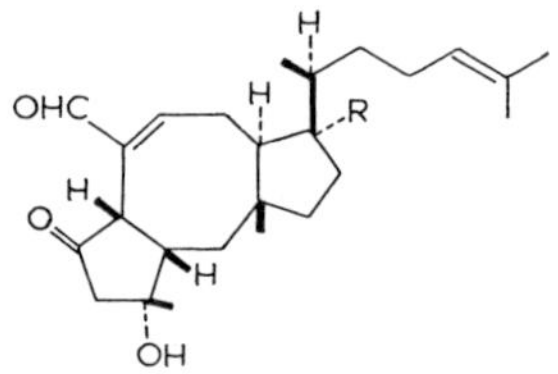

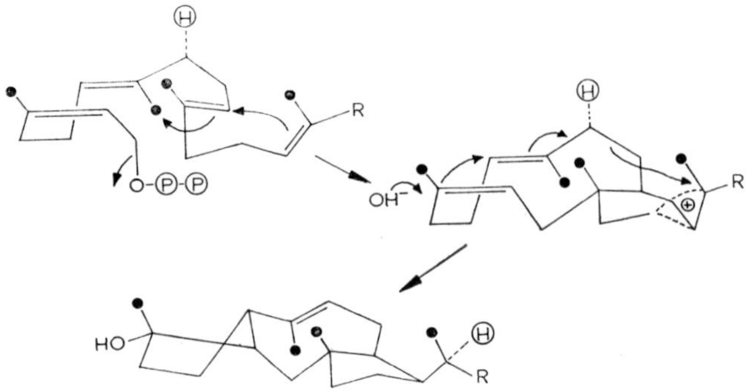

Ophiobolin

(I)

This is based on the observation of a 1,5-hydrogen migration from $C_{(8)}$ to $C_{(15)}$ when [2-$^{3}H_{1}$]MVA was administered to *Cochliobolus heterostrophus* (*idem, ibid.*, 1967, 3371). Further experiments with 2*S*- and 2*R*-[2-$^{3}H_{1}$]MVA demonstrated that the 8α hydrogen migrates (*idem, ibid.*, 1967, 4657). The oxygen at $C_{(14)}$ in ophiobolins A and B arises from free oxygen whilst that at $C_{(3)}$ arises from the medium (*S. Nozoe et al., ibid.*, 1967, 3371).

8. Triterpenes

(a) The basic unit

In the triterpenes a fundamentally different structure is encountered compared with the groups discussed previously. The six constituent isoprenoid residues are not arranged *head* to *tail* throughout the molecule, which can formally be derived from the *tail* to *tail* condensation of two C_{15} units formed by *head* to *tail* condensation of three isoprenoid units. One would expect that the simplest triterpene, the acyclic squalene (p. 55), would be the basic unit from which all triterpenoids are derived. The first experimental demonstration of this was by *R. G. Langdon* and *K. Bloch* (J. biol. Chem., 1953, **200**, 135) who showed that [^{14}C]squalene when fed to rats was converted into cholesterol. This and other work eventually led to the generalized conclusions that (a) all triterpenes are formed by cyclization of all-*trans*-squalene in various chair-boat conformational sequences; (b) the transformation of squalene follows the rules of anti-parallel cationic 1,2 additions, 1,2 shifts and 1,2 eliminations; and (c) the cyclization of squalene is concerted or "non-stop"; that is, no intermediates formed by neutralization of the cationic charges should occur (*L. Ruzicka*, Proc. chem. Soc., 1959, 341). Only the last conclusion has to be modified in the light of recent research. The first reaction is the production of squalene 2,3-epoxide (*E. J. Corey, W. E. Russey* and *P. R. Ortiz de Montellano*, J. Amer. chem. Soc., 1966, **88**, 4750; *E. E. van Tamelen et al., ibid.*, 1966, **88**, 4752; *P. D. G. Dean et al.*, J. biol. Chem., 1967, **242**, 3014) and it is this which undergoes concerted cyclization.

All-*trans*-squalene is biosynthesized from two molecules of *trans-trans*-farnesyl pyrophosphate (p. 85) by a particulate enzyme system from yeast in the presence of NADPH (*F. Lynen et al.*, Angew. Chem., 1958, **70**, 783; Biochem. Z., 1958, **330**, 269) and by liver microsomes in the presence of NADPH and ascorbate (*G. Popják*, Ann. Review Biochem., 1958, **27**, 583). The elegant studies of *Popják* and *J. W. Cornforth* (1966 *loc. cit.*) have revealed that in squalene biosynthesis from two molecules of farnesyl pyrophosphate the pro *R* hydrogen is lost from $C_{(1)}$ of one of the farnesyl units and is replaced

by the 4-pro *S* hydrogen from NADPH. The overall stereochemistry with [5-*R*-$^2H_1$]MVA is indicated in I (H_B is the hydrogen from NADPH).

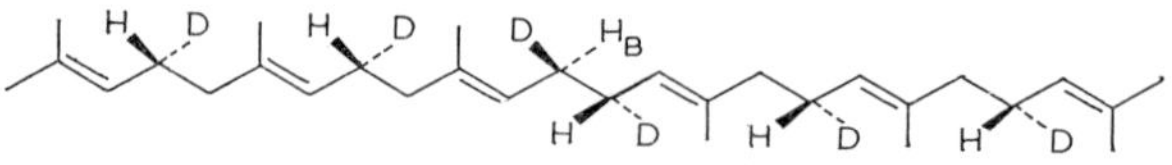

(I)

A possible mechanism for the formation of squalene is given in Scheme 9.

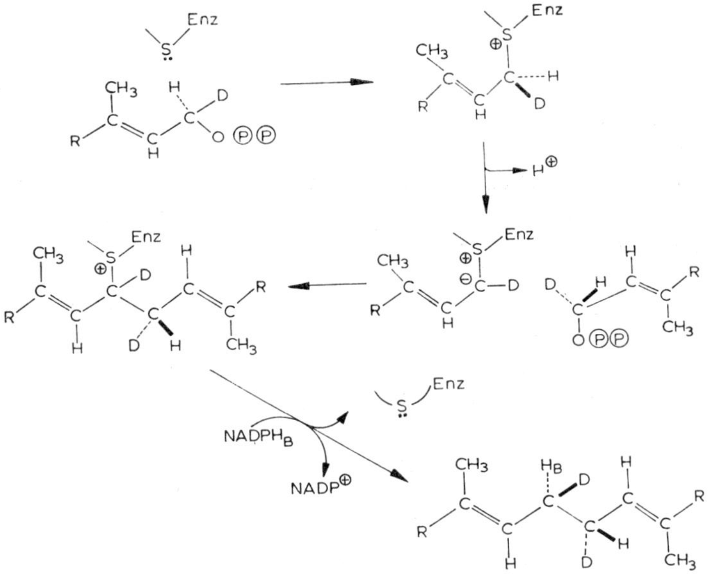

Scheme 9. A possible mechanism for the formation of squalene.

(b) *Biosynthesis of tetracyclic triterpenoids*

Because of the amount of experimental information available for this group, especially the sterols, it is appropriate to discuss them before the bicyclic and tricyclic compounds.

(i) *In animals*

Cholesterol synthesized in the presence of [1-^{14}C]acetate or [2-^{14}C]acetate was partly degraded and it was found that $C_{(7)}$, $C_{(13)}$, $C_{(18)}$ and $C_{(19)}$ arose from $C_{(2)}$ of acetate and $C_{(10)}$ from $C_{(1)}$ of acetate (see *Bloch* in "Biosynthesis of terpenes and sterols", Ed. *G. E. W. Wolstenholme* and *M. O'Connor*,

Churchill, London 1959, p. 4; *R. B. Clayton*, Quart. Reviews, 1965, **19**, 168). These observations strongly suggested a centre of symmetry about $C_{(11)}$ and $C_{(12)}$, and *R. B. Woodward* and *Bloch* (see *Bloch, loc. cit.*) suggested that squalene was folded in the all *trans* form II. This view implied (a) that only all *trans*-squalene would be a precursor of sterols; (b) a distribution of methyl (m) and carboxyl (c) carbons of acetate in cholesterol as indicated in

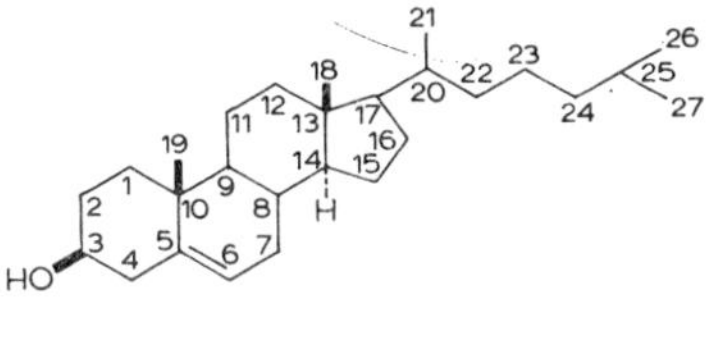

Cholesterol

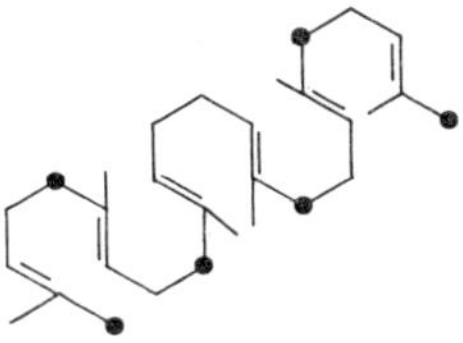

(II)

III, and (c) that following two 1,2-methyl shifts, from $C_{(14)}$ to $C_{(13)}$ and from $C_{(8)}$ to $C_{(14)}$, and hydrogen shifts indicated in Scheme 10 a true tetracyclic triterpene, lanosterol would be formed and would be the basic cyclic precursor of cholesterol. All these postulates were elegantly confirmed. Firstly, [^{14}C]-all-*trans*-squalene, synthesized by the unambiguous method of *D. W. Dicker* and *M. C. Whiting* (J. chem. Soc., 1958, 1994) is rapidly converted by liver preparations to cholesterol whilst various *cis* isomers are inactive (*Bloch, loc. cit.*); secondly, *Cornforth* and *Popják* and their colleagues devised an elegant carbon by carbon degradation of cholesterol and used it to demonstrate the incorporation pattern indicated in III (see *T. W. Goodwin*, "Recent Advances in Biochemistry", Churchill, London, 1959, for a summary and full references); thirdly, it has been shown that [^{14}C]lanosterol is formed from [^{14}C]acetate in rat liver homogenates (*Clayton* and *Bloch*, J. biol. Chem., 1956, **218**, 319) and in intact rats (*P. B. Schneider, Clayton* and *Bloch, ibid.*, 1957, **224**, 175). Kinetic studies by *Schneider et al.* (*loc. cit.*) indicated that lanosterol was a precursor of sterols but direct evidence was obtained by *Clayton* and *Bloch* (*loc. cit.*) who isolated [^{14}C]cholesterol from liver homogenates incubated with [^{14}C]lanosterol. Finally, *R. K. Maudgal, T. T. Tchen* and *Bloch* (J. Amer. chem. Soc., 1958, **80**, 2589) and *Cornforth et al.* (Tetra-

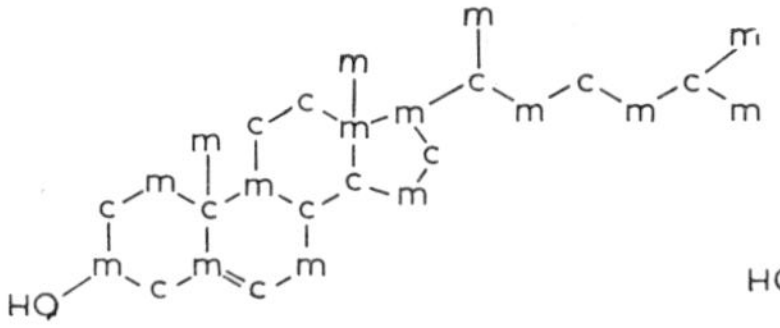

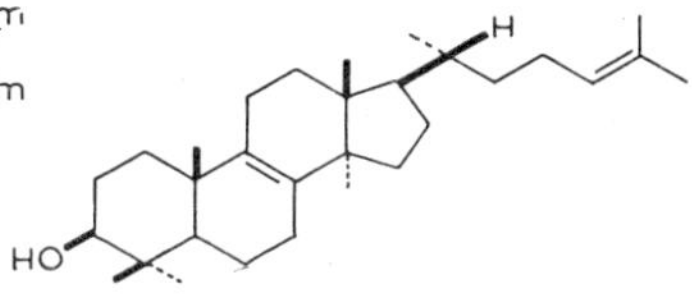

(III)

Lanosterol

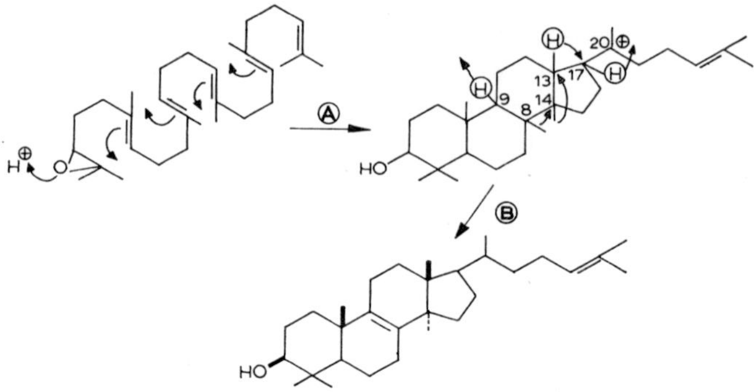

Scheme 10. The formation of lanosterol from squalene oxide.

hedron, 1959, **5**, 311) ingeniously demonstrated that two 1,2-methyl shifts occurred during the cyclization of squalene. For example, *Maudgal et al.* synthesized two species of geranylacetone from geranyl chloride and a mixture of [3-^{13}C] and [4-^{13}C]acetoacetic ester; these were converted into a mixture of four species of all-*trans*-squalene which was converted enzymically into lanosterol. This was oxidized by the Kuhn-Roth procedure and the distribution of ^{13}C between the carbons of the acetic acid produced measured in a mass spectrometer after the conversion of the acetic acid into ethylene. Only with two 1,2-shifts can ethylene with mass 30 arise and it should constitute one twenty-fourth of the total. This distribution was experimentally observed (Scheme 11) (next page).

The mechanism of the cyclization of squalene proposed by *Tchen* and *Bloch* (J. biol. Chem., 1957, **226**, 921) and *Cornforth et al.* (in "Biosynthesis of Terpenes and Sterols", Ed. *Wolstenholme* and *O'Connor*, Churchill, London 1959, p. 119) to cover the experimental facts began with an electrophilic attack by an oxidant formally represented as $OH^{\oplus}$. It is now known that the first step is the enzymic insertion of oxygen to form squalene 2,3-epoxide (Scheme 10). This, under the influence of squalene 2,3-oxide cyclase (*P. D. G. Dean et al.*, J. biol. Chem., 1964, **242**, 3014) undergoes prototropic attack which results in a forward cyclization leading to an electron deficiency at $C_{(20)}$ (**A,** Scheme 10) and the production of a transient carbonium ion or enzyme substrate complex. A backward rearrangement (**B,** Scheme 10) in which (a) one hydrogen moves from $C_{(18)}$ to $C_{(17)}$ and one from $C_{(17)}$ to $C_{(20)}$; (b) one methyl group moves to $C_{(13)}$ from $C_{(14)}$ and one to $C_{(14)}$ from $C_{(8)}$, and (c) the hydrogen at $C_{(9)}$ is expelled as a proton, leads to the formation of lanosterol.

To achieve the known stereochemistry of lanosterol the all-*trans*-squalene

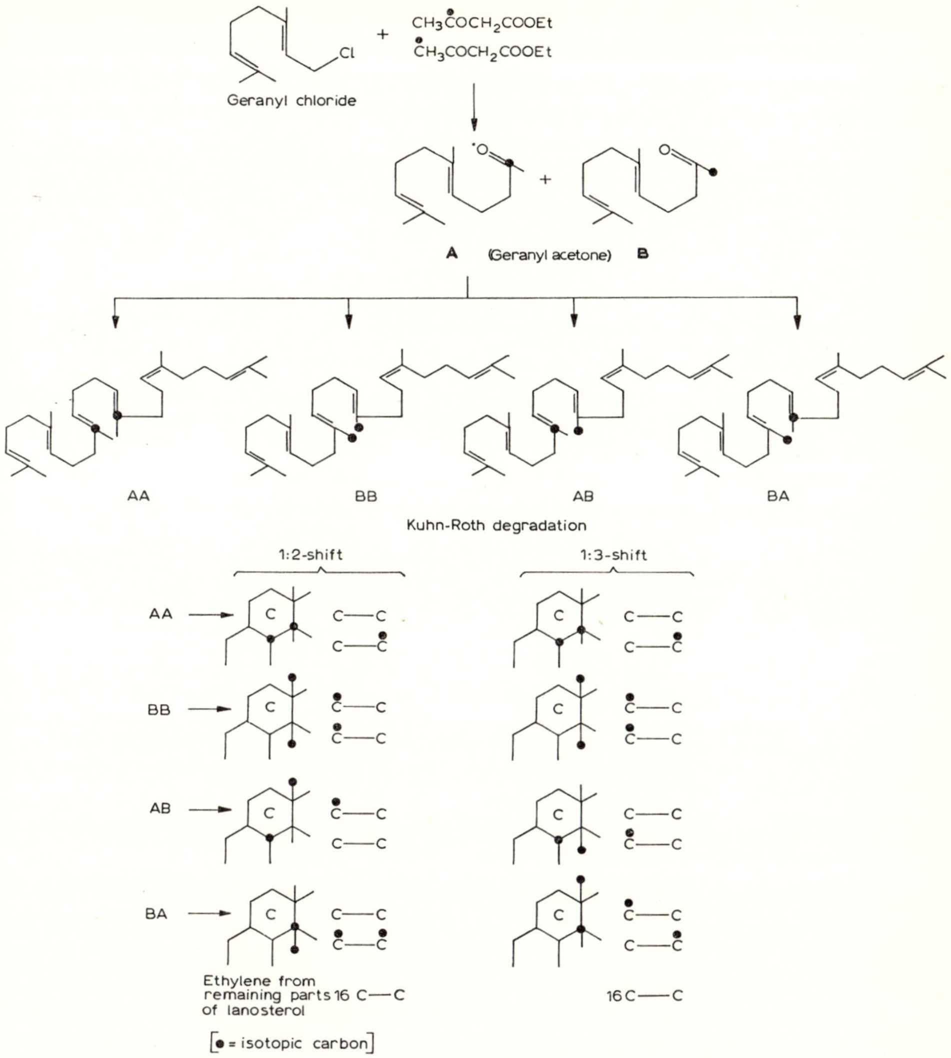

Geranyl chloride + $CH_3\overset{\bullet}{C}OCH_2COOEt$ / $\overset{\bullet}{C}H_3COCH_2COOEt$

A (Geranyl acetone) B

AA BB AB BA

Kuhn-Roth degradation

1:2-shift 1:3-shift

Ethylene from remaining parts of lanosterol 16 C—C

16 C—C

[● = isotopic carbon]

Scheme 11. Method for distinguishing between 1-2 and 1-3 methyl shifts in cyclization of squalene (*Maudgal et al.*, 1958, *loc. cit.*).

must be in the chair-boat-chair-boat conformation. Furthermore, the formation of a classical carbonium ion indicated in Scheme 10 would destroy the geometry of the system because the subsequent elimination or neutralization of the carbonium ion by an anion is non-stereospecific. This objection is

overcome chemically by assuming that bridged ions are formed from which, when an anion ($B^{\ominus}$) attacks the system, a specific conformation is achieved (*D. Arigoni*, Biochem. Soc. Symp., 1960, **19**, 32). In the cyclization of squalene 2,3-oxide ($B^{\ominus}$) is possibly the adjacent double bond and the conversion of squalene into lanosterol can be represented in full conformation (*Arigoni*, 1960, *loc. cit.*; angular methyl groups indicated "—") (Scheme 12).

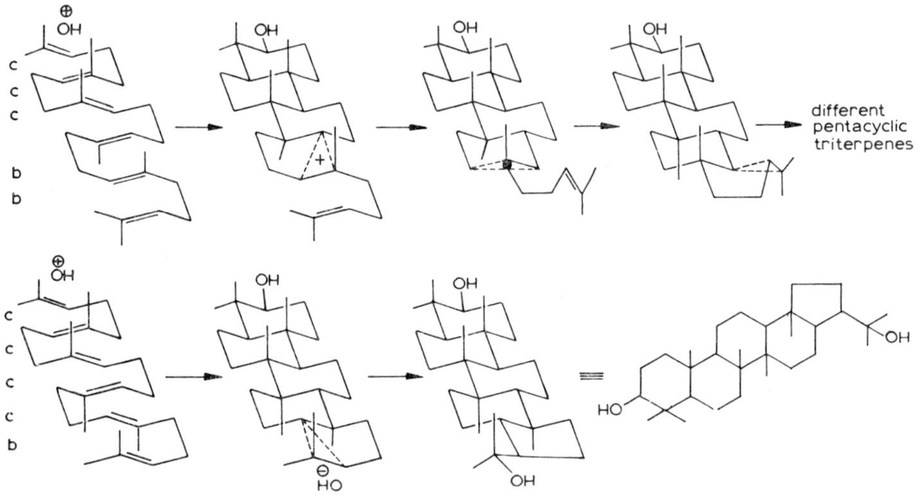

Scheme 12. General scheme for cyclization of squalene (*Arigoni*, 1959).

Biochemically it is more likely that an enzyme-substrate complex is formed rather than a carbonium ion, bridged or otherwise (see p. 67). With MVA stereospecifically labelled with tritium at $C_{(4)}$ *Popják* and *Cornforth* (1966, *loc. cit.*) have demonstrated that one hydrogen is lost from $C_{(9)}$, one moves from $C_{(17)}$ to $C_{(20)}$ and one from $C_{(18)}$ to $C_{(17)}$ [one is also lost at $C_{(3)}$ – this is discussed on p. 101]; this is indicated in Scheme 13.

The changes which lanosterol undergoes in its transformation into cholesterol involve (a) loss of three methyl groups; (b) movement of a double bond from Δ^8 to Δ^5 and (c) the saturation of the side chain [$C_{(24)}$]. There is no compelling evidence to conclude that one specific pathway occurs in all organisms and many of the considerable number of alternative possibilities have been demonstrated. To consider demethylations first, the removal of the three angular methyls as carbon dioxide was first demonstrated by *J. A. Olson, Jr. M. Lindberg* and *Bloch* (J. biol. Chem., 1957, **226**, 941). There were no indications that the 1-C unit was liberated at a lower oxidation level although the formation of carboxyl groups was probable. NADPH and

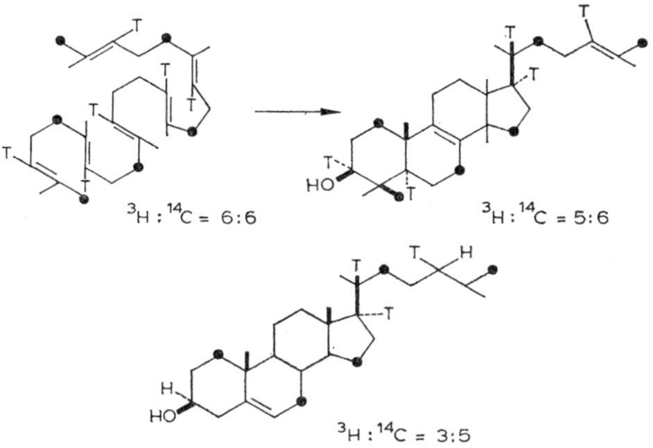

$^3H:^{14}C = 6:6$ $^3H:^{14}C = 5:6$

$^3H:^{14}C = 3:5$

Scheme 13. The location of tritium and carbon from [2-^{14}C-4*R*-^{3}H$_1$]mevalonic acid in squalene, lanosterol and cholesterol.
(● = ^{14}C; T = ^{3}H$_1$)

oxygen are obligatory co-factors, which indicated a direct attack on the methyl group by "active oxygen". Preliminary dehydrogenation is ruled out because of the absence of a hydrogen atom in the corresponding adjoining carbon atoms. Evidence exists for and against loss of the $C_{(14)}$-methyl group being the first step in the conversion. 4,4-Dimethylcholesta-8,24-dien-3β-ol has been identified as a metabolite and converted by a number of investigators into cholesterol (see *I. D. Frantz, jr.*, and *G. J. Shroepfer, jr.*, Ann. Review Biochem., 1967, **36**, 691). Furthermore no 14α-methyl sterols which lack methyls at $C_{(4)}$ have been observed in animals; however, macdougallin is present in cactus plants (*J. C. Knight, D. I. Wilkinson* and *C. Djerassi*, J. Amer. chem. Soc., 1966, **88**, 790) but is not converted into cholesterol in rat liver homogenates (*M. Slaytor* and *Bloch*, J. biol. Chem., 1965, **240**, 4598). Indications that removal of the $C_{(14)}$ methyl group is not a mandatory step include: (a) lanosta-7,24-dien-3β-ol accumulates in arsenite poisoning (*L. Gaylor*, Arch. Biochem. Biophys., 1962, **101**, 409) and is converted into cholesterol (*Idem*, J biol. Chem., 1963, **238**, 1649), although

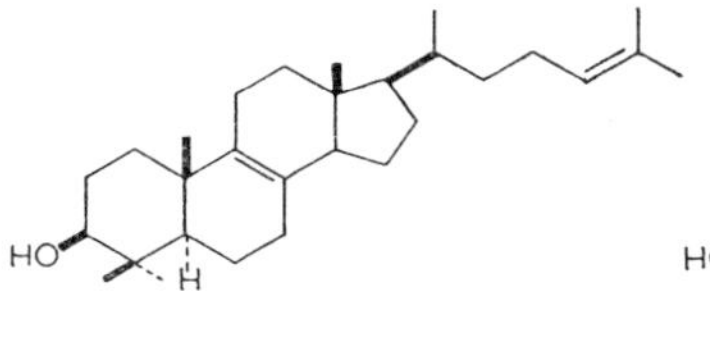

4,4-Dimethylcholesta-8,24-dien-3β-ol Macdougallin

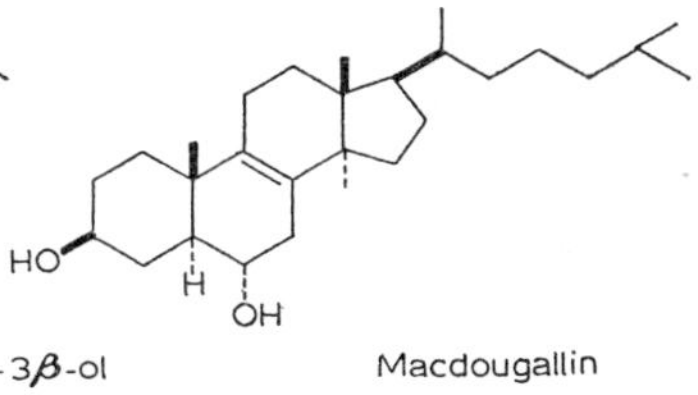

later experiments indicate that Δ^8-Δ^7 isomerization does not take place in compounds containing the 14α-methyl group (*Gaylor, C. V. Delwiche* and *A. C. Swindell,* Steroids, 1966, **8,** 353); (b) 14α-methylcholest-7-en-3β-ol is converted into cholesterol (*Knight, P. D. Klein* and *P. A. Szezepanik,* J. biol. Chem., 1966, **241,** 1502). The loss of the two methyl groups at $C_{(4)}$ is a stepwise process involving two separate demethylations (*Gaylor* and *Delwiche,* Steroids, 1964, **4,** 207) and evidence has been provided which indicates that it is the 4β-methyl group which is first removed. However, the natural occurrence of 4α-methyl derivatives in animal tissues (*e.g.* lophenol, cannot be taken as firm support for the conclusion that the first elimination involves the 4β group, because the enol form of the 3-oxo derivative is probably implicated in the oxidation (see p. 101) and on conversion back to the oxo form the remaining methyl group would take up the unhindered equatorial (α) configuration. Work by *Sharpless et al.,* (J. Amer. chem. Soc., 1968, 90, 6874) indicates that it is the 4α-methyl which is first removed. They found that 4α-hydroxymethyl-4β-methylcholestan-3β-ol was readily converted into cholestanol by rat liver homogenates whilst the 4β-hydroxy-4α-methyl derivative was not. This was directly demonstrated in the formation of 31-norcycloartanol by *Polypodium vulgare* (*E. L. Ghisalberti et al.,* Chem. Comm., 1969, in press).

As all known 4α-methylsterols have a double bond at either $C_{(7)}$ or $C_{(8)}$ and as synthetic 4,4-dimethylcholest-5-en-3β-ol is not a cholesterol precursor

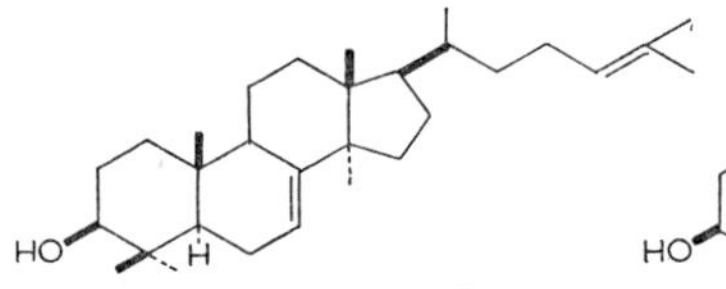

Lanosta-7,24-dien-3β-ol

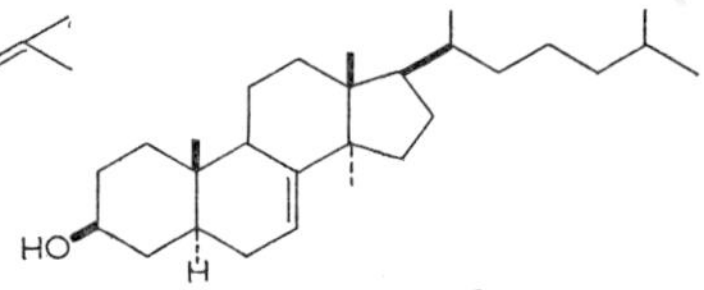

14α-Methylcholest-7-en-3β-ol

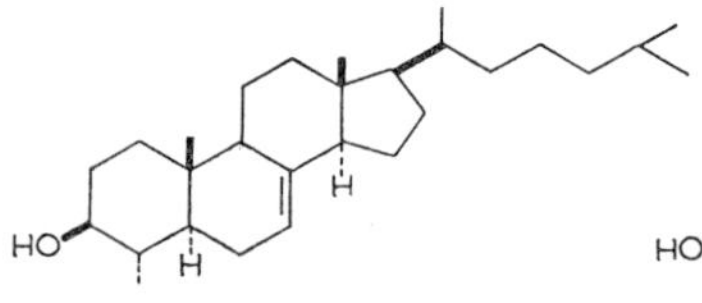

Lophenol

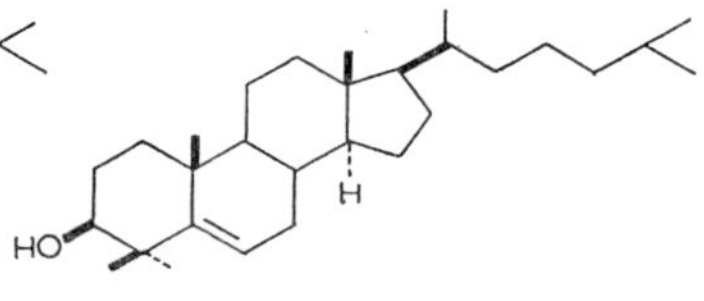

4,4-Dimethylcholest-5-en-3β-ol

in animal tissues (*F. Gautschi* and *Bloch,* J. Amer. chem. Soc., 1957, **79,** 684) it would appear that the demethylating enzyme is specific for compounds with a double bond at either $C_{(7)}$ or $C_{(8)}$: however, the presence or absence of a double bond at $C_{(24)}$ is apparently of little significance.

The demethylation is assumed to take place *via* the primary alcohol, aldehyde and carboxylic acid, because lanosterol gives stoicheiometric

amounts of carbon dioxide but no formaldehyde on conversion into cholesterol; semicarbazide inhibits the conversion and causes the accumulation of unidentified polar metabolites which in turn are converted into cholesterol (*Olson jr.*, *Lindberg* and *Bloch*, *loc. cit.*). Furthermore 4-hydroxymethylene-[^{14}C-2-$^{3}H_2$]-cholest-7-en-3-one is converted into [2-$^{3}H_2$]-cholest-7-en-3-ol and [^{14}C]carbon dioxide in the absence of oxygen (*J. Pudles* and *Bloch*, J. biol. Chem., 1960, **235**, 3417). This is in accord with the report that in the enzymic conversion of *p*-nitrotoluene into *p*-nitrobenzoic acid oxygen is required for the initial step only, subsequent oxidations being carried out by dehydrogenases (*J. R. Gillette*, *ibid.*, 1959, **234**, 139).

Decarboxylation of a carboxyl group at $C_{(4)}$ would be aided by the presence of a carbonyl function at $C_{(3)}$, and the conversion just mentioned supports this view. Furthermore, with [4-*R*-$^{3}H_1$]MVA as substrate there is no tritium present at $C_{(3)}$ of cholesterol (see p. 99) (*Popják* and *Cornforth*, 1966, *loc. cit.*) which indicates that the reaction has involved an oxo-intermediate.

It appears that a mixed function oxidase, requiring NADPH and O_2 is involved in the insertion of oxygen into the 4α-methyl group; the further reaction which involves a 3-oxo compound and results in loss of the methyl carbon as carbon dioxide can occur anaerobically. The 4-methyl-3-one derivative is then reduced to the 4α-methyl-3β-hydroxy derivative and the process is repeated thus (*A. C. Swindell* and *Gaylor*, J. biol. Chem., 1968, **243**, 5546):

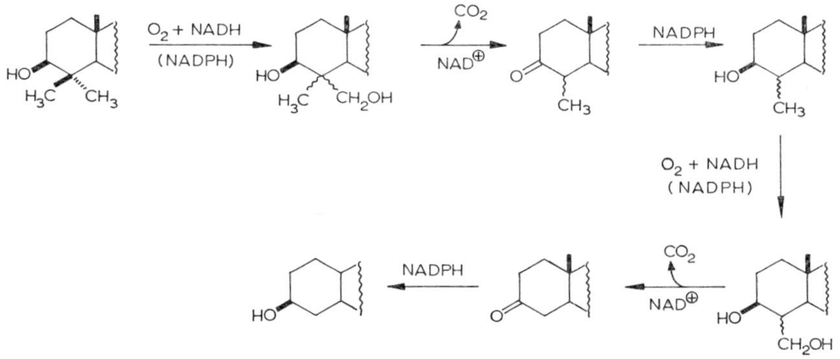

Work with stereospecifically labelled MVA has revealed a new intermediate in the removal of the 14α-methyl group from lanosterol. Cholesterol synthesized from [2*S*-2-$^{3}H_1$]MVA retained no tritium at $C_{(15)}$, indicating that the 15α-hydrogen had been removed (*L. Canonica et al.*, J. Amer. chem. Soc., 1968, **90**, 3597; *G. F. Gibbons et al.*, Chem. Comm., 1968, 1458); the 15β-

hydrogen which would arise from [$2R$-2-$^3H_1$]MVA, is retained (*Gibbons et al., loc. cit.*). Loss of an unspecified hydrogen at $C_{(15)}$ was also demonstrated by *M. Akhtar et al.* (Biochem. J., 1969, **111**, 757). A possible mechanism is:

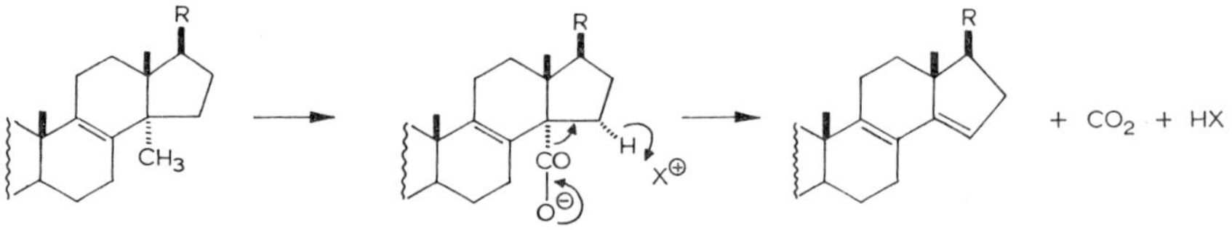

and this is supported by the demonstration that both 5α-cholesta-8,14-dien-3β-ol and 4,4-dimethyl-5α-cholesta-8,14-dien-3β-ol are converted into cholesterol by rat liver preparations (*Canonica et al.*, J. Amer. chem. Soc., 1968, **90**, 6532; *Akhtar et al., loc. cit.*; Chem. Comm., 1968, 1406; *B. N. Lutsky* and *G. J. Schroepfer*, Biochem. Biophys. Res. Comm., 1968, **33**, 492). Furthermore labelled 4,4-dimethyl-5α-cholesta-8,14-dien-3β-ol was isolated when labelled dihydrolanosterol was incubated anaerobically with a rat liver homogenate (*I. A. Watkinson* and *Akhtar*, Chem. Comm., 1969, 206). The reduction of the Δ^{14} bond involves NADPH with the hydrogen at 14α coming from NADPH and that at $C_{(15)}$ from water (*Akhtar et al., ibid.*, 1969, 149).

The next steps to consider are those involving ring B and which result in the conversion of a Δ^8 double bond into a Δ^5 bond. The pathway now appears to be well defined: $\Delta^8 \rightarrow \Delta^7 \rightarrow \Delta^{5,7} \rightarrow \Delta^5$ and the three steps will be considered without detailed reference to the structure requirements of the remainder of the molecule. However, other possibilities exist, *e.g.* cholest-8(14)-en-3β-ol is converted into cholesterol.

The isomerase carrying out the $\Delta^8 \rightarrow \Delta^5$ transformation is well known (see *Frantz* and *Schroepfer*, 1967, *loc. cit.*) and was considered not significantly reversible (*M. E. Dempsey*, J. biol. Chem., 1965, **240**, 4176); however, more recent work suggests that it is reversible (*D. C. Wilton et al.*, Biochem. J., 1969, **114**, 71). Experiments with [2-^{14}C-$2R$-$^3H_1$]MVA and [2-^{14}C-$2S$-$^3H_1$] MVA have shown that the pro-R hydrogen at $C_{(7)}$ arising from $C_{(2)}$ of MVA is retained in the formation of cholesterol in liver (*Gibbons et al.*, Chem. Comms., 1968, 1212; *Canonica et al.*, Steroids, 1968, **11**, 749; 1968, **12**, 445; *E. Caspi et al.*, Tetrahedron Letters, 1968, 3829) and of poriferasterol in *Ochromonas malhamensis* (*A. R. H. Smith et al., ibid.*, 1968, 1967). This means that the 7β-hydrogen is eliminated. The presence of a 14α-methyl group appears to prevent enzyme activity (*Gaylor et al.*, Steroids, 1966, **8**, 353).

The $\Delta^7 \rightarrow \Delta^{5,7}$ transformation has a mandatory requirement for oxygen but not for NADPH and the formation of the Δ^5 bond involves a *cis* elimi-

nation [5α(axial) and 6α(equatorial) hydrogens] (*A. M. Paliokas* and *Schroepfer*, Biochem. Biophys. Res. Comm., 1967, **26**, 736; *M. Akhtar* and *S. Marsh*, Biochem. J., 1967, **102**, 462; *L. J. Goad et al.*, *ibid.*, 1969, **114**, 885). The mechanism is not yet fully understood (*S. M. Dewhurst* and *Akhtar*, *ibid.*, 1967, **105**, 1187). The final step, the reduction of the Δ^7 bond is essentially irreversible (*Frantz jr.*, *A. T. Sanghvi* and *Schroepfer jr.*, J. biol. Chem., 1964, **239**, 1001) and is carried out by a soluble liver preparation in the presence of NADPH (*Dempsey et al.*, *ibid.*, 1964, **239**, 1381). The reaction involves a *trans*-addition of hydrogen 7α from NADPH (*Dempsey*, *ibid.*, 1965, **240**, 4176; *D. Dvornik*, *M. Kraml* and *J. F. Bagli*, Biochemistry, 1966, **5**, 1060) and 8β from $H_2O(H^{\oplus})$ (*D. C. Wilton et al.*, Biochem. J., 1968, 106, 803). Confirmation that the hydrogen introduced at $C_{(7)}$ occupies the 7α position comes from investigations with [2-^{14}C-2*R*-2-$^{3}H_1$]MVA (*Caspi et al.*, 1968, *loc. cit.*; *Gibbons et al.*, Chem. Comm., 1968, 1212).

The other reductive process required for cholesterol synthesis is the hydrogenation of the Δ^{24} bond; the enzyme responsible is present in cell microsomes, is NADPH-dependent and SH-dependent (*J. Avigan*, *D. S. Goodman* and *D. Steinberg*, J. biol. Chem., 1963, **238**, 1283; *Gaylor*, Arch. Biochem. Biophys., **101**, 108). The reduction is stereospecific (*K. A. Mitropoulos* and *N. B. Myant*, Biochem. J., 1965, **97**, 26C; *Caspi et al.*, Chem. Comm., 1969, 45).

The number of pathways possible for the later stages of cholesterol biosynthesis is very considerable and represents a three dimensional lattice of some complexity. If the simplifying assumption is made that a mandatory intermediate is 4α-methylcholesta-8,24-dien-3β-ol then the alternative possibilities are indicated in Scheme 14 (p. 104). The heavy arrows indicate steps which have been clearly demonstrated.

(*ii*) *Synthesis in higher plants and micro-organisms*

The main characteristics of the so-called phytosterols are (a) the presence of a supernumerary side chain at $C_{(24)}$, either methylene, methyl, ethylidene or ethyl; the methyl and ethyl derivatives can exist as epimers. However, cholesterol is widely distributed in higher plants in trace amounts (see *e.g.*, *H. J. Nicholas*, in *P. Bernfeld*, Ed. "Biogenesis of Natural Compounds", 2nd Edn. Pergamon Press, London, 1968). In red algae it is frequently the major sterol (*K. Tsuda et al.*, Chem. Pharm. Bull., Tokyo, 1958, **6**, 101, 724; 1959, **7**, 747; J. Amer. chem. Soc., 1960, **82**, 1442; *Gibbons et al.*, Phytochem., 1967, **6**, 677). Examples of alkylated sterols are chalinasterol (methylene), campesterol (α-methyl), 22-dihydrobrassicasterol (β-methyl), fucosterol and isofucosterol (ethylidene), β-sitosterol (α-ethyl), and clionasterol (β-ethyl); (b), the presence of an additional double bond at $C_{(22)}$, for example in

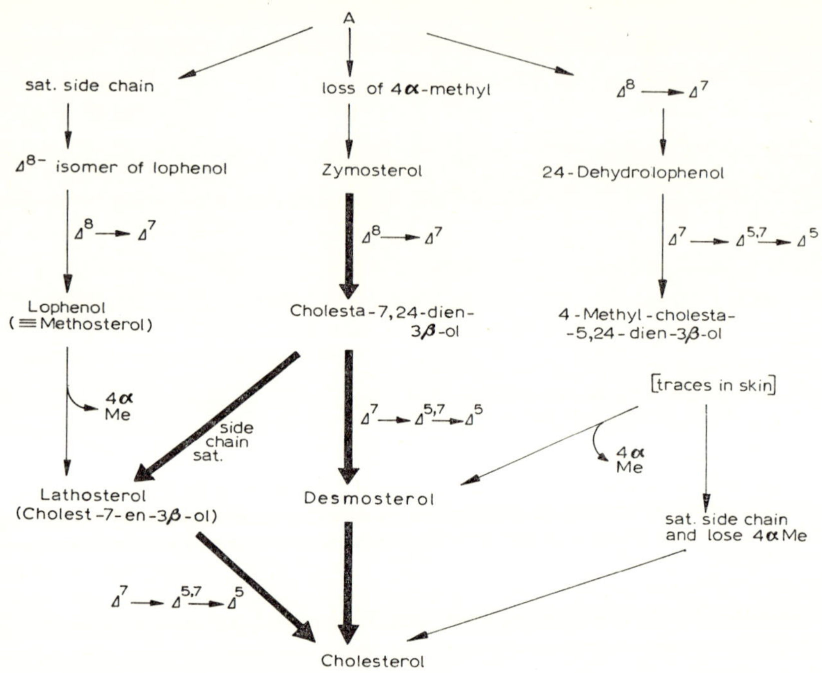

Scheme 14. Possible pathways for biosynthesis of cholesterol from 4α-methylcholesta-8,14-dien-3β-ol (A). (Heavy arrows indicate experimental demonstrations.)

α-spinasterol and brassicasterol, and (c) the existence of cycloartenol and its derivatives. These additional structural variables in combination with those already discussed under cholesterol biosynthesis in the previous section and with certain other occasional oxidative changes such as the appearance of an oxygen function at $C_{(6)}$ and $C_{(22)}$, make the possible number of sterols to be found in plants very great. Many of the possibilities have already been observed and many will undoubtedly be found as separation techniques improve (see *e.g.*, *L. J. Goad* in *J. B. Pridham*, Ed. "Terpenoids in Plants", Academic Press, London, 1967, p. 159).

There is no doubt that plant sterols arise from mevalonic acid *via* squalene (*E. Capstack et al.*, J. biol. Chem., 1965, **240**, 3258; *R. D. Bennett* and *E. Heftmann*, Phytochemistry, 1965, 4, 475), and that the overall labelling patterns are the same (*A. R. Battersby* and *G. V. Parry*, Tetrahedron Letters, 1964, 787). However there is still considerable doubt whether lanosterol is the first product of cyclization of squalene oxide in higher plants and algae and evidence is accumulating that it may be replaced by cycloartenol (see

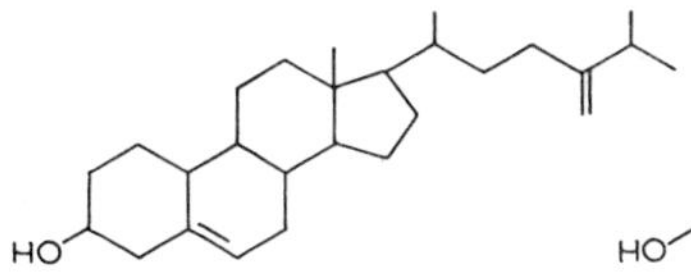

Chalinasterol 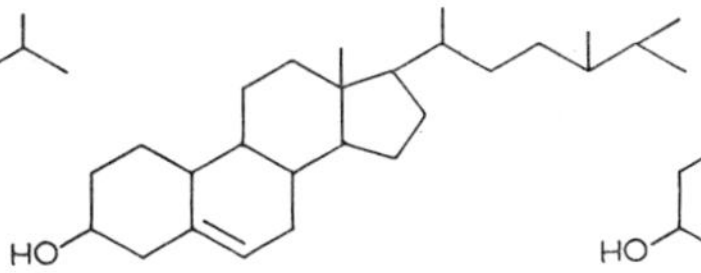 Campesterol 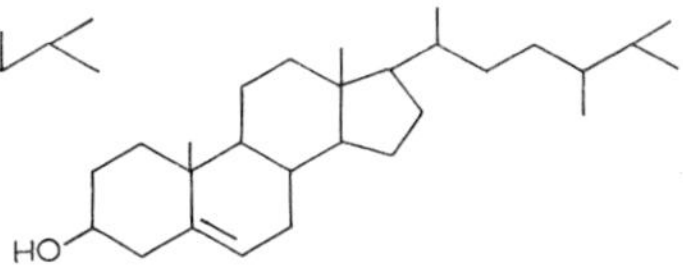 2-Dihydrobrassicasterol

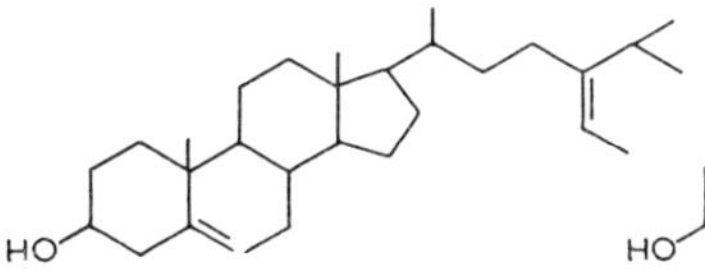

Isofucosterol 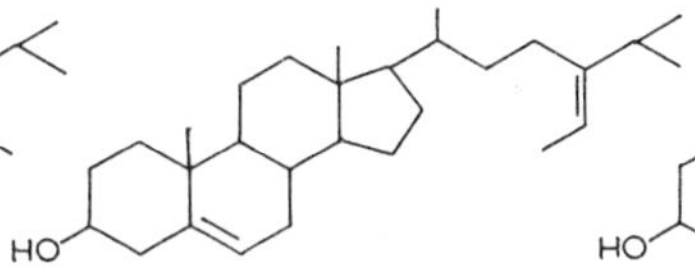 Fucosterol 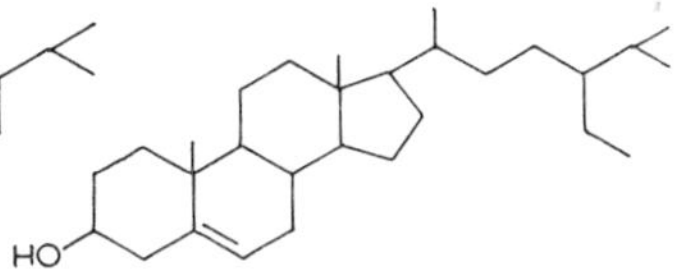 Clionasterol

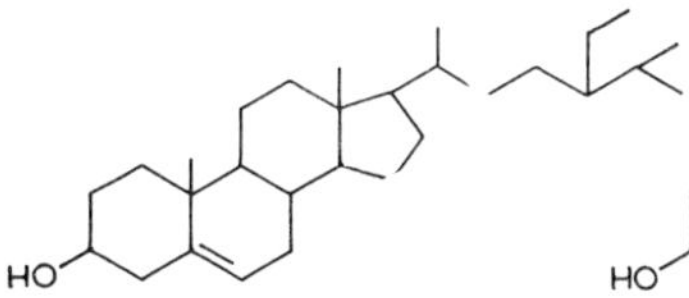

β-Sitosterol 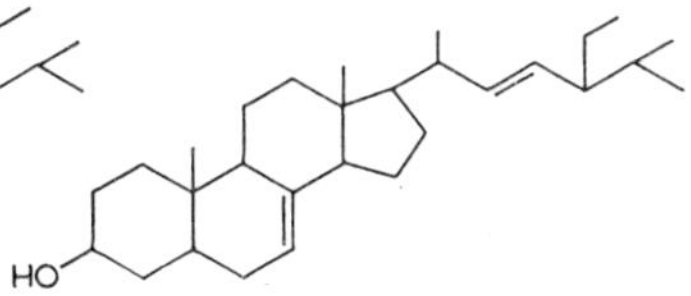 α-Spinasterol

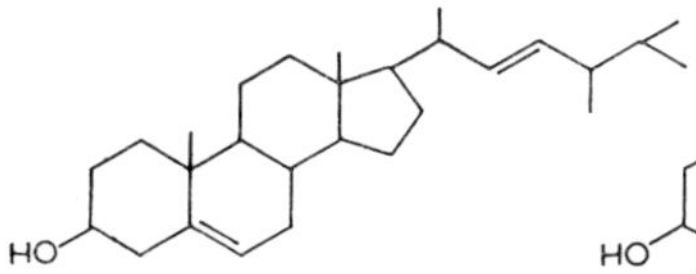

Brassicasterol 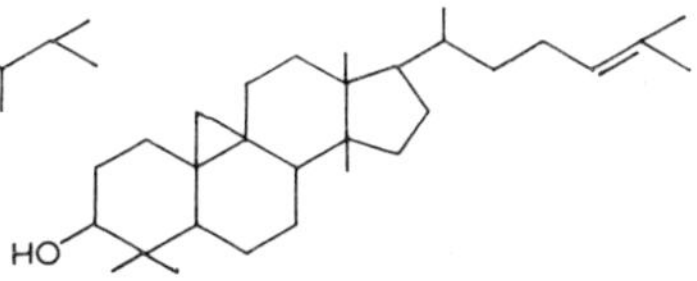 Cycloartenol

Goad, loc. cit.). Except in *Euphorbia* latex where it is formed from cycloartenol enzymically (*G. Ponsinet* and *G. Ourisson*, Phytochemistry, 1968, 7, 757), lanosterol has not been demonstrated unequivocally as a naturally occurring product in plants. An enzyme from bean leaves and from the alga *Ochromonas malhamensis* will convert [^{14}C]squalene oxide anaerobically into cycloartenol with the production of no detectable lanosterol (*H. H. Rees, Goad* and *Goodwin*, Tetrahedron Letters, 1968, No. 6, p. 723). Furthermore, studies with [2-^{14}C-4*R*-4-$^{3}H_1$]MVA demonstrate that cycloartenol is a primary product of cyclization of squalene oxide and that it is not formed from lanosterol (*Rees, Goad* and *Goodwin*, Biochem. J., 1967, **107**, 417). It has been shown recently that the algae *Ochromonas malhamensis* converts [2-$^{3}H_2$]-cycloartenol and [2-$^{3}H_2$]-24-methylenecycloartenol into poriferasterol (*J. Hall et al., ibid.*, 1969, **112**, 129) and that *Nicotiana tabacum* leaves convert

[24,25-$^3H_2$]cycloartanol into cholesterol (*M. Devys et al., Bull. Soc. chim. Biol.*, 1969, **51**, 133). However, labelled lanosterol is also effectively converted into phytosterols in *O. malhamensis* (*Hall et al., loc. cit.*), in *N. tabacum* (*M. J. E. Hewlins et al.*, European J. Biochem., 1969, **8**, 184) and in *Euphorbia peplus* (*D. J. Baisted et al.*, Phytochemistry, 1968, **7**, 945), as is 24-methylenelanosterol in *N. tabacum* (*A. Alcaide et al., ibid.*, 1968, **7**, 1773) and in *Spinacera oleracea* (*Devys et al., ibid.*, 1968, **7**, 613). This may be due to the possible lack of specificity of the enzyme concerned with the further transformations of the precursors. A similar lack of specificity has been noted in cholesterol biosynthesis (p. 104). A proposal has been made for the enzymic formation of cycloartenol, parkeol and lanosterol from a common enzyme substrate complex (Scheme 15).

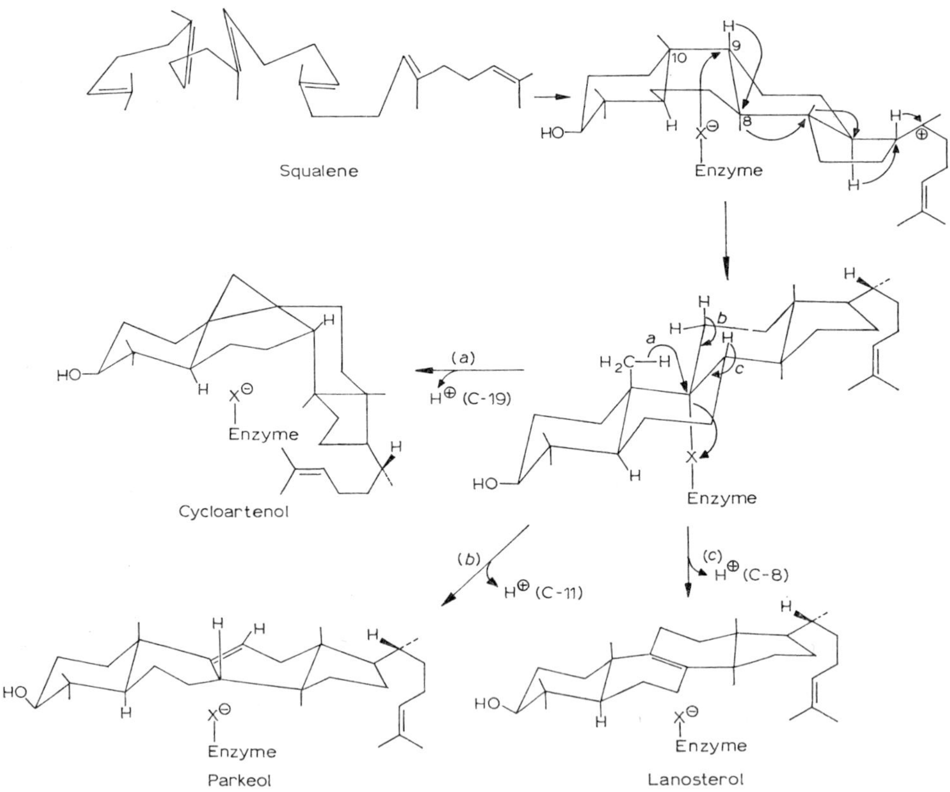

Squalene

Enzyme

(a)

$H^{\oplus}$ (C-19)

Enzyme

Cycloartenol

Enzyme

(b)

$H^{\oplus}$ (C-11)

(c)

$H^{\oplus}$ (C-8)

Enzyme

Parkeol

Enzyme

Lanosterol

Scheme 15. Proposed scheme for the enzymic formation of cycloartenol, parkeol and lanosterol.

Recent detection of possible intermediates in higher plants make the pathway indicated in Scheme 16 (below), a plausible means of synthesizing phytosterols from cycloartenol (*B. L. Williams, Goad* and *Goodwin,* Phytochem. 1967, **6,** 1137; *Goad, Williams* and *Goodwin,* European J. Biochem., 1967, **3,** 232; *Goad* and *Goodwin, ibid.,* 1967, **1,** 357), and it has recently been shown that in tobacco leaves 24-methylenelanosterol is incorporated into β-sitosterol whereas 24-methylenecholesterol is only reduced to campesterol (*A. Alcaide et al.,* Phytochemistry, 1968, **7,** 1773). Experiments on fungi suggests that lanosterol is the precursor of ergosterol, and on purely biogenetic studies the pathway in *Phycomyces blakesleeanus* and *Psalliota campestris* but not baker's yeast is indicated in Scheme 17 on p. 108 (*G. Goulston, Goad* and *Goodwin,* Biochem. J., 1967, **102,** 15C). A further intermediate, ergosta-5,7,24(28)-trien-3β-ol, consistent with this scheme has been found in *P. blakesleeanus* (*Goulston* and *E. I. Mercer,* Phytochemistry, 1969, **8,** 1945).

In baker's yeast, however, alkylation can occur apparently at a much

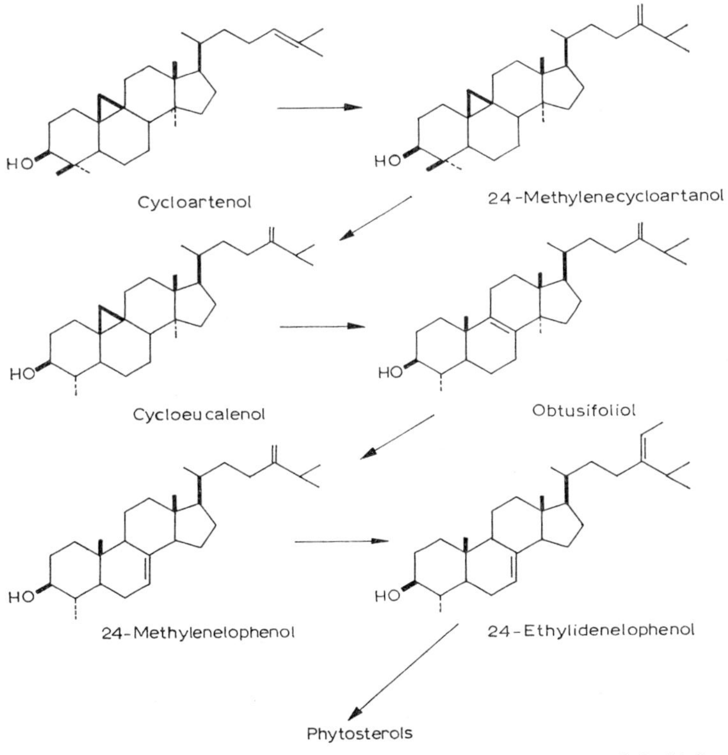

Cycloartenol
24-Methylenecycloartanol
Cycloeucalenol
Obtusifoliol
24-Methylenelophenol
24-Ethylidenelophenol
Phytosterols

Scheme 16. Possible biosynthetic sequence for the production of sterols in higher plants.

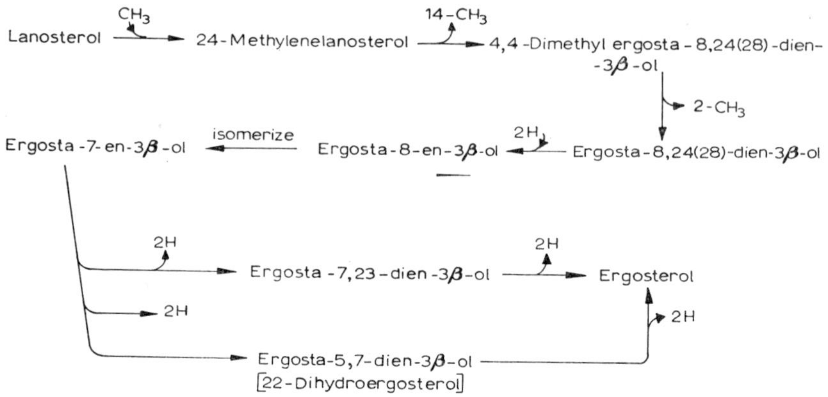

Lanosterol →(CH_3) 24-Methylenelanosterol →(14-CH_3) 4,4-Dimethyl ergosta-8,24(28)-dien-3β-ol

↓ (2-CH_3)

Ergosta-7-en-3β-ol ←(isomerize) Ergosta-8-en-3β-ol ←(2H) Ergosta-8,24(28)-dien-3β-ol

Ergosta-7-en-3β-ol →(2H) Ergosta-7,23-dien-3β-ol →(2H) Ergosterol

Ergosta-7-en-3β-ol →(2H) Ergosta-5,7-dien-3β-ol [22-Dihydroergosterol] →(2H) Ergosterol

Scheme 17. Possible pathway of biosynthesis of ergosterol in *Phycomyces blakesleeanus.*

later stage (*H. Katsuki* and *K. Bloch,* J. biol. Chem., 1967, **242,** 222) (Scheme 18) although 24-methylenelanosterol is also converted into ergosterol by intact yeast cells (*M. Akhtar, M. A. Parvez* and *P. F. Hunt,* Biochem. J., 1966, **100,** 38C; *D. H. R. Barton, D. M. Harrison* and *G. P. Moss,* Chem. Comm., 1966, **17,** 595).

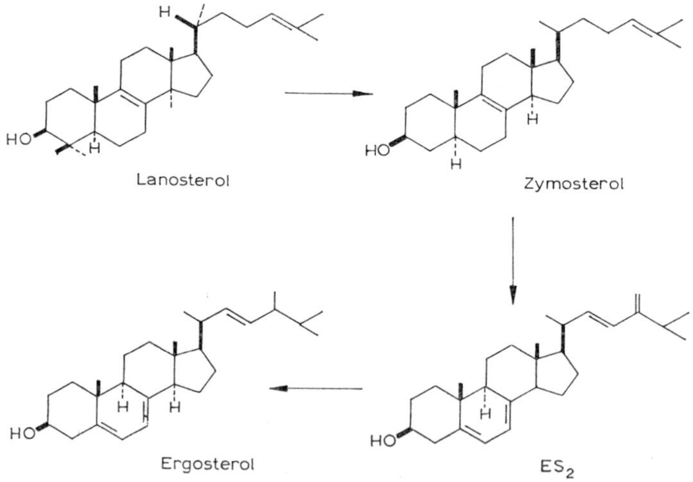

Lanosterol → Zymosterol → ES_2 → Ergosterol

Scheme 18. Possible pathway of ergosterol biosynthesis in yeast. (*Bloch*)

(*iii*) *Mechanism of alkylation*

The 24-methyl group of ergosterol arises from *S*-adenosylmethionine by transmethylation (see *e.g. E. Lederer,* Biochem. J., 1964, 93, 449). One of the

hydrogen atoms of the methyl group is lost during the transfer (*G. Jaurégui-berry et al.*, Biochemistry, 1965, **4**, 347) and the methylene derivative of lanosterol is a precursor of ergosterol (see above) and has been demonstrated in a number of fungi (*Goulston et al., loc. cit.*). The 24-ethyl and ethylidene groups arise by two successive methylations (*M. Castle, G. Blondin* and *W. R. Nes*, J. Amer. chem. Soc., 1963, **85**, 3306; *S. Bader, L. Guglielmatti* and *D. Arigoni*, Proc. chem. Soc., London, 1964, 16; *V. R. Villaneuva, M. Barbier* and *Lederer*, Bull. Soc. chim. Fr., 1964, 1423) and in the case of poriferasterol in *Ochromonas malhamensis* only four hydrogens of the entering methyl groups appear in the sterol (*A. R. H. Smith et al.*, Biochem. J., 1967, **104**, 56C). However, in the slime mould *Dictylostelium discoideum*, five hydrogens appear in the major sterol, stigmast-22-en-3β-ol (*M. Lenfant, E. Zissman* and *E. Lederer*, Tetrahedron Letters, 1967, No. 12, 1049; *Lenfant et al.*, European J. Biochem., 1969, **7**, 159). Mechanisms to explain these possibilities are given in Scheme 19. The reality of the hydrogen shift from $C_{(24)}$ to $C_{(25)}$ has been demonstrated in the case of ergosterol (*K. J. Stone* and *F. W. Hemming*, Biochem. J., 1967, **104**, 43; *Akhtar, P. F. Hunt* and *M. A. Parvez*, *ibid.*, 1967, **103**, 616) and fucosterol (*Goad* and *Goodwin*, *ibid.*, 1965, **96**, 79P; European J. Biochem., 1967, **1**, 357).

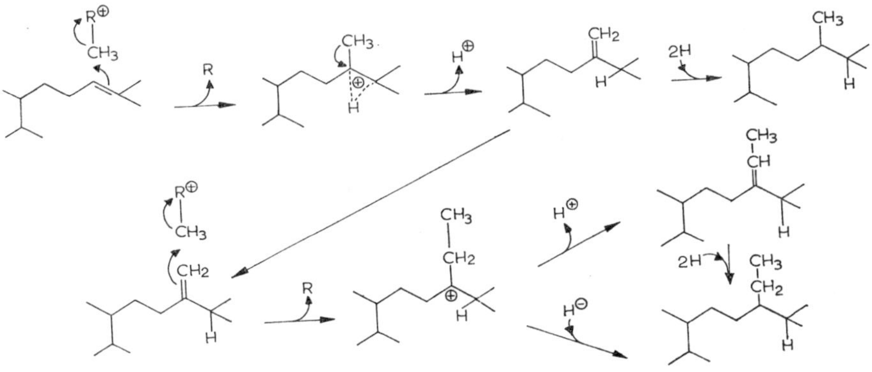

Scheme 19. Mechanisms for alkylation of plant sterols.

(iv) Formation of $C_{(22)}$ double bond in plant sterols

An outstanding problem is whether or not the double bond at $C_{(22)}$ in some phytosterols is inserted before or after alkylation.

In baker's yeast a compound with the probable structure IV is suggested as an intermediate (*H. Katsuki* and *K. Bloch*, J. biol. Chem., 1967, **242**, 772). However, as mentioned above 24-methylenelanosterol is a precursor of ergosterol in yeast (*D. H. R. Barton, D. M. Harrison* and *G. P. Moss*, Chem.

Comm., 1966, **3**, 595) as is 24-methyldihydrolanosterol (*Akhtar, Parvez* and *Hunt*, Biochem. J., 1968, **106**, 623). It is the pro-*R*-hydrogens at $C_{(22)}$ and $C_{(23)}$ which are removed during the introduction of this double bond into poriferasterol in *Ochromonas malhamensis* (*A. R. H. Smith, Goad* and *Goodwin*, Chem. Comm., 1968, 926, 1968). However, in *Aspergillus nidulans* the situation is reversed (*T. Bimpson et al., ibid.*, 1969, 297). The occurrence of 22-dehydrocholesterol in a red alga (*K. Tsuda et al.*, J. Amer. chem. Soc., 1958, **80**, 921) and the apparent complete absence of Δ^{22} sterols from pine bark (*J. W. Rowe*, Phytochemistry, 1965, **4**, 1) also suggest that the prior insertion of an alkyl group at $C_{(24)}$ is not necessary for the formation of a Δ^{22} double bond and *vice versa*. In fact, they may well be two processes which are in no way biosynthetically linked.

In concluding this section on alkylated sterols it might be emphasized that many of the structural differences encountered between fungal sterols and higher plant sterols could be ascribed to the absence of a Δ^{7}-hydrogenase from the fungi so far examined.

(*v*) *$C_{(21)}$ Sterols in plants*

(1) *Pregnane derivatives.* The compounds in this group include pregnenolone from *Xysmalobrium undulatum*, progesterone from *Holarrhena floribunda*, and their derivatives including the alkaloids such as holaphylline (see *e.g. R. Tschesche* in "Terpenoids in Plants", Ed. *J. B. Pridham*, Academic Press,

London, 1967). In animals these compounds are formed from cholesterol *via* 20α,22-dihydroxycholesterol with the liberation of isocaproic aldehyde thus:

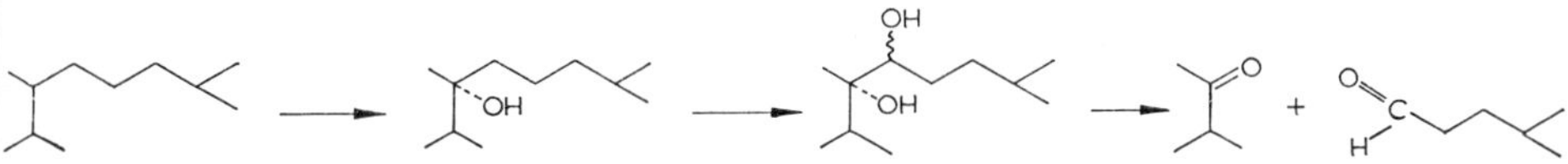

A 22-oxo derivative is not involved because [22-^{14}C-22-$^{3}H_2$]cholesterol yields isocaprylaldehyde with one tritium still attached (*F. Constantopoulus et al.*, Biochemistry, 1966, **5**, 1650). In plants [^{14}C]cholesterol is converted into pregnenolone in golden rod (*Haplopappus heterophyllus*) (*R. D. Bennett* and *E. Heftmann*, Phytochemistry, 1966 **5**, 747) and *Digitalis purpurea* (*E. Caspi et al.*, Experientia, 1966, **22**, 506) but the mechanism involved is unknown. In *Holarrhena* leaves [4-^{14}C]pregnenolone is converted into progesterone and into the alkaloids holaphylline and holaphyllamine. Progesterone is apparently not an intermediate in the synthesis of the alkaloids (*Bennett* and *Heftmann*, Science, N.Y., 1965, **149**, 652). The conversion of pregnenolone into progesterone in *Digitalis lanata* has also been demonstrated (*Caspi* and *D. O. Lewis*, *ibid.*, 1967, **156**, 519; *H. H. Sauer*, *Bennett* and *Heftmann*, Phytochemistry, 1967, **6**, 1521). Holaphyllamine itself can be converted into pregnenolone (*Bennett*, *Heftmann* and *S.-T. Ko*, *ibid.*, 1966, **5**, 517).

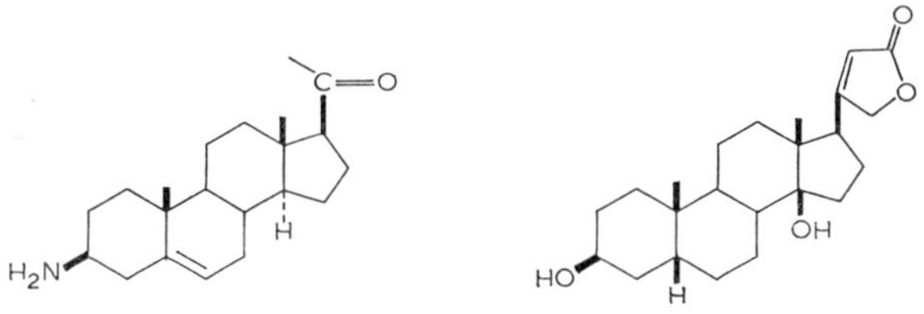

Holaphyllamine

Digitoxigenin

(2) *Cardenolides.* A typical cardenolide is digitoxigenin found in *Digitalis* spp. Gitoxigenin has an additional hydroxyl at $C_{(16)}$ and digoxigenin an additional hydroxyl at $C_{(12)}$. Only two carbons of the side chain arise from MVA (*J. von Euw* and *T. Reichstein*, Helv., 1964, **47**, 711) which suggested that pregnenolone was a precursor, a view that has been demonstrated by *Tschesche* and *G. Lilienweiss* (Z. Naturforsch., 1964, **19b**, 295), *Caspi* and *Lewis* (Science, 1967, **156**, 519) and *Bennett et al.* (Phytochemistry, 1968, **7**, 41). This lends support to the view that during formation of digitoxigenin a 5β-hydrogen is inserted to give rings A/B *cis* fusion and that similarly the 14α-hydrogen is replaced by a 14β-hydroxyl. It would appear that 14α-hydroxyprogesterone is not an intermediate (*Caspi* and *Lewis*, Phytochemistry, 1968, **7**, 683). 5β-Pregnan-3β,14β-diol-20-one is converted into cardenolides, but [5-^{14}C]-deoxydigitoxigenin is not, which is consistent with

the view that the addition of "acetate" is a late stage in the biosynthetic sequence. Oxidation at $C_{(3)}$ appears to be a necessary step in the conversion of pregnenolone into digitoxigenin (*Caspi* and *G. M. Hornby, ibid.*, 1968, 7, 423). Uzarigenin is formed from 5α-pregnan-3β-ol-20-one in *Strophanthus kombe* (*Sauer et al., ibid.*, 1969, **8**, 839)

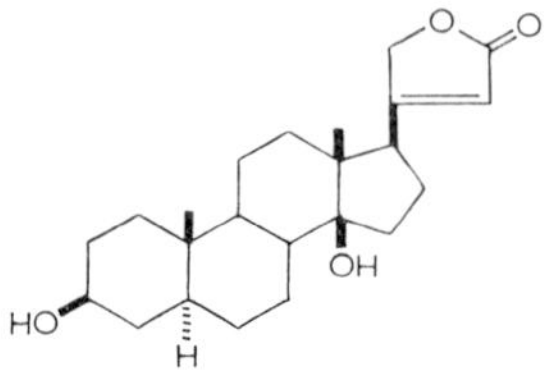

Uzarigenin

(vi) $C_{(18)}$ Sterols in plants

Estrone is found in date palm and pomegranate seeds (*Heftmann, T.Ko* and *Bennett*, Phytochemistry, 1965, **5**, 231, 1337), the latter contains 17 mg/kg. Nothing is known of the biosynthesis of oestrones in plants, but their formation in animals is better understood (see *Brener*, Vitamins and Hormones, 1962, **20**, 285; *E. Staple*, in *P. Bernfeld*, Ed. "Biogenesis of Natural Compounds", 2nd Ed., Pergamon Press, London, 1968).

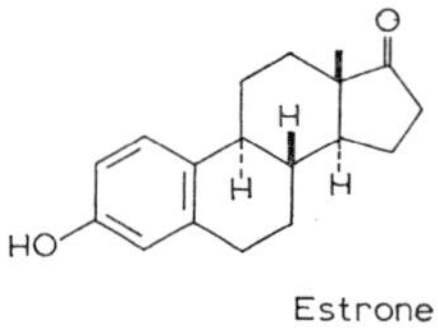

Estrone

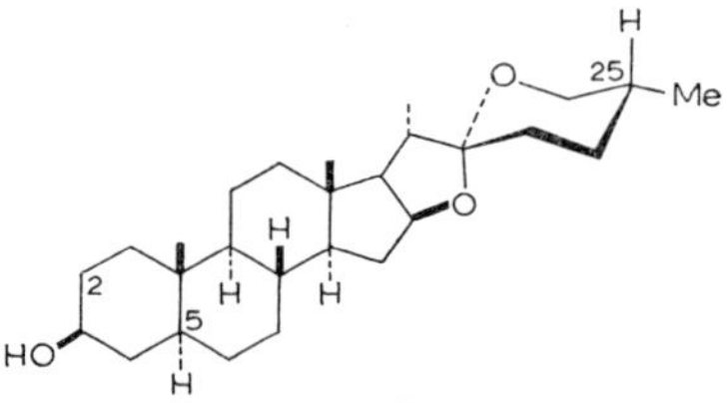

Tigogenin

(vii) Spirostanols $C_{(27)}$ in plants

Typical members of this group are found in the leaves of *Digitalis lanata* (see *e.g. R. Tschesche* in "Terpenoids in Plants", Ed. *J. B. Pridham*, Academic Press, London, 1967). The parent appears to be tigogenin; gitogenin has a α-hydroxyl at $C_{(2)}$; diosgenin a double bond at $C_{(5)}$ and neotigogenin has the opposite configuration to tigogenin at $C_{(25)}$. Rings A/B and C/D are *trans*-fused, as in 5α-cholestanol. [^{14}C]-Cholesterol is incorporated into tigogenin and gitogenin in *D. lanata* and the appearance of diosgenin suggested that it might be an intermediate (*Tschesche* and *H. Hulpke*, Z. Naturforsch.,

1966, **21b**, 494). Similar results were obtained in *Dioscorea spiculiflora* with label and also in kryptogenin which is a natural possibility as an intermediate (*Bennett* and *Heftmann*, Phytochemistry, 1965, **4**, 873). In *Digitalis* 22-oxo-cholesterol and 22ξ-hydroxycholesterol are not converted into tigogenin or gitogenin. As [3α-$^{3}H_{1}$]desmosterol does not label the spirostanols whilst [4-^{14}C]cholestenone does, it is concluded that a 3-oxo compound is involved in spirostanol biosynthesis (*Tschesche et al.*, *ibid.*, 1967, **7**, 2021). As experiments with [2-^{14}C]MVA and [3-^{14}C]MVA have shown that $C_{(26)}$ and $C_{(27)}$ are not equivalent in tigogenin as they are in cholesterol, it has been concluded that cholesterol is not a direct precursor but that previously reported incorporations were due to prior oxidation of cholesterol to cholest-24-enol (*R. Joly* and *C. Tamm*, Tetrahedron Letters, 1967, 3535).

The steroid alkaloids with 27 carbon atoms may be formed from the spirostanols because they occur together in many plants (see *e.g. K. Schreiber*, Kulturpflanze, 1963, **XI**, 422). An enzyme in *Hosta* spp. converts $\Delta^{25,27}$-gitogenin into both neogitogenin [*S*-configuration at $C_{(25)}$] and gitogenin [*R*-configuration at $C_{(25)}$]; 25-hydroxygitogenin is apparently not an intermediate.

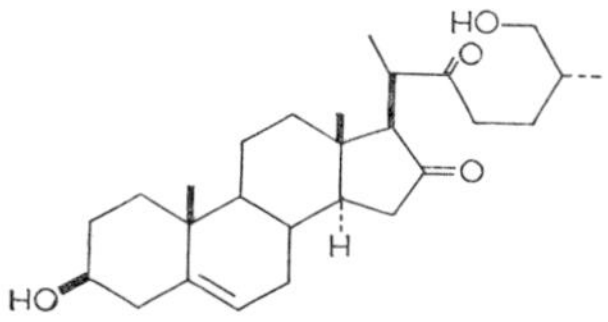

Kryptogenin

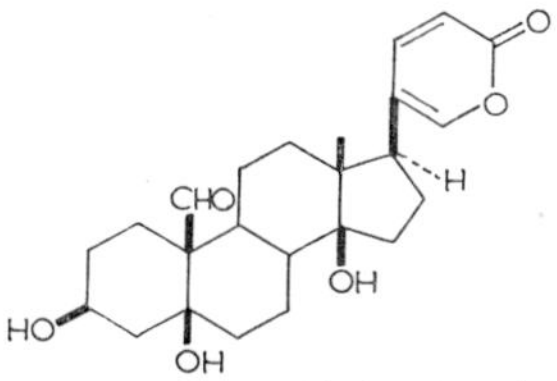

Hellebrigenin

(*viii*) *Bufadienolides*

Hellebrigenin is the aglycon of hellebrin which occurs to the extent of 2% in the roots of *Helleborus atrorubens*. It is derived from pregnenolone (*Tschesche* and *B. Brassat*, Z. Naturforsch., 1965, **20b**, 707). The source of the remaining three carbon atoms of the side chain is still unknown.

(*ix*) *Ecdysones*

Insects appear to be unique animals in the sense that no species so far examined can convert either acetate or squalene into sterols (*A. J. Clark* and *Bloch*, J. biol. Chem., 1959, **234**, 2578, 2583, 2689). However a variety of insects have a nutritional requirement for sterols which is frequently satisfied by their symbiotic micro-organisms. The moulting hormone, ecdysone is a steroid derived from cholesterol (*P. Karlson* and *H. Hoffmeister*, Hoppe-Seyler's Z., 1963, **331**, 298). It and related compounds such as ecdysterone,

pterosterone, the ponasterones A, B, C and polypodine B are now known to be widely distributed in plants, where they can be synthesized from cholesterol, but not from cholestenone (*H. H. Sauer et al.*, Phytochemistry, 1968, 7, 2027). It has been shown that ponasterone A and ecdysterone can arise from mevalonic acid in *Taxus baccata* (*N. J. de Souza et al.*, Biochem. J., 1969, **114**, 895).

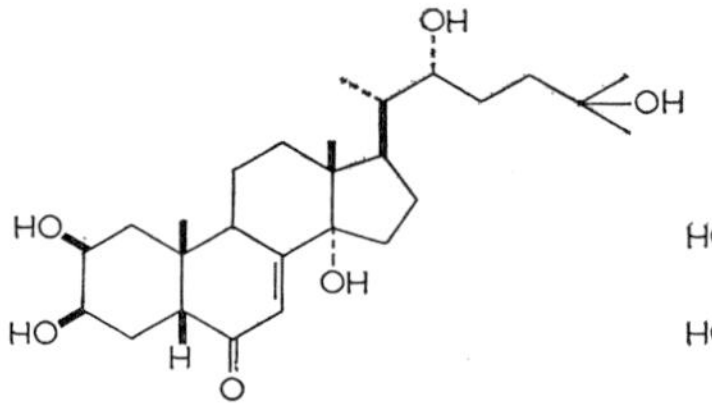

Ecdysone

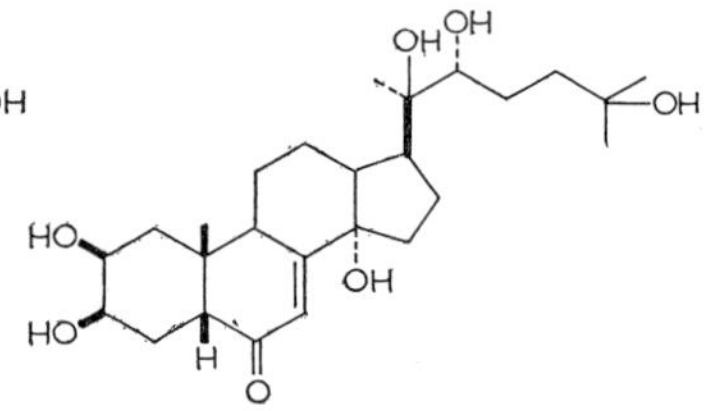

Ecdysterone

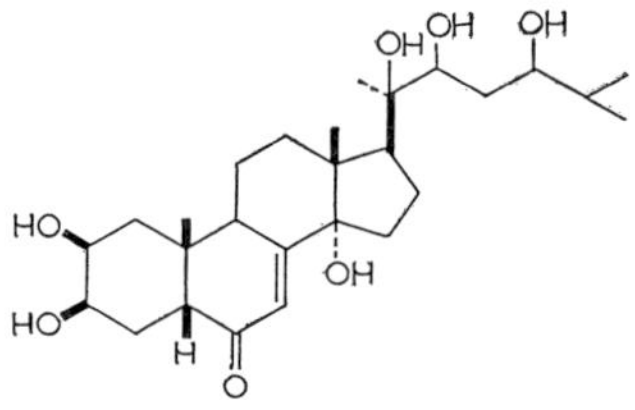

Pterosterone

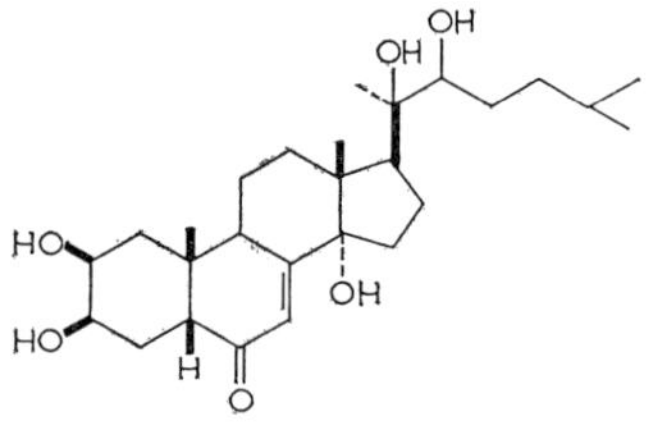

Ponasterone A

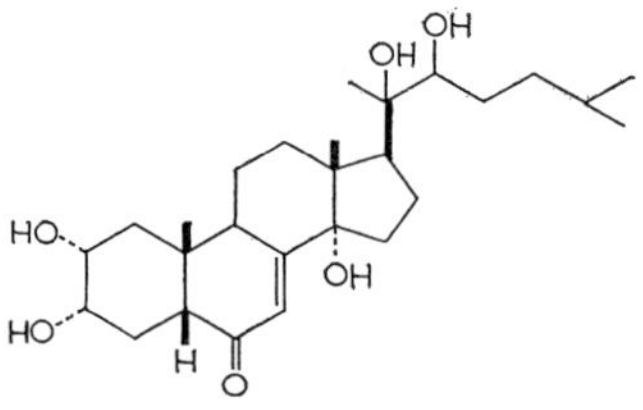

Ponasterone B

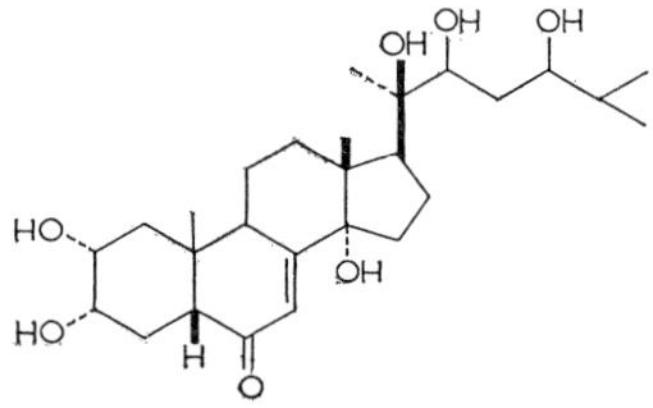

Ponasterone C

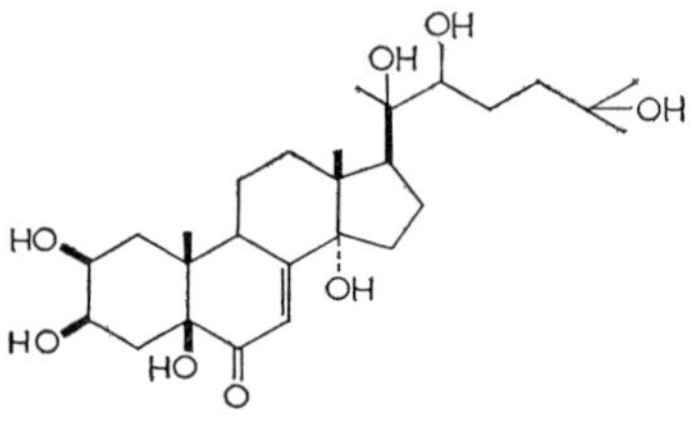

Polypodine B

(x) Compounds other than sterols with the cyclopentenophenanthrene nucleus

Two main groups are known (a) the tirucallol (20α-H) and euphol (20β-H) group (V) and (b) the dammarenediol group, in which the OH at $C_{(20)}$ can have the α- or β-configuration (VI):

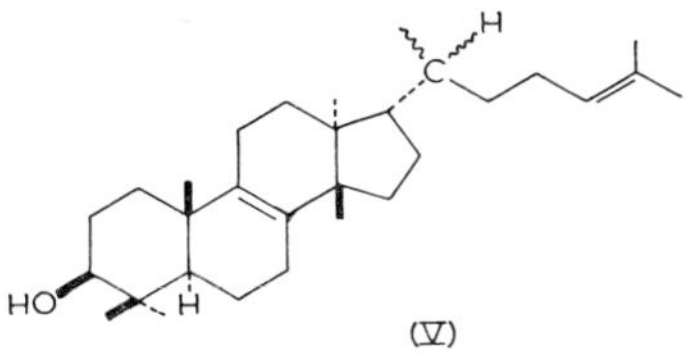

(V) 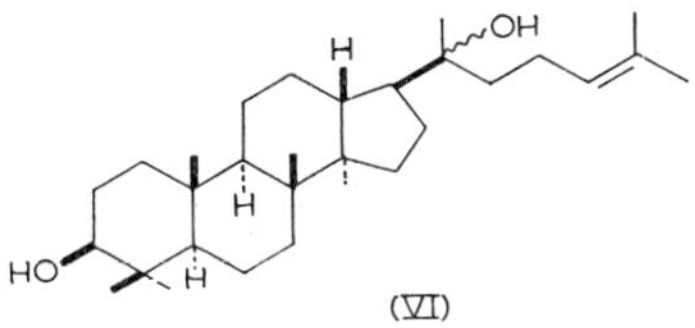 (VI)

The dammarenediols are probably formed by the cyclization of squalene (oxide) in the chair, chair, chair, boat conformation (*Ruzicka*, 1959, *loc. cit.*) (Scheme 20):

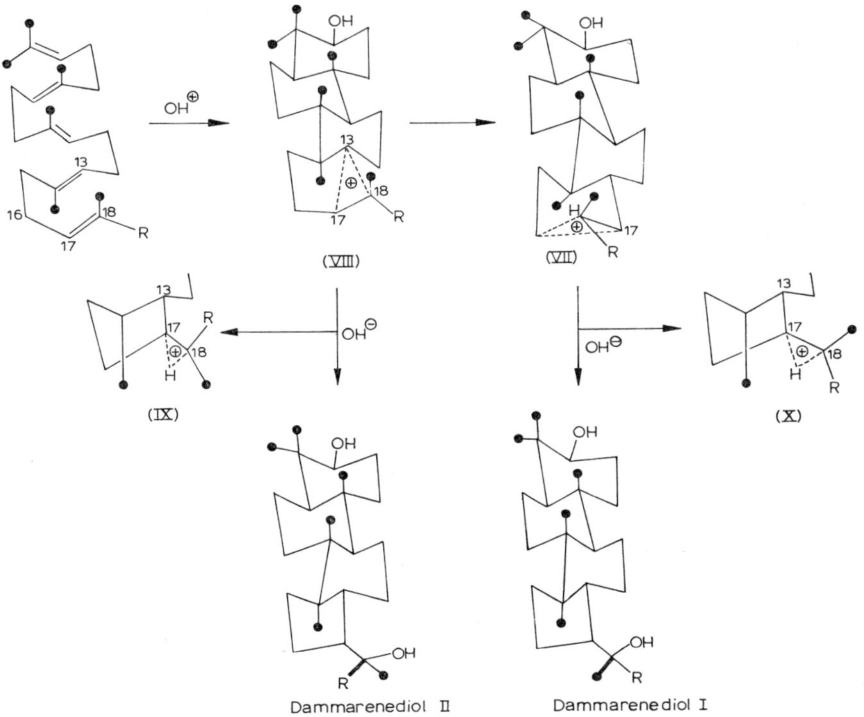

Scheme 20. The probable biogenetic formation of dammarenediols (*Ruzicka*, 1959, *loc. cit.*).

The euphol and tirucallol group can also be formed from the same bridged cations (VII and VIII, Scheme 20) or enzyme substrate complex as the dammarenediols, via IX and X, by a series of 1,2-methyl shifts similar to those proposed for lanosterol (p. 96). No direct demonstration of these shifts exists. Unless a bridged cation or enzyme substrate complex is invoked, the stereochemistry at $C_{(20)}$ is not determined by the conformation involved.

Ruzicka (1959) has given numerous examples of the mechanism of formation of tetracyclic triterpenoids similar to those just discussed.

A third group of tetracyclic triterpenoids contains only a few members; a typical example is α-onocerine

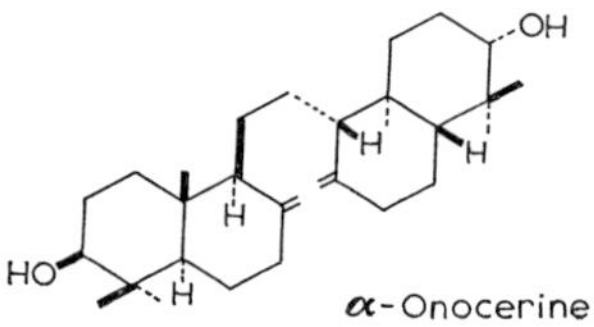

α-Onocerine

Ruzicka (1959) suggests that it may be derived by a simultaneous attack by $OH^{\oplus}$ at both ends of squalene folded in the chair, chair conformation thus:

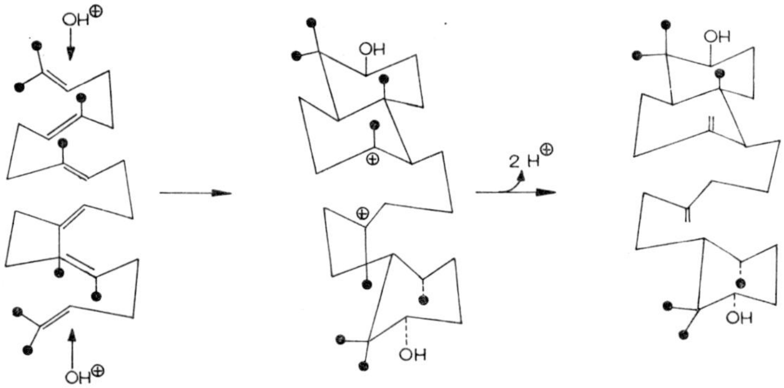

On more modern views a squalene diepoxide might be involved.

(c) Biosynthesis of pentacyclic triterpenes

A number of groups can be distinguished – in particular the amyrane, lupane, gammaoleanane, lopane, aborane, oleanane and ursane types. Two suitable examples for the present purpose are β-amyrin and lupeol. Both can be considered to arise by cyclization of squalene oxide in the chair, chair,

chair, boat conformation; lupeol directly following the formation of the bridged ion *via* XI (Scheme 21). For the formation of the six-membered

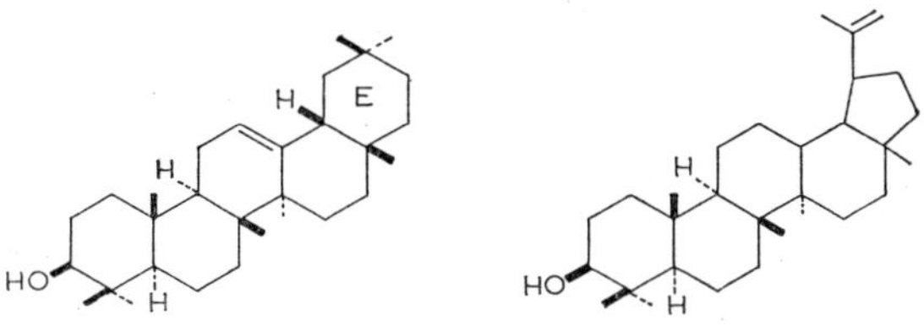

β-Amyrin

Lupeol

ring E of β-amyrin and related compounds, the rearrangement to a chair conformation (XII) is required (*Ruzicka*, 1959, *loc. cit.*) and this would be followed by a hydrogen shift from 17α to 21α and one from 13β to 17β with an elimination of a proton from 12α.

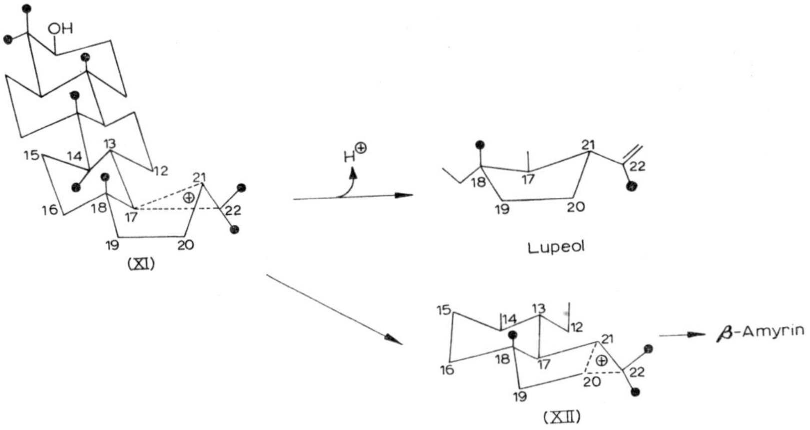

Scheme 21. Formation of lupeol and β-amyrin from squalene.

[2-^{14}C]MVA is incorporated into β-amyrin in pea seedlings (*D. Baisted et al.*, Biochemistry, 1962, **1**, 537) and maize seedlings (*E. I. Mercer* and *Goodwin*, Biochem. J., 1963, **88**, 46P) and by the use of [4*R*-4^{3}H$_1$]MVA the reality of Ruzicka's proposals have been demonstrated (Scheme 22) (*Rees, Mercer* and *Goodwin*, *ibid.*, 1966, **99**, 726; *Rees, Britton* and *Goodwin*, *ibid.*, 1968, **106**, 659). Squalene epoxide is also transformed into β-amyrin by crude cell-free homogenates from peas (*E. J. Corey* and *P. R. O. de Montellano*, J. Amer. chem. Soc., 1967, **89**, 3362).

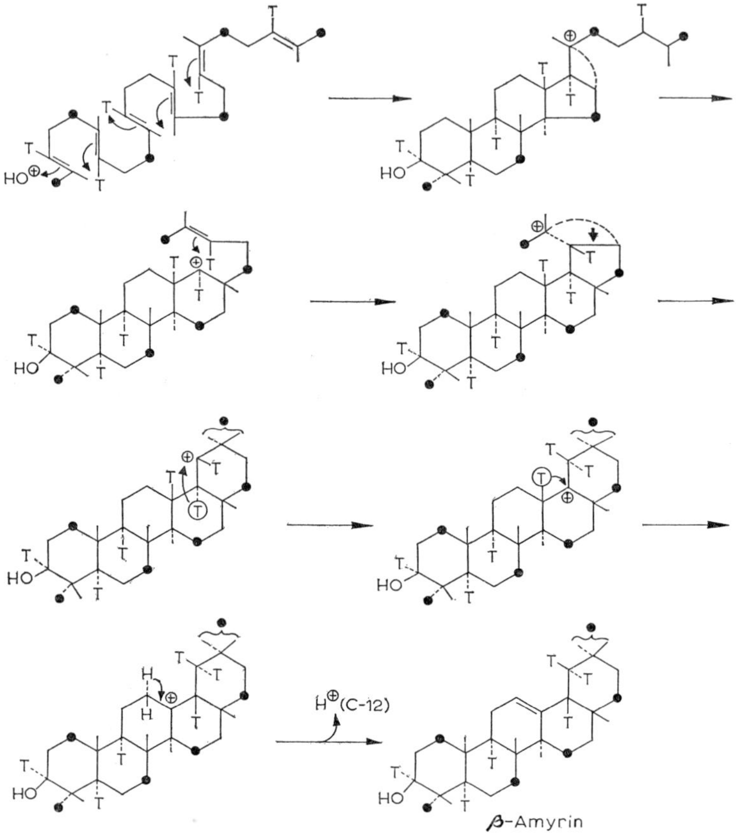

Scheme 22. The mechanism for the biosynthesis of β-amyrin.

The stepwise oxidation of the $C_{(28)}$ methyl group in β-amyrin is strongly suggested by the isolation of four structurally related triterpenoids from *Machaerium incorruptible* (*H. M. Alves et al.*, Phytochemistry, 1966, 5, 1327). They are β-amyrin, (XIII, $R{=}CH_3$) erythrodiol-3-acetate (XIII, $R{=}CH_2OH$) *O*-acetyloleanolic aldehyde (XIII, R=CHO) and *O*-acetyloleanoic acid (XIII, $R{=}CO_2H$).

D. Arigoni (Experientia, 1958, 14, 153) demonstrated that when [2-^{14}C]-mevalonate is incorporated into soyasapogenol A then $C_{(24)}$ is specifically derived from the methyl group of mevalonate and that the label from $C_{(2)}$ appears exclusively in the $C_{(23)}$ (equatorial) methyl group. Although it is known that the isomerization of IPP to DMAPP is stereospecific (p. 60), this important observation indicates (a) that two terminal methyl groups

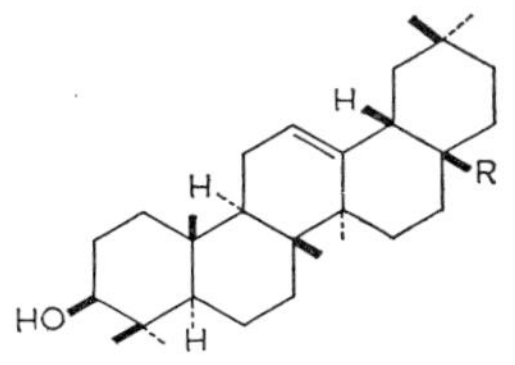

(XIII)

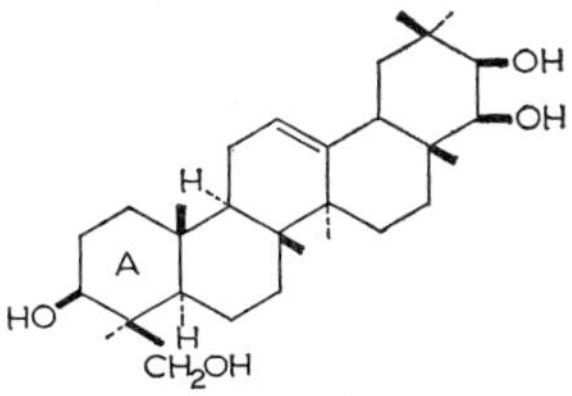

Soyasapogenol A

remain distinct throughout the cyclization of squalene and (b) that $C_{(2)}$ in MVA becomes the equatorial 4-methyl group; this confirms the view that ring A is formed by a chair-type folding of the chain and that at no time during the process has the configurational stability of the cation structure (Scheme 21) been lost (*Arigoni*, 1960, *loc. cit.*). This type of result has also been observed with gibberellic acid (p. 87) and rosenolactone (p. 88), lanosterol (p. 95), cycloartenol (p. 105) and torularhodin (p. 121).

The specific identity of the *gem*-dimethyl groups can be maintained at the end away from the initiation of cyclization. In lupeol, for example, the methyl group indicated (Scheme 21) is labelled from [2-^{14}C]MVA whilst the methylene group is unlabelled (*Ruzicka*, Pure appl. Chem., 1963, 6, 493). Other pentacyclic triterpenes, which have interesting biosynthetic problems although no experimental work has yet been reported, include friedelin which also has no hydroxyl at $C_{(3)}$ although it has lost a methyl group from $C_{(4)}$, which has migrated to $C_{(5)}$ and phyllanthol, which has a cyclopropane ring in an unexpected position and in which one methyl group at $C_{(4)}$ has migrated to $C_{(5)}$.

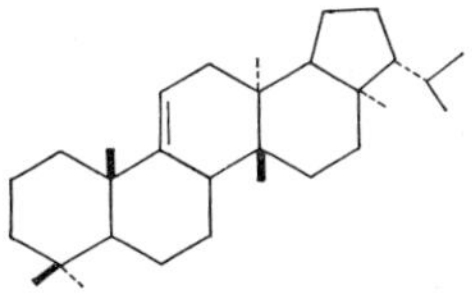

Fernene

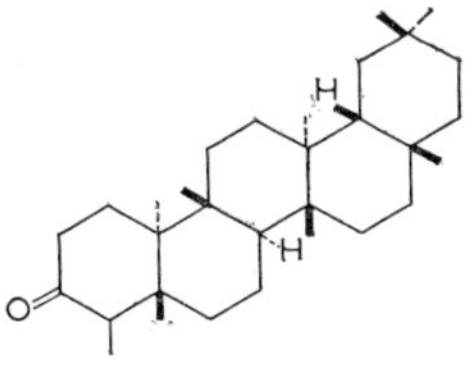

Friedelin

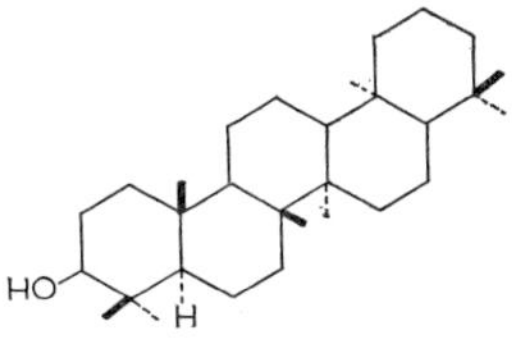

Tetrahymanol

Tetrahymanol synthesized by the protozoan *Tetrahymena pyriformis* is synthesized from squalene *via* a protonated cyclization and not *via* squalene oxide (*E. Caspi et al.*, J. Amer. chem. Soc., 1968, 90, 3563; Biochem. J., 1968, 109, 931). Fernene is also formed by a proton-initiated cyclization (*Barton et al.*, Chem. Comm., 1969, 184).

(d) Degraded triterpenes

[2-^{14}C]MVA was used to show that glaucarubolone and its congeners from *Simaruba glauca* are degraded triterpenes rather than modified diterpenes. No activity was found at $C_{(12)}$, although $C_{(1)}$ (and other) positions were labelled; if it arose from a diterpene it should have been labelled at $C_{(12)}$ (XIV), but this would not have been labelled if a lanosterol-like compound (XV) were an intermediate (*J. Moron, J. Rondest* and *J. Polonsky*, Experientia, 1966, **22**, 511).

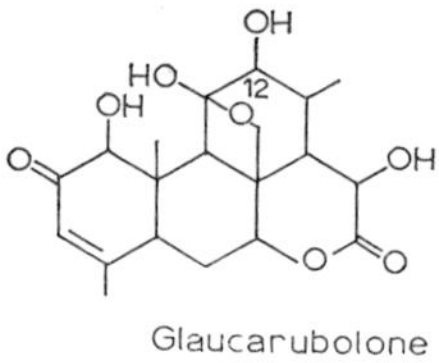

Glaucarubolone 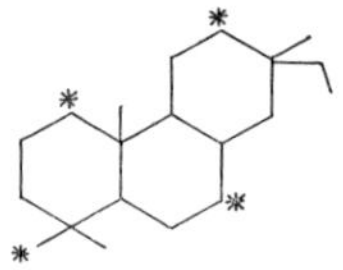 (XIV) 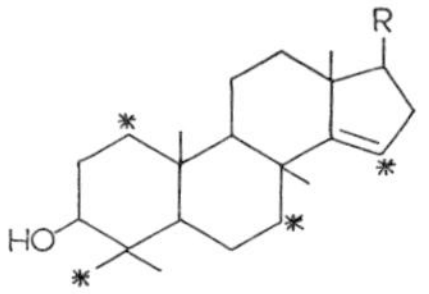 (XV)

9. Tetraterpenoids

The carotenoids (Vol. IIB, p. 231 *et seq.*) represent the only naturally-occurring tetraterpenoids; they are divided into *carotenes* (hydrocarbons) and *xanthophylls* (oxygen-containing carotenoids). They are widely distributed in nature (*T. W. Goodwin*, Ed., "Chemistry and Biochemistry of Plant Pigments", Academic Press, London, 1965) and are synthesized *de novo* by higher plants and protista. The basic carotene structures encountered are exemplified by α-carotene, β-carotene, γ-carotene and lycopene. They can all be formally represented by the tail to tail condensation of two $C_{(20)}$ units formed by the head to tail condensation of four isoprenoid units. Variations on these basic structures include (a) the addition of oxygen in the form of (1) an oxo-group (*e.g.* echinenone); (2) an hydroxyl group at $C_{(1)}$ (rhodopin), $C_{(3)}$ (lutein) or one of the *gem* dimethyl groups (lycoxanthin); (3) a carboxyl group (torularhodin); (4) a 5,6-epoxy group (antheraxanthin); and (5) a 5,8-furanoid oxide (auroxanthin); (b) the presence of acetylenic linkages in the polyene chain (alloxanthin); (c) the presence of allenic double bonds (neoxanthin); (d) the presence of pentacyclic rings (capsanthin); (e) the presence of aromatic rings (isorenieratene); (f) the presence of retro structures (eschscholtzxanthin).

(a) Biosynthesis of tetraterpenoids

The early stages in carotenoid biosynthesis have been fully reviewed (*J. W. Porter* and *D. G. Anderson*, Ann. Reviews Biochem., 1967, **18**, 1973;

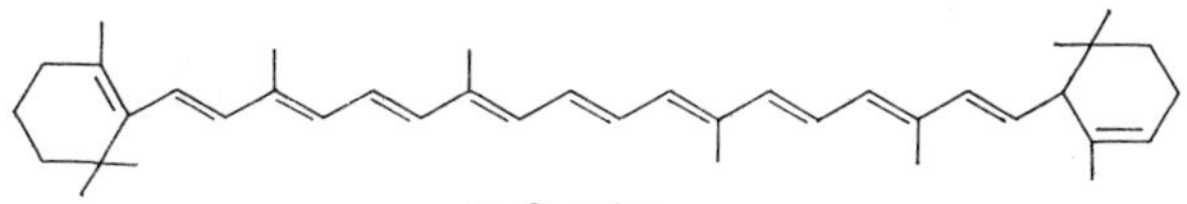

α-Carotene
β-Carotene 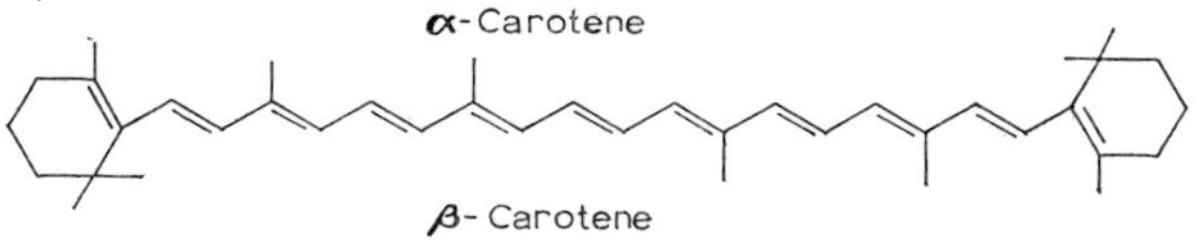
γ-Carotene 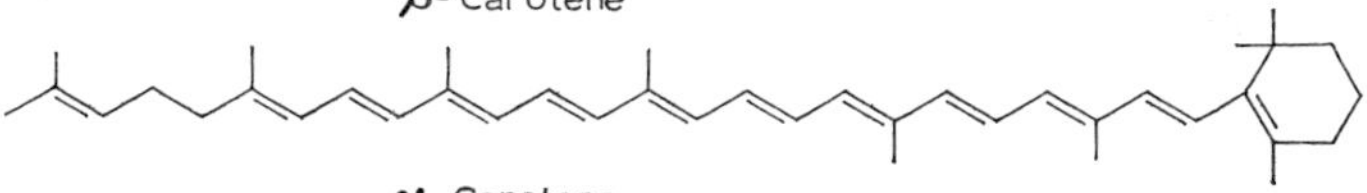
Lycopene 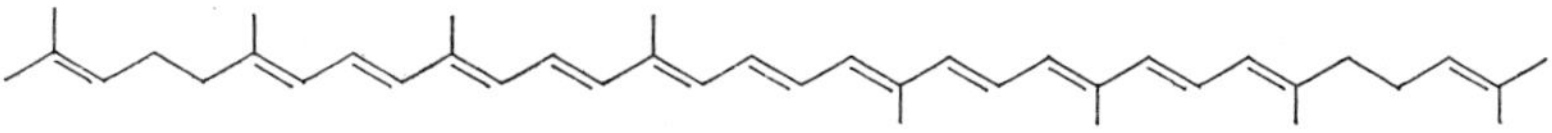 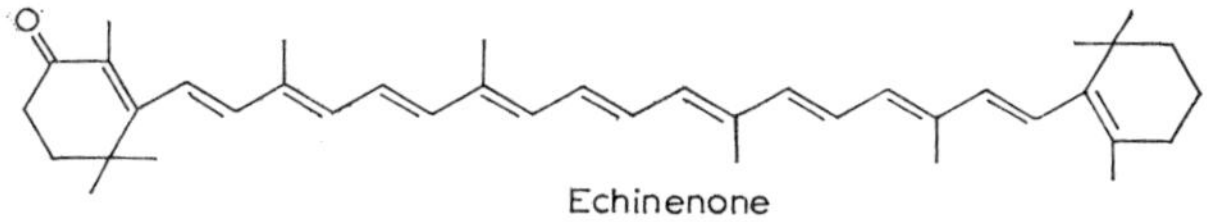
O
Echinenone 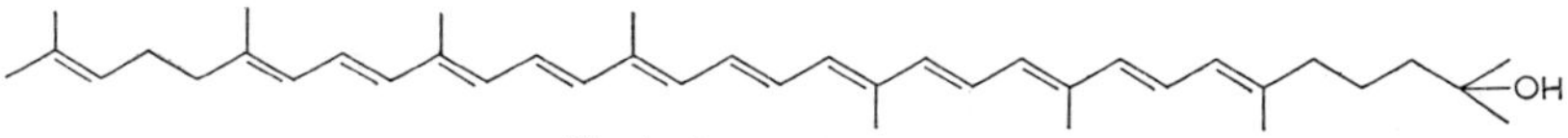
OH
Rhodopin 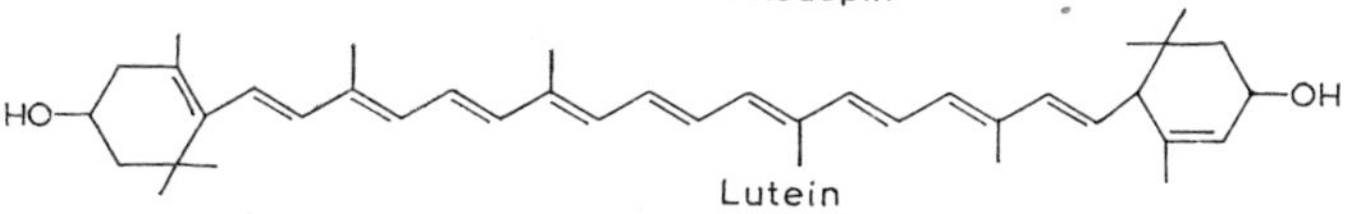
HO
OH
Lutein 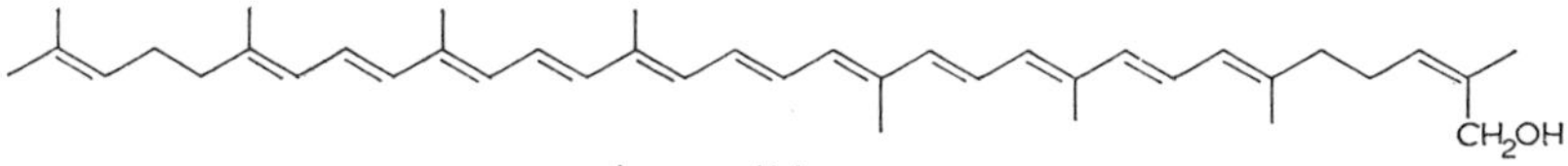
CH_2OH
Lycoxanthin 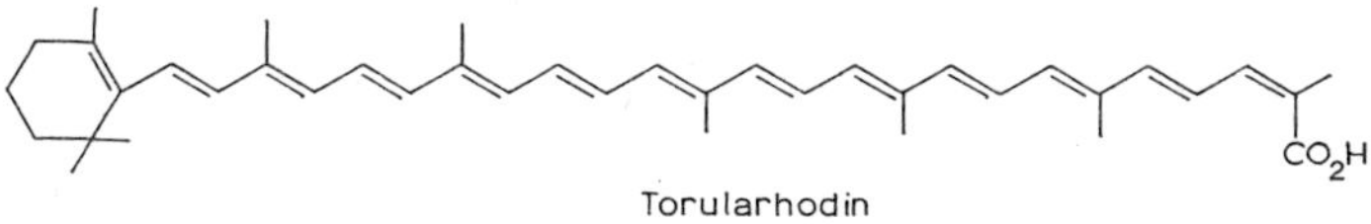
CO_2H
Torularhodin 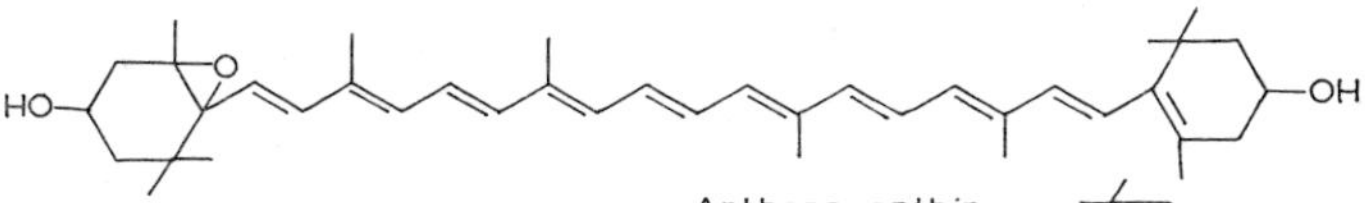
HO
O
OH
Antheraxanthin 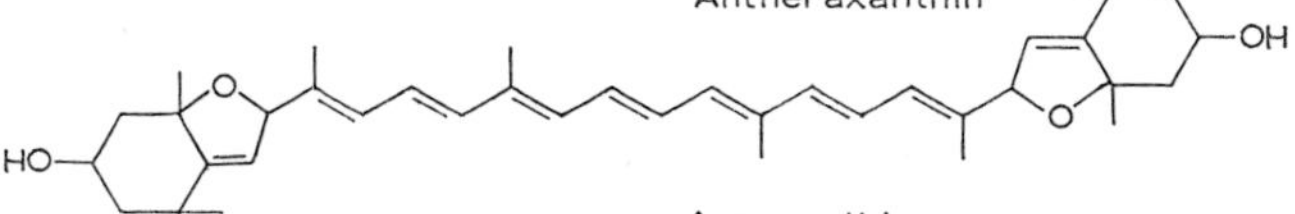
OH
O
HO
O
Auroxanthin

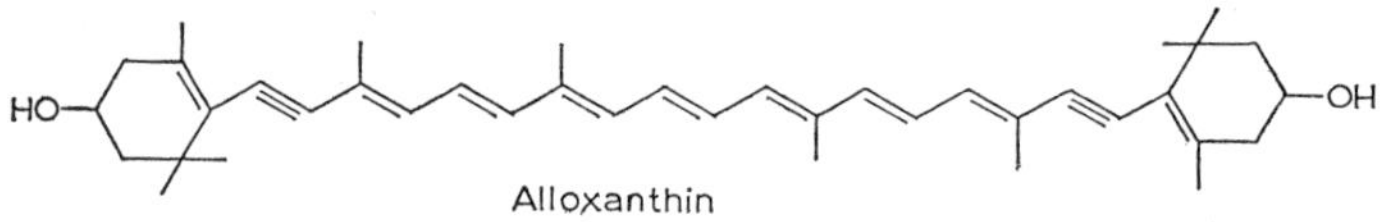

HO— —OH

Alloxanthin

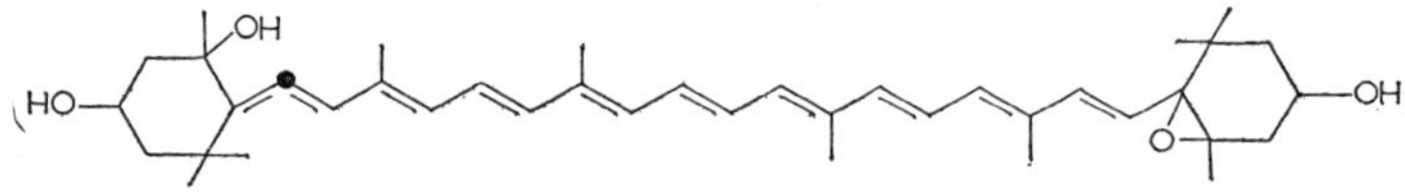

OH HO— —OH O

Neoxanthin

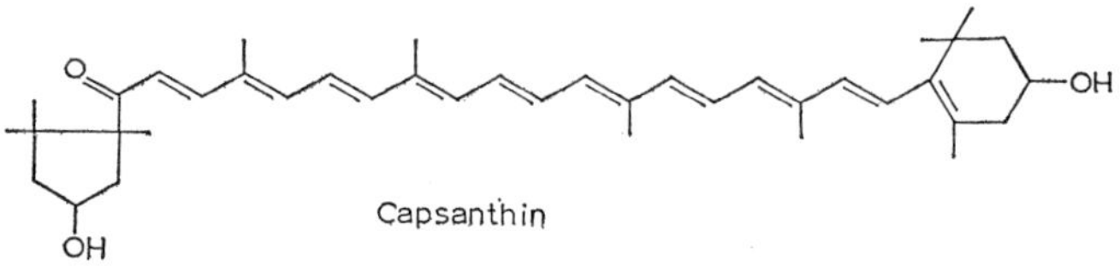

O —OH OH

Capsanthin

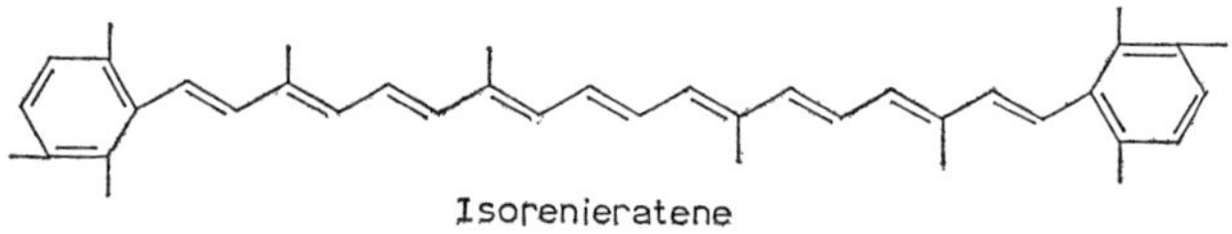

Isorenieratene

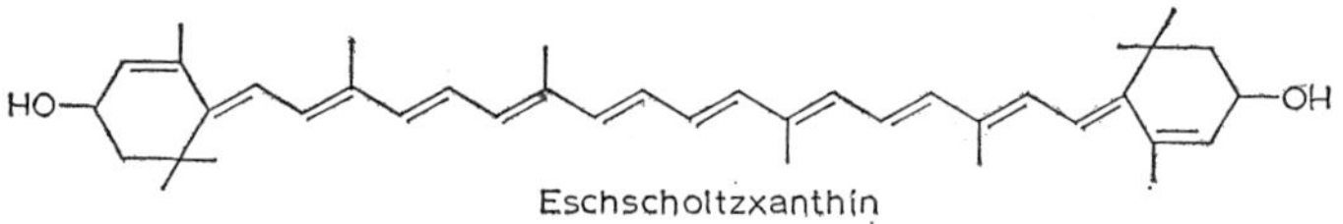

HO— —OH

Eschscholtzxanthin

Phytoene

T. W. Goodwin, "Chemistry and Biochemistry of Plant Pigments". Academic Press, London, 1965). Geranylgeranyl pyrophosphate is converted into phytoene by enzyme preparations from carrot roots (*F. B. Jungalwala* and *Porter*, Arch. Biochem., 1967, **119**, 209; *D. V. Shah et al.*, Arch. Biochem. Biophys., 1968, **127**, 124 and in preparations from a mutant of *Phycomyces blakesleeanus* (*T. Ching-Lee* and *C. O. Chichester*, Phytochemistry, 1969, **8**, 603); no NADPH is required, in contrast to squalene biosynthesis (*J. M. Charlton, K. J. Treharne* and *Goodwin*, Biochem. J., 1967, **105**, 205; *J. E. Graebe*, Phytochemistry, 1968, **7**, 2003). Experiments with [2-^{14}C-4*R*-$^{3}H_1$]-MVA and [2-^{14}C-4*S*-$^{3}H_1$]MVA demonstrated that the pro-4*S* hydrogen is stereospecifically eliminated from each of the eight MVA molecules incorporated into one phytoene molecule which indicates that all-*trans* geranylgeranyl pyrophosphate is the $C_{(20)}$ precursor (*R. J. H. Williams et al.*,

ibid., 1967, **104**, 767). The mechanism of phytoene biosynthesis must account for the fact that its central double bond is *cis*. Experiments with [2-^{14}C-5R-^{3}H$_1$]MVA indicate that the pro-R hydrogen at $C_{(1)}$ of geranylgeranyl pyrophosphate is retained at $C_{(15)}$ and $C_{(15')}$ in phytoene and a mechanism which takes this observation into consideration has been suggested (*Williams et al., loc. cit.*) (Scheme 23). In this scheme (based on that of *J. W. Cornforth et al.*, Proc. roy. Soc., 1966, **163B**, 492, for squalene biosynthesis) the geranylgeranyl residues are connected *via* a sulphonium ylide, which requires the presence of a thio-ether group (*e.g.* a methionine residue) at the active centre of phytoene synthase. The thio-ether group then displaces the pyrophosphate from a molecule of geranylgeranyl pyrophosphate by an S_N2 substitution reaction which involves inversion of configuration at $C_{(1)}$ of the geranylgeranyl group to give a sulphonium ion. The hydrogen atom at $C_{(1)}$, situated between a double bond and $S^{\oplus}$, now has a tendency to ionize with the formation of an ylide. This alkylates a second molecule of geranylgeranyl pyro-

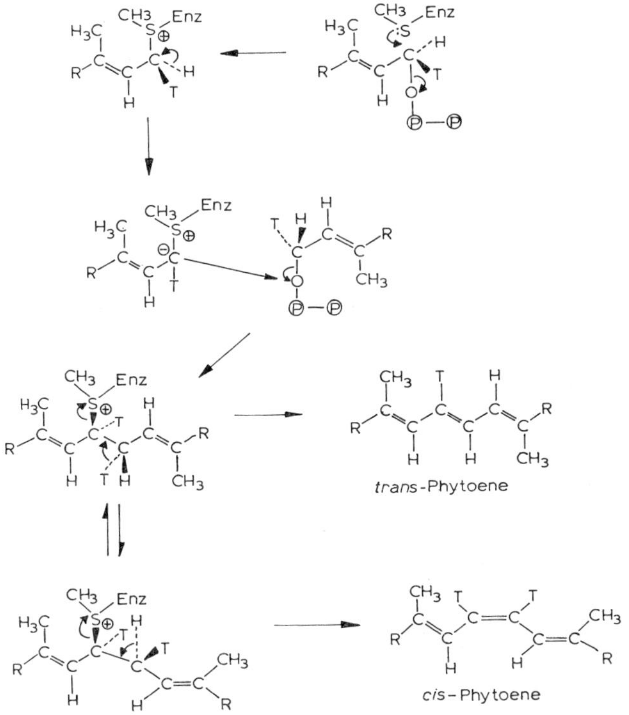

Scheme 23. Mechanism of phytoene formation from geranyl geranyl pyrophosphate.

phosphate [again with inversion at $C_{(1)}$] to give a lycopersylsulphonium ion (lycopersene is the C_{40} homologue of squalene). Finally, the incipient positive charge on the carbon atom from which the *S*-enzyme is lost is neutralized by loss of a proton from the adjacent methylene group, thus introducing the central double bond of phytoene. Both pro-*R* hydrogen atoms at the centre are retained only if the configuration of this double bond is *cis*.

The stepwise desaturation of phytoene to lycopene (Scheme 24) has considerable circumstantial support (see *e.g. Porter* and *Anderson*, 1967, *loc. cit.*); for example (a) the chemical structure of the proposed inter-

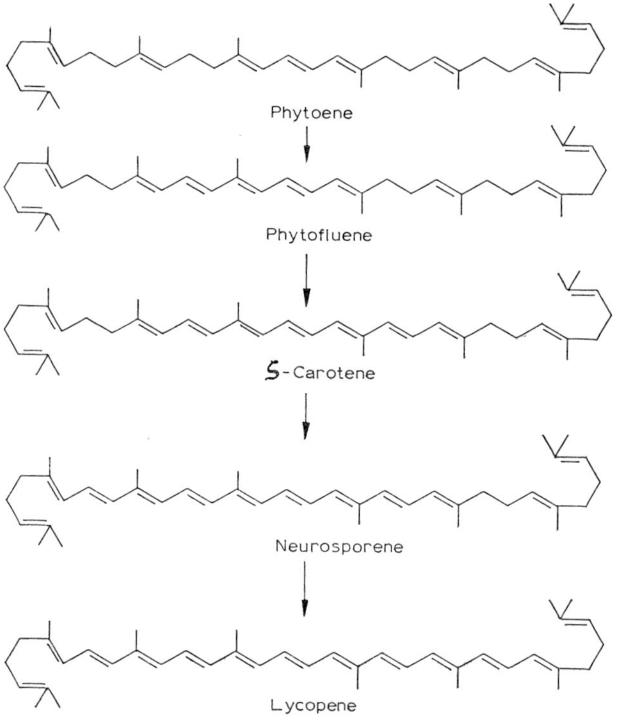

Phytoene

Phytofluene

ζ-Carotene

Neurosporene

Lycopene

Scheme 24. Conversion of phytoene into lycopene.
Phytoene and phytofluene are indicated in their *trans* configuration for ease of presentation.

mediates allow them to be arranged in a logical biogenetic sequence; (b) the intermediates accumulate in various mutants of algae, fungi and higher plants and in cultures in which carotenogenesis has been inhibited by compounds such as diphenylamine and (c) kinetic studies of synthesis of the fully unsaturated carotenes and disappearance of the most saturated

compounds are compatible with a precursor/product relationship. Although there are isolated claims for the enzymic conversion of phytoene into more unsaturated precursors there are no fully convincing reports of any conversion achieved with purified enzyme systems, and no indication of how the double bond which is *cis* in phytoene and phytofluene changes to *trans* at the ξ-carotene stage (Scheme 24).

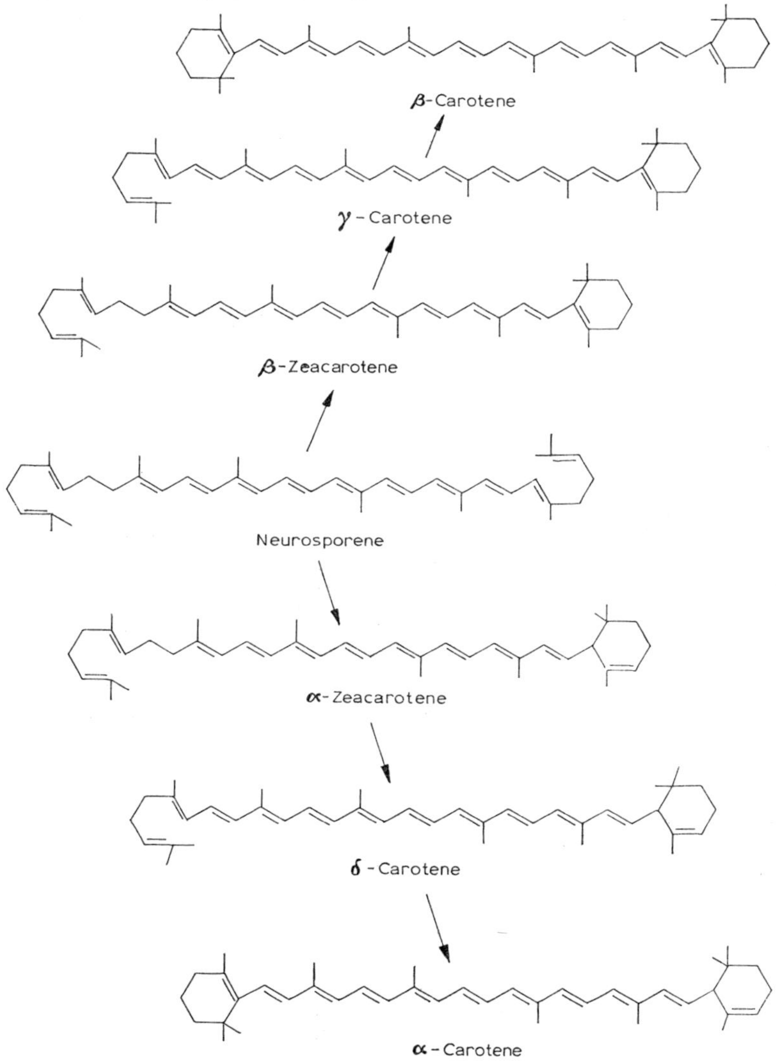

β-Carotene

γ-Carotene

β-Zeacarotene

Neurosporene

α-Zeacarotene

δ-Carotene

α-Carotene

Scheme 25. Conversion of neurosporene into α- and β-carotene.

(i) Formation of cyclic carotenoids

Cyclization of acyclic precursors takes place either at the neurosporene level or at the lycopene level (see Scheme 26). The neurosporene pathway requires the existence of α- and β-zeacarotenes (Scheme 25). These have been found in maize (*W. J. Rabourn* and *F. W. Quackenbush*, Arch. Biochem. Biophys., 1959, **44**, 159), and β-zeacarotene is present in certain *Rhodotorula* spp. (*K. L. Simpson et al.*, Biochem. J., 1964, **92**, 508) and in certain *Chlorella* mutants (*H. Claes*, Z. Naturforsch., 1958, **13b**, 222) and diphenylamine-inhibited micro-organisms (*Goodwin*, 1965 *loc. cit.*; *Williams et al.*, Phytochem., 1965, **4**, 759). Support for the lycopene pathway (Scheme 26) is based on reports of its conversion into cyclic carotenoids in chloroplasts (see *Porter* and *Anderson*, 1967, *loc. cit.*; *H. M. Hill* and *L. J. Rogers*, Biochem. J., 1969, **113**, 31P). However, whatever the immediate precursor, experiments with stereospecifically labelled MVA's have shown that the α- and β-ionone rings arise separately and that there is no interconversion of the two ring structures (*Williams et al.*, Biochem. J., 1967, **105**, 99). A mechanism consistent with the stereospecific results is given in Scheme 26.

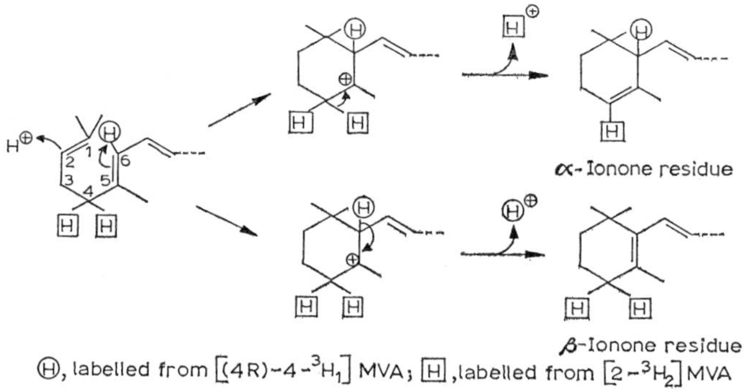

α-Ionone residue

β-Ionone residue

Ⓗ, labelled from [(4R)-4-$^3H_1$] MVA; [H], labelled from [2-$^3H_2$] MVA

Scheme 26. Mechanism for formation of α- and β-ionone rings of cyclic carotenes.

(ii) Insertion of oxygen

The weight of evidence supports the view that the insertion of oxygen into xanthophylls is a late step in the biosynthetic sequence. A mutant (5/520) of *Chlorella vulgaris* which synthesizes mainly phytoene when grown heterotrophically in the dark will on anaerobic illumination convert phytoene into coloured carotenes. If the cultures are then returned to darkness and allowed access to oxygen then xanthophylls are formed with the simultaneous disappearance of carotenes (*H. Claes*, Z. Naturforsch., 1958, **126**, 401). Anaerobic cultures of the non-sulphur purple photosynthetic bacterium

Rhodopseudomonas spheroides are yellowish-brown owing to the presence of spheroidene; on exposing the cultures to air they rapidly turn purplish red owing to the formation of spheroidenone (*C. B. van Niel*, Lieuwenhoek ned. Tijdschr., 1947, **12**, 156; *Goodwin et al.*, Biochem. J., 1956, **64**, 486; *J. B. Davis et al.*, Proc. chem. Soc., 1961, 261).

CH_3O

Spheroïdene

CH_3O

O

Spheroïdenone

Zeaxanthin can be stepwise and reversibly oxidized to antheraxanthin (5,6-epoxyzeaxanthin) and violaxanthin (5,6,5′,6′-diepoxyzeaxanthin). In *Euglena* the epoxidation reaction is considered to be non-enzymic, whilst the de-epoxidation is a dark enzyme reductive reaction (*N. Krinsky*, 1966, in "Biochemistry of Chloroplasts", Ed. *T. W. Goodwin*, Academic Press, London, Vol. 1, p. 377). In higher plants the reactions are considered to be photo-stimulated in both directions (Scheme 27) (*K. H. Lee* and *H. Y. Yamamoto*, Photochem. Photobiol., 1968, **7**, 101).

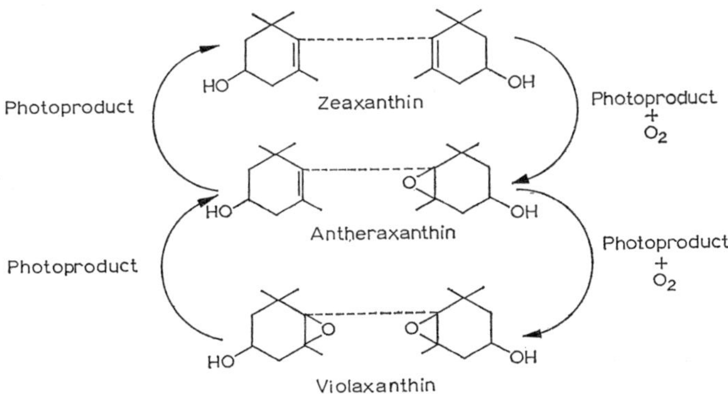

Scheme 27. Mechanism for epoxidation and de-epoxidation of xanthophylls in leaves.

Gaseous oxygen is the source of the oxygen in all these reactions as shown with experiments with $^{18}O_2$ and $H_2{}^{18}O$ on *Chlorella* [lutein (3,3′-dihydroxy-α-carotene) and zeaxanthin] (*Yamamoto et al.*, Arch. Biochem. Biophys., 1962, **97**, 168), on higher plants (antheraxanthin) (*Yamamoto* and *Chichester*, Biochim. Biophys. Acta, 1965, **100**, 303) and *Rsp. spheroides* (spheroidenone)

(*E. A. Shneour, ibid.*, 1962, **65**, 510). Furthermore, one oxygen atom of the carboxyl group of torularhodin comes from $^{18}O_2$ which is consistent with the reaction being catalysed by a mixed function oxidase (*Simpson et al.*, Biochem. J., 1963, **88**, 1688). Experiments with stereospecifically labelled MVA have shown that in maize seedlings lutein and zeaxanthin are not interconverted (*Goodwin et al.*, Plant Physiol., 1968 **43**, S 46) in agreement with the earlier observations on α- and β-carotene. Furthermore, in the insertion of the oxygen function an oxo compound is not involved, for only one hydrogen is lost from $C_{(3)}$; the hydrogen which is lost corresponds to the pro-*R* hydrogen of $C_{(5)}$ of MVA (*T. J. Walton et al.*, Biochem. J., 1969, **112**, 383). The mechanism of insertion of the carbonyl function into carotenoids such as spheroidenone and fucoxanthin is not known; however, in the case of okenone (*K. Schmidt et al.*, Arch. Microbiol., 1963, **46**, 117; *S. L. Jensen*, in "Biochemistry of Chloroplasts", Ed. *T. W. Goodwin*, 1966, vol. 1, p. 437), which is synthesized by an obligate anaerobe, the oxygen atom cannot arise

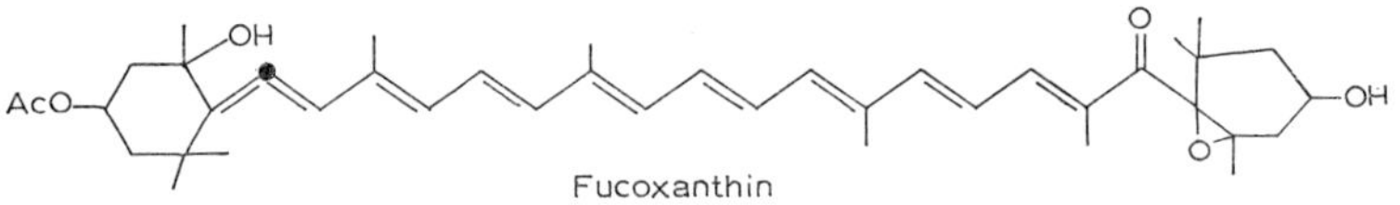

Fucoxanthin

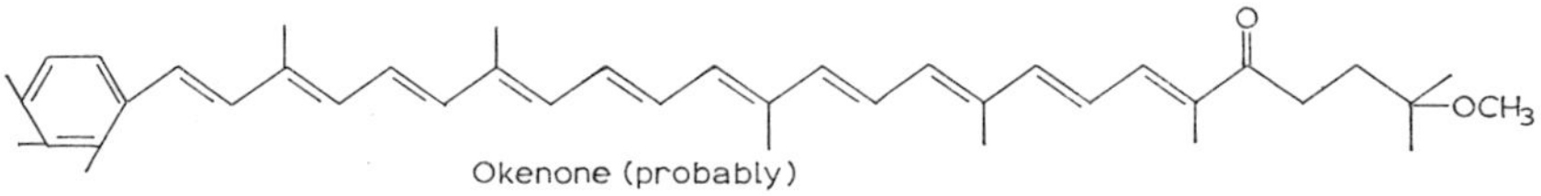

Okenone (probably)

from O_2. With echinenone the possibility is that it is formed by oxidation of the corresponding hydroxy-compound. 5,6-Epoxides are easily converted into 5,8-epoxides by traces of acids and the accumulation of acids in ripening fruit may account for the considerable amounts of 5,8-epoxides found in many fruit, especially citrus fruit (*Goodwin*, in *T. Swain*, Ed. "Comparative Phytochemistry", Academic Press, London, 1966).

The hydroxyl groups found in the photosynthetic bacteria would appear to have been formed by addition of water across a double bond (Scheme 28). The general pattern of synthesis in these organisms has been worked out from structural, kinetic, mutant and inhibitor studies (see *e.g. S. Liaaen-Jensen*, Ann. Rev. Microbiol., 1965, **19**, 163) but later investigations suggest that other pathways may arise which involve a new intermediate 7,8,11,12-tetrahydrolycopene (*B. H. Davies*, Biochem. J., 1970, **116**, 93). For example in *Rhodospirillum rubrum*, spheroidene, which is formed in the presence of the inhibitor diphenylamine, is probably formed as shown on p. 130.

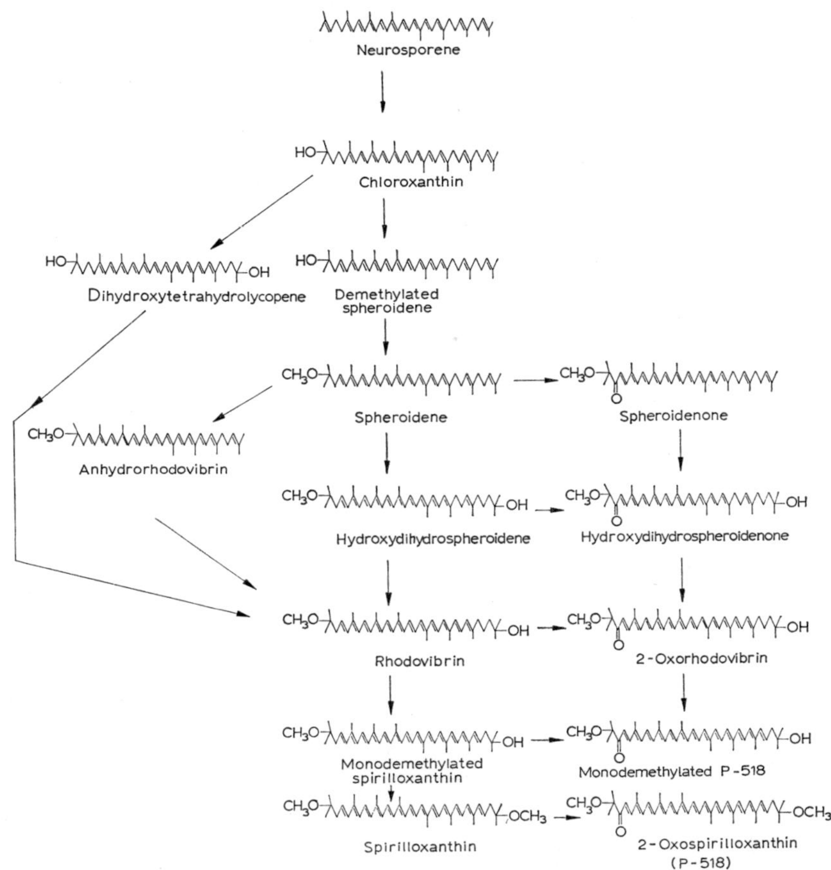

Scheme 28. Probable pathway for biosynthesis of carotenoids in photosynthetic bacteria.

The existence of methylated carotenoids is confined to the photosynthetic bacteria. The methyl groups arise by conventional 1-C transfer (*G. D. Braithwaite* and *Goodwin*, Nature, 1958, **182**, 1304) and incorporation of the methyl group of *S*-adenosylmethionine into spirilloxanthin has been achieved in isolated chromatophores from *Rhodospirillium rubrum* (*L. D. Beckman et al.*, Plant Physiol., 1965, **40**, 54).

A possible mechanism for the synthesis of the retro-carotenoid eschscholtzxanthin, indicated in Scheme 29, is supported by experiments with [2-^{14}C-4R-$^{3}H_1$]MVA in which the tritium atoms present in β-carotene and zeaxanthin are retained in eschscholtzxanthin (*Williams et al.*, Biochim. Biophys. Acta, 1966, **124**, 200). Plausible mechanisms for the synthesis of

the allenic grouping of neoxanthin and for the pentacyclic ring of capsanthin are indicated in Scheme 30. No experimental evidence for these mechanisms is yet available. Neither is anything known of the formation of the aromatic carotenoids, isorenieratene (≡ leprotene) (*Liaaen-Jensen*, Acta Chem. Scand., 1964, **18**, 1562), renieratene (p. 132) and chlorobactene found in the green

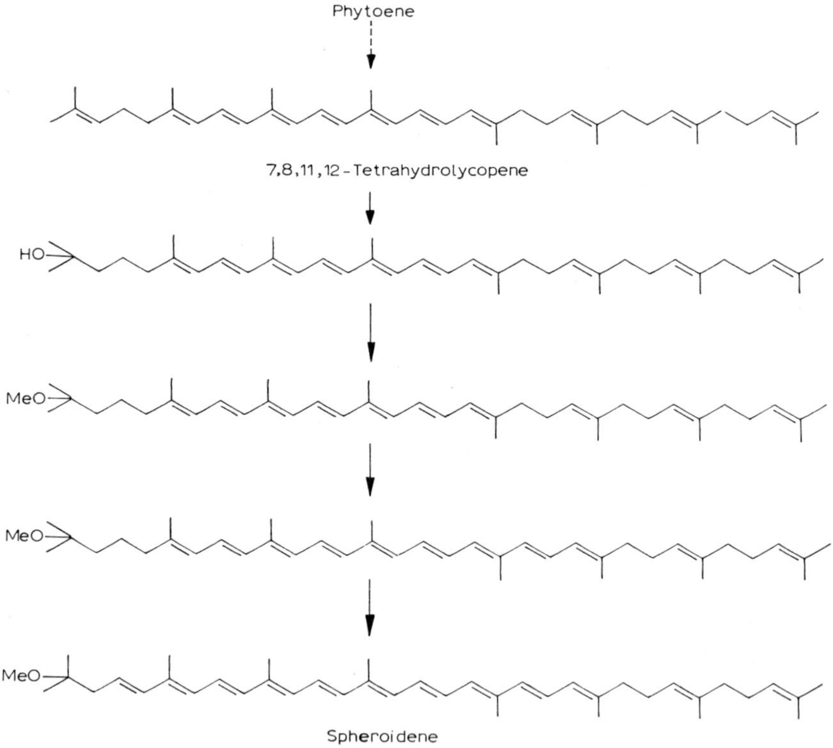

Phytoene

7,8,11,12-Tetrahydrolycopene

HO

MeO

MeO

MeO

Spheroidene

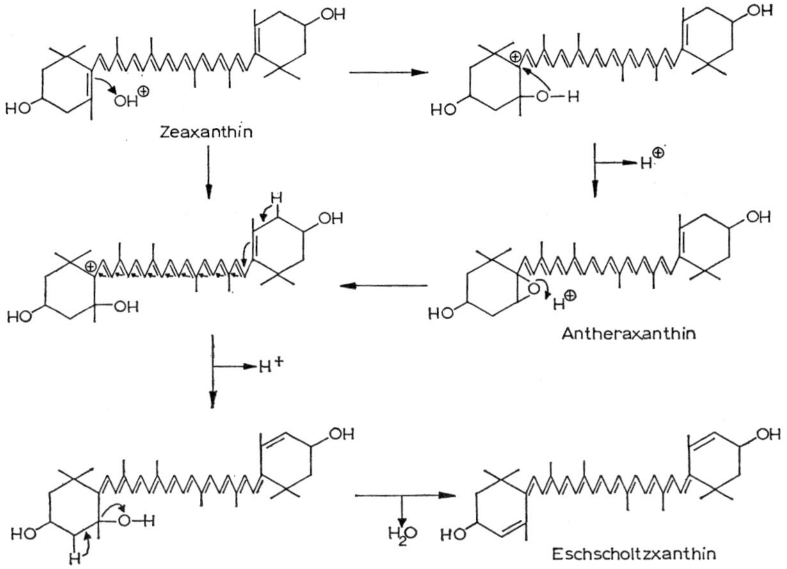

Scheme 29. Mechanism for the biosynthesis of eschsholtzxanthin.

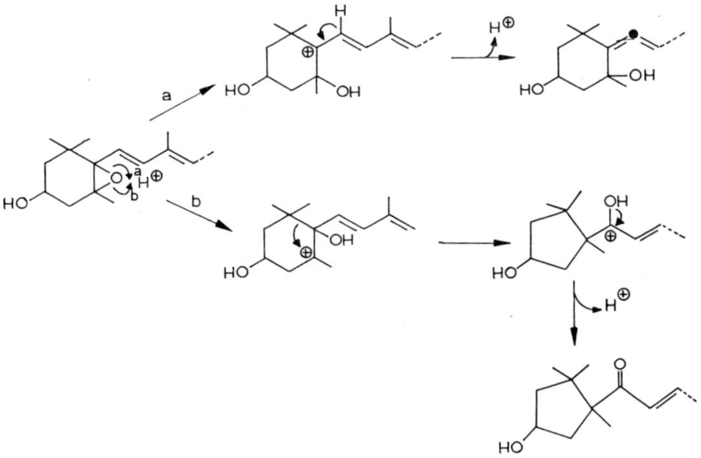

Scheme 30. Possible mechanisms for the biosynthesis of neoxanthin and capsanthin.

photosynthetic bacteria (*Chlorobacteriaceae*) (*idem, ibid.*, 1964, **18**, 1703):

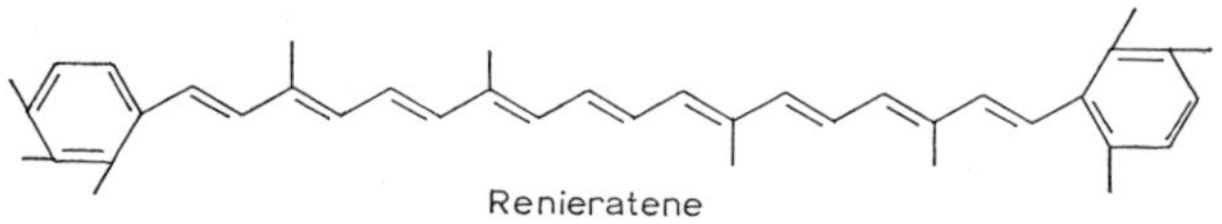

Renieratene

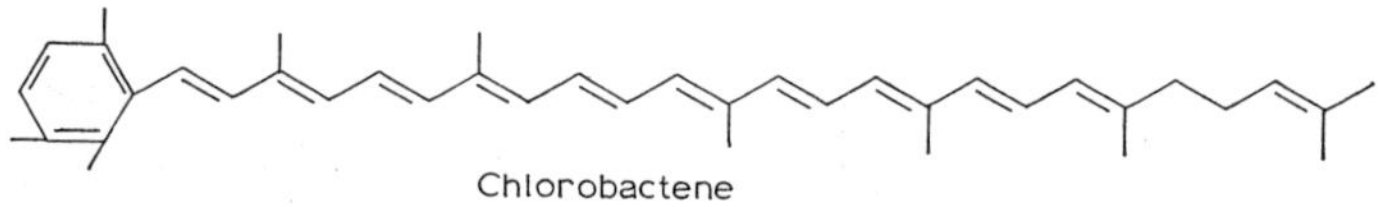

Chlorobactene

or of how the acetylenic linkages are formed in compounds such as alloxanthin (p. 122). The C_{45} carotenoid (nonaprenoxanthin) (*O. B. Weeks et al.*, Nature, 1969, **217**, 224) and the C_{50} carotenoid (decaprenoxanthin) from *Flavobacterium dehydrogenans* and other gram-positive bacteria (*Liaaen-Jensen et al.*, Nature, 1967, **214**, 379; *Weeks et al., loc. cit.*) also pose interesting biosynthetic problems.

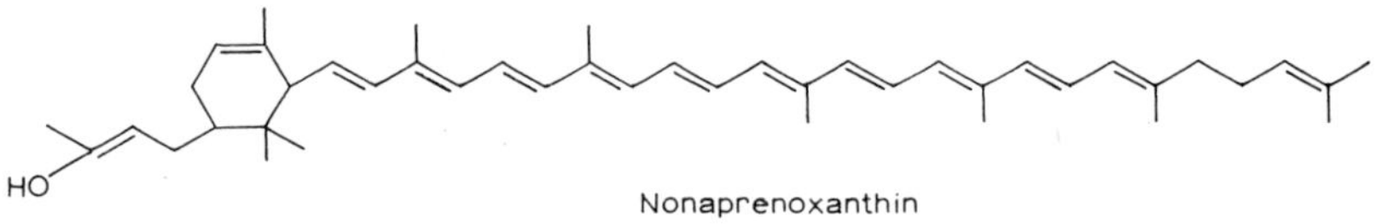

HO

Nonaprenoxanthin

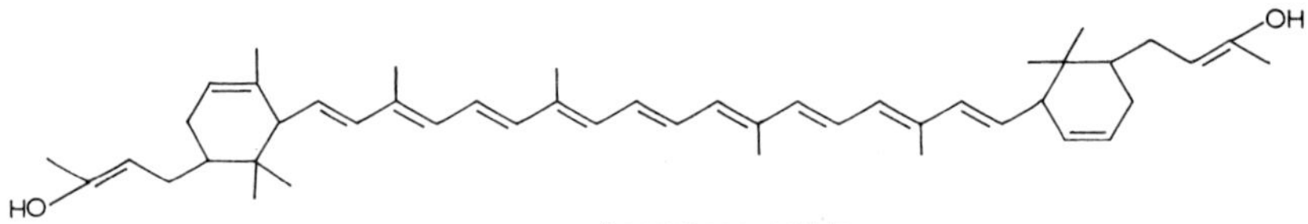

OH

HO

Decaprenoxanthin

(iii) Degradation products of β-carotene

A number of apo-carotenoids which occur in traces in plants (*A. Winterstein, A. Studer* and *R. Rüegg*, Ber., 1960, **93**, 2951) are almost certainly degradation products of C_{40} carotenoids formed by unilateral oxidation. A typical example is β-citraurin (3-hydroxy-β-apo-8′-carotenal). Perhaps the most important example of a "degraded" carotenoid is vitamin A which can formally be considered as resulting from the hydrolytic cleavage of β-carotene at its central double bond. Intensive study has not yet revealed the exact mechanism involved biologically (see *J. Glover*, Ann. Reports, Chem. Soc., 1960, **56**, 331).

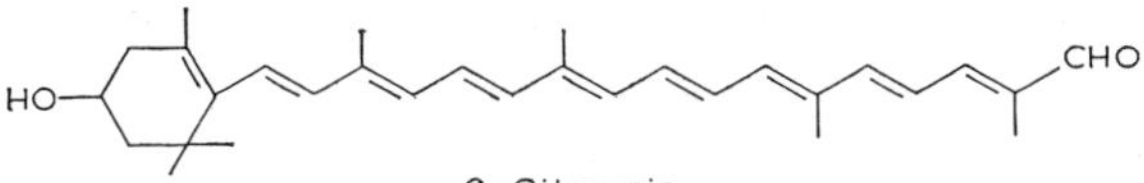

β-Citraurin

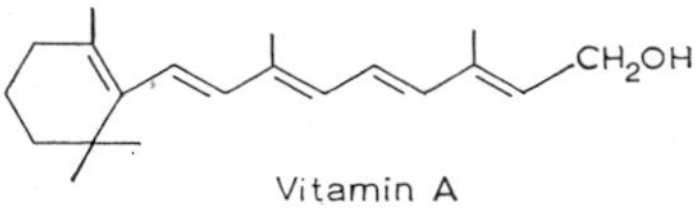

Vitamin A

Bixin from *Bixa orellana* and crocetin from saffron appear to result from bilateral oxidation of a C_{40} carotenoid, but no experimental evidence is available to support this view.

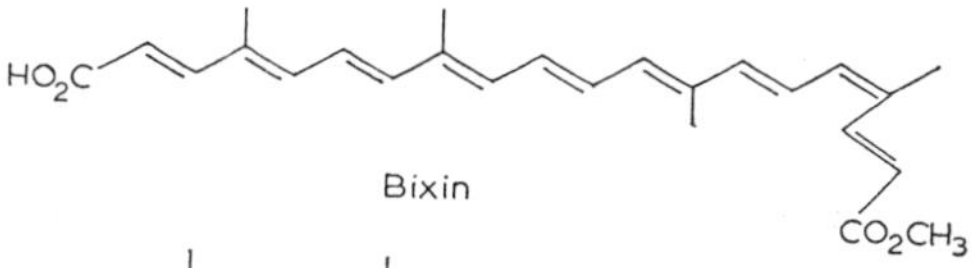

Bixin

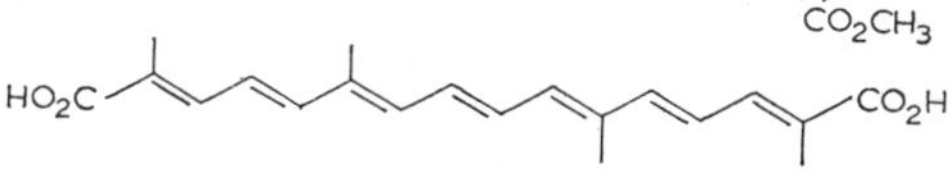

Crocetin

10. Polyterpenoids

Polyprenols

Long chain polyprenols with 6–13 isoprene residues with mixed *cis* and *trans* configurations (castaprenols, betulaprenols and ficaprenols) occur in leaves mainly in the chloroplasts (*F. W. Hemming*, in "Terpenoids in Plants", Ed. *J. B. Pridham*, Academic Press, London, 1967). Solanesol (9 isoprene residues, all in *trans* configuration) is also widespread in plants in traces but occurs in large amounts in tobacco leaves (*R. L. Rowland et al.*, J. Amer.

Solanesol

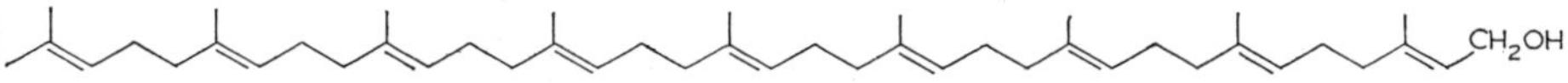

chem. Soc., 1956, **78**, 4680), where it is present mainly in the chloroplast (*J. Stevenson et al.*, Biochem. J., 1963, **88**, 52; *W. T. Griffiths et al*, European J. Biochem., 1968, **5**, 124). A C_{50} homologue is present in *Arum* spadices (*Stevenson et al.*, 1963, *loc. cit.*). [2-^{14}C]MVA is incorporated into these compounds only with difficulty because of the comparative impermeability of the chloroplast outer membrane to the substrate (*Hemming*, 1967 *loc. cit.*). Even longer polyprenols with mixed *cis-trans* configuration are found in

fungi and animal tissues. They contain from 13 to 23 isoprene residues and are named dolichols, after the C_{100} compound isolated from liver (*Hemming*, 1967 *loc. cit.*). Recently a prenol diphosphogalactose has been implicated in the biosynthesis of antigenic lipopolysaccharide in *Salmonella newingtonii* (*A. Wright et al.*, Proc. natl. Acad. Sci. Wash., 1967, **57**, 1798) and a similar compound is concerned in peptidoglycan biosynthesis in *Micrococcus lysodeikticus* (*Y. Higashi et al.*, *ibid.*, 1967, **57**, 1878). In polymannan biosynthesis in *M. lysodeikticus* a prenol monophosphomannose is involved (*M. Scher et al.*, *ibid.*, 1968, **59**, 1313). In all cases it is a prenol with 11 isoprene residues which is present. Experiments with stereospecifically labelled MVA confirmed the mixed *cis-trans* biogenetic origin (*K. J. Stone* and *Hemming*, Biochem. J., 1967, **104**, 43; 1968, **109**, 877).

11. Terpenoid quinones

The major quinones with long isoprenoid side chains are ubiquinones, plastoquinones (*e.g.* plastoquinone A, which is the major component), the plastoquinones B, which have a hydroxyl group esterified with a fatty acid in the side chain (*W. T. Griffiths*, Biochim. Biophys. Acta, 1966, **25**, 569), PQC's which are chemically related to PQA but have one hydroxy substituent in the side chain and PQZ's which have more than one substituent in the side chain (*J. C. Wallwork* and *J. F. Pennock*, Proc. int. Cong. Photosyn. Research, 1968, in press); phylloquinone (vitamin K_1) and the tocopherol quinones (*e.g.* α-tocopherolquinone) have already been discussed (p. 87) (see *Pennock* in "Terpenoids in Plants", Ed. *J. B. Pridham*, Academic Press,

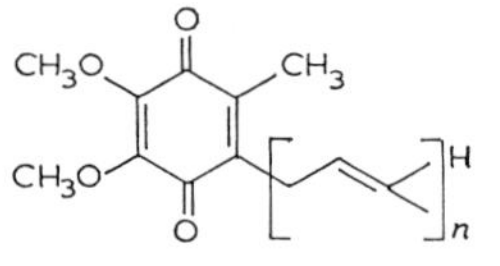

Ubiquinones

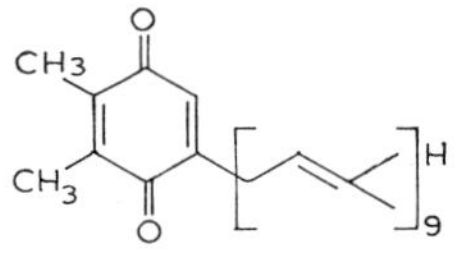

Plastoquinone A

London, 1967). Related compounds include the chromenols derived from the tocopherols (Vitamin E, α-tocopherol) and the chromanol, plastochromanol (*K. J. Whittle et al.*, Biochem. J., 1965, **96**, 17C).

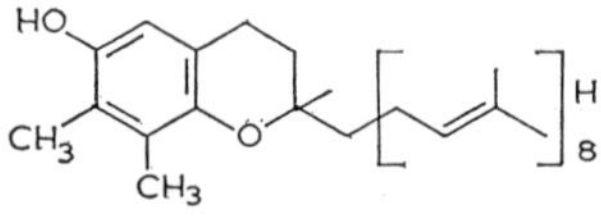

Plastochromanol

As expected the isprenoid side-chain is mevalonate-derived (see *D. R. Threlfall*, in "Terpenoids in Plants," Ed. *J. B. Pridham*, Academic Press, London, 1967), but this is frequently difficult to demonstrate because all the quinones mentioned except ubiquinone are synthesized in chloroplasts which, as indicated previously, are relatively impermeable to exogenous MVA. Experiments with stereospecifically labelled MVA showed that in *Aspergillus fumigatus* the side chain of ubiquinone is biogenetically all-*trans* (*Stone* and *Hemming*, Biochem. J., 1965, **96**, 14C).

Rubber

Rubber is a polyterpenoid with a molecular weight which varies from 20,000 to several millions according to source and location in the plant. It is present in small amounts in many dicotyledons but is absent from monocotyledons and gymnosperms (see *B. Arreguin*, In "Handbuch der Pflanzenphysiologie", Vol. X, Springer, Heidelberg 1958). Rubber from *Hevea* spp. is an all *cis*-isomer (I), whilst gutta percha is an all-*trans* isomer (II).

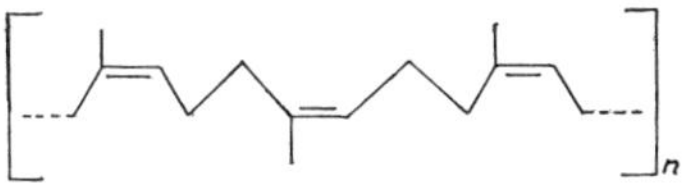

(I) 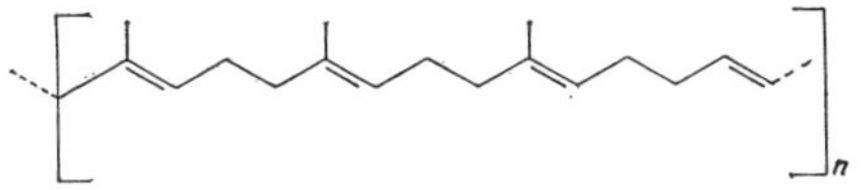 (II)

[2-^{14}C]MVA is effectively incorporated into rubber by isolated rubber latex (see *e.g. B. L. Archer* and *B. G. Audley*, Adv. Enzymol., 1967, **29**, 221). Experiments with [2-^{14}C-4*S*-$^{3}H_1$]MVA and [2-^{14}C-4*R*-$^{3}H_1$]MVA show that the pro-*S* hydrogen is retained in rubber and the pro-*R* lost, thus demonstrating that rubber is biogenetically *cis*, and is not formed as a *trans* compound followed by isomerization (*Archer et al.*, Proc. roy. Soc., 1966, **B163**, 359).

12. Regulation of terpenoid synthesis in seedlings

Ungerminated seeds contain sterols, but insignificant amounts of carotenoids and other terpenoid derivatives which are concerned in photosynthesis, including the chlorophylls (phytyl side chain), plastoquinones, phylloquinone (vitamin K_1) and tocopherol quinones. If the seedlings are germinated in the dark, sterols are synthesized and traces of highly oxidized carotenoids appear. If the dark-grown seedlings are illuminated then there is a massive synthesis of carotenoids, chlorophylls, plastoquinone, phylloquinone and tocopherol quinones (*i.e.* of the "photosynthetic terpenoids") but no corresponding stimulation of steroid synthesis. Thus on illumination

there is a channelling of isoprenoid precursors away from sterol synthesis to the synthesis of photosynthetic terpenoids.

If, for example, etiolated maize seedlings are excised from their roots, placed in a solution of [2-^{14}C]mevalonate and illuminated, the photosynthetic terpenoids are only weakly labelled, although they are rapidly synthesized during this period; on the other hand, squalene sterols and the pentacyclic triterpenoid, β-amyrin, are strongly labelled (*T. W. Goodwin*, Biochem. J., 1958, **70**, 612; and *Goodwin* and *E. I. Mercer*, in "The Regulation of lipid metabolism", Academic Press, London, 1963). On the other hand, if a similar experiment is carried out in the presence of $^{14}CO_2$, the photosynthetic terpenoids are strongly labelled whilst the sterols are only slightly labelled.

These, and other similar observations, have led to the view that it is possible that this regulation of biosynthesis is achieved by a combination of compartmentalization and enzyme segregation, in which sterols and pentacyclic triterpenoids are synthesized extraplastidically and the "photosynthetic terpenoids" are formed in the chloroplast (*Goodwin* and *Mercer*,

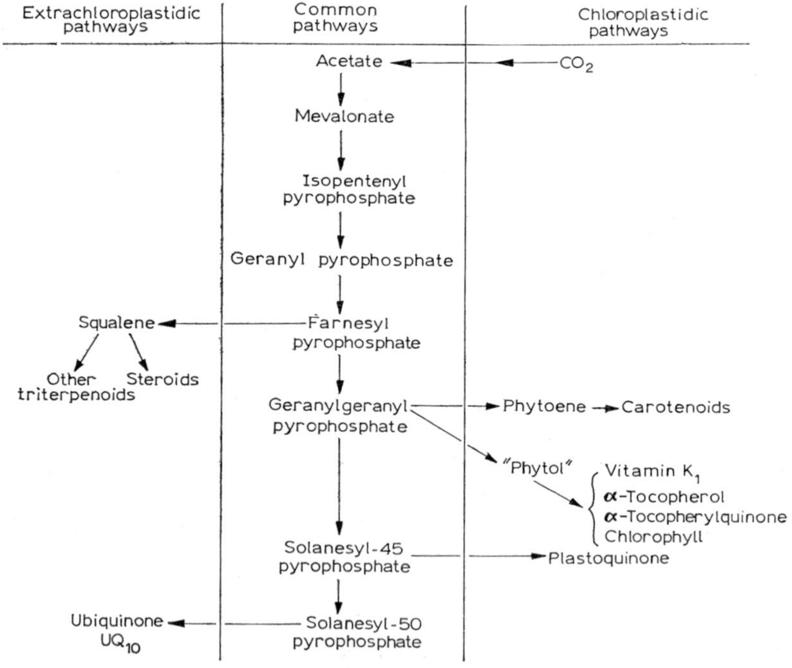

Scheme 31. Possible mechanism for the synthesis of terpenoids inside and outside the chloroplast.

1963, *loc. cit.*). Both sites of synthesis (chloroplastidic and extrachloroplastidic) are regarded as having a common basic biosynthetic backbone by which acetyl-CoA (derived either from storage material in the seed or from carbon dioxide by photosynthesis) is converted *via* mevalonate and IPP into geranyl, farnesyl, geranylgeranyl, solanesyl-45 and solanesyl-50 pyrophosphates (Scheme 31). Reactions specific to the extrachloroplastidic region of the cell are the formation of squalene and its conversion into cholesterol and other triterpenes (p. 94) and the incorporation of solanesol-50 into ubiquinone-50. Specific chloroplast reactions appear to implicate geranylgeranyl pyrophosphate; in the first reaction it dimerizes tail to tail to form phytoene, the C_{40} precursor of the carotenoids (p. 123), and in the second it is probably incorporated into the terpenoid side chains of the chlorophylls, tocopherol quinones, phylloquinone, presumably *via* an "active phytol" (? phytol pyrophosphate) as an intermediate; finally, the C_{45} side chain is incorporated into plastoquinone.

Evidence for this hypothesis includes (1) the presence of mevalonic acid kinases both inside and outside the chloroplast (*L. J. Rogers et al.*, Biochem. J., 1966, **96**, 7P; **100**, 14C): (2) the ability of chloroplasts isolated by non-aqueous methods to synthesize phytoene actively from MVA whilst only little squalene is synthesized (*J. M. Charlton et al.*, *ibid.*, 1967, **105**, 205); on the other hand crude extracts of plant tissue make large amounts of squalene and undetectable amounts of phytoene: (3) plastoquinone synthesized from carbon dioxide has the same specific activity in the carbons of the aromatic ring as in those of the side chain (*Threlfall et al.*, 1968, *loc. cit.*); they would be expected to be different if the side chain were coming from outside the chloroplast.

GUIDE TO THE INDEX

This index is constructed in a similar manner to the volume indexes of the first edition of the Chemistry of Carbon Compounds. However, to make the index easier to use, more descriptive entries have been made for the commonly occurring individual, and groups of chemicals.

The indexes cover primarily the chemical compounds mentioned in the text, and also include reactions and techniques, where named, and some sources of chemical compounds such as plant and animal species, oils, etc.

Chemical compounds have been indexed alphabetically under the names used by authors, editing being restricted to ensuring uniformity of entries under the same heading. In view of the alternative nomenclature that can often be used, a limited amount of cross-referencing has been done where it is considered to be helpful, but attention is particularly drawn to Convention 2 below.

For this and the succeeding volumes, the indexing conventions listed below have been adopted.

1. *Alphabetisation*

(a) The following prefixes have not been counted for alphabetising:

n-	*o-*	*as-*	*meso-*	D	*C-*
sec-	*m-*	*sym-*	*cis-*	DL	*O-*
tert-	*p-*	*gem-*	*trans*	L	*N-*
	vic-				*S-*
		lin-			*Bz-*
					Py-

Some prefixes and numbering have been omitted in the index, where they do not usefully contribute to the reference.

(b) The following prefixes have been alphabetised:

Allo	Epi	Neo
Anti	Hetero	Nor
Cyclo	Homo	Pseudo
	Iso	

(c) A letter by letter alphabetical sequence is followed for entries, firstly for the main entry, followed by the descriptive entry. The only exception

to this sequence is the placing of plural entries in front of the corresponding individual entries to prevent these being overlooked by a strict alphabetical sequence which could lead to a considerable separation of plural from individual entries. Thus "butanes" will come before *n*-butane, "butenes" before 1-butene, and 2-butene, etc.

2. *Cross references*

In view of the many alternative trivial and systematic names for chemical compounds, the indexes should be searched under any alternative names which may be indicated in the main body of the text. Only a limited amount of cross-referencing has been carried out, where it is considered that it would be helpful to the user.

3. *Esters*

In the case of lower alcohols esters are indexed only under the acid, *e.g.* propionic methyl ester, not methyl propionate. Ethyl is normally omitted *e.g.* acetic ester.

4. *Derivatives*

Simple derivatives are not normally indexed if they follow in the same short section of the text.

5. *Collective and plural entries*

In place of "— derivatives" or "— compounds" the plural entry has normally been used. Plural entries have occasionally been used where compounds of the same name but differing numbering appear in the same section of the text.

6. *Main entries*

The main entry of the more common individual compounds is indicated by heavy type. Where entries relate to sections of three pages or more, the page number is followed by "ff".

7. *Italicised entries*

Italicised entries other than names of species refer to the Appendix (I.U.P.A.C. Tentative Rules for Nomenclature of Steroids).

INDEX

Abelmoschus moschatus, C 289
Abienol, C 382
Abies balsamea oil, C 188
Abieslactone, C 422
Abies sachalinensis, C 293
Abies sibirica, C 193
Abietadiene, C 378
Abietanes, physical properties, C 392
Abietane group, C 378
Abietic acid, B 134; C 378, 379, 392; E 86
Abnormal guaianolide, C 357, 362
Abobioside, D **405**
Abogenin, D **405**
Abomonoside, D **405**
Aboranes, E 116
Aborescin, C **354**
Abscisic acid, E 84
Absorption spectra, carotenoids, B 253
Acacia, D 166
Acacic acid, C 440
Acenaphthene, C 74
17-Acetamido-3β-acetoxyandrost-5-ene, D 357
17α-Acetamido-3β-acetoxyandrost-5-ene-17β-carboxylic acid, D 358
17β-Acetamido-3β-acetoxyandrost-5-ene-17α-carboxylic acid, D 358
N-Acetamidoadamantane, C 131
1-Acetamido-3-hydroxyadamantane, C 127
m-Acetamidophenol, hydrogenation, B 83
o-Acetamidophenol, hydrogenation, B 82
γ-Acetamidotropolone, B 372
Acetates, pyrolysis, B 18
Acetic acid, B 75
—, complexes with deoxycholic acid, D 176
Acetoacetic acid, B 246
Acetoacetic ester, reaction with αω-dibromoalkanes, A 9
—,— with dihalides, A 22, 104
—,— with formaldehyde, B 144
[^{13}C]Acetoacetic ester, E 96
Acetoacetyl-ACP, E 57
Acetobacter suboxydans, B 60, 106
α-Acetobromorhamnose, D 382
γ-Aceto-α-isopropylbutyric acid, 316
Acetone, C 158, 259, 261
—, from pulegone, B 199
—, reaction with cyanoacetic ester, A 73
Acetonedicarboxylic ester, A 187, 201 B 151
Acetonitrile, C 131
Acetonylacetone, A 108
Acetophenone, B 3
Acetothiokinase, E 57
o-Acetotoluidide, hydrogenation, B 81
α-Acetoxyacrylonitrile, C 95
3β-Acetoxyallopregn-16-en-20-one, D 307
17-Acetoxyandrostanes, D 260
3β-Acetoxyandrostane-17-one, D 357
10β-Acetoxyandrost-4-ene-3,17-dione, D 204
3β-Acetoxyandrost-5-en-17-one, D 83
Acetoxybalchanolide, C 368
5-Acetoxybicyclo[2.2.1]heptane 2,3-dione, C 85
5-Acetoxybicyclo[2.2.1]heptan-2-one, C 94
Acetoxybicyclo[2.1.1]hexane carboxylic acids, C 85
2-Acetoxybicyclo[2.2.2]-oct-5-ene, C 111
1-Acetoxybicyclopropyl, C 7
3α-Acetoxybisnorchol-11-enic esters, D 313
3β-Acetoxybisnorchol-11-enic esters, D 313
3β-Acetoxybisnorchol-5-enoyl-chloride, D 109
2-Acetoxy-5-bromocyclopentanone, A 166
3-Acetoxy-5α-bromo-6β-hydroxy-pregnan-20-one, D 244
17β-Acetoxy-3-bromo-4-methoxy-A-homoestra-2,4,5(10)-triene, D 228
3α-Acetoxy-11α-bromo-12-oxochol-anoic methyl ester, D 188
3α-Acetoxy-11β-bromo-12-oxo-5β-cholanoic methyl ester, D 24
1-Acetoxybutadiene, B 90, 147, 150

3β-Acetoxy-5α-carda-14,16,20(22)-trienolide, D 397
3β-Acetoxy-5α-cholanoic acid, D 25
3β-Acetoxychol-5-en-22-al, D 154
3α-Acetoxychol-11-enic esters, D 313
3β-Acetoxychol-11-enic esters, D 313
3α-Acetoxychol-11-enic methyl ester, D 314
3α-Acetoxychol-9(11)-enic methyl esters, D 315
3α-Acetoxy-5β-chol-11-enoic acid, D 23
3α-Acetoxy-5β-chol-11-enoic methyl ester, D 55
3β-Acetoxychol-5-enoyl chloride, D 152, 158
3-Acetoxycholesta-3,5-diene, D 126, 129
2α-Acetoxy-5α-cholestan-3-one, D 119
3β-Acetoxy-5α-cholestan-6-one, D 17, 88
3β-Acetoxy-5α-cholestan-7-one, D 88
3β-Acetoxycholest-4-enes, D 105, 115
4-Acetoxycholest-5-enes, D 44
2α-Acetoxycholest-4-en-3-one, D 34, 127
3α-Acetoxycholest-5-en-7-one, D 42
3β-Acetoxycholest-4-en-7-one, D 109
4α-Acetoxycholest-5-en-3-one, D 125
5-Acetoxy-5β-cholest-3-en-7-one, D 109
17β-Acetoxy-6-chloro-9β,10α-androsta-4,6-dien-3-one, D 341
17β-Acetoxy-6-chloro-9β,10α-androsta-1,4,6-trien-3-one, D 341
17α-Acetoxy-6-chloro-1,6-bisdehydroprogesterone, D 347
17α-Acetoxy-6α-chloro-1-dehydroprogesterone, D 346, 347
17α-Acetoxy-5-chloro-6-dehydroprogesterone, D 347
17α-Acetoxy-6-chloropregna-1,6-diene-3,20-dione, D 240
3β-Acetoxy-5β,19-cycloandrost-6-en-17-one, D 335
6-Acetoxy-3α,5-cyclo-5α-cholest-6-ene, D 80
3-Acetoxycyclohexanone, B 105
4-Acetoxycyclohexanone, D 147
3-Acetoxycyclohex-1-ene, B 42
3-Acetoxycyclohex-4-ene-1,2-dicarboxylic anhydride, B 147
3-Acetoxycyclo-octene, B 403
Acetoxycyclopentadiene, C 93
2-Acetoxycyclopentanone, A 170
17β-Acetoxy-2α,3α-dibromomethylene 3β-methoxyestr-5(10)-ene, D 228
17β-Acetoxy-2β,3β-dibromomethylene-3α-methoxyestr-5(10)-ene, D 228
3β-Acetoxy-7α,11α-dibromo-8,9-oxido-5α,8α-lanostane, D 71
17α-Acetoxy-3α,4α-dichloro-methylenepregn-5-en-20-one, D 351
17α-Acetoxy-3α,4α-difluoro-methylenepregn-5-en-20-one, D 351
17α-Acetoxy-6,6-difluoroprogesterone, D 348, 349
21-Acetoxy-17α,20β-dihydroxypregn-4-en-3-one, D 311
2-Acetoxy-2,6-dimethyl-cyclohexa-3,5-dien-1-one, B 122
3β-Acetoxy-5,7-dioxo-5,6-secoandrostane-6-carboxylic acid, D 284
3β-Acetoxy-14β,15β-epoxy-5α-carda-16,20(22)-dienolide, D 397
3β-Acetoxy-14β,15β-epoxy-5β-card-20(22)-enolide, D 390
(25*R*)-11α-Acetoxy-2β,3β-epoxy-5β-spirostan, E 33
3β-Acetoxy-5α-ergosta-9,22-diene-5,8α-diol, D 136
3β-Acetoxy-5α-ergosta,7,9-dien-5-ol, D 136
3β-Acetoxyergosta-8,22-dien-11-one, D 319
3β-Acetoxyergosta-6,8(14),9(11),22-tetraene, D 318
3β-Acetoxyergosta-6,8,14,22-tetraene-11-one, D 319
17β-Acetoxyestr-5(10)-en-3-one, D 284
3β-Acetoxy-5,8β-etheno-5β-ergost-22-ene-6β,7β-dicarboxylic anhydride, D 134
3β-Acetoxy-5α-etianic methyl ester, D 58
3β-Acetoxy-5α-eti-14,16-dienoic methyl ester, D 58
3β-Acetoxyeti-5-enic acid, D 305, 306
3-Acetoxyeti-11-enic esters, D 313
3β-Acetoxyeti-5-enoyl chloride, D 153
17α-Acetoxy-6α-fluoro-1-dehydroprogesterone, D 347
17α-Acetoxy-6-fluoro-6-dehydroprogesterone, D 347
17α-Acetoxy-6-fluoro-9β,10α-pregn-4,6-diene-3,20-dione, D 348
3-Acetoxy-4-formylcyclohex-1-ene, B 90
3β-Acetoxy-5-formyloxy-6-oxacholestan-7-one, D 109

2-Acetoxyfuran, B 141
17β-Acetoxy-A-homoestra-2,4a,10(1)-trien-4-one, D 228
21-Acetoxy-17α-hydroxyallopregnane-3,11,20-trione, D 326
3α-Acetoxy-11β-hydroxycholanic methyl ester, D 316
3α-Acetoxy-6β-hydroxy-5α-cholanoic methyl ester, D 188
3α-Acetoxy-6β-hydroxy-5β-cholanoic methyl ester, D 188
3α-Acetoxy-12α-hydroxy-5β-cholanoic methyl ester, D 24
3α-Acetoxy-12α-hydroxy 5β-chol-9(11)-enoic methyl ester, D 57
3α-Acetoxy-12β-hydroxy-5β-chol-9(11)-enoic methyl ester, D 57
3β-Acetoxy-5-hydroxy-5β-cholestan-6-one, D 46
6β-Acetoxy-5-hydroxy-5α-cholestan-3-one, D 48
3β-Acetoxy-19-hydroxycholest-5-ene, D 204
3β-Acetoxy-14-hydroxyergosta-6,8,22-trien-11-one, D 318
3β-Acetoxy-14-hydroxyergosta-6,8,22-trien-11-one, D 318
3β-Acetoxy-14-hydroxy-5β,14β-etianic acid, D 12
3β-Acetoxy-14β-hydroxy-5β-etianic acid, D 389
11β-Acetoxy-3β-hydroxy-5β-etianic acid, D 39
3α-Acetoxy-14β-hydroxy-14-isoetianic acid, D 372
17α-Acetoxy-3β-hydroxy-6-methylpregn-5-ene-20-one, D 240
3β-Acetoxy-5β-hydroxy-17-oxo-B-nor-etio-6β-cholanic acid lactone, D 284
20β,21-Acetoxy-17β-hydroxy-17α-pregnane, D 60
3α-Acetoxy-17α-hydroxypregnane-11,20-dione, D 281
21-Acetoxy-17α-hydroxypregnane-3,20-dione, D 310, 311
21-Acetoxy-17α-hydroxypregnane-3,11,20-trione, D 327
12β-Acetoxy-14β-hydroxy-pregnan-20-one, D 404
17α-Acetoxy-19-hydroxyprogesterone, D 245
3β-Acetoxy-5-isoandrostan-17-one, D 393
3β-Acetoxy-14-iso-17-isoeti-5-enic acid, D 300
Acetoxykaurenic acid, C 395
Acetoxyketones, D 66
9-Acetoxy-9-methylbicyclo[4.2.1]-nonatriene, D 128
2-Acetoxy-2-methylcyclohexa-3,5-dien-1-one, B 122, 125
4-Acetoxy-4-methylcyclohexa-2,5-dien-1-one, B 125
2-Acetoxymethylcyclopropanecarboxylic esters, A 56
3-Acetoxy-4-methyl-4,5-oxidocholestanes, D 82, 83
5-Acetoxy-4-methylpentanoic acid, E 9
17α-Acetoxy-6-methylpregna-4,6-diene-3,20-dione, D 240
3β-Acetoxy-6-methylpregna-5,16-dien-20-one, D 240
3α-Acetoxy-16-methylpregna-16-ene-11,20-dione, D 352
17α-Acetoxy-6α-methylprogesterone, D 240
3β-Acetoxy-B-norandrost-5-en-17-one, D 284
3β-Acetoxynor-5α-cholanoic acid, D 12
17β-Acetoxy-A-nor-B-homoestr-5(10)-ene-3,6-dione, D 284
3β-Acetoxy-B-norpregn-5-en-20-one, D 284
3α-Acetoxy-5,6-oxido-5α-cholestane, D 82
3β-Acetoxy-5,6α-oxido-5α-cholestane, D 82
3β-Acetoxy-5,8α-oxido-5α-ergosta-9,22-diene, D 136
3-Acetoxy-11-oxobisnorcholanic acid, D 324, 325
3α-Acetoxy-11-oxocholanic methyl ester, D 316, 317, 323
3α-Acetoxy-11-oxo-5β-cholanoic methyl ester, D 55
3α-Acetoxy-12-oxocholanic methyl ester, D 314, 315
3α-Acetoxy-12-oxo-5β-cholanoic methyl ester, D 24, 55
3α-Acetoxy-12-oxochol-9(11)-enic methyl ester, D 315
3α-Acetoxy-12-oxo-5β-chol-9(11)-enoic methyl ester, D 56
3β-Acetoxy-7-oxochol-5-enoic methyl ester, D 53
3β-Acetoxy-11-oxo-5α-conamine, D 302
3α-Acetoxy-11-oxoetianic acid, D 324
3β-Acetoxy-11-oxoetianic acid, D 321

3β-Acetoxy-5-oxo-5,7-seco-B-norcholestan-7-oic acid, D 109
21-Acetoxy-20-oxosteroids, D 407
3-Acetoxy-9β,10α-pregna-3,5-dien-20-one, D 347
3β-Acetoxypregna-5,16-dien-20-one, D 357; E 10, 11
Acetoxypregnan-3,20-dione, D 248
3α-Acetoxy-5β-pregnane-11β,20β-diol-18-oic acid, D 303
3α-Acetoxy-5β-pregnane-20α-ol-11-one, D 303
21-Acetoxypregnane-3,11,20-trione, D 321, 322, 323, 324
3β-Acetoxy-5β-pregnan-20-one, D 40, 102
1β-Acetoxy-14α,17α-pregn-5-ene-3β,17β-diol-15,20-dione, D 404
12β-Acetoxy-14β,17β- pregn-5-ene-3β,14β-diol-15,20-dione, D 404
17α-Acetoxypregnenolone acetate, D 345, 348
3β-Acetoxypregn-5-en-20-one-oxime, D 357
2α-Acetoxyprogesterone, D 345
6α-Acetoxyprogesterone, D 345
6β-Acetoxyprogesterone, D 238, 239
17α-Acetoxyprogesterone, D 345
17α-Acetoxyprogesterone-3-enol ether, D 346
7-Acetoxyroyleanone, C 392
Acetoxysandaracopimaric acid, C 388
(25S)-3β-Acetoxy-5β-spirostan, D 437
(25*R*)-11α-Acetoxy-5β-spirostan, E 33
Acetoxysteroids, D 32
3β-Acetoxy-Δ^4-steroids, D 95
16β-Acetoxystrophanthidin, D 384
Acetoxytrachylobanic acid, C 400
17α-Acetoxy-5α,6,6-trifluoropregna-3,20-dione, D 349
2-Acetoxy-2,4,6-trimethyl-cyclohexa-3,5-dien-1-one, B 122
1,1′-Acetoxyvinylcyclohexene, B 129
α-Acetoxyvinylcyclopropane, C 7
Acetylacetone, reaction with methyl vinyl ketone, B 116
Acetylacetone chelates, A 129
β-Acetylacrylic acid, A 103
5-Acetyl-7-alkyl-4-hydroxy-2-methoxytropones, B 368
1-Acetylaminocyclopropane-1-carboxylic ester, A 21
17-Acetylamino-16-methylpregna-5,16-dien-3β-ol acetate, D 343
16-Acetyl-Δ^{14}-anhydrogitoxigenin, D 394
Acetylbalchanolide, C 365
2-Acetylbicyclo[2.2.0]hexanes, C 43
γ-Acetylbutyric ester, B 111
α-Acetyl-γ-butyrolactone, A 50
N-Acetylcarvomenthylamine, B 229
6β-Acetylcholest-4-en-3-one, D 126
Acetyl-CoA, E 55, 56, 57
Acetyl-co-enzyme A, see Acetyl-CoA.
Acetylcyclobutane, A 94
1-Acetylcycloheptene, B 348, 353
1-Acetylcyclohexa-1,4-dienes, B 128, **130**
Acetylcyclohexane, B 128, **129**; C 15
1-Acetylcyclohexan-1-ol, B 129
2-Acetylcyclohexan-1-one, B **130**
Acetylcyclohexenes, B 128
1-Acetylcyclohex-1-ene, B 66, **129**
1-Acetylcyclohex-3-ene, B **129**
6-Acetylcyclohex-2-enones, B 116
Acetylcyclopentadienylmesityleneiron, A 143
Acetylcyclopentadienyltetracarbonylvanadium, A 142
Acetylcyclopentadienyltricarbonylmanganese, A 140
Acetylcyclopentane, A 161, 175, 190
3-Acetylcyclopentane-1-carboxylic acid, A 192,
Acetylcyclopentan-1-one, A 189
1-Acetylcyclopent-1-ene, A 157, 162, 175
Acetylcyclopropane, A 50; C 13
1-Acetylcyclopropane-1-carboxylic ester, A 57
Acetyldigoxin, D **385**
α-Acetyldigoxin, D 362
β-Acetyldigoxin, D 362
1-Acetyldimethylcyclohexa-1,4-dienes, B 128
1-Acetyl-6,6-dimethylcyclohex-1-ene, B 128
2-Acetyl-3,3-dimethylcyclopropanecarboxylic acid, A 57
2-Acetyl-4,4-dimethyl-6-methylenecyclohexane-1,3-dione, B 130
Acetylenes, cyclic, B **388**
—, reaction with allene, B 28
—,— with carbenes, A 33
—,— with diazoacetic ester, A 55
Acetylene, D 268, 306
—, conversion to tropones, B 363
—, polymerisation, B 380, 392
—, reaction with cyclopentadiene, B 349
—,— with cyclopentadiene, C 89
—,— with methylene, A 33

Acetylenecarboxylic acid, B 141
Acetyldenedicarboxylic acid, Diels-Alder adducts, A 118
—, reaction with butadiene, B 147, 148
Acetylenedicarboxylic ester, B 126, 354; C 43
Acetylenedicarboxylic methyl ester, A 82, 173; B 356; C 110, 261
Acetylenedimagnesium bromide, B 248, 270, 299; C 12, 16
Acetylenic alcohols, cyclic, B **65**
Acetylenic aldehydes, B 90
Acetylenic carbinols, C 64
Acetylenic ketones, B 128
Acetylferrocene, A 132, 133, 134, 136, 143, 147
23-Acetylfurost-20(22)-ene, E 21
3-Acetylgitoxigenin, D 394
16β-Acetylgitoxigenin, D 394
1-Acetyl-1-halogenocyclohexane, B 134
Acetylides, B 65
1-Acetyl-4-isopropenylcyclopent-1-ene, B 208, 209
γ-Acetyl-α-isopropylbutyric acid, B 202
1-Acetyl-3-isopropylcyclopentane, B 209
2-Acetyl-1-isopropylcyclopropane-1-carboxylic acid, C 160
2-Acetyl-1-isopropylcyclopropylacetic acid, A 58
1-Acetyl-4-isopropylidenecyclopent-1-ene, E 69
N-Acetylmenthylamine, B 228
5-Acetyl-2-methoxy-1,4-benzoquinone B 368
17-Acetyl-3-methoxygona-1,3,5(10), 16-tetraene, D 248, 262
2-Acetyl-1-methyl-1,1′-bicyclopropyl, C 8
1-Acetyl-2-methylcyclohexa-1,4-diene, B 128
Acetylmethylcyclohexanes, B 128, 348
1-Acetyl-4-methylcyclohexane, B 219
4-Acetyl-1-methylcyclohexan-2-one, B 200
Acetylmethylcyclohexenes, B 128, 130
4-Acetyl-1-methylcyclohexene, C 291
1-Acetyl-1-methylcyclopentane, A 175
2-Acetyl-1-methylcyclopentane, A 175, 190
1-Acetyl-3-methylcyclopentane-1-acetic ester, C 28
2-Acetyl-3-methylcyclopentan-1-one, A 189
1-Acetyl-2-methylcyclopent-1-ene, A 175, **176**
1-Acetyl-2-methylcyclopent-3-ene, A 176
1-Acetyl-3-methylcyclopentene, B 209
Acetylnerrifolin, D 362, **386**
N-Acetyl-*N*-nitroso-(2′-oxocyclopentyl)butylamine, C 126, 129
O-Acetyloleanoic acid, E 118
O-Acetyloleanolic aldehyde, E 118
1-Acetylpentamethylcyclopent-1-ene, A 176
Acetyl perchlorate, A 55
1-Acetyl-2-phenylcyclohexane, B 128
1-Acetyl-1-phenylcyclopentane, A 175
2-Acetyl-1-phenylcyclopentane, A 175
2-(2′-Acetylpropyl)-4,4-dimethylcyclohexanone, B 119
Acetylquinones, B 368
23-Acetylspirostans, E 21
2-Acetylthiotropone, B 378
2-Acetyl-4,4,6-triethylcyclohexane-1,3-dione, B 130
1-Acetyl-2,6,6-trimethylcyclohex-1-ene, B 128
Acetyltropolones, B 377
Achillan C **354**, 367
Achillea lanulosa, C 354
Achillea millefolium, C 365
Acinitrocamphor, C 212
Acokanthera longiflora, D 405
Acokanthera venenata, D 405
Acolongiflorosides E, G, H, J, K, D **405**
Aconitic ester, A 201
Acorenone, C **312**, 335
Acorone, C **312**, 313, 314, 335
Acorus calamus, C 262, 270
Acoschimperoside P, D 386
Acovenose, D 405
Acovenosides A, B and C, D 364, **405**
Acovenosigenin, D 364, **405**
Acovenosigenin-acovenoside-[glucoside]2, D 405
Acraldehyde, C 79
Acroleins, A 47
Acrolein, B 91; C 112, 120, 121, 122, 124, 125
—, reaction with butadiene, B 90
—,— with cyclopentadiene, C 89
Acrylic acid, A 76; B 146; C 114
—, reaction with cyclopentadiene, C 96
Acrylic esters, A 202
Acrylic ester, C 114
—, adduct with butadiene, A 200

Acrylonitiiles, A 202
Acrylonitrile, A 76; C 90, 104, 114, 123
—, dimerisation, A 76
Acrylyl chloride, C 126
ACTH (adrenocorticotrophin), D 287
Actinamine, B **84**
Actinidia polygama, B 210, 212
Actinidine, B 212
Actinospectacin, B **84**
Acyclic hydroxycarotenoids, B 306
Acyclic monoterpenoids, B 153ff, 156ff
Acylacetic esters, A 57
1-Acyl-3-chloropropanes, A 50
Acylcyclohexanediones, B 130, 131
Acylcyclohexanetriones, B 130, 131
2-Acylcyclohexanones, A 190; B 96
1-Acylcyclohex-1-enes, B 20
Acylcyclopentadienes, A 124
1-Acylcyclopentan-1-ols, B 105
2-Acylcyclopentanones, A 189
Acylcyclopentenes, A 114
Acylcyclopropanes, A 50
Acylcyclopropanecarboxylic acids, A 57
Acylferrocenes, A 133
Acyloins, dehydration, B 410
—, macrocyclic, B **411**, 412
—, reduction to cyclanones, B 406
Acyloin condensation reaction, A 106
Acyloin oximes, B 418
Acyloxycyclopropanecarboxylic acids, A 56
Adamantanes, 1-substituted, C 130
Adamantane, C 103, 124, 127, *129*, 131, 132
Adamantane-1-carbaldehyde, C 131
Adamantane-1-carbamic methyl ester, C 131
Adamantane-2-carbamic methyl ester, C 131
Adamantane-1-carbo-*p*-toluene-sulphonohydrazide, C 131
Adamantane-1-carboxamide, C 131, 132, 133
Adamantane-1-carboxanilide, C 131
Adamantane-1-carboxylic acid, C 130, 131, 132
Adamantane-2-carboxylic acid, C 132
Adamantane-1-carboxylic acid hydrazide, C 131
Adamantane-1,3-dicarboxylic acid, C 132
Adamantane-1,3-diol, C 131
Adamantane-1,3,5,7-tetracarboxylic acid, C 132
Adamantane-1,3,5-triol, C 131
Adamantan-1-ol, C 127, 130, 131
Adamantan-2-ol, C 131
Adamantan-2-one, C 131
1-Adamantyl bromide, C 130
1-Adamantyl carbamate esters, C 131
Adamantyl carbonium ions, C 130, 131, 132
Adamantyl-1-isocyanate, C 131
N-(2-Adamantyl)-*p*-toluenesulphon-amide, C 131
Adenium boehmianum, D 386, 405
Adenium honghel, D 391
Adenium somalense, D 386
Adenocarpus species, A 101
Adenosine, diphosphate = ADP, E 56
Adenosine triphosphate = ATP, B 155; E 56, 57, 60
S-Adenosylmethionine, E 108, 129
Adhumulone, B **131**
Adianane group, triterpenoids, C 456
Adianene, C 417, 418, 456
Adianenediol, C 456
Adiantone, C 452
Adiantoxide, C 456
Adienolide, D 7
Adipaldehyde, A 163
Adipaldehydic acid, B 99
Adipic acids, conversion to cyclo-pentanones, A 105
—, cyclisation, A 165
—, esters, A 105
Adipic acid, A 198; B 46, 49, 74, 75, 98, 99, 110, 355
—, conversion to cyclopentanone, A 164
—, from cyclohexane, B 15
Adipic anhydride, conversion to cyclopentanones, A 105
Adipic dialdehyde, B 75
Adipic ester, A 199
Adipic hemialdehyde, B 98
Adipic methyl esters, B 103
Adipoin, B **103**
Adiponitriles, conversion to cyclo-pentanones, A 106
Adiponitrile, cyclisation, A 200
Adonirubin, B 332
Adonis annua, B 332
Adonis vernalis, D 396
Adonitoxigenin, D 364, **396**
Adonitoxin, D 364, **396**
Adonitoxol, D **396**
Adonitoxologenin, D **396**
Adonixanthin, B 332
ADP = Adenosine diphosphate, E 56

Adrenocortical hormones, D 2, 197ff, **286**
—, halogenation, D 351
Adrenocortical steroids, D 289
—, syntheses, D 303
Adrenocorticotrophin (ACTH), D 287
Adrenosterone, D 289, 294, **302**, **324**, 325, 329
Adrenosterone acetate, D 325
Adynerigenin, D 364, **394**, **395**
Adynerin, D 364, **394**, **395**
Aegiceradiol, C 415, 416
Aegicerin, C 441
Aescinidine, C 439
Aetianic acid, D 169
Aetiocholane, D 9
Aetiocholanic acid, D 169
Aetiocholan-17α-ol acetate, D 253
Agapanthagenin, E 27, **39**, 40
Agapathus sp., E 27
Agarofurans, C 335
α-Agarofuran, C **304**
β-Agarofuran, C **303**, 304
Agarol, C **300**, 334
Agathadiol, C 382
Agathic acid, C 384
Agatholal, C 383
Agatholic acid, C 383
Agave gracilipes, E 26
Agave huachucensis, E 27
Agave roezliana, E 27
Agave sisalana, E 26
Agave species, E 25, 26, 27
Agave striata, E 27
Agavogenin, E 27, **36**
Agliaol, C 425
Aglycons, D 360, 363
—, steroid, constitution, D 369
Agnosterol, C 407, 408, 421, 473; D 3, 75
—, ultraviolet absorption spectra, D 68
Agroclavine, E 65
Ailanthone, C 433
Alantolactone, C 301, **337**, 338, 366
Albaspidin, B 113, **131**
Albigenic acid, C 440
Albigenin, C 442
Albopetasin, C **308**, 335
Alcohols, acetylenic, B 65
—, cyclohexanes, B 68, 69
—, macrocyclic, B 402, 404
—, optically active from *cis*-olefins, C 173
Alcovenose, D 365
Aldehydes, alicyclic, B 86
—, condensation with cyclopentadienes, A 120
—, macrocyclic, B 406
—, reaction with dienes, B 86
—, terpenoid, B 195
Aldehydocamphors, C 223
Aldehydocarotenoids, B 312
Aldoketens, dimerisation, A 75
Aldol condensation, A 107
Aldosterone, D 1, 7, 87, 287, 289, **302**, 303, 333
—, *nomenclature*, D *439*
Aldosterone acetate, D 303
Aldosterone diethylene acetal, D 302
Aldosterone-21-monoacetate, D 303
Aldrin, C **105**
Aleprestic acid, A 196, 197, 198
Alepric acid, A 196, 197, 198
Aleprolic acid, A 191, 196, 198
Aleprylic acid, A 196, 198
Aleuria aurantia, B 332
Aleuritic acid, B 415
Aleurixanthin, B 332
Alfalfa, D 156
Alicyclic compounds, chemical properties, A 11
—, isomerism, A 16
—, nomenclature, A 2
—, physical properties, A 11
Alicyclic ketones, B 92ff
Alicyclic rings, classification, A 3
—, formation, A 9
—, strain, A 3
—, structure, A 2
—, systems, A 1
Alkaloids, D 2
—, diterpene, C 380, 470
—, indole, E 70
—, *nomenclature*, D *425*
—, with hemiterpene residues, E 64
αω-Alkanedicarboxylic acids, A 2
n-Alkanoic acids, complexes with deoxycholic acid, D 176
Alkenes, dimerisation, A 74, 75
Alkenoic acids, B 116
—, cyclisation, A 108, 169
Alkenoyl chlorides, cyclisation, A 108, 169
Alkenylcyclohexanes, B 23
Alkenylcyclohexenes, B 28
Alkenylcyclopentadienides, A 128
Alkenylferrocenes, A 133
Alkenyl-lithiums, B 395
2-Alkoxalyl-Δ^4-3-oxtosteroids, B 102
2-Alkoxybutadienes, B 91
1-Alkoxycyclohexa-1,4-dienes, B 116

Alkoxycyclohexanes, autoxidation, B 75
3-Alkoxycyclohex-1-enes, B 42
3-Alkoxycyclohex-2-en-1-ones, B 115
Alkoxycyclopropanecarboxylic acids, A 56
3-Alkoxy-3,5-dienes, D 92
1-Alkoxy-2,3-epoxycyclohexanes, B 55
1-Alkoxy-4-formylcyclohex-1-enes, B 91
2-Alkoxymethylenecyclohexan-1-ones, B 87
3-Alkoxypropylmagnesium halides, A 21
Alkoxytropones, B **377**
1-Alkyl-2-alkylcyclopentanols, A 116
β-Alkylallyl chlorides, A 20
3β-Alkylamino-5α-cholestanes, D 119
Alkylandrostane-17β-yl mixed acetals, D 274
Alkylation, mechanism, E 108
Alkyl boranes, C 173, 178
Alkylcarbenes, A 26
4-Alkylcholesteryl sulphonates, D 78
24-Alkylcorprostanols, D 159
Alkyl cyclobutyl ketones, A 91
5-Alkylcyclohepta-1,3-dienes, B 349
Alkylcycloheptanones, B 355
Alkylcycloheptatrienes, B 360
5-Alkylcyclohexa-1,4-dien-1-ols, B 64
Alkylcyclohexanes, B 13, 347
—, autoxidation, B 74
2-Alkylcyclohexane-1,3-diones, B 111
Alkylcyclohexanols, B 17, 46
2-Alkylcyclohexan-2-ol-1-ones, B 105
2-Alkylcyclohexanones, B 93, 94
1-Alkylcyclohex-1-enes, B 17, 19
3-Alkylcyclohex-1-enes, B 19
6-Alkylcyclohex-2-en-1-ols, B 62
2-Alkylcyclohex-2-enone, B 116
6-Alkylcyclohex-2-en-1-ones, B 62
2-Alkylcyclohexylboranes, B 19
α-Alkylcyclohexylglycidic esters, B 127
Alkylcyclopentadienes, A 120, 121
Alkylcyclopentanes, A 13
1-Alkylcyclopentanols, A 115
2-Alkylcyclopentanols, A 116
1-Alkylcyclopentanol acetates, A 115
Alkylcyclopentanones, A 166, 200
1-Alkylcyclopentenes, A 115
3-Alkylcyclopentenes, A 116
Alkyl cyclopentyl ketones, A 175
2-Alkylcyclopentyl tosylates, A 116
Alkylcyclopropenes, A 33
Alkyl cyclopropyl ketones, A 43, 50, 51
1-Alkylcyclopropyl methyl ketones, A 50
9-Alkyldecalins, C 63
Alkylferrocenes, A 132, 133
Alkyl formates, B 97
22-Alkylfurostan-26-ols, E 20
Alkyl halides, B 93
Alkylidene-bis-malononitriles, A 62
Alkylidenecarbenes, A 34
3-Alkylidenecyclohex-1,4-dienes, B 29
Alkylidencyclohexanes, B 23
Alkylidenecyclohexenes, B 28
1-Alkylidenecyclopenta-2,4-dienes, A 123
2-Alkylidenecyclopentanones, B 116
Alkylidenecyclopropanes, A 34
Alkylidene di-iodides, reaction with acetoacetic esters, B 113
Alkylidenephosphoranes, B 247
2-Alkyl ketone effect, B 12
Alkyllithiums, A 120
17α-Alkyl-19-nor-D-homotestosterones, D 281
24-Alkylsterols, D 29
Alkylsulphonium salts, B 380
Alkyltetracyanocyclopropanes, A 62
Alkynes, dimerisation, A 75
1-Alkynylcyclohexan-1-ols, B 94
Alkynylferrocenes, A 133
Allenes, A34, 39; B 388
—, dimerisation, 75
Allene, A 33; B 329; C 49
—, adduct with maleic anhydride, A 96
—, dimerisation, A 83
—, polymerisation, B 25
—, reaction with acetylene, B 28
—,— with alkenes, A 76
Allenic carotenoids, B 328
Alloaglycons, D 372
Alloaromadendrene, C **277**, 288, 320, 321
Δ^2-Allobetulene, C 437
Allobetulin, C 413, 414, 415, 437
Allobetulin group, triterpenoids, C 437
Allobetulone, C 437
"Allo" cardiac glycosides, D 29
Allocholane, D *428*
Allocholanic acid, D 179, 413
Allocholic acid, D 173, 189
Allocyclocitrals, B 173, 174
Allocyclogeranic acid, B 174
Allocyclogeraniol, B 173
Allocymarin, D 385
Allodeoxycholic acid, D 172
Alloetianic acid, D 370

Alloglycosides, D 372
Alloisolithobilianic acid, D 10
Allolithobilianic acid, D 10
D-Allomethylose, D 365
Allomicymarin, D 385
Allo-ocimeme, B **160**, 161, 172; C 172
Alloperiplogenin, D 385
Allopregnanes, D 248, 290
—, stereoisomers, D 237
Allopregnane, D 232, 248, 296
—, *nomenclature*, D *428*
Allopregnane-3,20-diols, D 249, 250, 253
Allopregnane-3α,20α-diol diacetate, D 250
Allopregnane-3β,20β-diol diacetate, D 250
Allopregnane-3β,21-diol-11,20-dione, D 289
Allopregnane-3β,17α-diol-20-one, D 289
Allopregnane-1,20-dione, D 249
Allopregnane-3,20-dione, D 232, 236, 237, 249, 250, 253, 298
Allopregnane-3β,11β,17α,20β,21-pentol, D 289
Allopregnane-3β,17α,20β,21-tetrol, D 289
Allopregnane-3α,11β,17α,21-tetrol-20-one, D 289
Allopregnane-3β,11β,17α,21-tetrol-20-one, D 289
Allopregnane-3,16,20-triols, D 251, 253
Allopregnane-3β,17α,20α-triol, D 289
Allopregnane-3β,17α,20β-triol, D 289
Allopregnane-3β,17α,21-triol-11,20-dione, D 289
Allopregnane-3β,11β,21-triol-20-one, D 289
Allopregnane-3β,17α,21-triol-20-one, D 289
Allopregnane-3,11,20-trione, D 297
Allopregnan-3α-ol, D 248
Allopregnan-3β-ol, D 248
Allopregnan-20α-ol, D 248, 249
Allopregnan-20β-ol, D 248
Allopregnanolone, D 232, 236, 237, 253
Allopregnan-3β-ol-20-one, D 249, 250, 263
Allopregnan-3β-ol-20-one, D 250, 263, 289, **302**
Allopregnan-20α-ol-3-one, D 248, 249, 250
Allopregnan-20β-ol-3-one, D 248
Allopregnanolone acetate, D 237
Allopregnan-3-one, D 249
Allopregnan-20-one, D 248, 249, 253, 260
Allopregn-1-en-3,20-dione, D **241**, 254
Allopregn-16-en-3,20-dione, D 249
Allopregn-16-en-3-ol-20-ones, D 253
Allopregn-16-en-3-one, D 248, 249
Allosantenic acid, D 190
Allospirostanes, D 321
Allostrophanthidin, D *431*
Allotestosterone, D 340
Alloxanthin, E 120, 122, 132
Allyl acetate, A 56
Allylbenzene, A 54
Allyl bromide, B 217
Allyl carbinol, A 78
Allylcarbinyl bromide, A 64
Allyl chlorides, A 56
Allylcyclohexane, B 21
1-Allylcyclohexan-1-ol, B 91
2-Allylcyclohexan-1-one, B 127
1-Allylcyclohexan-2-one-carboxylic ester, B 127
3-Allylcyclohex-1-enes, B 42
Allyldicyclopropylcarbinol, C 9
Allyl Grignard reagents, C 9
Allyl halides, reaction with sodium phenoxide, B 120
Allylic alcohols, B 157, 405
Allyl iodide, B 127
Allylmagnesium bromide, A 140; D 308
Allylmagnesium halides, B 42
Allylpalladium chloride, A 129
Alnusfoliendiolone, C 425
Alphitolic acid, C 436
Amarilin, C **360**, 368
Amarolide, C 433
Amarolide acetate C 433,
Ambergris oil, B 176
Amboside, D **402**
Ambrein, C 419, 457
Ambreinolide, C 468
α-Ambrinol, B 176
Ambrosia species, C 357
Ambrosia maritima, C 357, 359
Ambrosin, C **357**, 368
American cedaı leaf oil, C 145
American peppermint oil, C 144
Amines, exocyclic, B 86
—, macrocyclic, B **418**
1-Aminoadamantane, C 131
2-Aminoadamantane, C 131
α-Aminoalcohols, B 411
2-Aminoalcohols, macrocyclic, B 418
Aminoaldehydes, cyclohexanes, B 91

ω-Aminoalkanoic acids, B 409
17-Aminoandrost-4-en-3-one hydrochloride, D 357
Aminobenzoic acids, hydrogenation, B 142
m-Aminobenzoic acid, hydrogenation, B 143
1-Aminobicyclo[2.2.2]octane, C **110**
2-Aminobicyclo[2.2.2]octane, C 111, 118
2-Aminobicyclo [2.2.2]oct-5-ene, C 111
5-Aminobicyclo [2.2.2] oct-2-ene, C 109, 119
2-Aminobicyclopentyl, C 11
2-Aminobornanes, C 232
3-Amino-5-*tert*-butylcyclohexane-1-carboxylic acids B 143
α-Aminocamphor, C 221, **233**
3β-Aminocholest-5-ene, D 43
6β-Amino-3α,5-cyclo-5α-cholestane, D 43, 80
Aminocyclodecane, B 420, **422**
Aminocyclododecane, B 420
6-Aminocyclododecanol, B 411
Aminocycloheptadecane, B 420
1-Aminocycloheptanecarboxylic acid, B **357**
2-Aminocycloheptan-1-ol, B 352
Aminocycloheptatrienethione, B 378, 379, 380
Aminocyclohexanes, B **77, 80**
Aminocyclohexanecarboxylic acids, B 142, 143
2-Aminocyclohexan-1-ol, B **82**, 83
3-Aminocyclohexan-1-ol, B 83
Aminocyclohexanones, B **108**
2-Aminocyclohexan-1-one, B 83
3-Aminocyclohex-1-ene, B **82**
Aminocyclohex-1-ene-1-carboxylic acid, B 143
Aminocyclohexenetetrols, B 86
Aminocyclononane, B 420
Aminocyclo-octadecane, B 420
Aminocyclo-oct-4-ene, B 420
Aminocyclopentadecane, B 420
Aminocyclopentadienyltricarbonylmanganese, A 141
1-Aminocyclopentanecarboxylic acid, A 199; D 358
2-Aminocyclopentanecarboxylic acid, A 199
Aminocyclopentenes, A 149, 153
Aminocyclopropanecarboxylic acids, A 57
1-Aminocyclopropanecarboxylic acid, A 70
α-Aminocyclopropylacetic acid, A 57
Aminocyclotetradecane, B 420
Aminocyclotridecane, B 420
Aminocyclo-undecane, B 420
9-Aminodecalins, C 66
2-Aminodecanol, B 419
3-Aminodicyclopent-1-ene, A 153
2-Aminoepicamphor, C 233
2-Amino-3-ethoxycarbonyl-4,5-dihydrofuran, A 58
1-Aminoethyladamantane, C 133
α-Aminoethylcyclopropane, A 46
Aminoferrocene, A 134
17-Aminoheptadecanoic acid, B 414
2-Aminohexadecanol, B 419
3-Amino-1-hydroxyadamantane, C 127
4-Amino-7-iminocyclohepta-1,3,5-trienylamine, B 378
17-Aminomethylandrostanes, D 357
2-Aminomethylbicyclo[2.2.1]heptane, C 83, 112, 118
6-Aminomethylbicyclo[3.2.2]nonane, C 123
2-Aminomethylbicyclo[2.2.2]octane, C 122
1-Amino-1-methylcyclohexane, B 78, 81
2-Amino-1-methylcyclohexane, B 81
3-Amino-1-methylcyclohexane, B 81
4-Amino-1-methylcyclohexane, B 81
1-Aminomethylcyclohexan-1-ols, B 86
1-Amino-4-methylestra-1,3,5(10)-trien-17-one, D 356
5-Aminomethyl-5-hydroxyhydrindane, C 60
2-Aminomethylnorbornane, C 112
3-Amino-6-methyltropolone, B 375
2-Amino-6-methyltropone, B 375
o-Aminophenol, A 171
2-Amino-1-phenylcyclohexan-1-ol, B 83
p-Aminophenylcyclopropane, A 36
20-Aminopregnane, D 324
Aminospirostans, E 52
2-Aminotridecanol, B 419
Aminotropolones, B 373, 375
Aminotropones, B 368, 375, 376, **377**, 378, 379
2-Aminotropone, B 363
Aminotroponeimines, B 378, 379
Ammoresinol, E 81
Amolonin, E 2
Amylcyclohexane, B 14
1-*n*-Amylcyclopent-1-ene, A 116
2-*n*-Amyl-3-methylcyclopentan-1-one, A 167

Amylsodium, A 41, 134
Amyranes, E 116
Amyrenones, C 439
α-Amyrenone, C 450
α-Amyrin, C 406, 415, 416, 450
β-Amyrin, C 406, 413, 414, 415, 438, 475, 476; E 116, 117, 136
—, biosynthesis, E 118
δ-Amyrin, C 413, 414, 415, 438
α-Amyrin acetate, C 450
β-Amyrin acetate, C 438, 476
Andirobin, C 411, 412, 431
Andirobin group, triterpenoids, C 431
Androgens, D 2, 197, **254**
—, halogenation, D 339
—, isolation, D 254
Andrographolide, C 375, 376, 384
Andromedotoxin, C 380
Andropogon jwarancusa, C 153
5,14-Androstadiene, D *441*
8α,10α-Androsta-5,14-diene, D *441*
9β,13α-Androsta-5,14-diene, D *441*
9β,10α-Androsta-3,5-diene-3,17β-diol diacetate, D 341
Androsta-1,4-diene-3,17-dione, D 203, 338
Androsta-4,11-diene-3,17-dione, D 265
Androsta-3,5-diene-17β-ol acetate, D 344
Androsta-1,4-dien-17β-ol-3-one, D 213, 271
Androsta-4,6-dien-17β-ol-3-one, D 264
Androsta-4,11-dien-17β-ol-3-one, D 264
Androsta-1,4-dien-17β-ol-3-one acetate, D 262
5α-Androstan-19-al, D *438*
Androstanes, D 254, 263, 301
—, *nomenclature*, D *428*, *446*
—, steroids, D 357
—, thiol derivatives, D 275
Androstane, D 2, 5, 290
—, molecular rotation, D 63
5α-Androstane, D 9, 35
—, *nomenclature*, D *447*
5β-Androstane, D 6, 9, 35
—, *nomenclature*, D *448*
ψ-Androstane, D 61
Androstane-17-carboxylic acid, D 7, 191
—. *nomenclature*, D *439*, *440*
Androstane-3α,16α-diol, D 264
Androstane-3β,16β-diol, D 264
Androstane-3α,17-diols, D 264, 271
Androstane-3α,17α-diol, D 264
Androstane-3β,17α-diols, D 264, 268
Androstane-3β,17β-diols, D 264, 268, 290, 291
Androstane-6β,17β-diol, D 264
Androstane-3β,11β-diol-17-one, D 289, 290, 291, 301
5β-Androstane-1,6-dione, D 90
Androstane-3,17-dione, D 265, 291
5α-Androstane-3(R)*-spiro-2′-(4′*R*-methyl-1′,3′-oxathiolane)*, D *453*
(3S)-5α-Androstane-3-spiro-2′-oxiran, D *454*
Androsta-1,4,6-triene-3,17-dione, D 203, 228, 262
Androsta-1,4,6-trien-17β-ol-3-one acetate, D 262
Androsta-3,5,7-trien-17β-ol-propionate, D 261
Androsta-1,4,6-trien-3-one-17β-ol acetate, D 213
Androstane-3β 4β,17β-triol, D 265
Androstane-3β 11α,17β-triol, D 265
Androstane-3β 16α,17α-triol, D 265
Androstane-3β 16α,17β-triol, D 265
Androstane-3β 16β,17α-triol, D 265
5β-Androstane-1,6,17-trione, D 90
Androstane-3,11,17-trione, D 290, 291, 294
5α-Androstan-1α-ol, D 34
5α-Androstan-1β-ol, D 34
Androstan-3α-ol, D 263
Androstan-3β-ol, D 263
5α-Androstan-11-ols, D 54
5α-Androstan-15α-ol, D 60
5α-Androstan-15β-ol, D 60
5α-Androstan-16α-ol, D 60
5α-Androstan-16β-ol, D 60
Androstan-17α-ol, D 260, 263
Androstan-17β-ol, D 261, 263, 290
Androstan-17β-ol acetate, D 253, 260
Androstan-3α-ol-16-one, D 264
Androstan-3β-ol-7-one, D 264
Androstan-3β-ol-16-one, D 264
Androstane-3-ol-17-ones, 264, D 357
Androstan-17α-ol-3-one, D 264
Androstan-17β-ol-3-one, D 264
Androstan-17β-ol propionate, D 261
5α-Androstanones, D 90
5β-Androstanones, D 91
5α,14β-Androstanone, D 91
Androstan-3-one, D 263
5α-Androstan-3-one, D 337
5β-Androstan-3-one, D *438*
Androstan-11-one, D 263
Androstan-15-ones, D 268, 269
Androstan-17-one, D 60, 90, 263, 268, 359

5α-Androstan-3β-yl methyl ether, D *438*
5-Androsten-3β-amine, D *438*
ψ-Androstene, D 61
Androstene-1,2-diols, D 281
Androst-4-ene-3β,17α-diol, D 264
Androst-4-ene-3β,17β-diol, D 264
Androst-5-ene-3β,16α-diol, D 265
Androst-5-ene-3β,16β-diol, D 265
Androst-5-ene-3α,17β-diol, D 265
Androst-5-ene-3β,17α-diol, D 260, 265
Androst-5-en-3β,17β-diol, D 213, 258, 259, 260, 265, 271
Androst-5-ene-3β,17β-diol diacetate, D 344
Androst-5-ene-3β,17β-diol-diformate, D 258
Androst-5-ene-3β,17β-diol 3-monoacetate, D 258
Androst-4-ene-17β,19-diol-3-one diacetate, D 212
Androst-1-ene-3,17-dione, D 265
Androst-4-ene-3,16-dione, D 265
Androst-4-ene-3,17-dione, D 197, 230, 258, 259, 262, 265, 271, 289, 299, 301, 302
Androst-5-ene-3,17-dione, D 265
Androst-9(11)-ene-3,17-dione, D 265
Androst-4-ene-3,17-dione-17-monoethylene acetal, D 256
Androst-4-ene-1β,3β,17β-triol, E 31
Androst-5-ene-3β,4β,17β,-triol, D 265
Androst-5-ene-3β,16α,17β-triol, D 60, 265
Androst-5-ene-3β,16β,17β-triol, D 265
Androst-4-ene-3,11,17-trione, D 289, 302
Androst-16-en-3α-ol, D 262, 263
Androst-16-en-3β-ol, D 262
5α-Androst-16-en-15β-ol, D 60, 269
5α-Androst-1-en-16ξ-ol, D *428*
Androst-2-en-17β-ol acetate, D 344
Androst-4-en-11β-ol-3,17-dione, D 294
Androst-1-en-17β-ol-3-one, D 269
Androst-4-en-16β-ol-3-one, D 264
Androst-4-en-17β-ol-3-one, D 264
Androst-5-en-3α-ol-17-one, D 264
Androst-5-en-3β-ol-17-one, D 264
Androst-5-en-17β-ol-3-one, D 264
Androst-5-en-17β-ol-4-one, D 269, 271
5α-Androst-1-en-17β-ol-3-one-phenylpropionate, D 273
5α-Androst-3-en-17-one, D 256
5α-Androst-8(14)-en-15-one, D 269
Androst-16-en-3-one, D 262, 263
5α-Androst-16-en-15-one, D 269
5α-Androst-15-en-17-one ethylene acetal, D 269
5-Androsten-3β-ylamine, D *438*
Androsterone, D 2, 7, 41, 237, 254, **255**, 256, 258, 264, 269, 271, 279, 301, 302
—, isomerides, D 269
—, *nomenclature*, D *439*
—, total synthesis, D 102
Androsterone cyanohydrin, D 279
Androsterone trimethylsilyl ethers, D 200
Anemone species, A 102
Anemonin, A **72**, 102
Angelica lactones, D 371
Angelic acid, E 61
Angustione, B **130**, 178
Anhydroacetonebenzils, A 166, 176, 183
Anhydroacetonebenzil, A 170
Anhydroacovenosigenin A, D 405
Anhydro-α-angelicalactone, A 102
Anhydroazafrinone amide, B 328
Anhydrochimarol, D 195
11,12-Anhydrocorticosterone, D 299
Anhydrocorticosterone acetate, D 299
α-Anhydrodigoxigenin, D 386
β-Anhydrodigoxigenin, D 386
Δ^{16}-Anhydrodigoxigenin, D 386
α-Anhydrodigitoxigenin, D 389
β-Anhydrodigitoxigenin, D 389
β-Anhydrodigitoxigenin acetate, D 389
Anhydrodihydrostrophanthidin, D 381
Anhydrodihydrostrophanthidinic acid, D 381
Anhydroeschscholtzxanthin, B 305
Anhydrogamabufogenin, D 420
Anhydrogeigerin, C 351
Δ^{14}-Anhydrogitoxigenin, D 395
Δ^{16}-Anhydrogitoxigenin acetate, D 392
16-Anhydrogitoxigenin-α-L-acofriopyranoside, D 391
Δ^{16}-Anhydroglycosides, D 391
Anhydroinositols, B 58, 68, 85
Anhydro-α-isoperiplogenic methyl ester, D 387, 388
4,5-Anhydromarinobufagone, D 419
Δ^{14}-Anhydro-oleandrigenin, D 394, 395
Δ^{16}-Anhydro-oleandrin, D 391
Anhydro-ouabain hepta-acetate, D 401
Anhydroparthenin, C 356
Anhydropartheniol, C 316, 317
Anhydrorhodovibrin, B 336, 338; E 130
Anhydrosarsasapogenoic acid, E 8
Anhydroscillaridin A, D 409
Anhydroscymnol, D **194**, 195

Anhydrostrophanthidin, D 378, 379
Anhydrostrophanthidin benzoate, D 382
Anhydrostrophanthidinic ester, D 377
Anhydrostrophanthidone, D 381
Anhydro-2-sulphocyclohexyl hydrogen sulphate, B 77
Anhydrotetrahydropartheniol, C 316, 317
Anhydrouzarigenin, D 368, 370, 396, 397
Anhydrovitamin A_1, B **291**
Anhydrovitamin A_2, B **292**
Anilines, hydrogenation, B 80
1-Anilinocyclopentane-1-carboxylic acid, A 191
Anisole, B 3
—, reaction with diazomethane, B 361
Anisoxide, E 62, 63
Annulenes, B 347, **390**, 397
—, higher, B **396**
—, physical properties, B 400
[6]-Annulene, B 390
[8]-Annulene, B 390
[16]-Annulene, B 397
[22]-Annulene, B 398
[26]-Annulene, B 398
Antheraxanthin, B 294, 299, 325, 345; E 120, 121, 127, 131
Anthothecol, C 428
Anthracene, A 172; B 24, 30
Anthraergostapentaene, D 149
Anthranilic acid, reduction, B 143
Anthrasteroid rearrangement, D 149
Antiarigenin, D 364, 368, **400**, 401
Antiarigenin-antiaroside, D 400
Antiarigenin-rhamnoside, D 400
α-Antiarin, D 364, 368, **400**
β-Antiarin D 364, **400**
Antiaris toxicaria, D 384, 400
α-Antiarol, D 400, 401
Antiarose, D 361, 365
Anti-7-tosyloxybicyclo[2.2.1]hept-2-ene, A 13
Anymol, C 292, 334
Aphanicin, B 317
Aphanin, B **317**
Aphanizomenon-flosaquae, B 317
Aphanizophyll, B 331
Apium graveolens, C 273
Aplopappus, D 166
Aplysia kurodai, C 333
Aplysin, C **333**, 336
Aplysinol, C **333**, 336
Apoallobetulin, C 437
Apoaromadendrone, C 277, 321
π-Apoborneol-7-carboxylic acid, C 228
Apocamphoric acid, C 235, 238, 247
β-Apocarotenals, B **260**
β-Apo-4′-carotenal, B 280
α-Apo-8′-carotenal, B 259
β-Apo-7′-carotenal, B 231, 236, 249, 259, 273, 312, **313**, 316, 325
—, structure, B 233
β-Apo-10′-carotenal, B **313**
β-Apo-12′-carotenal, B 259
β-Apo-8′-carotenoic ester, B 249
Apocarotenoids, B 233, 236; E 132
Apocholic acid, D 175, 186, 187, 188
Apocrustacyanin, B 321
Apocynaceae, D 360
Apofenchene, C **235**
Apofenchocamphoric acid, C 246
Apo-6′-lycopenal, B 313
Apo-8′-lycopenal, B 313
Apo-oxyallobetulin, C 437
Apo-8′-zeaxanthinal, B 259
Aquila agallocha, C 300
L-Arabinose, E 3
Aramadendrene, C 319, 320
Araucarenolone, C 388
Araucarol, C 388
Araucarolone, C 388
Araucarone, C 388
Arborescin, C 355, 367
Arborinane group, triterpenoids, C 457
Arborinol, C 418, 419, 457
Arctiolide, C **363**
Arctiopicrin, C **363**, 366, 368
Arctium minus, C 363
Arenobufagin, D **421**
Arenobufogenin, D 414
Arenobufotoxin, D 414
Argentinogenin, D **421**
Arginine, D 413
Aristoic acid, C 308, 309
Aristolochia debilis, C 308
Aristolochia indica, C 320
Aristolachia reticulata, C 364
Aristolactone, C **364**, 368
Aristolene, C **274**, 288
Aristolol, C 308, 309
Aristolone, C 274, **308**, 309, 335
Arjunolic acid, C 440
Arndt-Eistert reaction, D 101, 204
Arnidiol, C 449
Aromadendrene, C **277**, 278, 288, 321
Aromadendrene oxide, C 319, 321
Aromaticin, C **360**, 368
Aromatin, C **360**, 368
3-Aroylpropionic acids, A 169
Artabsin, C **352**, 355, 367

Artebufogenin, D 414, **419**
Artemazulene, C 353, 356
Artemisia aborescens, C 354
Artemisia absinthium, C 141, 352
Artemisia alcohol, B 158, **164**, 165
Artemisia annua, B 164
Artemisia austriaca, C 353
Artemisia balchanorum, C 347, 363, 643, 365
Artemisia finita, C 347
Artemisia ketone, B 165; E 71
Artemisia leucodes, C 353
Artemisia maritima, C 345; E 83
Artemisia mexicana, C 354
Artemisia porrecta, C 226
Artemisia species, C 340
Artemisia taurica, C 347
Artemisia tilesii, C 353
Artemisia tridentata, C 199
Artemisia vulgaris, C 347
Artemisin, C 338, **345**, 346, 351, 363, 367
Arthrobacter simplex, D 186
Artificial camphor, C 193
Artilesin, C **353**, 354
Arum spadices, E 133
Arundoin, C 455
β-Aryllallyl chlorides, A 20
α-Arylcarbinols, A 133
1-Arylcyclohexan-1-ols, B 17
2-Arylcyclohexan-2-ol-1-ones, B 105
2-Arylcyclohexanones, B 94
1-Arylcyclohex-1-enes, B 17,19, 53
Arylcyclopentadienes, A 120
2-Arylcyclopentanols, A 115
1-Arylcyclopentenes, A 115
Arylcyclopentyl ketones, A 175
2-Arylcyclopentyl tosylates, A 115
Arylcyclopropanes, A 36
Arylcyclopropanecarboxylic acids, A 53
Arylcyclopropyl ketones, A 50, 51
Arylferrocenes, A 134
Aryllithiums, B 377
1-Aryl-2-nitrocyclohexanes, B 79
2-Aryloxycyclopropanecarboxylic acids, A 56
Aryl 2-phenylcyclopropyl ketones, A 27
5-Arylpyrazolines, A 36
Arylthiocarbenes, A 44
2-Aryltropones, B 377
Asarum sieboldi, E 69
Ascardic acid, B 207
Ascaridole, B 76, 194, **206**, 207
Ascaridole glycol, B 207
Ascaridole glycol anhydride, B 207
Ascosterol, D **150**, 165
Asiatic acid, C 450
Asparagus umbellatus, E 26
Aspergillus, E 110
Aspergillus carneus, D 210
Aspergillus fumigatus, D 162; E 135
Aspergillus niger, D 312, 257
Asperillin, C **340**, 366
Asperuloside, B **213**
Aspidin, B **131**
Astacene, B 242, 243, 249, **253**, 320, 321, 322
Astacene diacetate, B 321
Astacene dipalmitate, B 321
Astacin, B 320
Astaxanthin, B 238, 241, 266, **320**, **321**, **322**, **332**
Astaxanthin-protein complex, B 321
Athiorhodaceae, B 238
Atisine, C 381, 470, 471
Atisirane diterpenoids, C 400
Atisirene, C 379, 380, 381, 400, 470
Atisirene group, C 381
Atlantones, C 291, 334
ATP = Adenosine triphosphate, E 56, 57, 60
Atractylis ovata, C 322
Atractylodes japonica, C 302
Atractylodes lancea, C 322
Atractylol, C 322
Atractylon, C 302, 303, 335
Aucuba japonica, B 214
Aucubin, B **214**
Auraptene, E 63
Aureomycin, A 185
Aurochrome, B **303**
Auroglaucin, E 65
Auroxanthin, B 253, 299, **300**; E 120, 121
Austricin, C **353**
Auwers-Skita rule, B 4; C 177, 181, 186; D 38
Avenasterol, D **153**, 166
Aza-azulenes, B 379
Azaferrocene, A 138
Azafrin, B 233, 234, **328**
Azelaic acid, pyrolysis of salts, B 407
Azelaoin, B 412
Azelaone, B 407
6β-Azido-3α,5-cyclo-5α-cholestane, D 81
Azodicarboxylic ester, reaction with buta-1,3-dienes, A 74
Azotropolones, B 377
Azulenes, C 314, 324, 350

Azulene, A 128; C 59
Azulene-1,2-dicarboxylic methyl ester, B 350
Azulene group, C 277
Azurin, D **396**

Bacharis genistelloides, B 193
Bacillariophyceae, B 238
Bacillus lepreae, A 196
Backhousia angustifolia, B 130
Backhousia citriodora, B 166
Bacteria, carotenoid-rich, B 240
Bacterioruberin, B **310**
Baeyer strain theory, A 4, 5, 110; B 380
Baeyer-Villiger oxidation, A 42, 51; B 98, 410; C 43, 205, 277, 309, 313; D 119; E 19
Balbiano's acid, C 203
Balchanin, C **347**, 367
Balchanolide, C **365**, 366, 368
Balduilin, C 355, **359**, 361, 368
Balduina angustifolia, C 359
Balduina species, C 358
Balduina uniflora, C 359
Balsamorrhiza sagittata, C 299
Bambacic acid, A 70
Bamberger oxidation, B 120
Bamboo gibberellin, C 398
Bambord-Stevens reaction, D 106
Barbier-Wieland degradation, C 266, 277, 469; D 22, 25, 56, 191, 231, 380
Barosma species, B 202
Barrelene, C **109**
A-Barrigenol, C 439
Barringtogenic acid, C 442
Barringtogenol, C 438, 439
Bassic acid, C 441
Batatic acid, B **171**
Bauerenol, C 413, 415, 416, 451, 476, 477
Bayogenin, C 425, 440
Beckmann rearrangement, A 46; B 100, 118; C 237; D 27, 343
Benzaldehyde, A 109, 184, 186, B 375, 394
1-Benzamido-2-chlorocyclohexane, B 82
2-Benzamidocyclohexan-1-ol, B 82, 418
2-Benzamidocyclopentanol, B 418
2,3-Benz-5α-androst-2-ene, D 451, 452
Benzene, A 84; B 27
—, catalytic hydrogenation, B 2, 15
—, cyclopentadienylmetal complexes, A 143
—, fluorination, B 37
—, reaction with bromine, B 37
—, — with carbenes, B 359
—, — with diazoactic esters, B 356
—, — with diazomethane, B 349
—, reduction, B 3
—, to cyclohexadiene, B 27
Benzene compounds, catalytic hydrogenation, B 1, 4
—, reduction to cyclohexanes, B 1
5-Benzeneazotropolones, B 373
Benzeneglycol, B 65
Benzenehexacarboxylic acid, B 152
Benzene hexachlorides, B **36**, 37
Benzenesulphonic acid, from cyclohexane, B 15
Benzenetetracarboxylic acids, B 152
Benzene-1,2,3,4-tetracarboxylic acid, D 379
Benzene-1,2,3-tricarboxylic acid, B 152
Benzil, condensation with ketones, A 183
Benzilic acid rearrangement, D 20
Benzoheptafulvenes, B 350
Benzoic acids, B 136, 368, 375
Benzoic acid, B 374, 394
—, hydrogenation, B 133
—, reduction, B 137, 138
Benzonitrile, reaction with isobutene, C 49
4,5-Benzo-octalene, C 73
Benzophenone, C 266
Benzoquinones, C 50
Benzoquinone, A 173; B 28, 54, 112; C 76; D 328
—, cyclohexadiene, adducts, C 116
—, Diels-Alder adducts, A 118
—, reaction with cyclopentadiene, C 103
Benzotropones, B 364
2,3-Benzotropone, B 362
12-Benzoxycholanic acid, D 312
Benzoylalanine, C 342
Benzoylaminomethylcyclopropane, A 39
N-Benzoylcarvomenthylamine, B 229
Benzoylcyclobutane, A 94
Benzoylcyclopentadienyltricarbonylmanganese, A 140
Benzoylcyclopentane, A 175
2-Benzoylcyclopentan-1-one, A 189
1-Benzoylcyclopent-1-ene, A 175
1-Benzoylcyclopropane-1-carboxylic acid, A 57
1-Benzoyl-2,2-diphenyl-1-methyl-

cyclopropane, A 64
N-Benzoylmenthylamine, B 228
1-Benzoyl-2-nitro-3-phenylcyclopropane, A 50, 63
4-Benzoyloxycyclohexanone, B 355
1-Benzoyloxyethene, C 49
3α-Benzoyloxy-12-oxocholanic methyl ester, D 316
3-Benzoyl-2-phenylcyclopropanecarboxylic acids, A 58
β-Benzoylpropionic ester, A 121
2,3-Benzsuberone, C 72
Benzylamine, hydrogenation, B 86
Benzylcycloheptane, C 18
2-Benzylcyclohexane-1,3-dione, B 109
2-Benzylcyclohexan-1-ols, B 48
Benzylcyclohex-1-enes, B 21
2-Benzylcyclopentan-1-ols, A 155
2-Benzylcyclopropanecarboxamides, A 54
2-Benzylcyclopropanecarboxylic esters, A 54
Benzyl halides, B 93
—, reaction with sodium phenoxide, B 120
Benzylhydroquinone, B 112
Benzylideneacetophenone, A 94
5-Benzylidene-2-benzylcyclopent-2-en-1-one, A 164
2-Benzylidene-bicyclo[2.2.2]octane, C 112
Benzylidenecyclohexanones, B 96, 100
Benzylidenediphenylmaleide, A 186
Benzylidenemethylamine, B 228
2-Benzyl-4-methylcyclohexanone, B 95
Benzyloxyacetone, A 109, 184
N-Benzyloxyamine, C 213
3-Benzyltropolone, B 369
α-Bergamotene, C 267
β-Bergamotene, C **266,** 288
Betamethasone, D 352
Bethogenin, E **44,** 45
Betula alba, C 331
Betulafolienetetrol, C 425
Betulafolienetriol, C 425
Betulaprenols, E 133
Betulenols, C **331,** 332, 336
Betulin, C 413, 435, 461, 467
Betulinal, C 413, 435
Betulinal acetate, C 436
Betulinic acid, C 413, 414, 436, 461, 467
Betulonic acid, C 436
Beyerol, C 399
Bicyclcic compounds, bridged cyclohexanes, C 263
Bicyclic diterpenoids, C 375
Bicyclic hydrocarbons, C 108
Bicyclic monoterpenes, E 79
—, absolute configuration, C 249
Bicyclic monoterpenoids, C 136 ff
Bicyclobutane, A 48
Bicyclo[1.1.0]butanes, C **32,** 88
Bicyclo[1.1.0]butane-1-carbonitriles, C 33
Bicyclo[1.1.0]butane-1-carboxylic ester, C 32
Bicyclobutonium ions, A 24
Bicyclo[5.5.0]decadienes, C 72
Bicyclo[5.3.0]decal-2-ol, B 405
Bicyclo[4.4.0]decanes, C 5, **60**
Bicyclo[5.3.0]decanes, C **59**
Bicyclo[6.2.0]decanes, C **50**
Bicyclo[7.1.0]decanes, C **42**
Bicyclo[6.2.0]decan-1-ol-10-one, C 51
Bicyclo[5.3.0]decan-2-one, C 60
Bicyclo[5.3.0]decan-3-one, C 60
Bicyclo[5.3.0]decan-4-ones, C 59
Bicyclo[5.3.0]decan-9-one, C 60
Bicyclo[6.2.0]deca-1,3,5,7-tetraene, C 50
Bicyclo[5.3.0]dec-1(7)-ene-2-one, C 60
Bicyclo[5.3.0]dec-1(7)-ene-8-one, C 60
Bicyclo[5.3.0]dec-7-ene-9-one, C 60
Bicyclo[5.5.0]dodecanes, C **72**
Bicyclo[5.5.0]dodecan-2-one, C 73
Bicyclo[6.4.0]dodecene, B 387
Bicyclo[5.5.0]dodec-1(7)-ene-2-one, C 73
Bicycloheptadiene, C 55
Bicyclo[2.2.1]heptadiene, C 38, 85, 89, 99, 105, 118
Bicyclo[3.2.0]hepta-2,6-diene, B 350; C 45
Bicyclo[4.1.0.]hepta-2,4-diene, C 38
Bicyclo[2.2.1]heptadiene-2,3-dicarboxylic acid, C 91
Bicyclo[3.2.0]hepta-3,6-dien-2-one, C 46
Bicyclo[2.2.1]heptan-*endo*-2-*exo*-5-diol, C 94
Bicyclo[2.2.1]heptan-*exo*-2-*syn*-7-diol, C 94
Bicyclo[2.2.1]heptan-2,5-dione, C 95
Bicycloheptanes, C 136
Bicyclo[2.2.1]heptanes, C 78, **89,** 99
Bicyclo[2.2.1]heptane, C 79, 82, 89, 206
Bicyclo[3.1.1]heptanes, C 4, **88**
Bicyclo[3.2.0]heptanes, C 4, **45,** 46
Bicyclo[4.1.0]heptanes, C **38**
Bicycloheptanecarbaldehydes, C 95

Bicyclo[2.2.1]heptanecarboxylic acids, C 96, 97
Bicyclo[3.1.1]heptane-3-carboxylic acid, C 88
Bicyclo[4.1.0]heptane-7-carboxylic acid, C 39
Bicyclo[2.2.1]heptane-2,2-dicarboxylic acid, C 99
Bicyclo[2.2.1]heptane-2,3-dicarboxylic acid, C 97, 98, 100, 101
Bicyclo[2.2.1]heptane-2,3- dicarboxylic ester, C 98
Bicyclo[2.2.1]heptane-2,5-(or 2,6)-dicarboxylic acid, C 99
Bicyclo[3.1.1]heptane-3,3-dicarboxylic acid, C 88
Bicyclo[2.2.1]heptane-2-ols, C 91, 97
Bicyclo[2.2.1]heptan-7-ol, C 92
Bicyclo[3.2.0]heptan-2-ol, C 46
Bicyclo[4.1.0.]heptan-2-ol, C 38
Bicyclo[2.2.1]heptan-2-one, C 83, 84, 95
Bicyclo[2.2.1]heptan-7-one, C 95
Bicyclo[3.1.1]heptan-6-one, C 88
Bicyclo[3.2.0]heptan-3-one, C 42, 46
Bicyclo[4.1.0.]heptan-2-one, C 38
Bicyclo[3.1.1]heptan-6-one-1-carboxylic acid, C 89
Bicyclo[2.2.1]hept-2-ene, A 192; C 89, 118, 120, 121
Bicyclo[3.2.0]hept-1(5)-ene, A 191; B 350; C 45
Bicyclo[3.2.0]hept-2-ene, C 34, 45
Bicyclo[3.2.0]hept-6-ene, A 191; B 350; C 45
Bicyclo[4.1.0]hept-3-ene, C 37, 38
Bicyclo[2.2.1]hept-2-ene-1-carboxylic acid, C 97
Bicyclo[2.2.1]hept-2-ene-7-carboxylic acid, C 97
Bicyclo[2.2.1]hept-5-ene-2-carboxylic acid, C 96
Bicyclo[2.2.1]hept-2-ene-2,3-dicarboxylic acid, A 192 C 98
Bicyclo[2.2.1]hept-5-ene-2,2-dicarboxylic acid, C 99
Bicyclo[2.2.1]hept-5-ene-2,3-dicarboxylic acids, C 98
Bicyclo[2.2.1]hept-5-ene-2,3-dicarboxylic ester, C 98
Bicyclo[2.2.1]hept-5-en-2-ol, C 92
Bicyclo[3.2.0]hept-2-en-6-ol, C 45
Bicyclo[3.2.0]hept-3-en-2-ol, C 46
Bicyclo[3.2.0]hept-6-en-3-ol, C 46
Bicyclo[2.2.1]hept-2-en-5-ol acetate, C 46
Bicyclo[2.2.1]hept-2-en-7-one, C 95
Bicyclo[2.2.1]hept-5-en-2-one, C 93, 95
Bicyclo[3.2.0]hept-2-en-6-one, C 45
Bicyclo[3.2.0]hept-6-en-2-one, C 46
Bicyclo[3.2.0]hept-6-en-3-one, C 46
Bicyclohept-1-enyl, C 18
Bicyclohepten-5-yl halides, C 91
Bicycloheptyl, C **18**
Bicyclo[2.2.1]heptyl carbonium ion, C 83, 97
Bicyclo[2.2.1]heptyl derivatives, C 162
Bicycloheptylidene, C 18
Bicyclo[2.2.0]hexa-2,5-diene, C 42, 43
Bicyclo[2.2.0]hexa-2,5-diene-2-carboxylic ester, C 43
Bicyclo[2.2.0]hexa-2,5-diene-2,3-dicarboxylic ester, C 43
Bicyclo[2.1.1]hexanes, C 34, **84**, 85, 86
Bicyclo[2.2.0]hexanes, C **42**, 43
Bicyclo[3.1.0]hexanes, B 123, 161; C **36**
Bicyclo[2.2.0]hexane-2-carboxylic acid, C 43
Bicyclo[3.1.0]hexane-1-carboxylic acid, C 37
Bicyclo[3.1.0]hexane-6-carboxylic acid, C 38
Bicyclo[2.2.0]hexane-1,4-dicarboxylic acid, C 42
Bicyclo[2.2.0]hexane-2,3-dicarboxylic acid, C 43
Bicyclo[2.1.1]hexan-2-ol, C 85
Bicyclo[2.1.1]hexan-5-ols, C 86
Bicyclo[2.2.0]hexan-2-ol, C 43
Bicyclo[3.1.0]hexan-2-ol, C 36
Bicyclo[3.1.0]hexan-3-ol, C 36, 37
Bicyclo[2.1.1]hexan-2-one, C 85
Bicyclo[3.1.0]hexan-2-one, C 37
Bicyclo[3.1.0]hexan-3-one, C 36
Bicyclo[2.2.0]hex-2-ene, C 43
Bicyclo[3.1.0]hex-2-ene, C 36
Bicyclo[3,1,0]hex-2-ene-6-carbaldehyde, C 38, 91
Bicyclo[3.1.0]hex-2-ene-6-carboxylic acid, C 38
Bicyclo[2.2.0]hex-2-ene-1,4-dicarboxylic acid, C 42
Bicyclo[3.1.0]hex-3-en-2-one, B 123
Bicyclohexenyls, hydrogenation, C 14
Bicyclohex-1-enyl, C 15
Bicyclohex-2-enyl, B 42; C 15
Bicyclohexyl, C 9, **14**
Bicyclohexyl-4-carboxylic acid, C 17
Bicyclohexyl compounds, C 6

Bicyclohexyl-2,2′-dicarboxylic acids, C 17
Bicyclohexyl-4,4′-dicarboxylic acid, C 17
Bicyclohexyl-4,4′-dicarboxylic methyl ester, C 17
Bicyclohexyl-1,1′-diol, C 15
Bicyclohexyl-4,4′-diol, C 16
Bicyclohexylidene, C **14**
Bicyclo[3.3.1]nona-3,7-diol, C 125
Bicyclo[3.3.1]nona-2,6-dione, C 124, 126, 132
Bicyclo[3.3.1]nona-2,9-dione, C 126
Bicyclo[3.3.1]nona-3,7-dione, C 125, 126
Bicyclo[3.2.2]nonanes, C **121,** 122
Bicyclo[3.3.1]nonanes, C **124,** 129
—, transannular reactions, C **127**
Bicyclo[4.2.1]nonanes, C **128**
Bicyclo[4.3.0]nonanes, C 5, **55**
Bicyclo[5.1.1]nonanes, C 129
Bicyclo[5.2.0]nonanes, C **50**
Bicyclo[6.1.0]nonanes, C **41**
Bicyclo[3.2.2]nonane-6-carbaldehyde, C 122
Bicyclo[3.2.2]nonane-6-carbonitrile, C 123
Bicyclo[6.1.0]nonane-9-carboxylic acid, C 41
Bicyclo[3.2.2]nonane-6,7-dicarboxylic acid, C 123
Bicyclo[3.3.1]nonane-2,6-dione-1,3,5,7-tetracarboxylic acid, C 132
Bicyclo[3.3.1]nonane-2,6-dione-1,3,5,7-tetracarboxylic methyl ester, C 125, 126
Bicyclo[3.3.1]nonan-2-en-9-one, C 126
Bicyclo[3.2.2]nonan-1-ol, C 122
Bicyclo[3.2.2]nonan-2-ol, C 122
Bicycle[3.2.2]nonan-3-ol, C 122
Bicyclo[3.3.1]nonan-2-ol, C 125, 128
Bicyclo[3.3.1]nonan-9-ol, C 125
Bicyclo[3.2.2]nonan-2-one, C 122
Bicyclo[3.2.2]nonan-3-one, C 122
Bicyclo[3.2.2]nonan-6-one, C 122
Bicyclo[3.2.2]nonan-7-one, C 123
Bicyclo[3.3.1]nonan-2-one, C 125
Bicyclo[3.3.1]nonan-3-one, C 125
Bicyclo[3.3.1]nonan-9-one, C 125, 126
Bicyclo[4.2.1]nonan-2-one, C 128
Bicyclo[4.2.1]nonan-3-one, C 128
Bicyclo[4.2.1]nonan-9-one, C 128
Bicyclo[5.1.1]nonan-8-one, C 129
Bicyclo[5.2.0]nonan-4-one, C 50
Bicyclo[3.3.1]nonan-2-one-1-carboxylic acid, C 126
Bicyclo[6.1.0]nonatriene, C 58
Bicyclo[6.1.0]nona-2,4,6-triene-9-carboxylic acid, C 41
Bicyclo[3.2.2]non-2-enes, C 122
Bicyclo[5.2.0]non-8-ene, C 50
Bicyclo[3,2.2]non-8-ene-6,7-dicarbonitrile, C 123
Bicyclo[3.2.2]non-8-ene-6,7-dicarboxylic acid, C 123
Bicyclo[3.2.2]non-6-ene-2,3-dione-8,9-dicarboxylic acid, B 370
Bicyclo[3.2.2]non-8-ene-6,6,7,7,-tetracarbonitrile, C 123
Bicyclo[3.2.2]non-6-en-2-one, C 123
Bicyclo[3.3.1]non-2-en-9-one ethylene acetal, C 125
Bicyclo[3.2.2]nonyl 1-chloride, C 122
Bicyclo[2.2.2]octa-2,5-diene, C 108, 109, 110
Bicyclo[3.2.1]octa-2,6-diene, C 117
Bicyclo[4.2.0]octa-2,4-diene, A 82
Bicyclo[4.3.0]octa,4,6-diene, C 58
Bicyclo[2.2.2]octa-2,5-diene-2-carboxylic acid, C 114
Bicyclo[3.3.0]octa-1,4-diene-3-carboxylic acid, C 54
Bicyclo[3.3.0]octa-2,6-diene-4,8-dione, C 54
Bicyclo[5.1.0]octa-3,5-diene-2-one, C 40
Bicyclo[5.1.0]octadienyl hexachloroantimonate, B 394
Bicyclo[2.2.2]octa-2,3-dione, C 112
Bicyclo[2.2.2]octa-2,5-dione, C 112
Bicyclo[2.2.2]octa-2,6-dione, C 112
Bicyclo[3.3.0]octa-2,6-dione, C 54
Bicyclo[3.3.0]octa-2,8-dione, C 54
Bicyclo[2.2.2]octa-2,3-dione-1,4-dicarboxylic acid, C 114
Bicyclo[2.2.2]octa-5,7-dione-2,3-dicarboxylic anhydride, C 115
Bicyclo[3.3.0]octa-2,6-dione-3,7-dicarboxylic ester, C 54
Bicyclo[2.2.2]octan-2,3-epoxide, C 111
Bicyclo-octanes, C 109
Bicyclo[2.2.2]octanes, C 78, 108, 122
Bicyclo[2.2.2]octane, C 4, 51, 79, 108, 112, 114
Bicyclo[3.2.1]octane, C 5, 51, **117**
Bicyclo[3.3.0]octanes, C **51,** 117
Bicyclo[3.3.0]octane, B 384
Bicyclo[4.2.0]octanes, C **47,** 50
Bicyclo[4.2.0]octane, B 395
Bicyclo[5.1.0]octanes, C **39**
Bicyclo[2.2.2]octane-2-carbaldehyde, C 108, **112**

Bicyclo[2.2.2]octane-1-carboxylic acids, C 113
Bicyclo[2.2.2]octane-2-carboxylic acid, C 114, 117
Bicyclo[3.3.0]octane-3-carboxylic acid, C 53
Bicyclo[2.2.2]octane-1,2-dicarboxylic acid, C 114
Bicyclo[2.2.2]octane-1,3-dicarboxylic acid, C 114
Bicyclo[2.2.2]octane-2,3-dicarboxylic acid, C 115
Bicyclo[3.2.1]octane-3,3-dicarboxylic acid, C 121
Bicyclo[4.2.0]octane-7,8-dicarboxylic acids, C 49
Bicyclo[4.2.0]octane-7,8-dicarboxylic esters, C 49
Bicyclo[3.3.0]octane,-2,4,6,8-tetracarboxylic acid, C 55
Bicyclo[2.2.2]octan-1-ol, C 111
Bicyclo[2.2.2]octan-2-ol, C 111, 112, 119
Bicyclo[3.2.1]octan-2-ol, C 109, 117, 118, 119, 120
Bicyclo[3.2.1]octan-3-ol, C 118, 119
Bicyclo[3.2.1.]octan-8-ol, C 120
Bicyclo[3.3.0]octan-2-ol, B 405
Bicyclo[3.3.0]octan-3-ol, C 53
Bicyclo[4.2.0]octanols, C 48
Bicyclo[5.1.0]octanols, C 39, 40
Bicyclo[2.2.2]octan-2-ol acetate, C 112
Bicyclo[4.2.0]octan-7-ol-8-one, C 47
Bicyclo[2.2.2]octan-2-one, C 112
Bicyclo[3.2.1]octan-2-one, C 112, 119, 120
Bicyclo[3.2.1]octan-3-one, C 112, 119, 120
Bicyclo[3.2.1]octan-6-one, C 121
Bicyclo[3.2.1]octan-8-one, C 120, 121
Bicyclo[3.3.0]octan-2-one, C 51, 53
Bicyclo[3.3.0]octan-3-one, C 51, 53
Bicyclo[4.2.0]octan-2-one, C 47, 48, 49
Bicyclo[4.2.0]octan-3-one, C 48
Bicyclo[4.2.0]octan-7-one, C 48
Bicyclo[3.2.1]octan-6-one-7-carboxylic ester, C 121
Bicyclo[3.3.0]octan-3-ylamine, C 53
Bicyclo[3.3.0]octan-3-yl magnesium bromide, C 53
Bicyclo[2.2.2]octa-2,5,7-triene, C 109, 110
Bicyclo[4.2.0]octatriene, B 396
Bicyclo[2.2.2]octa-2,6,7-trione, C 113
Bicyclo[2.2.2]octenes, C 110
Bicyclo[2.2.2]oct-2-enes, B 148
Bicyclo[2.2.2]oct-2-ene, C 108, 109, 118
Bicyclo[3.2.1]oct-2-ene, C 117
Bicyclo[3.2.1]oct-3-ene, C 118
Bicyclo[3.3.0]oct-2-ene, B 383
Bicyclo[3.3.0]oct-2-ene, C 52
Bicyclo[4.2.0]oct-7-ene, C 49
Bicyclo[5.1.0]oct-3-ene, B 384
Bicyclo[2.2.2]oct-5-ene-2-carbaldehyde, C 112
Bicyclo[2.2.2]oct-5-ene-2-carbonyl chloride, C 111
Bicyclo[2.2.2]oct-5-ene-2-carboxylic acid, C 114, 133
Bicyclo[2.2.2]oct-5-ene-2-carboxylic ester, C 114
Bicyclo[2.2.2]octenedicarboxylic acids, B 145, 151
Bicyclo[2.2.2]oct-2-ene-1,4-dicarboxylic acid, C 114
Bicyclo[2.2.2]oct-5-ene-1,2-dicarboxylic acid, C 114
Bicyclo[2.2.2]oct-2-ene-2,3-dicarboxylic anhydride, C 114
Bicyclo[2.2.2]oct-5-ene-2,3-dicarboxylic anhydride, C 115
Bicyclo[4.2.0]oct-3-ene-1,6-dicarboxylic methyl ester, C 49
Bicyclo[3.2.1]oct-3-en-2-ol, C 119
Bicyclo[2,2.2]oct-5-en-2-ol, C 111, 112
Bicyclo[3.2.1]oct-3-en-2-one, C 119 120
Bicyclo[3.2.1]oct-2-en-8-one, C 121
Bicyclo[3.3.0]oct-1(5)-en-2-one, C 54
Bicyclo[3.3.3]oct-1-en-3-ones, C 54
Bicyclo[4.2.0]octen-7-one, C 47
Bicyclo[5.1.0]oct-1(7)-en-8-one, C 39
Bicyclo-oct-1-enyl, C 20
Bicyclo[2.2.2]oct-5-en-2-yl tosylate, C 119
Bicyclo-octyl, C 18
Bicyclo[2.2.2]octyl 2-brosylates, C 83
Bicyclo[2.2.2]octyl carbonium ion, C 83
Bicyclo-octylium ions, C 118
Bicyclo[2.2.2]octyl-2-methyl *p*-toluenesulphonate, 122
Bicyclopentadienyl metals, A 126
Bicyclo[1.1.1]pentanes, C 84
Bicyclo[2.1.0]pentanes, C 35
Bicyclo[2.1.0]pent-2-ene, C 35
2,2-Bicyclopentenyl, C 10
Bicyclopentyl, C 3, 9
Bicyclopentylidene, C 10
Bicyclopropyl, C 6

4-Bicyclopropyl-*n*-butanol, C 7
Bicycloprop-2-ylcyclohexane, C 13
Bicyclopropyl-hexacarboxylic acid, C 8
Bicyclo steroids, D 336
Bicyclo[6.6.0]tetradecaheptaene, C 73
Bicyclo[6.6.0]tetradecanes, C **73**
Bicyclo[5.4.0]undecanes, C **72**
Bicyclo[5.40]undecane group C 281
Bicyclo[5.4.0]undec-1(7)-ene, C 72
Bicyclo[5.4.0]undecan-1-ol, C 72
Bicyclo[5.4.0]undecan-2-one, C 72
Bicyclo[5.4.0]undecan-9-one, C 72
Bicyclovetivene, C 323
Bicyclovetivenol, C **323**, 336
Biferrocenyl, A 135
Biformene, C 383
Bigelovin, C **360**, 368
Bijvoet's Anomalous Dispersion Method, C 481
Bile acids, D 1 ff, 2, 18, 168 ff
—, analysis and characterisation, D 177
—, chromatographic analysis, D 178
—, conjugated, D 176, 177
—, degradation, D 24, 185
—, from cholesterol, D 169
—, isolation, D 170
—, labelled, D 185
—, nomenclature, D 169
—, physical properties, D 171
—, structure, D **8**ff
—, unsaturated, D 186ff
—, unsubstituted, D 179
—, with more than 24C, D 192
Bile alcohols, D 193ff
—, physical properties, D 195
Bilianic acids, D 20, **169**
Biloidanic acid, D 20
Biogenetic Isoprene Rule, B 234; C 369, 370, 371, 406, 460, 464; E 54
Biosynthesis, diterpenoids, C 460
—, gibberellic acid, C 461
—, lanosterol, C **464**
—, mutilin, C 463
—, rosenonolactone, C 460
—, sesterperpenoids, C 463
—, terpenoids, C 460ff
—, triterpenoids, C 464
Biphenyl, hydrogenation, C 14
—, reduction, C 6
Birch reduction B 2, 3; C 58; D 213, 242, 244, 246, 247, 262, 274
Birdcage alcohol, C 107
Birdcage hydrocarbons, C **106**
Bisabolane, C **291**
Bisabolene, C **259**, 250, 288, 291; E 85
Bisabolene trihydrochloride, C 259
Bisabol myrrh, C 259
Bisabolol C **292**, 334; E 80
2,3-Bisacetoxymethylbicyclo [2.2.1]-heptane, C 91
1α,7α-Bisacetylthio-17α-methyl-testosterone, D 275
1,4-Bis(2-aminoethyl)cyclohexane, B **86**
1,1-Bisbromomethylcyclohexane, C 24
4,4-Bisbromomethylcyclohex-1-ene, C 24
1,1-Bis(bromomethyl)cyclopropane, C 21
2,3-Bisbromomethylmethylene-cyclopropane, A 34
Bisbutadiene-quinone, C 76
Bis-2-butyl-3-methylborane, B 179
Bis(carbonylcyclopentadienenickel), A 130
Bis(cyclopentadienyldicarbonyliron), A 146
Bis(cyclopentadienyliron), A 124
Bis(cyclopentadienylmanganese), A 125
Bis(cyclopentadienylrhenium), A 142
Bis(cyclopentadienyltitatium), A 126
Bis(cyclopentadienyltricarbonyl-tungsten), A 146
Bis(cyclopentadienylvanadium), A 142
Bisdehydro-annulenes, B 397
Bisdehydro-[14]-annulene, B 398
Bisdehydro-β-carotene, B 253
Bisdehydrodoisynolic acids, D 224, 226
8,14-Bisdehydro-17α-estradiol 3-methyl ether 17-acetate, D 207
Bisdehydroestrolic acids, D 227
Bisdehydroestrolic lactones, D 227
Bisdehydro-estrone, D 206, 207
8,14-Bisdehydro-estrone 3-methyl ether, D 207
9,11-Bisdehydro-17α-ethynylnortes-tosterone, D 247
Bisdehydrolycopene, B 249, **280**, 336
Bisdehydromarrianolic acids, D 224
9,11-Bisdehydro-17α-methyl-19-nortestosterone, D 273
1,6-Bisdehydro-17α-methyltestoste-rone, D 275
Bisdehydronorprogesterones, D **247**
Bisdehydronortestosterones, D **247**
Bis-desmethyl-lanostane derivatives, C 408

1,1-Bis(dialkylboryl)-3-chloropropanes A 41
Bis(2,2-dibromo-3,3-dimethylcyclopropyl) ether, A 43
Bis(2,2-dichlorocyclopropyl) ether, A 43
Bis(2,2-dicholoro-3,3-dimethyl cyclopropyl) ether, A 43
Bisdihydropentalenyliron, C 52
1,6-Bis(dimethylamino)cycloocta-2,4-diene, B 420
1,2-Bis (dimethylaminomethyl)-cyclohexane, B 24
1,1-Bis(*N*,*N*-dimethylaminomethyl)-ferrocene, A 133
Bis(dimethylfulvene) C 103
1,2-Bis(diphenylhydroxymethyl)-cyclopropane, A 44
Bis-1,3-diphenylmethylene-cyclobutane, A 84
Bisergostatrienol, D 135
2,2-Bis-ethoxycarbonylbicyclo-[2.2.1]heptane, C 99
2,2-Bis-ethoxycarbonylbicyclo-[2.2.1]hept-5-ene, C 99
3,3-Bis-ethoxycarbonylbicyclo-[3.3.1]nonane, C 124
1,4-Bis (*N*-ethoxycarbonyl-*N*-nitrosoamino) butane, C 128, 129
1,3-Bis (*N*-ethoxycarbonyl-*N*-nitrosoamino)propane, C 129
3,20-Bisethylenedioxy-11α-hydroxy-5α,11β,19-cyclopregnane, D 334
3,20-Bisethylenedioxy-11-oxo-5α,9β,19-cyclopregnane, D 334
3,20-Bisethylenedioxy-11-oxo-5α-pregnane, D 334
Bishomoestrone, D 280
α, ω-Bis(1-hydroxycyclohexyl)-alkanes, C 16
Bis-(1-hydroxycyclohexyl)ethene, C 16
Bis-(1-hydroxycyclohexyl)ethyne, C 16
1,4-Bis(1′-hydroxycyclopentyl)-buta-1,3-diyne, C 13
Bis-(1-hydroxycyclopentyl)diacetylene C 13
Bis(1-hydroxycyclopentyl)ethane, C 12
Bis(1-hydroxycyclopentyl)ethene, C 12
Bis(1-hydroxycyclopentyl)ethyne, C 12
Bis 1,3-(α-hydroxydiphenylmethyl)-cyclobutane, A 83
Bishydroxymethylcyclohexanes, B 16
1,3-Bis(hydroxymethyl)cyclohexane, C 124
1,4-Bis(hydroxymethyl)cyclohexane, B 86
1,3-Bis(hydroxymethyl)cyclohexyl di-*p*-toluenesulphonate, C 126
1,2-Bishydroxymethylcyclopentane, A 118, 162
1,3-Bis(hydroxymethyl)cyclopentane, C 117
1,3-Bis(hydroxymethyl)cyclopentane ditosylate, C 121
Bisindenylcobalticinium ion, A 139
Bisindenyliron, A 145
Bisindenylmetal complexes, A 128
1,1-Bisiodomethylcyclobutane, C 23
1,2-Bis(iodomethyl)cyclohexane, B 24
Bismethoxycarbonylbicyclo[1.1.0]-butane, C 33
2,3-Bismethoxycarbonylbicyclo-[2.2.1]hept-5-ene, C 98
2,3-Bis(methoxycarbony)norborn-adien-7-ylcyclopentadienylnickel, A 140
4,6-Bis(methylamino)cyclohexane-1,2,3,5-tetrol, B 84
3β,20α-Bis(methylamino)-14α-methyl-4-methylene-9β,19-cyclopregn-16α-ol, D 335
Bismethylenecyclobutanes, C 22
17α,20:20,21-Bismethylenedioxy-pregn-5-ene-3,19-diol 3-acetate, D 331
17α,20:20,21-Bismethylenedioxy-pregn-4-en-3-one, D 331
1,4:5,8-Bis-*endo*methyleneperhydro-anthracene, C 103
Bisnitrosocyclohexane, B 92
Bisnoradamantane, C 55
Bisnorchaulmoogric acid, A 197
Bisnorcholanic acids, E 21
Bisnorcholanoic acids, D 169, 179, 191
9β,10α-Bisnorchol-4-en-3-on-22-al, D 237
Bisnordeoxycholic acid, D 322, 323
18,19-Bisnordeoxycorticosterone, D 301
Bisnoronoceranediols, C 475
18,19-Bisnorprogesterone, D 247
18,19-Bisnortestosterone, D 262
Bis(pentaphenylcyclopentadienyl)-molybdenum, A 126
Bisphenylazocyclopentadienylmetal complexes, A 128

Bis-(2,2,3,3-tetrafluorocyclobutyl)-ethyne, C 9
1,3-Bistosylmethylcyclobutane, C 88
Bistrifluoromethylethyne, C 34
Bis(triphenylcyclopropenyl), C 8
Bis(triphenylphosphite)carbonylnickel B 25, **28**
Bitocholic acid, D 173
Bitter fennel oil, B 182
Bixa orellana, B 326; E 133
Bixin, B 244, 257, **326**; E 133
Bixindial, B 267, **327**
Blakeslea trispora, B 249; E 86
Blighia sapida, A 70
Bluensidine, B **85**
Bluensomycin, B **85**
Blumae balsamifera, C 199, 226
Blumae eriantha, B 197
Blumae malcomii, B 197
Boat conformation, cyclohexane ring, B 7
Boistroside, D 364, **398**
Boivinose, D 361, 365
Bolasterone, D 272
Bornane, C 136, 137, **206**, 207, 225, 229; E 69
Bornane bases, C 232
Bornane-3-carboxylic acid, C 224
Bornanediones, C **218**
Bornenes, C 231
Bornene, C 198, 206, 229, 230, 233
Bornene-3-carboxylic acid, C 224
Bornene-3-carboxylic azide, C 225
Bornene-3-carboxylic methyl ester, C 225
Bornene-3-isocyanate, C 225
Borneo-camphor, C 226
Borneols, C 218, 231, 241
—, relation to camphor, C 227
Borneol, C 89, 137, 164, 193, 197, 200, 218, 223, **226**, 227, 228, 229, 233, 251; E 69, 79
Borneol acetate, C 228
Borneol nitrite esters, C 228
Bornesitol, B **61**
Bornyl acetate, C 222, 223, 227
Bornyl alcohol, C 172
Bornylamine, C **232**, **233**
Bornylamine acetate, C 232
Bornylamine hydrochoride, C 232
Bornylane C 137
Bornyl chlorides, C 194, 227, 231
Bornyl chloride, C 160, 164, 171, 206, 207, 212, 229, 251, 252
β-Bornylene, C 192
Bornyl halides, C 230
Bornyl iodide, C 194, 206
Bornyltrimethylammonium hydroxide, C 230, 233
Boronia megastigma, B 174
Boswellia carterii, C 181; E 76
Boswellia serrata, C 138
α-Boswellic acid, C 439
β-Boswellic acid, C 450
Bothriochloa intermedia, C 301
Botogenin, E **26**, **34**
Boujotinolone A, C 426
Bourjotinolone B, C **426**
Bourjotone, C 426
Bouveault-Blanc method, A 43
Bovea volubilis, D 412
Bovogenin A, D 366, 412
Bovogenin E, D 412
Bovogen-19-ol, A, D 412
Bovosides A, B and C, D 366, **412**
Bovoside D, D 366, 412
Bovoside E, D **412**
Brassica campestris, D 156
Brassica rapa, D 155
Brassicasterol, D 3, 13, **155**, 165; E 104, 105
Brassylic ester, A 197
Brassyloin, B **412**
Bredemolic acid, C 440
Bredt's rule, C 79
Brein, C 450
Bridged carbonium ions, C **80**
Bridged cyclohexanes, C 263
Bridged ring compounds, C 77 ff
—, synthesis, C 78 ff
N-Bromoacetamide, B 35; D 312, 326; E 33
N-Bromoacetamide sodium, D 322
Bromoacetic acid, D 407
Bromoacetic ester, A 201; B 167; D 407
α-Bromoacrylic ester, A 61
1-Bromoadamantane, C 130, 131
3-Bromoadamantane-1-carbamic methyl ester, C 124
1-Bromoadamantane-3-carboxylic acid, C 126
ω-Bromoalkylamines, ring closure, A 9
4-Bromo-*n*-amylrethrone, A 183
2α-Bromoandrostan-17β-ol-3-one, D 269
14α-Bromo-5α-androstan-15-one, D 269
14β-Bromo-5α-androstan-15-one, D 269
Bromobenzenes, B 13

7-Bromobicyclo[2.2.1]heptane, C 97
1-Bromobicyclo[2.2.2]octane, C 111
2-Bromobicyclo[2.2.2]octane, C 108, 117
Bromobicyclo[3.2.1]octane, C 108, 117
3-Bromobicyclo[3.2.1]oct-2-ene, C 118, 120
8-Bromobicyclo[3.3.1]oct-2-ene, C 108
3-(4′-Bromobenzenesulphonyloxy)-cyclohexanecarboxylic methyl ester, C 37
2-Bromobornane-3-carboxylic acid, C 224, 230
1-Bromo-3-bromomethylcyclobutane, C 84
1-Bromo-2-(bromomethyl)cyclopentane, C 36
2-(ω-Bromobutyl)cyclopentanone, C 28
4-Bromo-*n*-butylrethrone, A 183
Bromocamphors, physical properties, C 211
α-Bromocamphor, C 208, 209, 212, 223, 264
α′-Bromocamphor, C 208, 209, 212
π-Bromocamphor, C 209, 210, 218
α-Bromocamphorcarboxylic acid, C 223
3-Bromocamphor-3-carboxylic acids, C 208
Bromocamphoronic acid, C 203
α-Bromocamphor-10-sulphonic acid, C 215
α-Bromocamphor-π-sulphonic acid, C 215
α-Bromocamphor-π-sulphonyl bromide, C 209
α-Bromo-α′-chlorocamphor, C 209
α′-Bromo-α-chlorocamphor, C 209
π-Bromo-α-chlorocamphor, C 210
Bromochlorocholestan-3-ones, D 122
1-Bromo-2-chlorocycloalk-1-enes, B 388
3-Bromo-1-chlorocyclobutane, C 32
2-Bromo-1-chlorocyclohept-1-ene, B 351
1-Bromo-1-chlorocyclohexane, B 33, 34
1-Bromo-2-chlorocyclohexane, B 33, 34, 35
1-Bromo-3-chlorocyclohexane, B 40
2-Bromo-1-chloro-1-fluorocyclohexane, B 35
3-Bromo-5-chlorotropolone, B 373
6α-Bromo-5α-cholestane, D 50
6β-Bromo-5α-cholestane, D 50
5-Bromo-5α-cholestane-3β,6β-diol, D 49
β-Bromo-5α-cholestan-3α-ol, D 121
Bromocholestan-3-ones, D 122
2α-Bromocholestanone, D 120, 123
2α-Bromo-5α-cholestan-3-one, D 120, 121, 129
2β-Bromo-5α-cholestan-3-one, D 120
4α-Bromo-5α-cholestan-3-one, D 120
4β-Bromo-5α-cholestan-3-one, D 120
4β-Bromo-5β-cholestan-3-one, D 124
6β-Bromo-5α-cholestan-3-one, D 50
4-Bromocholest-4-en-3-one, D 128, 131
6α-Bromocholest-4-en-3-one, D 128
6β-Bromocholest-4-en-3-one, D 128, 131
7β-Bromocholesterol, D 103
7β-Bromocholesteryl benzoate, D 54
7α-Bromocholesteryl chloride, D 41
7-Bromocholesteryl esters, D 54
Bromocyanoacetic ester, A 63
Bromocycloalkanes, B 401
Bromocyclobutane, A 88
3-Bromocyclobutanecarboxylic ester, C 32
2-Bromocyclobutanone, A 92
1-Bromocyclobutene, A 87, 88
3-Bromocyclobut-1-ene, A 82
7α-Bromo-3α,5-cyclo-5α-cholestan-6-one, D 80
Bromocyclodecane, B 402
2-Bromocyclodecanone, B 411
3-Bromocyclodecanone, B 411
1-Bromocyclodecene, B 402
3-Bromocyclodecene, B 402
9-Bromocyclododeca-1,5-diene, B 402
Bromocyclododecane, B 402
2-Bromocyclododecanone, B 410, 411
1-Bromocyclododecene, B 402
1-Bromocycloheptadecene, B 402
1-Bromocycloheptanecarboxylic acid, B 356, 357
2-Bromocyclo[2.2.1]heptane-1-carboxylic acid, C 97
2-Bromocycloheptanone, B 354, 355
1-Bromocyclohept-1-ene, B 351
7-Bromobicyclo[2.2.1]hept-2-ene, C 97
4-Bromocyclohexadienones, B 126
Bromocyclohexane, B 33
Bromocyclohexanecarboxylic acids, B 135, 136, 137
2-Bromocyclohexane-1,3-dione, B 111
2-Bromocyclohexan-1-ol, B 66
2-Bromocyclohexanones, B 102, 103, 104

6-Bromocyclohexan-1-one-2-carboxylic ester, A 191
1-Bromocyclohex-1-ene, B 34, **41**, 137
3-Bromocyclohex-1-ene, B 27, 41, **42**, 82
2-Bromo-3-cyclohexylprop-1-ene, B 31
2-Bromocyclopentadienone, C 44
Bromocyclopentanes, A 148, 152
2-Bromocyclopentanone, A 166, 184
1-Bromocyclopentene, A 149
3-Bromocyclopentene, A 149, 156, 162
4-Bromocyclopent-2-ene-1-one, A 168, 172
1-Bromocyclopentyl-1-acetic acid, A 192
Bromocyclopropanes, A 39, 40
Bromocyclononane, B 402
2-Bromocyclononanone, B 411
Bromocyclo-octane, B **402**
2-Bromocyclo-octanone, B 356, **411**
1-Bromocyclo-octene, B 402
3-Bromocyclo-octene, B 402
2-Bromocyclo-octyl acetate, B 403
Bromocyclopentadecane, B 402
2-Bromocyclopentadecanone, B 411
1-Bromocyclopentadecene, B 402
Bromocyclotetradecane, B 402
2-Bromocyclotetradecanone, B 411
3-Bromocyclotetradecanone, B 411
2-Bromocyclotridecanone, B 411
3-Bromocyclotridecanone, B 411
2-Bromocycloundecanone, B 411
23-Bromo-9-dehydrohecogenin acetate, E 35
Bromodeoxyinositols, B 85
2-Bromo-2,6-dibenzylcyclohexanone, B 104
6-Bromo-1,4-dien-3-ones, D 93
Bromodihydroisophotosantonic lactone, C 342, 344
7α-Bromo-3β,5-dihydroxy-5α-cholestan-6-one, D 46
21-Bromo-3α,17α-dihydroxypregnan-20-one, D 310
Bromodimedone, B 112
2-Bromo-1,4-dimethylcyclohexan-1-ol, B 66
2-Bromo-6,6-dimethylcyclohexanone, B 104
2-Bromo-3,3-dimethylglutaric ester, A 60
3-Bromo-4,6-dimethyl-2-pyrone-5-carboxylic ester, A 60
2-Bromo-epicamphor, C 226
2-Bromo-epicamphor-2-carboxylic acid, C 226
α-Bromoethylbenzene, B 395
4-(2-Bromoethyl)cyclohexanone, C 118
2(1′-Bromoethyl)cyclopentanone, C 23
3-(β-Bromoethyl)cyclopent-1-ene, A 149
Bromofenchone, C 236
1-Bromo-2-fluorocyclohexanes, B 33, 35
1-Bromo-2-fluorocyclohex-1-ene, B 31
10β-Bromo-5β-fluoroestrane-3α,17β-diol, D 229
Bromogeigerin acetate, C 351
Bromohelenalin, C 359
α-Bromoheptoic ester, A 180
1-Bromo-1-hydroxymethylcyclobutane, A 83
12-Bromo-3α-hydroxy-11-oxonorcholanic methyl ester, D 317
3-Bromo-5-hydroxytropolone, B 373
Bromoiodocholestan-3-ones, D 122
1-Bromo-2-iodocyclohexane, B 35
3-Bromo-5-iodotropolone, B 373
π-Bromoisobornyl formates, C 210
α-Bromoisobutyric ester, B 219
6-Bromoisofenchyl alcohol, C 241
Bromoisotenulin, C 358
2-Bromoketones, macrocyclic, B 410
Bromomalonic ester, A 47
Bromomalononitrile, A 62
3-Bromo-5-methoxytropolone, B 373
Bromomethylcyclobutane, A 88
3-Bromomethylcyclobutane-1-carboxylic acids, C 84, 88
3-Bromomethylcyclobutane-1,1-dicarboxylic acid, C 88
3-Bromo-3-methylcyclobutane-1-methyl bromide, C 84
Bromomethylcyclohexane, B 33, 77
2-Bromo-1-methylcyclohexan-1-ol, B 66
2-Bromo-2-methylcyclohexanone, B 104
2-Bromo-3-methylcyclohex-1-ene, B 41
Bromomethylcyclopropane, A 39, 40
2-Bromomethylcyclopropane-carboxylic acid, A 56
1-Bromo-1-methyl-2,2-diphenylcyclopropane, A 51
1-Bromo-4-methylestra-1,3,5(10)-trien-17-one, D 356
2-Bromomethyl glutaric esters, A 59
Bromonitrocamphors, physical properties, C 213
α-Bromo-α′-nitrocamphor, C 212
α′-Bromo-α-nitrocamphor, C 212
π-Bromo-α′-nitrocamphor, C 213

ω-Bromo-α′-nitrocamphor, C 213
3-Bromo-5-nitropolone, B 373
5α-Bromo-6β,19-oxidopregnane-3β, 17α-diol-20-one 3,17-diacetate, D 245
5α-Bromo-6β,19-oxidopregnan-3β-ol-20-one acetate, D 244
7α-Bromo-6-oxo-5α-cholestan-3β-yl acetate, D 105
9α-Bromo-11-oxoprogesterone, D 235
4-Bromo-3-oxosteroids, D 345
2(ω-Bromopentyl)cyclopentanone, C 29
2-Bromo-1-phenylcyclobutene quinone, A 93
2-Bromo-4-phenylcyclohexanones, B 104
Bromopinic acid, C 168
4-Bromo-9β,10α-pregna-4,6-diene-3,20-dione, D 348
4-Bromopregnandione, D 232
5-Bromopregnane-3β,6β,17α-triol-20-one, D 244
3β-Bromopregn-5-en17β-ol-20-one acetate, D 345
4-Bromoprogesterone, D **345**
6-Bromoprogesterone, D **345**
16-Bromoprogesterone, D 350
α-Bromopropionic ester, E 21
Bromoquercitols, B 58, 68
Bromoquercitol penta-acetates, B 64
2-Bromo-α-santonin, C 342
1-Bromospirohexane, C 23
2-Bromospiro[3.5]nonane, C 26
(24R)-Bromo-(25R)-5β-spirostan-3β-ol, D *434*
11β-Bromosteroids, D 11
α-Bromostipitatic acid, B 377
N-Bromosuccinimide, B 35, 261, 263, 277, 280, 281, 304, 318, 334, 367; D 319, 323, 407
—, reaction with cyclobutene, A 82
2-*exo*-Bromosulphonates, C 163
2α-Bromotestosterone, D 340
π-Bromotricyclene, C 210
2-Bromotropinone, B 362
Bromotropolones, B 373
α-Bromotropolone, B 372, 377
2-Bromotropone, B 376, 378
α-Bromo-*m*-xylene, B 28
Bruceine A, C 433
Bruceine B, C 433
Brucine succinate, D 100
Bryodulcosigenin, C 424
Bryogenin, C 424
Bryosigenin, C 424
Buccinum undatum, D 146
Bucherer synthesis, D 358
Bufadienolides, E 113
20,22-Bufadienolide, D *432*
Bufagins, D 413
Bufalin, D 2, 12, 377, 412, 413, 414, **416**
—, *nomenclature*, D *432*
Bufalin acetate, D 416
Bufanolides, nomenclature, D *426, 431, 432, 437*
Bufanolide, D 2, 7
14β-Bufanolide, D 5
5β,14β-Bufanolide, D *432*
Bufarenogenin, D 421
ψ-Bufarenogenin, D 421
Bufa arenarum, D 415, 418, 421
Bufo bufo gargarizans, D 412
Bufo formosus, D 415, 418, 419, 421
Bufo gargarizans, D 412, 415
Bufogenins, D 407, 413, 414, 416
Bufo japonicus, D 415
5β-Bufol, D 195
Bufo marinus, D 415, 417, 420
Bufonic acid I, D 193
Bufo paracnemis, D 415, 420
Bufo regularis, D 415
Bufotalidin, D 414
Bufotalien, D 416, 418
Bufotalienone, D 417
Bufotalin, D 13, 413, 414, **417**
Bufotalinin, D **418**
Bufotalone, D 417
Bufotenidine, D 412
Bufotenine, D 412
Bufothionine, D 412
Bufotoxins, D 413, 414, 416, 417
Bufotoxinone, D 417
Bufo vulgaris, D 413, 415, 417, 420
Bulnesia sarmentia, D 314, 315
Bulnesol, C **315**, 335
Bupleurum falcatum, D 166
Bursera diacid, C 414, 416, 442
Buta-1,3-dienes, from cyclobutenes, A 82
Butadiene, A 55; B 63, 129, 141, 151; C 30, 32, 49; D 102
—, adduct with acrylic ester, A 200
—, — with citraconic anhydride, D 148
—, conversion to cyclohexene, B 16
—, cyclopropene adduct, A 33
—, dimerisation, A 76; B 24, 28, 383
—, from cyclobutane, A 81
—, from cyclohexane, B 15
—, polymerisation, B 380
—, reaction with acetylenedicarboxylic acid, B 147, 148

Butadiene with acrolein, B 90
—, — with benzoquinone, C 76
—, — with cyclopentenone, C 58
—, — with diazoacetic ester, A 53
—, — with fumaric acid, B 146
—, — with maleic acid, B 6
—, — with maleic anhydride, B 146
—, — with methyl methacrylate, B 134
—, — with methyl vinyl ketone, B 129
—, — with β-nitrostyrenes, B 78
—, — with propargylic aldehyde, B 90
—, — with propiolic acid, B 138
—, trimerization, B 385
Butadienecyclopentadienylcarbonylmanganese, A 142
Buta-1,4-diene-1,4-diol, B 64
17α-Butadiynylestradiol 3-methyl ether, D 215
17α-Butadiynylestra-1,3,5(10)-triene-3,17β-diol 3-methyl ether, D 215
n-Butane, A 81
—, conformation, A 6
Butanetetracarboxylic ester, A 200; B 144
Butane-1,3,3-tricarboxylic acid, D 22
Butane-1,2,4-tricarboxylic ester, A 201
But-3-enol, A 64
Butenolides, D 371, **373**
2-Butenyldimethylcyclohexa-3,5-dien-1-ones, B 125
But-3-enyllithium, A 41
2-But-2′-enyl-3-methylcyclopent-2-en-4-ol-1-one, A 182
But-2-enylrethrin, A 180
7-*tert*-Butoxynorbornadiene, C 94
5-*tert*-Butyl-3-aminocyclohexane-1-carboxylic acids, B 143
tert-Butylbenzoic acids, hydrogenation, B 134
2-*n*-Butylbicyclopropyl, C 7
n-Butylboronic acid, A 159; B 54
n-Butylboroxine, A 159; B 54
1-*n*-Butyl-2-chlorocyclohexan-1-ol, B 66
Butyl-2-chlorocyclohexanone, B 104
tert-Butyl chromate, D 84, 85, 109
2-*n*-Butylcyclohexa-1,3-diene, B 26
Butylcyclohexanes, B 12, 14
Butylcyclohexanecarboxylic acids, B 32, 133, 134
4-Butylcyclohexane-1-thiol, B 76
Butylcyclohexanols, B 32, 47, **49**, 50
Butylcyclohexanone, B 49, 50, 95, 102
4-*tert*-Butylcyclohexan-1-one, B 46
Butylcyclohex-1-enes, B 21
Butylcyclohex-2-enone, B 117
Butylcyclohexyl bromide, B 32
tert-Butylcyclopentadiene, A 122
4-*tert*-Butyl-2-methylcyclohexanones, B 12
2-*n*-Butyl-1-methylcyclopent-1-ene, A 116
Butylcyclopentanes, A 113
Butylcyclopentan-1-ol, A 155, 157
Butylcyclopent-1-enes, A 116
Butylglycidic esters, B 129
Butylhydroxycyclohexanecarboxylic acids, B 138
Butyl-1-hydroxycyclopentyl acetate, A 117
Butyl hypochlorite, A 39
Butyllithium, A 34, 41, 134; B 395; C 23, 34, 52
n-Butylmalonic ester, A 27
Butyl-5-methylcyclohexane-1,4-diol, B 51
Butyl-1-methylcyclohexan-1-ol, B 17, 46
Butyl-2-methylcyclohexan-1-ols, B 47, 48
Butylmethylcyclohexanones, B 95
2-*n*-Butyl-3-methylcyclopent-2-en-4-ol-1-one, A 179
2-*n*-Butyl-3-methylcyclopent-2-en-1-one, A 179
n-Butyl nitrite, D 286
Butylphenol, B 49, 50
5-Butyltropolone, B 369
But-1-yne-1,4-dicarboxylic ester, A 60
n-Butyric acid, C 313
Butyric anhydride, D 233
Butyrophenone, A 64
Butyrospermol, C 411, 426
Buxenine G, C 408, 409
Buxus alkaloids, C 408
Buxus microphylla, D 336
Buxus sempervirens, D 335
Bystropogon mollis oil, B 208

Cacalea decomposita, C 304
Cacalol, C **304**, 335
Cacalone, C **304**, 335
Caccigenin, C 441
Cacogenin, E 27, **36**
Cactus sterols, D 3
Cadalene, C 256, 261, 268, 271, 281, 312, 314, 329
2-Cadalenol, E 84
Cadinane, C 297, 298
Cadinenes, C 256, 268, 270, 272, 288

β-Cadinene, C 268, **269**, 270
δ-Cadinene, C **269**
γ-Cadinene, C **269**; E 80
ε-Cadinene, C **270**
Cadinene dihydrochloride, C 268, 296
Cadinene hydrochloride, C 271, 296, 298
Cadinols, C **296**, 334
α-Cadinol, C **296**
δ-Cadinol, C 290, **296**, **297**
Cadmium alkyls, A 94
α-Caesalpin, C 391
β-Caesalpin, C 391
δ-Caesalpin, C 391
Cafestol, C 380, 396
Cage hydrocarbons, C **106**
Cajaput oil, B 204
Calacone, C **292**, 312, 334
Calactin, D 406
Calactinic acid, D 406
Calamenene, C **270**, 271, 288
Calamenol, C 270
Calarane, C 274, 308, 309
Calarane derivatives, C 308
Calarene, C **274**, 308
Calciferol, D 138, 140, 143, **144**
Calicene, C 13
Callitris glauca, C 314
Callitropis araucarioides, C 299
Calotoxin, D 406
Calotropagenin, D **406**
Calotropin, D **406**
Calotropis procera, D 406
Calycanthus occidentalis, B 60
Calycanthus oil, D 154
Calythrix tetragona, A 187
Calythrone, A 109, 187
Campanulin, C 446
Campestanol, D 6, 75, 156, 157
—, *nomenclature*, D 427
Campestanyl acetate, D 156
Campesterol, D 3, 13, 75, 154, **155** 156, 157, 165; E 103, 105
Camphane, C 137
Camphanic acid, C 202
Camphene, C 137, 160, 164, 165, 172, 193, 194, 195, 196, 197, 198, 199, 206, 207, 227, 229, 233, 251, 252; E 69
—, hydration, C 163
—, racemisation, C 164
Camphene[8-^{13}C], C 198
Camphene[8-^{14}C], C 198
Camphene glycol, C 199
Camphene hydrate, C 163
Camphene hydrochloride, C 160, 161, 163
Camphenic acid, C 194, 195, 199
Camphenilanaldehyde, C 196
Camphenilane, C 196, 199
Camphenilol, C 164
Camphenilone, C 188, 194, 195, 196, 197, 199
Camphenilone oxime, C 199
Camphenilonylacetic acid, C 265
Camphenylic acid, C 194, 195, 199
Campherol, C 241
Camphol, C 227
Campholene A 116
α-Campholene aldehyde, C 186
$\Delta^{1,0}$-Campholene aldehyde, C 228
α-Campholenic acid, C 169, 170
Campholides, C 204
α-Campholide, C 201, 205, 207
β-Campholide, C 201, 207
α-Campholonic acid, C 223
Camphonanic acid, C 267
Camphononic acid, C 203
Camphor, C 84, 89, 137, 162, 166, 166, **193**, 196, **199**, 201, 202, 203, 204, 205, 206, 210, 217, 219, 222, 228, 232, 238, 241, 242, 249, 250, 264; E 73, 77
—, applications, C 207
—, constitution, C 200
—, halogen derivatives, C 208
—, interconversion, C 165
—, nitro derivatives, C **212**
—, reduction to borneol, C 227
—, substitution products, C 207
—, sulphonation, C 214, 215
β-Camphor, C 224
Camphorcarboxylic acids, C **223**
Camphor-3-carboxylic acid, C 219 223, 224
Camphor-3-carboxylic ester, C 224
Camphor chlorosulphoxide C 216
Camphor enol acetate, C 206
Camphor hydrazone, C 192, 206
Camphoric acid, A 106; C 200, 201, 202, 203, 204, 206, 207, 208, 212, 219, 230
—, half esters, C 201
Camphoric acid *sec*-semialdehyde, C 216
Camphoric acid *tert*-semialdehyde, C 217, 219
Camphoric anhydride, C 201, 207
Camphoric dialdehyde, C 220
Camphorimide, C 207
Camphorimine, C 232

Camphor methanesulphonylhydrazone, C 206
Camphor oil, B 208
Camphoronic acid, B 324, 325; C 202
Camphor oxime, C 169, 170, 232
Camphorquinone, C 216, 218, 219, **220**, 221, 233
Camphorquinone dioximes, C 220, 221
Camphorquinone mono-oxime, C 220
Camphorquinone oximes, C 209, 221
Camphorsulphonic acids, C **213**, 214
Camphor-8-sulphonic acid, C 216
Camphor-π-sulphonic acid, C 215
Camphor-10-sulphonyl chloride, C 214
Camphor-π-sulphonyl chloride, C 216
Camphor-π-sulphonyl halide, C 209
Camphor-wood oil, C 180
Canaric acid, C 141, 436
Canarium luzonicum, C 295
Canarium samonense, C 304
Canarium strictum, C 302
Canarone, C **302**, 335
Canaryxanthophyll, B 238
Candogenin, D **403**
ψ-Candoside, D **402**
Cannabidiol, E 70
Cantharellus cinnabarinus, B 318
Cantharidin, E 71
Cantharis vesicatoria, E 71
Canthaxanthin, B 249, 253, 256, 262, 317, **318**, 319, 320, 321
Capitasterone, D 161, 162
n-Caproic acid, A 179; B 74
Caprolactam, B 100, 118
ε-Caprolactone, B 98, 99
Capsanthin, B 257, 299, **322**, 323, 324, 345; E 120, 122, 130
—, biosynthesis, E 131
Capsanthin diacetate, B 324
Capsanthin dipalmitate, B 324
Capsanthone, B 323
Capsicum annuum, B 322
Capsorubin, B 256, 257, **322**, 323, 345
—, stereochemistry, B 324
—, structure, B 233
Capsorubin diacetate, B 325
Capsorubin dipalmitate, B 325
Capsorubone, B 323
Carane, C 136, 137, 153; E 69
Caranecarboxylic acids, C **157**
Carane group, C **150**
Carane ketones, C **154**
Caran-2,3-epoxide, C 154
Caran-3-ol, C 153
Carapin, C 431
Caraway oil, B 191, 192, 200
N-Carbamoylcarvomenthylamine, B 229
1-Carbamoylcyclopropane-1-carboxylic ester, A 58
3-*O*-Carbamoyl-1-deoxy-1-guanidino-*scyllo*-inositol, B 85
N-Carbamoylmethylamine, B 228
Carbazole, reduction, C 10
Carbenes, C 8, 118
—, reaction with acetylenes, A 33
—, — with olefins, A 24
—, — with styrenes, A 36
Carbocamphenilone, C 195
Carbocyclic rings, strain, A 3
Carbonium ions, C 159, 160, 163
—, bridged, C 80
Carbonium ion rearrangements, C 83
Carbonyl bromide, A 74
Carboranes, cyclopentadienylmetal complexes, A 143
3-Carboxyandrostan-17-ones, D 271
N-*o*-Carboxybenzoylmenthylamine, B 228
1-Carboxybicyclo [3.3.1] nonan-2-one, C 125
γ-Carboxycamphenehydratolactone, C 230
4-Carboxycyclohexanonepropionic acid, C 123
Carboxycyclopentadienyltricarbonylmanganese, A 140
4-(3-Carboxycyclopentyl)-butyric acid C 128
1-Carboxy-3,3-dimethylcyclobutane-2-acetic acid, A 97
1-Carboxy-3,3-dimethylcyclobutane-2-propionic acid, A 97
2-(3′-Carboxy-2′,2′-dimethylcyclopropyl)-3-methylcyclopentanone, C 277
19-Carboxy-13-epi-*enantio*-manoyl oxide, C 383
Carboxylic acids, macrocyclic, B 417
3-(2′-Carboxyprop-1-enyl)-2,2-dimethylcyclopropane-1-carboxylic acid, A 67
Cardanolides, nomenclature, D *430*, *437*
Cardanolide, D 2, 7
14β-Cardanolide, D 5
5β,14β-Cardanolide, D *431*
Cardenolides, D 369; E 111
—, 17α-hydroxylation, D 389
—, *nomenclature*, D *426*, *430*
Card-20(22)-enolides, D **373**, 407
Cardiac aglycons, D 2, 191, 371
Cardiac glycosides, D 360 ff, 371; E 5

Cardiazole, B 100
Cardiotonic glycosides, D 360 ff
Carene, B 184
Car-2-ene, C 152, **153**, 154, 155; E 71, 79
Car-3-ene, C **150**, 151, 153, 252, 255; E 69
Car-4-ene, C 252
Car-5-ene, C 152
Δ^4-Carene, A 58
Car-3-en-5,6-epoxide, C 185
Car-3-en-5-ol, C 152
Car-3(10)-en-4-ol, C 152
Carinene dihydrobromide, C 270
Carissa lanceolata, C 301
Carissone, C **301**, 302, 335
Carnaubadiol, C 425
Carneagenin, E 49
Carone, A 19; E 78
—, oxidation, A 60
Car-2-one, C **154**, 155, 252
Carone hydrazone, C 153
Caronic acids, C 151
Caronic acid, A 19, 60, 65, 67; C 151, 152, 154, 156, 158
Caro's acid, C 201
β-C25-Carotenal, B 273
Carotenes, B 234, 237, 294; E 120, 124
—, biosynthesis, B 338
—, cyclic, E 126
—, separation from xanthophylls, B 242
—, synthetic analogues, B 281
α-Carotene, B 237, 250, 259, 263, 268, **271**, **272**, 295, 302, 338; E 120, 121, 125, 128
—, oxidation, B 259, 260
—, reaction with *N*-bromosuccinimide, B 261
—, structure, B 231, 232
—, synthesis, B 246
β-Carotene, B 231, 237, 238, 239, 249, 250, 253, 259, 263, 267, **268**, 269, 271, 281, 300, 302, 303, 312, 313, 316, 318, 328, 332, 338, 339, 340, 341, 342; E 120, 121, 125, 128, 129, 133
—, biological activity, B 283
—, conversion into Vitamin A, B 283
—, degradation products, E 132
—, manufacture, B 270
—, oxidation, B 259, 260, 263
—, pyrolysis, B 258
—, reaction with *N*-bromosuccinimide, B 261, **262**
β-Carotene, structure, B 232
—, synthesis, B 245, 246, 269, 270
β-Carotene[^{14}C], B 283
γ-Carotene, B 239, 253, 259, **272**, 273, 277, 304, 317, 326, 332, 338, 345; E 120, 212, 125
δ-Carotene, B 247, 253, 257, 259, **273**, 274, 277; E 125
ζ-Carotene, B **275**, 277, 333, 334, 335, 336, 339; E 124, 125
ε-Carotene, B **274**
θ, Carotene, B **275**
λ-Carotene, B 338
η-Carotene, B **275**
β-Carotene diepoxide, B 283, 293, 295, 303
α-Carotene epoxide, B 283, 302
β-Carotene monoepoxide, B 303
Carotenoids, B 153, 231 ff; E 135
—, aldehydes, B 312
—, allenic B 328
—, basic structure constitution, B 231
—, biosynthesis, B 333; E 120, 130
—, carboxylic acids, B 325
—, catalytic hydrogenation, B 259
—, chemical properties, B 258
—, chromatography, B 241, 243
—, classification and nomenclature, B 234
—, conformation, B 236
—, containing allylic hydroxyl (or alkoxyl) groups, B 264
—, conversion into Vitamin A, B 283
—, cyclic, formation, E 126
—, formation, E 126
—, from purple bacteria, B 307
—, function B 239
—, hydroxyaldehydes, B 312
—, hydroxyoxo derivatives, B 314
—, infra-red spectra, B 256
—, isolation from natural products, B 240
—, isomerisation, B 258
—, mass spectrometry, B 257
—, microbiological preparation, B 249
—, occurrence, B 236
—, of uncertain constitution, B 330
—, oxidation reactions, B 259
—, oxo derivatives, B 314
—, physical properties, B 252
—, properties, B 249
—, proton magnetic resonance, B 256
—, pyrolysis, B 258
—, reaction with boron trifluoride etherate, B 263
—, structure determination, B 243

Carotenoids, synthesis, B 245
Carotenoid acids, B 234
Carotenoid epoxides, B 242, 293 294
Carotenoid hydrocarbons, B 266
Carotenoid P518, B 345
Carotenoid pigment R, B 345
Carotenoid pigment Y, B 344
β-Carotenone, B 260
Carotol, C **323**, 324, 325, 336; E 81, 84, 85
Carpesia abrotanoides, C 351
Carpesia lactone, C **351**, 367
Carpotioche brasiliensis, A 198
Carquejol, B **193**
Carquejyl acetate, B 193
Carromenthol, B 220
Carrot oil, B 275, 333
Carr-Price reaction, 250, 283, 289, 291
Carvacrol, B 187; E 77
Carvanacetone, C 147
Carvenone, B 187, 200; C 147
Carvenone hydrazone, B 219
Carveol, B 179, **192**, 194
Carvestrene, B 184
Carvomenthene, B 179, **218**
Carvomenthol, B **187**, 197, 218, 230
Carvomenthone, B **197**, 218, 229; C 333
Carvomenthoxime, B 229
Carvomenthylamine, B **227**, 229, 230
Carvomenthyltrimethylammonium hydroxide, B 218, 230
Carvone, B 178, 192, **201**; C 84, 154, 251; E 74
Carvone-camphor, B 201; C 84
Carvone hydrate, B 194
Carvone hydrobromide, B 215; C 155
Carvone oxime, B 201
Carvotanacetol, B 198
Carvotanacetone, B **197**, 198
Caryophyllene, A 97; C 51, **284**, 285, 286, 287, 289, 331, 332; E 80, 85
α-Caryophyllene, C 283, 284
β-Caryophyllene, C **284**
α-Caryophyllene alcohol, C 287
β-Caryophyllene alcohol, C 287
Caryophyllenic acid, A 97; C 284
Cascarillin, C 377, 387
Cascarillin A, C 385, 387
Cassaic acid, C 378, 391
Cassaidic acid, C 391
Cassamic acid, C 391
Cassius abus, B 213
Castaprenols, E 133
Catalaposide, B **214**
Catalpa species, B 214
Catechol, A 171
—, hydrogenation, B 103
Catenanes, A 18; B **422**; C 2
Catharanthus roseus, C 75
Cativic acid, C 383
Catnip oil, B 209, 210
Caucasian petroleum, B 13
Ceanothenic acid, C 414, 436
Ceanothic acid, C 414, 436
Cedrelane group, triterpenoids, C 428
Cedrelanol, C 296
Cedrela odorata Brasiliensis, C 296
Cedrela toona, C 271
Cedrelone, C 410, 411, 428
Cedrene, C **279**, 289, 325
Cedrol, C 279, **325**, 326, 336
Cedrolic acid, C **325**, 336
Cedronine, C 434
Cedronine group, triterpenoids, C 434
Cedronyline, C 434
Cedrus deodara, C 281
Cedrus species, C 291
Celastrus scandens, B 304
Celastrol, C 415, 416, 446
Celaxanthin, B **304**
Celluloid, C 207
Cembrane diterpenoids, physical properties, C 401
Cembrane group, C 381
Cembrene, C 374, 381, 401; E 91
Centaura scabiosa, C 365
Centaura stoebe, C 365
Cephalonic acid, C 403, 405
Cephalosporin, C 406
Cephalosporin P_1, C 420; D **162**, 163
Cephalosporium, D 162
Cephalosporium caerulens, C 403
Ceratostomella fimbriata, C 291
Cerberin, D 386
Cerberose, D 386
Cerebrosterol, D 96, 108, 164
Cerevisterol, D 151, 165
Cerin, C 447
Cerotic acid, D 176
Cevanine nomenclature, D *449*, *450*
Chair conformation, cyclohexane, ring, B 7
Chaksine, B **212**
Chalkones, dimerisation, A 76
Chalinasterol, D **157**, 165; E 103, 105
Chamaecyparis formosensis, C 137, 180, 270, 301; E 77
Chamaecyparis lawsoniana, C **296**
Chamaecyparis nootkatensis, B **365**; C 157, 273
Chamaecyparis obtusa, B **365**

Chamaecyparis obtusa formosana, C 180
Chamaecyparis species, C 275
Chamaecyparis thyoides, C 267
Chamazulene, C 351
Chamic acid, C **157**, 158, 255
Chaminic acid, B 357; C **158**, 255
Chanoclavine I, E 65, 66
Ch'an Su, E 413, 415, 416, 419, 421
Chaparrin, C 433
Chaparrinone, C 433
Chasmanthin, C 385, 387
Chaulmoogra oil, A 196, 197
Chaulmoogric acid, A 104, 161, 193, 196, 197, 198
Cheiranthus cheiri, D 383
Cheirosides A and H, D 364, **396**
Cheirosides C, D, E, F and G, D 405
Cheirotoxin, D 362, **382**, **383**
Chenodeoxycholic acid, D 51, 52, 169, 170, 171, 180, 181, 182, **184**, 185, 186
Chenopodium ambrosioides, B 206
Chenopodium oil, B 76, 180, 206
Chiapagenin, E 26, **35**
Chiapagenin 3-mono-acetate, E 35
Chichipegenin, C 438
α-Chigadmarene, C **278**, 288
Chimonanthis fragrans, B 60
Chimaerol, D 195
Chionographis japonica, **E** 44
Chloranil, B 265, 266; C 261; D 240, 272
Chlorella species, E 126, 127
Chlorella vulgaris, B 334, 344; E 126
α-Chloroacetamide, A 61
6β-Chloro-17α-acetoxyprogesterone, D 346
Chloroacetyl chloride, A 201
α-Chloroacrylic esters, A 49
3α-Chloro-5α-androstan-17-one, D 51
5β-Chloro-5α-androstan-17-one, D 41
3-Chloroandrosta-1,3,5-trien-17-one, D 338
3β-Chloroandrost-5-en-17-one, D 42, 257
Chlorobactene, B 257, 279, 305, 307; E 130, 132
Chlorobacteriaceae, E **132**
1-Chlorobicyclo[2.2.1]heptane, C 97
1-Chlorobicyclo[2.2.1]hexane, C 85
6-Chlorobicyclo[2.1.1]hexane, C 87
Chlorobicyclo[2.2.0]hexanes, C 42
4-Chlorobicyclo[2.2.0]hexane-1-carboxylic acid, C 42
9-Chlorobicyclo [6.1.0] nona-2,4,6-triene, B **399**
1-Chlorobicyclopropyl, C 7
3β-Chlorobisnorallocholanic acid, D 250
3-Chlorobornane, C 192
Chlorobromocamphors, physical properties, C 211
1-Chloro-2-bromocycloheptene, B 349
1,1-Chlorobromocyclohexane, B 40
1,2-Chlorobromocyclohexane, B 40
1-Chloro-4-bromocyclohexane, B 36
4-(3-Chloro-2-butenyl)androst-4-ene-3,17-dione, D **230**
2-Chloro-1-*n*-butylcyclohexan-1-ol, B 66
3-Chloro-1-*n*-butylcyclohex-1-ene, B 41
α-Chlorocamphene, C **166**
Chlorocamphors, physical properties, C 211
α-Chlorocamphor, C 208, 209, 212
α'-Chlorocamphor, C 208, 209, 212
π-Chlorocamphor, C 209, 216
6-Chlorocamphor, C 212
α-Chlorocamphorcarboxylic acid, C 223
3-Chlorocamphor-3-carboxylic acids, C 208
Chlorocamphoric anhydride, C 212
Chlorocamphorquinone, C 212
α-Chlorocamphor-π-sulphonic acid, C 215
α-Chlorocamphor-10-sulphonic acid, C 215
π-Chlorocamphor-π-sulphonic acid, C 216
α-Chlorocamphor-π-sulphonyl chloride C 209
Chlorocamphor sulphoxide, C 214
Chlorocarbonylcyclopentadienyltricarbonylmanganese, A 141
1-Chloro-2-carboxamidohexafluorocyclopent-1-ene, A 151
2-Chloro-4-(1'-chlorovinyl)-cyclohex-1-ene, B 41
3β-Chlorochol-5-enoic methyl ester, D 188
3-Chlorocholestra-3,5-diene, D 126
2α-Chloro-5α-cholestane, D 36
3-Chlorocholestanes, D 40
3α-Chloro-5α-cholestane, D 40
3α-Chloro-5β-cholestane, D 41
3β-Chlorocholestane, D 256
3β-Chloro-5α-cholestane, D 40, 110, 115
3β-Chloro-5β-cholestane, D 41
5α-Chlorocholestane, D 15, 16
5-Chloro-5α-cholestane, D 50
6α-Chloro-5α-cholestane, D 50

6β-Chloro-5α-cholestane, D 50
7-Chloro-5α-cholestanes, D 53
6α-Chloro-5β-cholestane-3β,5-diol, D 97
6β-Chloro-5α-cholestane-3β,5-diol diacetate, D 45
3α-Chloro-5α-cholestan-6α-ol, D 50
3β-Chloro-5α-cholestan-6α-ol, D 51
Chlorocholestan-3-ones, D 122
2α-Chlorocholestanone, D 123
2α-Chloro-5α-cholestan-3-one, D 120
3α-Chloro-5α-cholestan-6-one, D 17, 18
3β-Chloro-5α-cholestan-6-one, D 78
3α-Chlorocholest-4-ene, D 105
3β-Chlorocholest-4-ene, D 105
3β-Chlorocholest-5-ene, D 15, 16, 41
4-Chlorocholest-4-en-3-one, D 128
7-Chlorocholesteryl esters, D 54
9α-Chlorocortisol acetate, D 329
9α-Chlorocortisone acetate, D 329
1-Chlorocycloalkenes, B 383, 412
6β-Chloro-3α,5-cyclo-5α-cholestane, D 90
1-Chlorocyclodecene, B 402
9-Chlorocyclododeca-1,5-diene, B 402
2-Chlorocyclododecanol, B 410
2-Chlorocyclododecanone, B 410
1-Chlorocycloheptanecarboxylic acid, B 357
2-Chlorocycloheptanone, B 354, 355
1-Chlorocycloheptene, B **351**
Chlorocyclohexane, B 33
1-Chlorocyclohexane-1-carboxylic acid, B **135**
3-Chlorocyclohexane-1-carboxylic acid, B 135
4-Chlorocyclohexane-1-carboxylic acid, B 135
3-Chlorocyclohexane-1,2-dione, B 110
2-Chlorocyclohexanethiol, B 77
2-Chlorocyclohexan-1-ol, B 35, 66
4-Chlorocyclohexan-1-ol, B 66
2-Chlorocyclohexan-1-ol acetate, B 67
2-Chlorocyclohexanone, A 190; B 40, 94, 103, 104, 110
3-Chlorocyclohexanone, B 104
1-Chlorocyclohex-1-ene, B 34, 35, **40**, 41, 94
3-Chlorocyclohex-1-ene, B 19, 41, **42**; C 15
3-Chlorocyclohex-2-en-1-one, B 111
4-Chloro-1-cyclohex-1-enyl-3,3,4-trifluorocyclobutene, A 88, 89
2-(1′-Chlorocyclohexyl)cyclohexanone, C 16
1-Chloro-1-cyclohexylethylene, B 31
1-Chlorocyclononene, B 402
1-Chlorocyclo-octene, B 402
1-Chlorocyclopentadiene, A 110
Chlorocyclopentane, A 148
3-Chlorocyclopentane-1,2-dione, A 184, 185
2-Chlorocyclopentan-1-ol, A 156
2-Chlorocyclopentanone, A 157, 166, 184
1-Chlorocyclopentene, A 118
3-Chlorocyclopentene, A 116, 149, 153, 155, 193, 194; C 10
2-Chlorocyclopent-1-ene-1-aldehyde, A 163
2-Chlorocyclopent-2-en-1-one, A 171, 184
Chlorocyclopropanes, A 38, 39, 40
Chlorodane, C **102**
16α-Chlorodeoxycorticosterone, D 354
1-Chloro-3-diazobicyclo[2.2.1]heptan-2-one, C 85
α-Chloro-α′, 10-dibromocamphor, C 211
3-Chloro-3,5-dienes, D 93
Chlorodifluoroacetic acids, D 344
2-Chlorodimethylcyclohexan-1-ols, B 66
2-Chlorodimethylcyclohexanones, B 104
3-Chloro-1,4-dimethylcyclohex-1-ene, B 41
2-Chloro-5,5-dimethyl-3,4-diphenylcyclopent-2-en-1-one, A 184
10-Chloro-10,10-dinitrocamphor, C 214
6-Chloroepicamphor, C 231
Chloro-1,2-epoxycyclohexanes, B 72
16α-Chloroestradiol 3-methyl ether, D 339
16α-Chloroestrone 3-methyl ether, D 339
17α-(2-Chloroethoxy)-3-methoxyestra-1,3,5(10)-triene, D 339
1-Chloro-4-ethoxy-2,3,3-trifluorocyclobutene, A 89
2-Chloro-1-ethylcyclohexan-1-ol, B 66
3-Chloro-1-ethylcyclohex-1-ene, B 41
6-Chloro-17α-ethynyl-6-dehydro-19-nortestosterone, D 346
6-Chloro-17α-ethynyl-6-dehydrotestosterone, D 346
17α-Chloroethynylestra- 4,9-dien-17β-ol-3-one, D 350
6α-Chloro-17α-ethynyltestosterone, D 346
6α-Chloro-17α-ethynyltestosterone acetate, D 346
ω-Chloro-α-fenchene, C 245

Chlorofluoralkenes, A 76
1-Chloro-1-fluorocyclohexane, B 33, 34, 40
Chlorofluoro-olefins, dimerisation, A 75
Chloroformic esters, reaction with cyclopentanone enamines, A 200
Chlorogalum pomeridianum, E 2, 25, 26
Chlorogenin, E 26, **33**, 36
β-Chlorogenin, E 33
Chlorogenone, E 33
3-Chlorohomoadamantane, C 133
12α-Chloro-3α-hydroxy-5β-chol-9(11)-enoic acid, D 38
1-Chloro-1-hydroxymethylcyclobutane, A 83
Chloroiodocamphors, physical properties, C 211
Chloroiridic acid, D 110
Chloroiosoborneol, C 166
4-Chloroisobornyl trichloroacetate, C 166
β-Chloroketones, reaction with sodioacetoacetic ester, B 114
Chloromercuricyclopentadienyltricarbonylmanganese, A 141
Chloromercuriferrocenes, A 135
1-Chloromercuri-2-methoxycyclohexanes, B 22
20α-*N*-Chloromethylaminopregn-4-en-3-one, D 355
18-Chloro-20α-methylaminopregn-4-en-3-one trifluoroacetate, D 355
7-Chloro-1-methylbicyclo[3.1.1] heptan-6-one, C 88
1-Chloro-1-methylcycloalkyl chlorides, B 401
1-Chloromethylcyclobutan-1-ol, A 83
2-Chloromethylcyclohexan-1-ols, B 66
2-Chloro-4-methylcyclohexanone, A 190
2-Chloro-5-methylcyclohexanone, B 103
Chloromethylcyclohex-1-enes, B 41
2-Chloro-1-methylcyclopentan-1-ol, A 157
2-Chloromethylcyclopropane-carboxylic acids, A 56
16α-Chloro-6-methyl-dehydroprogesterone, D 349
3-Chloro-1-methyl-1,2-epoxycyclohexane, B 72
1-Chloro-4-methylestra-1,3,5(1)-trien-17-one, D 338
5-Chloro-2-methylphenylphosphoric acid, B 126
2-Chloro-6α-methylpregn-4-ene-11β,17α,21-triol-3,20-dione 21-acetate, D 355
16α-Chloro-6α-methylprogesterone, D 349, 350
Chloronitrocamphors, physical properties, C 213
α-Chloro-α′-nitrocamphor, C 212
α′-Chloro-α-nitrocamphor, C 212
5-Chloro-6β-nitro-5α-cholestan-3β-yl acetate, D 97
2-Chloronorbornane, C 89
2-Chloropentanone, A 168
5-Chloropentan-2-one, A 50
2-Chloro-1-phenylcyclohexan-1-ol, B 66
4-Chloro-2-phenylquinazoline, D 356
1-Chloro-10-phosphonato-α-fenchene, C 238.
Chlorophyceae, B 238
Chlorophylls, B239; E 87, 91, 135, 136, 137
6-Chloro-9β,10α-pregna-4,6-diene-3,20-dione, D 348
20β-Chloro-5β-pregnane, D *426*
20α-Chloropregna-1,4,16-triene-11β, 21-diol-3-one 21-acetate, D 353
6β-Chloro-9β,10α-pregn-4-ene-3,20-dione, D 348
3β-Chloropregn-5-en-17β-ol-20-one acetate, D 345
Chloroprene, A 55
Chloroprene dimer, B 383, 402
4-Chloroprogesterone, D 345
16α-Chloroprogesterone, D **349**, 350
21-Chloroprogesterone, D 350
α-Chloropropionic ester, A 96
3-Chloropropionyl chloride, A 50
3-Chloropropylborane derivatives, A 20
3-Chloro-1-propylcyclohex-1-enes, B 41
3-Chloropropyl sulphoxides, A 45
Chloroquercitols, B 68
α-Chlorosabina ketone, C 139
Chlorospirohexanes, C **23**
Chlorospiropentane, C 22
N-Chlorosuccinimide, D 341
ω-Chlorosulphides, ring closure, A 9
2α-Chlorotestosterone, D 340
4-Chlorotestosterone, D **340**
6α-Chlorotestosterone acetate, D 341
6β-Chlorotestosterone acetate, D **341**
1-Chlorotetrafluorocyclobutene, A 88
3-Chlorothujane, C 138
1-(4′-Chloro-3′,3′,4′-trifluorocyclobutenylcyclohexene, C 13

Chlorotrifluoroethene, C 13
4-Chloro-3,3,4-trifluoro-1-phenyl-cyclobut-1-ene, A 93
2-Chloro-1,1,2-trifluorotriethylamine, D 339
Chlorotropolones, B 373
2-Chlorotropones, B 372, 376, 378
Chloroundecafluorocyclohexane, B 39
Chloroxanthin, B 257, **310**, 338, 344; E 130
Chola-1,4-dienoic acid, D 186
Chola-4,6-dienoic acid, D 186
5α-Cholan-24-aldehyde, D *438*
Cholanes, nomenclature, D *428*, *429*
Cholane, D 2, 5
—, molecular rotation, D 63
5α-Cholane, D 9
5β-Cholane, D 9, 35
Cholanic acid, D 179, 249, 408, 413
Cholanoic acids, D 191
5α-Cholanoic acid, D 15, 16, 24, 169, 179, 189
5β-Cholanoic acid, D 8, 15, 16, 22, 24, 169, 179
5β-Cholan-12-ols, D 54
5β-Cholanones, D 90
5β-Cholan-12-one, D 54
Cholatrienic acid, D 16, 17, **22**
5β-Chola-2,6-11-trienoic acid, D 186
Cholecalciferol, D *439*, *447*
Cholegenin, E **46**, 47, 48, 49
Cholegenin diacetate, E 46
Choleic acids, D **175**, 176
Cholenoic acids, physical properties, D 187
5β-Chol-3-enoic acid, D 186
Chol-4-enoic acid, D 186
Chol-5-enoic acids, D 192
Chol-11-enoic acid, D 312
5β-Chol-13(17)-en-24-oic-acid, D *448*
Chol-4-enoic methyl ester, D 189
Cholestadienes, D 67, **117**
—, physical properties, D 118
Cholesta-1,3-diene, D **117**
Cholesta-2,4-diene, D **117**
Cholesta-3,5-diene, D 43, 80, 105, **117**
Cholesta-4,6-diene, D **117**
Cholesta-5,7-diene, D **118**
5α-Cholesta-6,8-diene, D **118**
5α-Cholesta-7,9-diene, D **118**
5β-Cholesta-7,9(11)-diene, D *436*
5α-Cholesta-7,14-diene, D 118
5α-Cholesta-8,14-diene, D 118
Cholestadienols, D 106
Cholesta-4,6-dien-3β-ol, D 107
Cholesta-4,7-dien-3β-ol, D 107
Cholesta-5,7-dien-3β-ol, D 74, 106
Cholesta-5,20-dien-3β-ol, D 107
Cholesta-5,25-dien-3β-ol, D 107
Cholesta-7,24-dien-3β-ol, D 74; E 104
5α-Cholesta-6,8-dien-3β-ol, D 107
5α-Cholesta-6,8(9)-dien-3β-ol, D 106, 151
5α-Cholesta-7,9(11)-dien-3β-ol, D 107
5α-Cholesta-7,14-dien-3α-ol, D 107
5α-Cholesta-7,22-dien-3α-ol, D 107
5α-Cholesta-8,14-dien-3β-ol, D 107; E 102
5α-Cholesta-14,24-dien-3β-ol, D 151
Cholestadien-3-ones, D **130**
Cholesta-1,4-dien-3-one, D 120, 123, **129**, **130**
Cholesta-3,5-dien-4-one, D 263
Cholesta-3,5-dien-7-one, D 52, 109, 117
Cholesta-4,6-dien-3-one, D 105, 128, 130, **131**, **262**
Cholesta-4,7-dien-3-one, D **131**
Cholesta-4,22-dien-3-one, D 131, 154
Cholesta-4,24-dien-3-one, D **131**, 158
5α-Cholesta-8,24-dien-3-one, D **131**
Cholesta-4,7-dien-3β-yl acetate, D 106
Cholestanes, nomenclature, D *428*, *429*
Cholestane, D 2, 5, 35, E 7;
—, molecular rotation, D 63
5α-Cholestane, D 4, 9, 10, 15, 16, 24, 35, 45, 119
—, *nomenclature*, D *448*
5β-Cholestane, D 4, 6, 9, 10, 15, 16, 24, 35
5α-Cholestane-3β-carboxylic acid, D 43
Cholestanediols, D 87
—, physical properties, D 110, 111
5α-Cholestane-2,3-diols, D 34
5α-Cholestane-2α,5-diol, D 117
5α-Cholestane-3β,4α-diol, D 130
5α-Cholestane-3β,4β-diol, D 84
5α-Cholestane-3β,5-diol, D 45, 46
5β-Cholestane-3β,5-diol, D 47
5α-Cholestane-3β,6-diols, D 48
5α-Cholestane-3β,6α-diol, D 48, 50, 152
5α-Cholestane-3β,6β-diol, D 45, 46
5β-Cholestane-3α,6β-diol, D 48
5β-Cholestane-3β,6β-diol, D 48
5α-Cholestane-3β,7α-diol, D 53, 106
5α-Cholestane-3β,7β-diol, D 53
5β-Cholestane-3α,7α-diol, D 53
5β-Cholestane-3α,11β-diol, D *438*
5β-Cholestane-3α,12α-diol, D 54
5α-Cholestane-4α,5-diol, D 44
5β-Cholestane-4β,5-diol, D 44

5α-Cholestane-5,6β-diol, D 47, 48
5β-Cholestane-5,6α-diol, D 48
Cholestane-16β,26-diol, E 19
Cholestane-3α,5α-diol 3-acetate, D 31
5α-Cholestane-4β,5 diol 4-acetate, D 44
5α-Cholestane-3β,7α-diol 3-acetate 7 benzoate, D 105
5β-Cholestane-3α,12α-diol 3-acetate 12-benzoate, D *437*
5β-Cholestane-3β,6β-diol diacetate, D 49
5β-Cholestane-3α,12α-diol diacetate, D *437*
Cholestane-4α,5β-diol-3-one, D 127
Cholestane-4β,5β-diol-3-one, D 127
5α-Cholestane-3β,5-diol 3-tosylate, D 45, 104
5β-Cholestane-4β,5-diol 4-tosylate, D 47
Cholestanediones, D **124**, **131**
Cholestane-3,4-dione, D 84, **124**, 125
Cholestane-3,6-diones, D 90
5α-Cholestane-1,3-dione, D **124**
5α-Cholestane-2,3-dione, D **124**
5α-Cholestane-3,6-dione, D 17, 132, 134
5α-Cholestane-3α-thiol, D 110
5α-Cholestane-3β-thiol, D 110
5α-Cholestane-3β,5,6α-triol, D 45, 46
5α-Cholestane-3β,5,6β-triol, D 39, 45, 46, 96, 263
5β-Cholestane-3β,5,6β-triol, D 46
5α-Cholestane-3β,5,6β-triol, 3,6-diacetate, D 82
5β-Cholestane-3β,5,6β-triol diacetate, D 82
5α-Cholestane-4β,5,7β-triol 4,7-diacetate, D 82
5β-Cholestan-27-oic acid, D 2
Cholestanols, D 40, **109**
Cholestanol, B 32; D 17, 34, 38, 39, 40, 43, 74, **109**, 150, 165
Cholestan-1β-ol, D 30
Cholestan-3β-ol, D 44
Cholestan-12-ols, D 54
5α-Cholestanols, D 30
5α-Cholestanol, D 87; E 112
5α-Cholestan-1α-ol, D 34
5α-Cholestan-2-ols, D 40
5α-Cholestan-2β-ol, D 34
5α-Cholestan-3α-ol, D 34, 73, 74, **110**
5α-Cholestan-3β-ol, D 10, 17, 19, 24, 29, 73, 74, 79, 96, 102, 105, 106, **109**
5α-Cholestan-4α-ol, D 44
5α-Cholestan-4β-ol, D 44, 115
5α-Cholestan-5-ol, D 117
5α-Cholestan-6α-ol, D 47, 50, 78
5α-Cholestan-6β-ol, D 47, 78
5α-Cholestan-7α-ol, D 52, 115
5α-Cholestan-7β-ol, D 52, 53
5α-Cholestan-16β-ol, D 60
5β-Cholestanol, D 87
5β-Cholestan-2α-ol, D 36
5β-Cholestan-3α-ol, D 29, 36, 37, 73, 75, **110**
5β-Cholestan-3β-ol, D 10, 18, 73, 74, **110**, **159**
5β-Cholestan-5-ol, D 47
5β-Cholestan-6β-ol, D 47
5β-Cholestan-7α-ol, D 52
5β-Cholestan-7β-ol, D 52
5β-Cholestan-16β-ol, D 60
Cholestanol acetate, D 154
Cholestan-3β-ol-6-one, D 262
Cholestanones, D **119**
Cholestanone, D 151
i-Cholestanone, D 77, 80
Cholestan-1-one, D 30
Cholestan-3-ones, D 33
—, halogenated, D 122
Cholestan-4-ones, D 90
Cholestan-6-ones, D 90
5α-Cholestan-2-one, D 89
5β-Cholestan-3-one, D 10, 17, 38, 39, 40, 44, 87, 102, **119**, 120, 122, 124
5α-Cholestan-6-one, D 21, 47
5β-Cholestanones, D 91
5β-Cholestan-3-one, D 10, 38, 39, 87, **124**, 125, 126
5β-Cholestan-4-one, D 125
5β-Cholestan-6-one, D 46, 47, 78
5β-Cholestan-7-one, D 52
Cholestanone diethyl acetal, D 119
Cholestanone dimethyl acetal, D 119
Cholestanone enol ethyl ether, D 119
Cholestanone ethylene acetal, D 119
Cholestanone ethylene dithioacetal, D 119
Cholestan-4β,5β-oxido-3-one, D 127
Cholestanyl acetates, D 36
—, hydrolysis rates, D 31
Cholestanyl acetate, D 45, 185, 255
5α-Cholestan-3β-yl acetate, D 25, 110
—, *nomenclature*, D *437*
5α-Cholestan-4β-yl benzoate, D 115
5α-Cholestan-5α-yl benzoate, D 115
Cholestanyl bromide, D 43
α-Cholestanyl chloride, D 40
β-Cholestanyl chloride, D 40
5α-Cholestan-3β-yl methyl ether, D 110

5α-Cholestan-3β-yl-methyl sulphate, D 44
5α-Cholestan-3β-yl tosylate, D 110, 115
5α-Cholestan-7β-yl tosylate, D 115
5β-Cholestan-7β-yl tosylate, D 110
Cholesta-5,7,24-trien-3β-ol, D 158, 164
Cholesta-1,4,6-trien-3-one, D 130
Cholesta-4,6,24-trien-3-one, D 158
Cholestenes, D **115**
—, physical properties, D 116
—, reactions, D 116
Δ^5-Cholestenes, 3α-substituted, D 86
ψ-Cholestene, D 15
Cholest-2-ene, D 40
Cholest-4-ene, D 15, 16, 44, 47, **115**, 126
Cholest-5-ene, D 15, 16, 43, 47, 50, 77, **115**
Cholest-8(14)-ene, D 152
5α-Cholest-1-ene, D **115**
5α-Cholest-2-ene, D 34, **115**
5α-Cholest-3-ene, D 44, **115**, 126
5α-Cholest-6-ene, D 52, **115**
—, *nomenclature*, D *436*
5α-Cholest-7-ene, D **115**, 118
5α-Cholest-8-ene, D **115**, 118
5α-Cholest-8(14)-ene, D 115
5α-Cholest-14-ene, D 115
5β-Cholest-2-ene, D 41
5β-Cholest-6-ene, D **115**
Cholest-5-ene-3β-carboxylic acid, D 43
Cholestene-1,2-diols, D 281
Cholest-4-ene-3α,6α-diol, D 49
Cholest-4-ene-3β-6α,diol, D 49
Cholest-4-ene-3β,6β-diol, D 48, 49, 84, 262
Cholest-5-ene-3β,4α-diol, D 107, 130
Cholest-5-ene-3β,4β-diol, D 84
Cholest-5-ene-3,7-diols, D **53**
Cholest-5-ene-3α,7α-diol, D 53
Cholest-5-ene-3β,7α-diol, D 96
Cholest-5-ene-3β,7β-diol, D 263
Cholest-5-ene-3β,12α-diol, D 54, 108
Cholest-5-ene-3β,19-diol, D 78, 108
Cholest-5-ene-3β,20α-diol, D 108
Cholest-5-ene-3β,22-diols, D 108
Cholest-5-ene-3β,24α-diol, D 108
Cholest-5-ene-3β,24β-diol, D 96
Cholest-5-ene-3β,25-diol, D 83, 96, 108
Cholest-5-ene-3β,26-diol, D 108, 109
5α-Cholest-2-ene-4β,5-diol, D 117
5α-Cholest-7-ene-3β,6α-diol, D 49
5α-Cholest-7-ene-3β,6β-diol, D 50
5α-Cholest-8(14)-ene-3β,6α-diol, D 152
5α-Cholest-8(9)-ene-3β,6α-diol, D 151
Cholest-5-ene-3β,7β-diol benzoate, D 107
Cholest-4-ene-3,6-dione, D 105, 130, **131**
Cholest-4-ene-3,7-dione, D **132**
Cholest-5-ene-3,4-dione, D **131**
Cholest-5-ene-3α-thiol, D 110
Cholest-5-ene-3β-thiol, D **110**
α-Cholestenol, D 86, 106
β-Cholestenol, D 106
δ-Cholestenol, D 106
γ-Cholestenol, D 106
Cholesten-3-ols, isomeric, D 103
Cholest-14-enol, D 150
Cholest-4-en-3-ols, D **104**
Cholest-4-en-3β-ol, D 47, 73, 85
Cholest-4-en-6α-ol, D 47, 48
Cholest-4-en-6β-ol, D 47, 48
Cholest-4-en-7α-ol, D 52
Cholest-4-en-7β-ol, D 52, 81
Cholest-5-en-3α-ol, D 33, 74
Cholest-5-en-3β-ol, D 74
Cholest-5-en-4α-ol, D 44
Cholest-5-en-4β-ol, D 44
Cholest-5-en-7α-ol, D 52
Cholest-5-en-7β-ol, D 117
Cholest-7-en-3β-ol, E 104
Cholest-8(14)-en-3β-ol, E 102
Cholest-24-enol, E 113
5-Cholesten-3β-ol, D *439*
5α-Cholest-1-en-3α-ol, D 104
5α-Cholest-1-en-3β-ol, D 85, 104, 115, 129
5α-Cholest-2-en-1α-ol, D 34
5α-Cholest-3-en-5-ol, D 47
5α-Cholest-6-en-3β-ol, D 105
5β-Cholest-6-en-3β-ol, D **105**
5α-Cholest-7-en-3β-ol, D 106
5α-Cholest-7-en-3β-ol, D 86, 96, **106**, 164
—, ultraviolet absorption spectra, D 68
5β-Cholest-7-en-3β-ol, D 106, 164
5α-Cholest-8(9)-en-3β-ol, D 106, 151
5α-Cholest-8(14)-en-3β-ol, D 106, 151
5β-Cholest-8(14)-en-3β-ol, D 106
5α-Cholest-9(11)-en-3β-ol, D 106
5α-Cholest-14-en-3β-ol, D 106
5β-Cholest-14-en3β-ol, D 106
[2-$^3H_2$]-Cholest-7-en-3-ol, E 101
Cholestenones, D **126**
Cholestenone, D 124, 126; E 114
—, autoxidation, D 128
[4-^{14}C]Cholestenone, D 126; E 113
Cholest-4-en-3-one, D 18, 22, 38, 39, 40, 44, 95, 102, 104, 105, 115, 120, **126**, 129, 232, 262

Cholest-4-en-7-one, D 52
Cholest-5-en-3-one, D 17, 103, 126, **129**, 159
Cholest-5-en-4-one, D 120
Cholest-5-en-7-one, D 52
Cholest-7-en-3-one, D 151
5α-Cholest-1-en-3-one, D 104, 120, **129**
5β-Cholest-1-en-3-one, D **129**
5α-Cholest-6-en-3-one, D 124, **130**
5β-Cholest-6-en-3-one, D 115, **130**
5α-Cholest-7-en-3-one, D **130**
5α-Cholest-8-en-3-one, D **130**
5α-Cholest-8(14)-en-3-one, D **130**
Cholestenone enol acetate, D 126, 130
Cholestenone enol ethyl ether, D 126, 128
Cholest-5-en-3-one ethylene acetal, D 126
Cholest-4-en-3-one ethylene dithioacetal, D 126
Cholest-4-en-6β-yl acetate, D 47
5α-Cholest-7-en-3β-yl acetate, D 106
Cholest-5-en-7β-yl benzoate, D 118
5α-Cholest-1-en-3β-yl benzoate, D 117
5α-Cholest-14-en-3β-yl benzoate, D 25
Cholest-4-en-7β-yl tosylate, D 81
5α-Cholest-7-en-3β-yl tosylate, D 115
5α-Cholest-14-en-3β-yl tosylate, D 115
Cholesterilene, D 117
Cholesterol, C 164, 187, 373, 406, 408, 460, 464, 465; D 2, 7, 11, 15, 17, 36, 38, 41, 43, 44, 46, 48, 73, 74, 77, 79, **95**, 103, 106, 108, 117, 143, 145, 151, 154, 157, 159, 164, 185, 203, 213, 232, 233, 235, 255, 256, 257, 289, 302, 326, 413; E 1, 4, 19, 29, 30, 44, 55, 93, 98, 99, 100, 101, 102, 103, 106, 110, 137
—, biosynthesis, D 74; E 94
—, — from 4α-methylcholesta-8,14-dien-3β-ol, E 104
—, chlorination, D 97
—, conversion to bile acids, D 169
—, degradation, D 189; E 95
—, dehydrogenation, D 96
—, determination, D 40, 76
—, enzymic oxidation, D 25
—, hydroxylation, D 45
—, *nomenclature*, D *432*, *439*
—, oxidation, D 45, 53, 84, 96
—, ozonolysis, D 96
—, structure, D 8
—, synthesis, C 473; D 99; E 94
—, ultraviolet absorption spectra, D 68
ψ-Cholesterol, D 52
i-Cholesterol, D 43, 74, 77, 78, 80
[^{14}C]Cholesterol, D 51, 126; E 95, 111, 112
[22-^{14}C-22-$^{3}H_2$]Cholesterol, E 111
[7α-^{3}H]Cholesterol, D 52
[7β-^{3}H]Cholesterol, D 51
Cholesterol derivatives, physical properties, D 98
Cholesterol dibromide, D 17, 39, **97**
Cholesterol dichloride, D **97**
Cholesterol digitonide, D 39
Cholesterol esters, physical properties, D 98
i-Cholesterol methyl ether, D 77
Cholesterol α-oxide, D 103, 129
Cholesterol-5α,6α-oxide, D 44
Cholesterol-5β,6β-oxide, D 44
Cholesterol series, D 95 ff
Cholesterol tigonide, D 39
Cholesteryl acetate, D 17, 18, 25, 36, 37, 43, 49, 50, 77, 84, 97, 109, 204, 284
i-Cholesteryl [^{14}C]acetate, D 79
Cholesteryl acetate dibromide, D 83, 185, 191, 257
Cholesterylamine, D 40
Cholesteryl benzoate, D 54, 107
Cholesteryl bromide, D 43
Cholesteryl chloride, D 15, 17, 41, 42, 77, 80, 115
Cholesteryl chloroformate, D 42
Cholesteryl halides, D 77
Cholesteryl halogenoacetates, D 271
Cholesteryl hydrogen phthalate, D 53, 85
Cholesteryl iodide, D 27, 41
Cholesteryl isothiuronium tosylate, D 110
Cholesterylmagnesium bromide, D 43
Cholesterylmagnesium chloride, D 103
Cholesteryl methanesulphonate, D 42
Cholesteryl methyl ether, D 42, 77, 99
i-Cholesteryl methyl ether, D 42, 79
Cholesteryl thiocyanate, D 110
Cholesteryl tosylate, D 43, 77, 79, 81
Cholic acid, D 2, 16, 17, 20, 22, 51, 169, 170, 173, 175, 178, 181, **184**, 185, 186
—, *nomenclature*, D *439*
—, structure, D 8
Choloidanic acid, D 19, 20
Chondrilla nucula, D 158
Chondrillasterol, D **158**, 166
Christyoside, D 364, **398**
Chromanols, E 134
Chromatium okenii, B 332

Chromatium species, B 307
Chromatium vinosum, B 337
Chromatium warmingii, B 332
Chromatography, bile acids, D 178
—, carotenoids, B 240, 241, 242
—, steroids, D 32, 71
—, thin-layer, D 369
Chromenols, E 134
Chromium acetylacetonate dibromide, A 129
Chromium biscyclopentadienyls, A 131
Chromium hexacarbonyl, A 130; C 91
Chromocene, A 131
Chromoproteins, B 239, 320
Chrysanthemaxanthin, B 296, 297
Chrysanthemic acid, A 65, 178, 180, 192; B 214; C 252; E 71
Chrysanthemic acid chloride, A 183
Chrystanthemic anhydride, A 183
Chrysanthemum carboxylic acids, A 60, 65, 66
Chrysanthemum cinerarifolium, A 65, 177; E 69
Chrysanthemum cinneratiaefolium, C 362
Chrysanthemumdicarboxylic acid, A 65, 67, 178, 180; B 214; E 68
Chrysanthemum indicum, C 184
Chrysanthemum-monocarboxylic acid, A 65, 178
Chrysanthemum parthenium, C 363
Chrysanthemum sinense, C 184, 199
Crysanthenone, C **184**, 185
Chrysene, C 76; D 8, 17, 202, 413
Cichorium intybus, C 352
Cilianic acid, D 20
Cimigenol, C 408, 409, 423
Cinchona bark, B 139; D 154
Cincholic acid, C 414, 442
α-Cinenic acid, B 205
β-Cinenic acid, B 205
Cineole, E 71
1,4-Cineole, B **206**
1,8-Cineole, B **204, 205**
Cineolic acid, B 205
Cinerins, A 65, 176, 178, 183
Cinerolone, A 65, 176, 179, 182
Cinerone, A 181, 183
Cinnamic acid, dimers, A 98
—, photodimerisation, A 76, 98
O-Cinnamoyltaxicin, C 402
Cinnanocum camphora, C 199
Cinobufagin, D 413, **420**
Cinobufalidin, D 414
Cinobufogenin, D 414
Cinobufotalin, D 413, 414, **420**
Cinobufotoxin, D 414
Cinogenic acid, B **205**
Circular dichroism, steroids, D 64
—, structure determination, C 477, 479
Cistus ladaniferus, C 319
Citraconic anhydride, D 142
—, adduct with butadiene, D 148
Citral, B 158, 163, **166**, 168, 286
—, cyclization, B 173
Citral-a, B 162, 166, 167, 168
Citral-b, B 162, 166, 167, 168
Citral anil, B 174
Citranaxanthin, B 332
Citraurin, B **312**
α-Citraurin, B 295
β-Citraurin, B 259, **312**, 322; E 132
Citronellal, B **165**, 166, 191, 212, 223; C 292; E 72
Citronella oil, B **165**
Citronellic acid, B 199
Citronellol, B **161**, 166, 170; E 67
Citrostadienol, C 408, 422; D 3, 155, **156**, 157, 166
Citrostanol, D 157
Citrostanone, D 157
Citroxanthin, B **303**
Citrus reticulata, B 332
Citrylidenecyanoacetic acid, B 167, 168, 173
Citrylidenemalonic acid, B 167
Civet, B 414
Civetene, B 414
Civetone, A 1, 2, 8; B 380, 409, **414**; D 262, 263
—, physical properties, B 415
Claisen reaction, A 51, 105; B 7, 96, 111
Clanberg test, D 240
Clemmensen reduction, B 112; C 57, 74, 330; D 16, 21, 22, 108, 119, 232, 248, 249, 290, 296, 298, 370, 397; E 36
Clerodin, C 377, 387
Clionasterol, D 3, 155, **158**, 166; E 103, 105
Clostridia butyricum, A 69
Clovene, C **287**
Cnicin, C **365**, 366, 368
Co A = Coenzyme A, E 56
CoASH = Reduced coenzyme A, E 57
Cobaltocene, A 131, 139
Cobaltocene carborane analogue, A 144
Cobalticinium salts, A 128, 131

Cocaic acids, A 98
Cocaine, B 347
Cochalic acid, C 440
Cochliobolin A, C 403, 405, **463**, 464; E 92
Cochliobolin B, C 403, 405, 464; E 92
Cochliobolus heterostrophus, E 93
Cochliobolus miyabeanus, C 463; E 92
Codfish bile, D 169
Coenzyme A, B 155; E 56, 57, 58, 59
Cogeijerene, C **263**
Cohumulone, B **131**
Colchicine, B 347, 366
Colchium autumnale, B 366
Colensenone, C 384, 385
Coleone A, C 392, 394
Collidine, D 221, 259, 262, 279, 312, 324
γ-Collidine, D 348
Collinin, E 70
Colophony resin, C 167
Comlumbin, C 377, 387; E 88
Column chromatography, bile acids, D 178
—, carotenoid isolation, B 242
Colour reactions, stenols, D 76
—, sterols, D 76
Colupulone, B **132**
Commic acid C, C 440
Commic acid D, C 450
Commic acid E C 450
Communic acid, C 384
5α-Conanine, D *449*
Condensed cyclic systems, C 32 ff
Conduramines, B 86
Conduritols, B 61, **63**, 64, 85
Conduritol epoxides, B 64
5-Conenine, D *449*
Confertolin, C **347**, 348, 367
Conformational transmission, D 88
Congressane, C **134**, 135
Coniferae oils, C 167
Convallagenin-A, E 3, 27, **39**
Convallagenin-B, E 3, 27
Convallamaretin, D 368
Convallamarin, D 368
Convallamarogenin, E 48, 49, **51**
Convallaria keisukei, E 3, 27
Convallaria majalis, D 382; E 51
Convallasaponin-A, E 3
Convallasaponin-B, E 3, 5
Convallasaponin-C, E 3
Convallasaponin-D, E 3, 5
Convallatoxin, D 362, **382**, 383, 384
Convallatoxol, D 362
Convalloside, D 362, **382**, **383**
Copaene, C **271**, 272, 288, 298
Copalic acid, C 383
Copalol, E 90
Coprosma species, B 213
Coprostane, D 4, 6
—, *nomenclature*, D *428*
Coprostane (5β) series, E 7
Coprostanic acid, D 169
Coprostanol, D 18, 19, 34, 38, 39, 41, 74, **110**, **159**, 165, 369
Coprostanone, D 38, 110, **124**
—, ultraviolet absorption spectra, D 68
Coprostanyl acetate, D 255
Coprost-7-en-3β-ol, D 159
Coriamyrtin, E 83
Corner-Allen test, D 231, 240
Cornforth-Robinson synthesis, cholesterol, D 100
Coroglaucigenin, D 364, 398
Coroglaucigenin- δ-allomethyloside, D 398
Coronene, C 76
Coronopilic acid, C 357
Coronopilin, C **357**, 367
Corotoxigenin, D 364, **398**
Corotoxigenin-D-allomethyloside, D 398
Correllogenin, E 26, **34**, **35**
Correllogenin acetate, E 35
Cortical hormones, E 1, 10
Corticoids, fluorinated, biological activity, D 354
Corticosteroids, D 191
—, synthesis, D 54
Corticosterone, D 2, 7, 55, 287, 289, 291, **296**, 297, 298, 299, 312, **323**, **324**
—, *nomenclature*, D *439*
Corticosterone acetate, D 298, 303, 324
Cortisol, D 7, 287, **294**, 312, **329**
—, *nomenclature*, D *439*
Cortisol acetate, D *439*
Cortisol 3,20-disemicarbazone, D 331
Cortisol 21-methyl ether, D *438*
Cortisone, D 2, 7, 149, 287, **294**, **295**, **324**, 325, 332, 333; E 1, 33
—, homologues, D 331
—, *nomenclature*, D *439*
—, total synthesis, D 102, 328, 330
Cortisone acetate, D 325, 326, 327, 328, 329, 332
—, *nomenclature*, D *439*
Cortisone 3,20-disemicarbazone, D 331

Corymbol, C 395
Corynebacterium simplex, D 218, 332
Costol, C **301**, 334
Costunolide, C 337, **363**, 368
Costus lactones, C **351**, 367
Cotton effect, B 103
Cottonseed oil, D 154
Coumarins, photodimerisation, A 76
Countercurrent distribution, carotenoid isolation, B 241
Courmontosides, D **405**
Cram's rule, D 108
o-Cresol, hydrogenation, B 44, 49
m-Cresol, hydrogenation, B 49
p-Cresol, B 121
—, hydrogenation, B 44, 49
Crocetin, B 244, **327**, 328; E 133
—, structure, B 233, 234
all trans Crocetin, B 328
Crocetindial, B 248, 277, 278, 279, 308, 318, 322, 324, 327
Crocetin diesters, B 328
Crocin, B **327**, 328
Croconic acid, A 188, 189
Croconic acid dimethyl ether, A 188
Crotonic acid, C 308
Crotonic ester, A 59
α-Crustacyanin, B 321
Crustecdysone, D 161
Crypto-acorone, C 312, 313, 314, 335
Cryptoflavin, B **302**
Cryptograndoside A, D 364, **391**, 394
Cryptograndoside B, D 364, **391**
Cryptols, B 201
Cryptomeradol, C 299
Cryptomeria japonica, C 299, 301
Cryptomeridiol, C **301**, 334
Cryptone, B 117, **200**, 201; E 77
Cubane, C **44**
Cubane-1,4-dicarboxylic acid, C 44
Cubane monocarboxylic acid, C 44
Cubebs, C 298
Cubeb camphor, C 297, **298**, 334
Cucuberbitacins, C 424, 479
Cucurbitacin A, C 409
Cucurbitacin group, triterpenoids, C 424
5α-Cucurbitane, D *429*
Cumin alcohol, C 140, 143
Cunninghamella, D 328
Cuparene, C **267**, 288
Cuparenic acid, C 267, 288
Cuprenene, C 268, 288
Cupressaceae species, B 365; C 275
Cupressic acid, C 383
Cupressus diterpene A, C 393
Cupressus diterpene B, C 392
Cupressus macrocarpa, B 180, 365
Cupressus torulosa, B 189
Cupric tropolone, B 369
Curcuma aromatica, C 262
Curcuma longa, C 260
Curcuma zedoraris, C 260
Curcumenes, C 259, **262**, 288, 293
Curtius reaction, A 46, 53; B 80, 83; C 170; D 232, 324
Curvularia, D 328
Cyanaropicrin, C **355**, 367
Cyanoacetamide, A 108; B 149
Cyanoacetic acid, B 167; C 169
Cyanoacetic ester, A 162, 192; B 100, 358
—, reaction with acetone, A 73
—, reaction with dihalides, A 22
2-Cyanobicyclo[2.2.2]oct-5-ene, C 114
6-Cyanobicyclo[3.2.2]non-8-ene, C 123
α-Cyanocamphor, C 223, **224**
Cyanochogenin, D 285
1α-Cyano-5α-cholestan-3-one, D 129
5-Cyano-5α-cholestan-3-one, D 128
5-Cyano-5β-cholestan-3-one, D 128
1-Cyanocyclohex-1-ene, B 149
1-Cyanocyclopentane-1,2-dicarboxylic ester, A 191
2-Cyanocyclopentan-1-one, A 200
1-Cyanocyclopropanecarboxylic acid, A 58
1-Cyanocyclopropane-1-carboxylic ester, A 58
7-Cyanoheptan-2-one, B 74
Cyanohydrins, cyclanones, B 409
3-(2-Cyanomethylcyclopentyl)propionitrile, C 128
3-Cyano-2-methyldeca-2,4,6,8-tetraene, C 49
1-Cyano-4-phenylbuta-1,3-diene, A 99
1-Cyano-1-phenylcyclobutane, A 95
Cyanophyceae, B 238
20-Cyanopregn-17(20)-ene, D 325
16α-Cyanopregn-5-ene-3β,20β-diol, D 235
Cyanora cardunculus, C 355
Cyanora scolymus, C 255
6-Cyano-7,7,8-trimethylbicyclo-[4.2.0] octa-2,4-diene, C 49
Cyanotropolones, B 373
β-Cyanotropolone, B 377
Cyanotropones, B 377
Cyasterone, D 161, 162
Cyclamigenin B, C 441
Cyclanols, oxidation, B 408
Cyclanones, B 401, 406, 408

—, physical properties, B 409
Cyclic acetylenes, B 387, 401
Cyclic acetylenic ketones, B 408
Cyclic carotenoid epoxides, B 293
Cyclic dihydroxycarotenoids, B 295
Cyclic ethers, terpenoid, B 204
Cyclic Grignard reagents, conversion to polycarboxylic compounds, C 6
Cyclic halides, B 401
Cyclic ketones, conversion to polycarboxylic compounds, C 6
Cyclic olefins, B 382
Cyclic sulphonium salts, A 9
Cyclic terpenes, nomenclature, D *425*
Cyclisation reactions, C 78
Cycloalkadienes, A 2
Cycloalkadiynes, A 2
Cycloalkanes, B 381, 390
—, heats of combustion, A 4
—, strain, A 3
Cycloalkane carboxylic acids, B 417
Cycloalkanediols, A 16, 159
Cycloalkane-1,2-diones, A 16
Cycloalkanols, B 401
Cycloalkanones B 418
Cycloalkan-2-onecarboxylic acids, B 417
Cycloalkanone cyanohydrins, B 417
Cycloalkanone, oximes, B 418
Cycloalkenes, 2, 15; B 382, 388, 401, 419
Cycloalkenones, B 354, 410
Cycloalkylamines, B 382
Cycloalkylamine oxides, A 15
Cycloalkylboranes, A 154
Cycloalkylcyclopropyl ketones, A 50
Cycloalkylmagnesium halides, B 417
Cycloalkyltriethylammonium hydroxides, B 419
Cycloalkyltrimethylammonium hydroxides, A 15
Cycloalkynes, A 2; B 383, 388
9,19-Cyclo-5α,9β-androstane, D *430*
11β,19-Cyclo-5α-androstane, D *430*
5β,19-Cycloandrost-1-ene-3,17-dione, D 334
6β-19-Cycloandrost-4-ene-3,17-dione, D 335
Cycloartane, C 473
Cycloartane group, triterpenoids, C 423
Cycloartanol, C 423; E 106
Cycloartenol, C 408, 409, 423; E 104, 105, 106, 107, 119
Cycloartenone, C 423
Cyclobutadienes, A 84, 86
Cyclobutadiene-irontricarbonyl complex, C 43
Cyclobutadienetricarbonyliron, A 85
Cyclobutadienoquinones, A 93
Cyclobutanes, aldehydes, A 91
—, amines, A 91
—, general properties, A 78
—, halogen derivatives, A 87
—, infrared spectra, A 79
—, ketones, A 91
—, ring contraction, A 23
—, syntheses, A 72
Cyclobutane, A 2, 80, 164
—, alcohols, A 90
—, conformation, A 79
—, hydrogenation, A 81
—, structure, A 79
Cyclobutanecarboxylic acid, A 78, 87, 91, 95; C 9
Cyclobutanecarbonyl chloride, A 94
Cyclobutane-1,2-dicarbonyl chloride, A 94
Cyclobutane-1,1-dicarboxylic acid, A 95
Cyclobutane-1,2-dicarboxylic acid, A 96; C 46
Cyclobutane-1,3-dicarboxylic acids, A 1, 96
Cyclobutane-1,1-dicarboxylic ester, A 1
Cyclobutane-1,2-diol, A 90
Cyclobutane-1,2-diol-1,2-diacrylic acid dilactone, A 102
Cyclobutanediones, A 75, 93
Cyclobutanepentacarboxylic acid, A 97
Cyclobutanetetracarboxylic acid, A 96, 97
Cyclobutane-1,2,3,4-tetracarboxylic dimethyl ester, C 33
Cyclobutanetetranitriles, 97
Cyclobutane-1,2,3-tricarboxylic acid, A 97
Cyclobutanols, A 78
Cyclobutanol, A 64, 77, 90
—, oxidation, A 47
Cyclobutanol esters, A 23
Cyclobutanone, A 23, 78, 90, 92,; C 9
Cyclobutanone bases, A 76
Cyclobutanone pinacol, C 9, 26
Cyclobutenes, A 82, 92; B 384
—, halogen derivatives, A 87
Cyclobutene, A 80, 81, 82, 87; C 45
Cyclobutenecarboxylic acids, A 95
Cyclobutene-1,2-dicarboxylic acid, A 96

Cyclobutene-1,2-dicarboxylic methyl ester, C 49
Cyclobutenones, A 92
Cyclobutenylmetal complexes, A 86
Cyclobutylamine, A 23, 90, 91
Cyclobutyl bromide, A 87
Cyclobutylcarboxylic acid, A 90
Cyclobutylcyclohexane, C 13
2-Cyclobutylethanol, A 90
Cyclobutylmagnesium bromide, A 80
Cyclobutylmethanol, A 80, 82, 90, 91
Cyclobutylmethylamine, A 91
Cyclobutyltrimethylammonium hydroxide, A 82
Cyclobuxine, C 408, 409; D 335, 336
3,5-Cyclocampestanyl methyl ether, D 156
3,8-Cyclocamphor, C 210
(20R)-18,21-Cyclo-5α-cholane, D *430*
3α,5-Cyclo-5α-cholestane, D 6, 43, 77
3α,5-Cyclo-5α-cholestan-6β-ol, D 74, 77, 78
—, *nomenclature*, D *430*
3β,5-Cyclo-5β-cholestan-6β-ol, D 80
5,7α-Cyclo-5α-cholestan-4α-ol, D *430*
5,7β-Cyclo-5β-cholestan-3β-ol, D 81
1β,5-Cyclo-5β,10α-cholestan-2-one, D 128
3α,5-Cyclo-5α-cholestan-6-one, D 77
5,7-Cyclo-5β-cholestan-3-one, D 81
5,7α-Cyclo-5α-cholestan-4-one, D 81
3α,5-Cyclo-5α-cholestan-6β-yl acetate, D 77
3,5-Cyclocholestan-6-yl methyl ether, D 41
3α,5-Cyclo-5α-cholest-6-ene, D 80
3α,5-Cyclo-5α-cholest-7-ene, D 43, 80
5,7β-Cyclo-5β-cholest-3-ene, D 81
Cyclocitrals, B **173**, 174
α-Cyclocitral, B 88, 90
β-Cyclocitral, B 90
Cyclocitrylidenecyanoacetic acids, B 174
Cyclocolorenone, C **322**, 336
Cyclodeca-1,2-diene, B 383
Cyclodeca-1,3-diene, B 386
Cyclodecanes, A 3, 16
Cyclodecane, B 381, 384, 399; C 71
Cyclodecanecarboxylic acid, B 417
Cyclodecanediols, B 404, 405
Cyclodecane-1,6-diol, A 13
Cyclodecane-1,2-dione, B 413; C 51
Cyclodecane-1,6-dione, B 414; C 60, 69
Cyclodecane-1,6-dione dioxime, C 66
Cyclododecane-1,5,7-triol, B 404
Cyclodecanol, B 404, 420
Cyclodecanone, B 409, 420
Cyclodecan-2-onecarboxylic methyl ester, B 417
Cyclodecapentaene, B 399; C 71
Cyclododeca-1,5,9-triene, C 72
Cyclodecenes, B 403
Cyclodecene, A 13; B 382, 385, 386, 401, 420
Cyclodec-1-enecarboxylic acid, B 417
Cyclodecene oxide, B 405
Cyclodec-2-enol, B 404, 405
Cyclodec-2-en-1-one, B 410
Cyclodec-5-en-1-one, B 410
Cyclodec-1-enyllithium compounds, B 401
Cyclodex-1-enylmagnesium bromide, B 401
Cyclodecylamine, B 420
Cyclodecyl bromide, B 401
Cyclodecyl *p*-toluenesulphonate, B 420
Cyclodecyne, B 386
3α,5-Cyclo-4,4-diethyl-5α-cholestan-6β-ol, D 79
3α,5-Cyclo-4,4-dimethyl-5α-cholestan-6β-ol, D 79
Cyclodocosa-1,12-diene, B 386
Cyclodocosa-1,12-diyne, B 389
Cyclodocosane, B 381
Cyclodocosane-1,12-dione, B 413
Cyclodocosanone, B 409
Cyclodocosatetrayne, B 389
Cyclododecadienes, B 382
Cyclododeca-1,3-diene, B 386
Cyclododeca-1,7-diene, B 386
Cyclododeca-1,7-diyne, B 389
Cyclododecane, B 381, 421
Cyclododecanediols, B 404
Cyclododecane-1,2-dione, B 413
Cyclododecane-1,3-dione, B 412, 413
Cyclododecane-1,6-dione, B 387, 414
Cyclododecanol, B 404, 408
Cyclododecanone, B 387, 408, 409
Cyclododecan-2-onecarboxylic methyl ester, B 417
Cyclododeca-1,5,9-triene, B 380, 385, 387, 408
Cyclododecene, B 386
Cyclododec-1-enecarboxylic acid, B 417
Cyclododec-2-enol, B 404, 405
Cyclododec-2-en-1-one, B 410
Cyclododecyne, B 386
Cyclodotriacontane, B 381
Cyclodotriacontane-1,17-dione, B 413
Cyclodotriaconta-octayne, B 389

Cycloeicosa-1,11-diene, B 386
Cycloeicosa-1,11-diyne, B 389
Cycloeicosane, B 381
Cycloeicosane-1,10-dione, B 414
Cycloeicosane-1,11-dione, B 413
Cycloeicosanol, B 404
Cycloeicosanone, B 409
Cycloeicosatetrayne, B 389
3α,5α-Cyclo-6β,19-epoxy-steroids, D 256
3α,5-Cyclo-5α-ergosta-7,22-diene, D 43
3α,5-Cyclo-5α-ergosta-7,22-dien-3β-ol, D 78
3α,5-Cyclo-5α-ergosta-7,9(11),22-trienol, D 78
3α,5-Cyclo-5α-ergost-22-en-6α-ol, D 155
3α,5-Cyclo-5α-ergost-22-en-6-one, D 155
5,7β-Cyclo-4β-ethoxy-5β-cholestane, D 81
Cycloeucalenol, C 423; E 107
Cyclofenchene, C 240, 243, 244, 247, 248, 249
α-Cyclogeranic acid, B 174
Cyclogeraniols, B 172, 174; E 68,
α-Cyclogeraniol, B 173
β-Cyclogeraniol, B 173
γ-Cyclogeraniol, B 173
Cycloheneicosa-hexayne, B 389
Cycloheneicosane, B 381
Cycloheneicosanone, B 409
Cycloheptacosane, B 381
Cycloheptadeca-1,8-diyne, B 389
Cycloheptadecane, B 381
Cycloheptadecanol, B 404
Cycloheptadecanone, B 409, 415
Cycloheptadecatetrayne, B 389
Cycloheptadecene, B 386
Cycloheptadec-9-enone, A 2; B 414
Cycloheptadecyne, B 386
Cycloheptadienes, C 121
Cycloheptadiene, A 35; B 347, 348, **349**, 350, 367; C 34, 35, 122, 123
Cyclohepta-3,5-dienol, C 46
Cycloheptadienones, B 361, 362
Cycloheptadienone, B 370
Cycloheptadienone bishydrazone, B 349
Cycloheptanes, A 16; B 347 ff
—, alcohols and ethers, B 351
—, aldehydes, B 352
—, amines, B 358
—, carboxylic acids, B 356
—, ketones, B 353 ff
—, monoterpenes, E 78
Cycloheptane, B **347**, 348, 405
Cycloheptane-1-acetic-1-bromoacetic ester, C 25
Cycloheptane-1-acetic acid-1-carboxylic acid, B 358
Cycloheptanecarbaldehyde, B **352**, 353, 405
Cycloheptanecarboxamide, B 358
Cycloheptanecarboxylic acid, B **356**, 358
Cycloheptane-1,1-diacetic acid, B **358**
Cycloheptane-1,2-diacetonitrile, C 60
Cycloheptane-1,2-dicarboxylic acid, B **357**
Cycloheptane-1,2-diol, B **351**, 371
Cycloheptane-1,4-diol, B **352**
Cycloheptane-1,2-dione, B 355, 367
Cycloheptane-1,3-dione, B **355**
Cycloheptane-1,4-dione, B 355
Cycloheptanehexacarboxylic acid, B 357
Cycloheptanepentacarboxylic ester, B 357
Cycloheptane-1,3,5,6-tetracarboxylic acid, B 357
Cycloheptane-1,2,4-tricarboxylic acid, B 357
Cycloheptanol, B 348, 350, **351**, 358
Cycloheptanones, dehydrogenation to tropones, B 361
Cycloheptanone, B 347, 351, 352, **353**, 355, 357, 406, 407, 408
—, reaction with bromoacetic ester, B 357
—, reduction, C 18
—, self-condensation, C 18
Cycloheptanone-2-acetone, C 60
Cycloheptanone cyanohydrin, B 356, 357
Cycloheptanone oxime, B 358
Cycloheptanone pinacol, C 18, 31
Cycloheptatrienes, C 38, 45
—, 1,5-hydrogen shifts, B 349
Cycloheptatriene, B 347, 348, **349**, 359, 360, 367
—, cyclopentadienylmetal complexes, A 143
—, reaction with bromine, B 351
Cycloheptatrienecarboxylic acids, B 356, 359
Cycloheptatrienide anion, B 361
Cycloheptatrienols, B 359
Cycloheptatrienone, B 355, 359, 361
Cycloheptatrienyl acetate, B 359
Cycloheptatrienylmethylamine, B 350
Cycloheptenes, B 351, 355

Cycloheptenes, halogen compounds, B 350
Cycloheptene, B 347, **348**
—, adduct with 2,5-diphenyl-3,4-isobenzofuran, B 349
—, hydroxylation, B 352
—, reaction with bromine, B 350
Cycloheptene-1-*n*-butyric acid, C 73
Cyclohept-1-enecarbaldehyde, B 353
Cyclohept-1-enecarboxylic acid, B **356**; C 39
Cycloheptene oxide, B **352**
1,2-Cycloheptene-*O,O'*-thiocarbonate, B 348
Cycloheptenols, B 351
Cyclhepten-3-ol, C 39
Cyclohept-2-enone, B 353, **354**
Cyclohept-3-enone, B 353, 354
Cyclohept-1-enyl-1-acetic acid, B 357
Cyclohept-2-enyl-1-acetic acid, B 358
Cycloheptenylamine, B 349, 358
Cyclohept-1-enyl cyanide, B 356
Cycloheptenyl-lithium, B 351, 353
Cyclohept-2-enyl-1-malonic acid, B 358
β-(1'-Cycloheptenyl)-*β*-methoxy-carbonylpropionic acid, C 60
Cycloheptylacetic acid, B **357**
Cycloheptylamine, B 348, **358**
Cycloheptyl bromide, C 18
Cycloheptylcarbinol, B 351
Cycloheptyl chloride B **350**
Cycloheptyl halides, B 347
Cycloheptylideneacetic acid, B 357
Cycloheptylidenecyanoacetic ester, B 358
Cycloheptylidenecycloheptanone, C 18
Cycloheptylmagnesium bromide, B 356
Cycloheptylmalonic acid, B 358
Cycloheptylmethylamine, B 358, 402
Cycloheptyne, B **349**
Cyclohexacosane, B 381
Cyclohexacosane-1,14-dione, B 413
Cyclohexacosanone, B 409
Cyclohexacosatetrayne, B 389
Cyclohexadeca-1,9-diene, B 386
Cyclohexadecadiynes, B 389
Cyclohexadecane, B 381
Cyclohexadecane-1,2-dione, B 413
Cyclohexadecane-1,9-dione, B 413
Cyclohexadecanol, B 404
Cyclohexadecanone, B 409
Cyclohexadecan-2-onecarboxylic ethyl ester, B 417
Cyclohexadecatetrayne, B 389
Cyclohexadecene, B 386
Cyclohexadec-8-en-1-one, B 410
Cyclohexadienes, B 5, 25 ff, 57
—, benzoquinone addition products, C 116
—, physical properties, B **26**
—, reaction with oxygen, B 74
—, dimerisation, C 115
Cyclohexa-1,3-dienes, B 25, 40
Cyclohexa-1,3-diene, B 26, **27**, 42, 62, 75; C 36, 47, 78, 108, 111, 112, 114, 117
—, oxidation, B 76
—, reaction with maleic anhydride, C 115
Cyclohexa-1,4-dienes, B 3, 25, 26
Cyclohexa-1,4-diene, B 2, 26, **27**, 36, 63; C 38
Cyclohexadiene-acetylenedicarboxylic ester adduct, C 114
Cyclohexadiene-benzoquinone adduct, C 116
Cyclohexa-2,5-dienecarbinol, B 367
Cyclohexadienecarboxylic acids, B 136, 138, 141
Cyclohexa-1,3-diene-1-carboxylic ester, C 113, 114
Cyclohexadiene-dibromomaleic anhydride adduct, C 114
Cyclohexadienedicarboxylic acids, B 146, 147, 148
Cyclohexa-1,3-diene-1,4-dicarboxylic ester, C 114
Cyclohexa-1,3-diene-1,6-dicarboxylic ester, C 114
Cyclohexa-1,3-diene-1,4-dicarboxylic methyl ester, C 42
Cyclohexa-3,5-diene-1,2-diol, B 60, 64 86
Cyclohexa-1,3-diene-5,6-diol, B 65
Cyclohexa-1,3-diene peroxides, B 75
Cyclohexadienols, B 29, **64**
Cyclohexadienones, B **120**, 123
—, conversion to phenols, B 124
—, reactions, B 122
Cyclohex-1,4-dienyl-1-methanol, B 29
1-(Cyclohexa-1,4-dienyl) methyl ether, B 64
Cyclohexa-1,3-dione-5-acetic acid, C 113
Cyclohexa-1,3-dione dioxime, B 79
Cyclohexanes, A 16; B 1 ff, 12 ff; C 36
—, alcohols, B 44 ff
—, alicyclic aldehydes, B **86**
—, amino-aldehydes, B 91
—, carbaldehydes, B 91

Cyclohexanes, conversion to alkylcyclopentanes, B 13
—, dehydrogenation to benzene derivatives, B 13
—, exocyclic alcohols, B 68
—, exocyclic amines, B 86
—, formation, B 1 ff
—, halogen compounds, B 32 ff
—, hydroxycarbaldehydes, B 90
—, nitration, B 13
—, peroxides, B **74**
—, physical properties, B 13
—, polycarboxylic, acids B 151
—, reaction with oxygen, B 74
—, sulphinic acids, B 77
—, sulphonic acids, B 77
—, sulphur containing, B **76**
—, tetracarboxylic acids, B 151
—, tricarboxylic acids, B 151
Cyclohexane, B **15,** 405
—, acetylation, A 176
—, conformation, A 14
—, cracking, A 112
—, dehydrogenation, B 4
—, fluorination, B 33
—, free energy differences, B 9
—, from benzene, B 3
—, nitration, B 78, 79
—, oxidation, B 74
—, photonitrosation, B 15, 98
—, physical properties, B 13
—, reaction with carbon monxide, B 134
—, reaction with formaldehyde, B 68
—, reaction with nitrosyl chloride, B 92
—, reaction with sulphur dioxide and oxygen, B 77
—, reaction with sulphuryl chloride, B 77
Cyclohexane-1-acetic acid, B 24
Cyclohexane-1-acetic ester-1-bromoacetic acid, C 25
Cyclohexane-1-acetic-2-propionic acid, B 150; C 62
Cyclohexanecarbaldehyde, B 70, **87,** 88, 89
Cyclohexanecarbonyl chloride, B 135
Cyclohexanecarboxylic acids, B 80; C 14, 480
—, halogenated, B 135
—, physical constants, B 132
Cyclohexanecarboxylic acid, B 89, **133**
—, esters, B 68
Cyclohexane-1-carboxylic-1-acetic acid, B 149
Cyclohexane-1-carboxylic-2-acetic acid, B 149; C 56
Cyclohexane-1-carboxylic-3-acetic acid, B 149; C 120
Cylcohexane-1-carboxylic-4-acetic acid, C 112
Cyclohexane-1-carboxylic-2-propionic acid, C 58, 62, 66, 69
Cyclohexane compounds, B 1 ff, 12 ff
Cyclohexane-1,1-diacetic acid, B 24, 149
Cyclohexane-1,2-diacetic acid, B 149; C 57, 58, 62, 70
—. ester, C 67
Cyclohexane-1,3-diacetic, acid, C 125
Cyclohexane-1,4-diacetic acid, B 150
Cyclohexanedicarboxylic acid, A 2; B **144**
Cyclohexane-1,2-dicarboxylic acid, B 35
—, ester, C 47
Cyclohexane-1,3-dicarboxylic acid, B 82, **145,** 149
—, dimcthyl ester, C 121
Cyclohexane-1,4-dicarboxylic acid, B **145**; C 17, 116
Cyclohexanediols, B 50, 51
—, dehydration, B 25, 27, 62
Cyclohexane-1,2-diols, B 51, 94, 352
Cyclohexane-1,2-diol, A 163; B **50, 52,** 75, 103, 110
Cyclohexane-1,3-diols, B 51
—, monomethyl ethers, B 54
Cyclohexane-1,3-diol, B **53,** 54, 105
Cyclohexane-1,4-diols, B 51
Cyclohexane-1,4-diol, B **54,** 62, 73, 105
Cyclohexane-3,4-diol-1-carboxylic acids, B **138**
Cyclohexane-1,2-diol diacetate, B 25
Cyclohexane-1,3-diol-1,3-dicarboxylic acid, B 150
Cyclohexane-1,4-diol-1,4-dicarboxylic acid, B 150
Cyclohexanediones, B **108**
—, physical properties, B 109
Cyclohexane-1,2-dione, A 198; B 98, 99, 105, 108, 109, **110**
Cyclohexane-1,3-diones, B 7, 108
Cyclohexane-1,3-dione, B 1, 109, **111**
Cyclohexane-1,4-dione, B 8, 54, 99, 105, 109, **112**
Cyclohexane-1,3-dione dicyanohydrin, B 150
Cyclohexane-1,4-dione dicyanohydrin, B 150
Cyclohexane-1,2-dione dioxime, B 81

Cyclohexane-1,4-dione dioxime, B 79, 82
Cyclohexane-1,3-dione monoethylene acetal, B 355
Cyclohexane-1,2-disulphonic acid, B 77
Cyclohexane-1,2-halohydrins, B 72
Cyclohexanehexacarboxylic acid, B 152
Cyclohexanehexols, B **58**
Cyclohexane hydrocarbons, B 1 ff, 12 ff
Cyclohexane monoethers, B 52
Cyclohexanepentols, B **57**
Cyclohexane peroxides, B 74
Cyclohexane ring, configuration, B 7
—, conformations, D 14
Cyclohexanesulphinic acid, B **77**
Cyclohexanesulphonic acid, B **77**
Cyclohexanetetracarboxylic acids, B 144, 151, 152
Cyclohexanetetrols, B 28, **56**, 60, 64
Cyclohexanethiol, B **76**
Cyclohexane-1,3,5-tricarboxylic methyl ester, B 151
Cyclohexanetriols, B 50, 51
Cyclohexane-1,2,3-triol, B **55**, 75
Cyclohexane-3,4,5-triol-1-carboxylic acid, B 139
Cyclohexanetriones, B 112
Cyclohexane-1,2,3-trione dihydrate, B 112
Cyclohexane-1,2,3-trione 1,3-dioxime, B 100
2β,3α-Cyclohexano-5α-androstane, D 451
Cyclohexanols, B 2, **44**
—, dehydration to cyclohexenes, B 46
—, halogenated, B **66**
—, oxidation, B 46, 92
—, physical constants, B 47
—, properties, B 46
—, reaction with halides, B 32
—, tertiary, B 46
Cyclohexanol, A 2; B **46**, 47, 75, 98; C 15
—, conversion to cyclohexene, B 49
—, dehydration, B 20
Cyclohexan-1-ol-1-acetic acid, B 137, B 142
—, esters, B 94
Cyclohexan-2-ol-1-acetic acid, B 142
Cyclohexan-3-ol-1-acetic acid, B 142
Cyclohexan-4-ol-1-acetic acid, B 142
Cyclohexanolcarboxylic acids, B 136
Cyclohexan-1-ol-1-carboxylic acid, B 129, **138**
—, ester, B 70
Cyclohexan-2-ol-1-carboxylic acids, B 138
Cyclohexan-3-ol-1-carboxylic acid, B 135, 137, 138
Cyclohexan-4-ol-1-carboxylic acids, B 55, 138
Cyclohexanoldicarboxylic acids, B 150
Cyclohexan-2-ol-1-ones, B 103, 113
Cyclohexan-2-ol-1-one, B 98, **103**, 105
Cyclohexan-3-ol-1-one, B **105**
Cyclohexan-4-ol-1-one, B **105**
Cyclohexan-1-ol-2-one oxime, B 82
Cyclohexan-2-ol-1-sulphonic acid, B 77
Cyclohexanones, B 2, 6, 17, 20, 40, 65, 92 ff, 93
—, dehydrogenation, B 98
—, halogenated, B 102
—, oxidation, B 98
—, physical constants, B 95
—, polyhalogenated, B 113
—, properties, B 94
—, reaction with amines, B 97
—, reduction with oxalic esters, B 143
—, reduction, B 45, 98
Cyclohexanone, A 190; B 3, 15, 24, 34, 40, 49, 66, 70, 83, 91, 95, **98**, 105, 142, 151; C 129
—, autocondensation, C 17
—, condensation with cyclopentadiene, C 14
—, conformation, B 12
—, oxidation, B 99, 110
—, reaction with ammonium cyanide, B 142
—, reaction with bromoacetic ester, B 142
—, reaction with ethyl acetate, B 130
—, reaction with hydrogen cyanide, B 138
—, reaction with succinic ester, B 150
—, reduction, B 44, 46
—, reduction with sodium, C 15
—, ring expansion, B 353
Cyclohexan-1-one-3-acetic acid, C 112
Cyclohexan-2-one-1-acetic acid, B 142, **143**
Cyclohexan-3-one-1-acetic acid, B 143
Cyclohexanone acetylenic glycol, C 15
Cyclohexanone azine, B 81
Cyclohexanone bishydroperoxide, B 99
Cyclohexanone-3-*n*-butyric ester, C 67
Cyclohexanonecarboxylic acids, B 143
—, oximes, B 94, 142
Cyclohexan-1-one-2-carboxylic acid, B 108, 150

Cyclohexan-1-one-3-carboxylic acid, B 150
Cyclohexan-2-one-1-carboxylic acid, B 102, 138, 143
Cyclohexan-3-one-1-carboxylic acid, B 143
Cyclohexan-4-one-1-carboxylic acid, B 143
Cyclohexanone-2-carboxylic ester, C 68
Cyclohexanone cyanohydrins, B 86, 137, 138
Cyclohexan-2-one-1,3-dicarboxylic ester, B 150
Cyclohexan-2-one-1,4-dicarboxylic acid, B 150
Cyclohexan-4-one-1,2-dicarboxylic methyl ester, B 151
Cyclohexanone dimethyl acetal, B 62
Cyclohexanone enamines, B 127
Cyclohexan-1-one-2-glyoxylic acid, B 89
Cyclohexan-1-one-2-glyoxylic esters, B 96
Cyclohexan-2-one-1-glyoxylic esters, B 143
Cyclohexanonehydrazone, B 81
Cyclohexanone hydroxyhydroperoxyperoxide, B 99
Cyclohexanone morpholine enamine, C 124, 125, 126
Cyclohexanone pinacol, C 15, 31
2-(2-Cyclohexanone)-propionic ester, C 126
Cyclohexanone-3-propionic ester, C 59
Cyclohexanone oxime, B 78, 80, 81
Cyclohexatriacontane, B 381
Cyclohexatriaconta-octayne, B 389
Cyclohexenes, B 5, 16 ff, 40, 67, 80
—, autoxidation, B 19
—, disproportionation, B 20
—, halogen derivatives, B 41
—, hydroxylation, B 50
—, physical properties, B 21
—, oxidation, B 113
—, properties, B 19
—, reaction with acyl chlorides, B 127
—, reaction with diborane, B 20
—, reaction with hydrogen halides, B 32
—, reaction with oxygen, B 74, 75
—, thermal decomposition, B 20
Cyclohexene, A 2; B **20**, 27, 42, 129, 349; C 49, 50
—, autoxidation, B 19
—, conformations, B 11
Cyclohexene, cracking, B **22**
—, fluorination, B 33
—, from benzene, B 3
—, hydroformylation, B 87
—, hydroxylation, B 52
—, oxidation, B 22, 92
—, physical properties, B 21
—, reaction with ammonium bisulphite, B 77
—, reaction with carbon monoxide, B 22, 69, 87, 134
—, reaction with diazoacetic ester, C 39
—, reaction with dinitrogen tetroxide, B 78, 79
—, reaction with formaldehyde, B 70, 71
—, reaction with halogens, B 35
—, reaction with hydrogen peroxide, B 22
—, reaction with oxalyl bromide, B **137**
—, sulphonation, B 77
Cyclohexenecarbaldehydes, B 70, 87, 88, 90
Cyclohex-1-enecarbaldehyde, B **89**, 90
Cyclohex-3-enecarbaldehyde, B **90**
Cyclohexenecarboxylic acids, B **136**, 137
Cyclohex-1-ene-1-carboxylic acid, B 90, 135, 143
Cyclohex-3-ene-1-carboxylic acids, B 136, 139
Cyclohexenecarboxylic ester, B 69, 71, 149
Cyclohexenedicarboxylic acids, B **145**, 146, 151
Cyclohex-4-ene-1,2-dicarboxylic acid, B 6
Cyclohex-4-ene-1,2-dicarboxylic anhydride, B 144
Cyclohexenediols, B 57
Cyclohex-4-ene-1,2-diol, B **63**
Cyclohex-1-ene-1-methyl acetate, B 28
Cyclohex-1-ene 3-nitronate anion, B 80
Cyclohexene oxide, A 161, 163; B 142
Cyclohexene peroxides, B 75
Cyclohex-4-enespirocyclobutan-2′-one-3-carboxylic acid, C 4
Cyclohex-1-ene-1-sulphonic acid, B 77
Cyclohex-2-ene-1-sulphonic acid, B **77**
Cyclohex-4-ene-tetracarboxylic ester, B 151
Cyclohexenetetrol, B 61, 63, 68
Cyclohex-2-ene-4,5,6-triol-1-ones, B 64

Cyclohexenols, B **61,** 67
—, dehydration, B 25
Cyclohex-2-enol, C 38
Cyclohexen-3-ol, C 36
Cyclohex-1-en-1-ol, B **62,** 73
Cyclohex-2-en-1-ol, B 22, 42, 46, 55, **62,** 75
Cyclohex-3-en-1-ol, B 28, **62,** 63
Cyclohexenones, B 61, **113**
—, hydrogenation, B 118
—, physical constants, B 117
—, properties, B 118
—, reactions, B 118
Cyclohex-2-enones, A 169
Cyclohex-2-en-1-ones, B 102
—, 2-substituted, B 116
—, 3-substituted, B 115
Cyclohex-2-en-1-one, B 62, 144; C 49, 286
Cyclohex-1-en-3-ones, C 17
Cyclohex-3-en-2-one-1-carboxylic acid, B **144**
Cyclohexenonedicarboxylic esters, B 113
Cyclohexenyl-acetaldehydes, B 91
3-Cyclohex-1-enyl acetate, B 75
Cyclohex-1-enyl-acetic acids, B 137
Cyclohexenylacetoacetic ester, B 100, 130
Cyclohexenylacetones, B 128
Cyclohex-1-enylacetone, B **130**
Cyclohex-1-enylacetonitrile, B 100, 149
Cyclohexenyl-alkanecarboxylic acids, B 137, 142
2-(Cyclohex-1′-enyl)cyclohexanone, C 16
2-(Cyclohex-1′-enyl)ethanol, B 24
2-(Cyclohexenyl)ethyl alcohols, B **71**
Cyclohex-2-enyl hydroperoxide, B 19
Cyclohexenyl-lithium, B 89
Cyclohex-3-enylmagnesium bromide, B 90, 137
Cyclohexenylmagnesium chloride, B 40
Cyclohexenylmalonic ester, B 149
Cyclohex-1-enylmethanesulphonic acid, B 24
(Cyclohex-1-enylmethyl)succinic acid, B 24
Cyclohexenyl peroxyester, B 19
α-Cyclohexenylpropionaldehyde, B 92
Cyclohex-1-enylsuccinic acid, B 150
1-(Cyclohex-1′-enyl)succinic ester, B 100
Cycloheximide, B 100, 101
Cyclohexylacetamide, B 86
Cyclohexylacetic acid, B 91, **135,** 149
Cyclohexylacetone, B 128
Cyclohexylacetonitrile, B 86
Cyclohexylacetylene, B **31**
Cyclohexylalkanecarboxylic acids, B 135
Cyclohexylalkylmagnesium halides, B 68
Cyclohexylamine, B 46, 78, **80,** 81, 122
Cyclohexylboranes, B 20
β-Cyclohexyl-$\Delta^{\alpha\beta}$-butenolide, D 371
Cyclohexylcarbaldehyde, C 13
Cyclohexylcarbinol, B **68,** 69, 89
Cyclohexyl chloride, B 16
Cyclohexyl cyanide, B 86
Cyclohexylcycloheptane compounds, C **18**
Cyclohexylcycloheptylmethane, C 18
2-Cyclohexylcyclohexanol, C 14
2-Cyclohexylcyclohexanone oxime, C 16
1-Cyclohexylcyclohexene, C **14**
3-Cyclohexylcyclohexene, C 14
2-Cyclohexylcyclohexylamine, C 16
Cyclohexylcyclopentenylmethane, C 14
1-Cyclohexyl-2-cyclopropylcyclopropane, C **13**
1-Cyclohexyl-3-cyclopropyl-2-methylpropane, C 3
5-Cyclohexyl-3-cyclopropyl-2-pyrazoline, C 13
N-Cyclohexyldimethylamine, B 81
Cyclohexyldimethylcarbinol, B 69
Cyclohexyl ethers, B 48, 49
2-Cyclohexylethyl alcohol, B 69, **70**
β-Cyclohexylethylamine, B **86**
N-Cyclohexylethylamine, B 81
Cyclohexyl fluoroformates, B 33
Cyclohexyl Grignard reagent, C 14, 15
Cyclohexyl halides, B 18, 22, 78
Cyclohexyl hydroperoxide, B **74**
Cyclohexylideneacetaldehyde, B 66, 91
Cyclohexylideneacetic acids, B 24, 135, 137
Cyclohexylideneacetone, B 128, **130**; C 31
Cyclohexylidene-alkanecarboxylic acids, B 142
Cyclohexylidenecyanoacetic ester, B 100, 149
Cyclohexylidenecyclohexanone, B 100; C 15, 16
β-Cyclohexylidene-ethanol, B 91
α-Cyclohexylidene-ethylcyclohexane, C 15

Cyclohexylideneketen, C 27
Cyclohexylidenesuccinic anhydride, B 150
Cyclohexyl iodide, C 14
S-Cyclohexylisothiouronium bromide, B 77
Cyclohexylketones, B 68
Cyclohexylmagnesium bromide, B 69
Cyclohexylmagnesium halides, B 46, 68, 70, 77
Cyclohexylmalonic acid, B **149**
Cyclohexylmercuric nitrate, A 163
Cyclohexylmethanesulphonic acid, B 77
Cyclohexylmethanol, B 24
Cyclohexylmethylamine, B 81, **86,** 348
Cyclohexylmethyldimethylamine oxide, B 23
Cyclohexylmethylene acetophenone B 89
Cyclohexyl methyl ether, B 48
Cyclohexylmethyl iodide, B 24
Cyclohexyl methyl ketone, B **129**
Cyclohexylmethylmagnesium iodide, B 70
Cyclohexylmethyl sulphide, B 76
Cyclohexylphenols, B 20
o-Cyclohexylphenyl C 15
2-Cyclohexylpropane-1-carboxylic acid, B 136
Cyclohexylpropyl alcohols, B 69
3-Cyclohexyprop-1-yne, B 31
Cyclohexylsulphamic acid, B 81
β-Cyclohexylvinyl cyclopropyl ketone, C 13
Cyclolaudenol, C 423
Cyclolavandulol, B **173**
3*α*,5-Cyclo-6*β*-methoxy-5*α*-steroids, D 80
3*α*,5-Cyclo-4*α*-methyl-5*α*-cholestan-6*β*-ol, D 78
Cyclomicrobuxine, D 336
Cyclomicrophylline-B, D **336**
Cyclomyretin A, 440
Cyclononacosane, B 381
Cyclononacosanone, B 409
Cyclononadeca-1,10-diene, B 386
Cyclononadecadiynes, B 389
Cyclononadecane, B 381
Cyclononadecanol, B 402, 404
Cyclononadecanone, B 409
Cyclononadecatetrayne, B 389
Cyclononadecatriyne, B 389
Cyclonona-1,2-diene, B 383
Cyclonona-1,3-diene, B 384; C 50
Cyclonona-1,5-diene, A 118,
Cyclononanes, A2, 16; B 381
Cyclononanecarboxylic acid, B 410, 417
Cyclononanediols, B 404, 405
Cyclononane-1,2-dione, B 413
Cyclononane-1,3-dione, B 412
Cyclononane oxide, B 405
Cyclononanol, B 402, 404
Cyclononanone, B 408, 409
Cyclononan-2-onecarboxylic methyl ester, B 417
Cyclononatetraenide, B 399
Cyclononatriene, B 385; C 42, 58
Cyclononene, B 382, 386
Cyclonon-3-ene-1,6-diol, B 385
Cyclononylamine hydrobromide, B 421
Cyclononyl bromide, B 401
Cyclononyne, B 386
3*α*,5-Cyclo-6-nor-5*α*-cholestan-7-one, D 81
Cyclo-octacosane, B 381
Cyclo-octacosane-1,15-dione, B 413
Cyclo-octacosanone, B 409
Cyclo-octacosaoctayne, B 389
Cyclo-octacosatetrayne, B 389
Cyclo-octadeca-1,3-diene, B 386
Cyclo-octadeca-1,10-diene, B 386
Cyclo-octadeca-1,10-diyne, B 389
Cyclo-octadecahexaenetriyne, B 390, 397
Cyclo-octadecahexayne, B 390, 396
Cyclo-octadecane, B 381, 397
Cyclo-octadecane-1,10-dione, B 413
Cyclo-octadecanol, B 404
Cyclo-octadecanonaene, B 397
Cyclo-octadecanone, B 409
Cyclo-octadecatetrayne, B 389
Cyclo-octadec-9-en-1-one, B 410
Cyclo-octadiene, B 408; C 128
Cyclo-octa-1,3-diene, B 383, 384, 386; C 49
Cyclo-octa-1,5-diene, B 380, **383,** 384, 386, 392; C 52, 85, 126
Cyclo-octanes, A 2; B 347 ff
Cyclo-octane, B 381, 383, 421, 422
Cyclo-octanecarbaldehyde, B **406**
Cyclo-octanecarboxylic acid, B **417**
Cyclo-octane-1,2-dicarboxylic acid, B 422
Cyclo-octane diols, B 403, 404, 405
Cyclo-octane-1,4-diol, C 54
Cyclo-octane-1,2-dione, B 413
Cyclo-octane-1,3-dione, B **412,** 413
Cyclo-octane-1,4-dione, B 414
Cyclo-octanol, B 358, 394, 402, 404, 408, 418

Cyclo-octan-1-ol-4-one, C 54
Cyclo-octanone, B 405, 406, 407, 408, 409
Cyclo-octanone pinacol, C 20
Cyclo-octan-2-onecarboxylic methyl ester, B 417
Cyclo-octatetraene, A 96; B 347, 352, 380, 383, 384, 390, **392**, 395, 396; C 36, 41, 47, 128
—, oxidation, B 360
—, reactions, B 393
Cyclo-octatetraene dimer, B 397; C 135
Cyclo-octatetraenide, C 41
Cyclo-octatetraenyl-lithium, B 395
Cyclo-octatrienes, B 393
Cyclo-octa-1,2,6-triene, C 49, 58
Cyclo-octa-1,3,5-triene, B **384**, 395
Cyclo-octa-1,3,6-triene, B 385
Cyclo-octa-2,4,6-trienone, B 396
Cyclo-octene, A 8, 16; B 382, 386, 393, 401, 419; C 20, 41, 128
Cyclo-oct-1-enecarboxylic acid, B 417
Cyclo-octene oxide, B 353, 403; C 53
Cyclo-octenols, B 404
Cyclo-oct-2-enol, B 410
Cyclo-oct-4-en-1-ol, B 405
Cyclo-octen-4-ol benzoate, C 39
Cyclo-oct-2-en-1-one, B 410
Cyclo-oct-5-enyldimethylamine, B 383
Cyclo-oct-1-enyllithium compounds, B 401
4-Cyclo-octenyl *p*-toluenesulphonate, C 120, 125
Cyclo-octylamine, A 16; B 418, **420**
Cyclo-octyl bromide, B 401
Cyclo-octylcyclo-oct-1-ene, C 18
Cyclo-octylmagnesium bromide B 406
Cyclo-octylmethylamine, B 402
Cyclo-octyltrimethylammonium bromide, B 419
Cyclo-octyne, B 386
Cyclo-olefins, B 401
3α-5-Cyclo-6β,19-oxido-5α-cholestane, D 78
Cyclopentacosane, B 381
Cyclopentacosanone, B 409
Cyclopentacosatetrayne, B 389
Cyclopentadeca-1,8-diene, B 386
Cyclopentadeca-1,8-diyne, B 389
Cyclopentadecane, A 5; B 381
Cyclopentadecanol, B 402, 404
Cyclopentadecanone, B 409
Cyclopentadecan-2-onecarboxylic methyl ester, B 417
Cyclopentadecene, B 386
Cyclopentadec-2-en-1-one, B 410
Cyclopentadecyne, B 386
Cyclopentadiene, A 2, 104, 112, 114, 119, 120, 127, 129, 156, 159, 172; B 29; C 23, 27, 35, 36, 46, 55, 78, 90, 91, 92, 95, 96, 97, 98, 99, 105, 266
—, adduct with ketene, C 45
—, adducts with tetrafluoroethylene, B 368
—, condensation with cyclohexanone, C 14
—, cyclopropene adduct, A 33
—, dimerisation, C 100
—, oxidation, A 159
—, polymerisation products, C **99**
—, reaction with acetylene, B 349
—, — with acrolein, C 89
—, — with diazoacetic ester, C 38
—, — with ethylene, C 89
—, — with formaldehyde, A 123
—, — with fumaric acid, A 192
—, —with halogens, A 149
—, — with hydrogen chloride, A 149
—, — with maleic acid, A 192
Cyclopentadiene acetals, A 174
Cyclopentadien-benzoquinone adduct, C 103
Cyclopentadienecarboxylic acids, A 121
Cyclopentadiene–maleic anhydride adduct, C 97
Cyclopentadiene mono-epoxide, A 159
Cyclopentadienides, A 127, 128
Cyclopentadienones, A 131; C 96
—, substituted, A 173
Cyclopentadienone, A 172
Cyclopentadienone acetals, A 173
Cyclopentadienone dimer, A 172, 173
Cyclopentadienonehydrazones, A 123
Cyclopentadienone metal complexes, A 131
Cyclopentadienyls, metal derivatives, A 124
Cyclopentadienylallylnickel, A 140
Cyclopentadienylbenzene, A 144
Cyclopentadienylbenzeneirons, A 143
Cyclopentadienylcyclopentenylnickel, A 130
Cyclopentadienyldicarbonylcobalt, A 139
Cyclopentadienyldicarbonylnitrosylmanganese cation, A 142
Cyclopentadienylexo-1-benzoylcyclopentadienecobalt, A 139
Cyclopentadienyl Grignard reagents, A 127

Cyclopentadienyl-lithium, A 129
Cyclopentadienylmagnesium bromide, A 122, 142, 149
Cyclopentadienylmetalcarbonyls, A 130
—, binuclear, A 137
Cyclopentadienylmetalcarbonyl complexes, A 126
Cyclopentadienylmetal complexes, A 126
Cyclopentadienylmetal compounds, structure, A 145
Cyclopentadienylnickelnitrosyl, A 140
Cyclopentadienyl-*N*-piperidine-dicarbonylmanganese, A 142
Cyclopentadienyl radicals, A 122
Cyclopentadienylsodium, A 120, 121, 129, 130
Cyclopentadienyltetracarbonyl-vanadium A 142
Cyclopentadienyltricarbonyl-manganese, A 140, 142
Cyclopentadienyltricarbonyl rhenium, A 142
Cyclopentadienyltricarbonyl-technetium, A 142
Cyclopentadienyl-1-trichloromethyl-cylopentadienecobalt, A 139
Cyclopentanes, aldehydes, A 163
—, carboxylic acids, A 189
—, exocyclic monoketones, A 175
—, halogen derivatives, A 148
—, hydroxyketones, A 176
—, ketones, A 164
—, phenyl substituted, A 114
—, physical constants, A 113
—, polyketones, A 176
—, ring contraction, A 77
—, synthesis, A 104
Cyclopentane, A 5, 112, 113; C 35
—, autoxidation, A 164
—, geometry, A 110
—, pseudorotation, A 110
—, reaction with acetyl chloride, A 175
—, strain, A 8
Cyclopentane-1-acetic-1-carboxylic acid, A 194
Cyclopentane-1-acetic-2-carboxylic acids, A 194
Cyclopentane-1-acetic-1-glyoxylic acid, C 23
Cyclopentanealdehyde, A 161, 163
Cyclopentane-1,1-bisbromoacetic acid, C 23
Cyclopentane-1,1-bisbromoacetic ester, C 23
Cyclopentane-1,2-bis(trimethyl-ammoniomethyl) dihydroxide, A 118
Cyclopentane-1-butyric-1-carboxylic acid, C 27
Cyclopentanecarbaldehyde, B 83
Cyclopentanecarboxylic acids, B 103
Cyclopentanecarboxylic acid, A 189
Cyclopentanecarboxylic ester, A 161
Cyclopentane-1-carboxylic-2-propionic acid, C 53
Cyclopentane-1,1-diacetic acid, A 194
Cyclopentane-1,2-diacetic acid, A 194, C 53
Cyclopentane-1,3-diacetic acid, A 194
Cyclopentane, dicarboxylic acids, C 95
Cyclopentane-1,1-dicarboxylic acid, A 105, 189, 191
Cyclopentane-1,2-dicarboxylic acid, A 191
Cyclopentane-1,3-dicarboxylic acid, A 192; C 100, 188
Cyclopentanediols, A 121, 158, 159, 160
1,2-Cyclopentanediol, A 158
1,3-Cyclopentanediol, A 159
Cyclopentane-1,2-dione, A 118, 170, 184
Cyclopentane-1,3-dione, A 185, 186
Cyclopentane-1,2-dione-3,5-dicarb-oxylic esters, A 201
Cyclopentane-1,3-dione-2,4-carboxylic esters, A 201
Cyclopentane-1,2-dione dioxime, A 153
Cyclopentane-1,2-dipropionic acid, A 194
Cyclopentane-1,3-dipropionic acid, A 194
Cyclopentane-1,2-dipropionitriles, C 59
Cyclopentane enamine, A 202
Cyclopentane terpenoids, B 208
Cyclopentanetetracarboxylic acids, A 192
Cyclopentanetetracarboxylic ester, A 1
Cyclopentanethione, A 168
Cyclopentane-1,2,3-tricarboxylic acid, A 192
Cyclopentane-1,2,4-tricarboxylic acid, 192
1,3,4-Cyclopentanetrione, A 187
Cyclopentan-1-imino-2-nitrile, A 200
Cyclopentanols, A 148, 154, 155
—, 2-substituted, 154
—, tertiary, A 156

Cyclopentanol, A 114, 153, 154, 155
Cyclopentan-1-ol-2-acetic acid, A 199
Cyclopentan-1-ol-3-acetic acid, A 199
Cyclopentan-1-ol-1-acetic ester, A 199
Cyclopentan-3-ol-1-aldehyde, B 73
Cyclopentan-1-ol-2-carbonitrile, A 198
Cyclopentan-1-ol-1-carboxylic acid, A 198; B 110
Cyclopentan-1-ol-2-carboxylic acid, A 198
Cyclopentan-1-ol-3-carboxylic acid, A 199
Cyclopentan-1-ol-2-one, A 92
Cyclopentanones, C 121
—, from adipic acids, A 105
—, α-methylated, A 165
—, physical properties, A 167
Cyclopentanone, A 2, 91, 104, 108, 111, 112, 115, 117, 153, 154, 157, 164, 167, 199, 200; C 12
—, acylation, A 189
—, halogenation, A 166
—, photolysis, A 81, 164
—, reaction with acetone, A 170
—, — with cyanoacetic ester, A 193
—, — with hydrogen peroxide, A 164
—, — with malonic esters, A 193
—, — with succinic ester, A 194
—, reduction, C 10
—, ring contraction A 78
—, selfcondensation, C 11
Cyclopentan-1-one-2-acetic acid, A 199, 201
Cyclopentan-1-one-3-acetic acid, A 202
Cyclopentan-1-one-2-butyric acids, A 202
Cyclopentan-1-one-3-carboxylic acid, A 201
Cyclopentan-1-one-2-carboxylic ester, A 154, 162, 199
Cyclopentan-1-one-3-carboxylic ester, A 199, 200
Cyclopentan-2-one-1-carboxylic ester, C 11
Cyclopentanone cyanohydrin, A 190, 198
Cyclopentan-1-one-3,4-dicarboxylic acid, A 201
Cyclopentan-1-one-2,3-dicarboxylic ester, A 201
Cyclopentan-1-one-2,4-dicarboyxlic ester, A 201
Cyclopentan-1-one-2,5-dicarboxylic ester, A 201
Cyclopentanone enamines, A 200, 202
Cyclopentanone ethylene acetal, A 168
Cyclopentanone morpholine enamine, C 120, 121
Cyclopentanone oxime, A 153
Cyclopentanone pinacol, C 10, 29
Cyclopentan-1-one-2-propionic esters, A 202
Cyclopentan-1-one-2-propionitriles, A 202
Cyclopentanone pyrrolidine enamine, A 165; C 129
Cyclopentan-1-one-2,3,5-tricarboxylic ester, A 201
Cyclopentan-1-thione-2-carboxylic acid, A 168
Cyclopentatetracontadecayne, B 389
Cyclopentatetracontane, B 381
Cyclopentenes, A 112, 148, 154
—, physical constant, A 116
—, reaction with halogens, A 148
Cyclopentene, A 35, 110, 114, 116, 163, 175; C 36, 46, 49
—, chemical properties, A 114
—, oxidation, A 155
—, physical properties, A 115
—, reaction with diazoacetic ester, A 193
Cyclopent-1-ene-1-acetic acid, A 193, 194
Cyclopent-2-ene-1-acetic acid, A 193, 197
Cyclopent-3-ene-1-acetic acid, A 193
Cyclopent-2-ene-1-acetic-2-carboxylic acid, A 194
Cyclopent-1-ene-1-acetonitrile, A 193
Cyclopent-1-ene-1-carbaldehyde, A 163, 176; B 75
Cyclopent-1-ene-carbonitrile, A 190
Cyclopent-1-ene-1-carboxylic acid, A 190, 199
—, ester, A 191
Cyclopent-1-ene-4-carboxylic acid, A 153
Cyclopent-2-ene-1-carboxylic acid, A 191
Cyclopent-3-ene-1-carboxylic acid, A 191
Cyclopent-3-ene-1,1-dicarboxylic acid, A 192
Cyclopentenediols, A 159, 168
1,2-Cyclopentenediol, A 159
Cyclopent-1-ene-3,4-diol, A 161
Cyclopent-2-ene-2,3-diol-1-one, A 185
Cyclopent-1-ene-1,2-diol-3,4,5-trione, A 188
Cyclopent-4-ene-1,3-dione, A 172, 185, 186; C 59

Cyclopent-2-ene-1-malonic esters, A 194
Cyclopentene nitroso-chlorides, A 169
Cyclopentene oxides, A 154, 161, 199
Cyclopentene sulphide, A 161
Cyclopentene-3,4,5-trione, A 186
Cyclopentenols, A 121, 149
Cyclopent-2-en-1-ol, A 155
Cyclopent-3-en-1-ol, A 156, 193; C 37
Cyclopentenolone, A 176, 182
Cyclopentenones, A 104, 108, 110, 114, 156, 168, 169, 180; B 116; C 46
Cyclopentenone, C 58
Cyclopent-2-en-1-one, A 114, 170, 202
Cyclopent-3-en-1-one, A 156, 171
Cyclopentenone oximes, A 169
1,2-Cyclopentenophenanthrene, D 1
Cyclopentenophenanthrene group, E 115
Cyclopentenylacetaldehyde, A 163
Cyclopentenylacetic acid, A 176, 193
Cyclopentenylacetone, A 176
Cyclopent-1-enylacetophenone, A 176
4-Cyclopent-1′-enylbut-3-en-2-one, A 176
Cyclopent-3-enylcarbinol, A 162
10-Cyclopent-2′-enyldecane-1-carboxylic acid, A 197
2-Cyclopent-3-enylethan-1-ol, A 162
β-(Cyclopent-3-enyl)ethyl *p*-toluene sulphonate, C 91
Cyclopentenyl halides, A 149
Cyclopent-1-enylmethylcarbinol, A 162
Cyclopenten-3-yl peroxyacetate, A 114
Cyclopent-1-enylsuccinic acid, A 194
Cyclopent-3-enyl tosylate, A 162
Cyclopentylacetic acid, A 192
Cyclopentylacetic ester, A 161
Cyclopentylacetone, A 175
Cyclopentylamine, A 153
Cyclopentylboranes, A 114
Cyclopentyl bromide, A 192; C 9, 35
4-Cyclopentylbutan-4-one-1-carboxylic acid, C 11
Cyclopentylbutyric acids, A 193
Cyclopentylcarbinol, A 161
Cyclopentyl chloride, A 29
Cyclopentylcycloheptane compounds, C **17**
Cyclopentylcyclohexane, C **14**
Cyclopentylcyclohexane types, C 267
Cyclopentylcyclopentane, C 3
2-Cyclopentylcyclopentanol, C **10**
Cyclopentylcyclopentanone, C 9
1-Cyclopentylcyclopent-2-ene, C 10
Cyclopentylcyclopentylidenemethane, C **12**
Cyclopentyldimethylcarbinol, A 161
1-Cyclopentylethanol, B 73
2-Cyclopentylethan-1-ol, A 161
2-Cyclopentylethyl bromide, A 193
Cyclopentyl Grignard reagents, C 10, 12
Cyclopentyl halides, A 148
Cyclopentylideneacetic acid, A 176, 193
Cyclopentylidenecyanoacetate-2-acetic ester, A 194
Cyclopentylidenecyanoacetic ester, A 194
2-Cyclopentylidenecyclopentanol, C 10
Cyclopentylidenecyclopentanone, A 164; C 10, 11
Cyclopentylidenemalonic esters, A 193, 194
Cyclopentylidenesuccinic acid, A 194
Cyclopentyl iodide, A 153
Cyclopentyl ketones, B 67
Cyclopentylmagnesium bromide, A 161
Cyclopentylmagnesium chloride, A 190
Cyclopentylmethylamine, A 161
Cyclopentylmethylcarbinol, A 161
Cyclopentylmethyl iodide, C 12
Cyclopentyl nitrite, A 153
1-Cyclopentylpentene, C 10
Cyclopentylpropionic acid, A 193
Cyclopentylpropyl bromide, A 193
α-Cyclopentylsuccinic acid, A 193
Cyclopentyl *p*-toluenesulphonate, A 148
Cyclopentylvaleric acid, A 193
Cyclopentyne, A 118
14β,18-Cycloprogesterone, D **335**
Cyclopropanes, A 19, 31
—, chemical properties, A 26
—, from carbenes, A 24
—, from quaternary ammonium compounds, A 21
—, halogen derivatives, A 40
—, hydrogenation, A 26
—, physical properties, A 27
—, preparation, A 19
—, pyrolysis to cyclopentenes, A 109
—, ring enlargement, A 77
—, ring fission, A 31
—, sulphur derivatives, A 44
Cyclopropane, A 2, 32
—, alcohols, A 42
—, aldehydes, A 47
—, electron diffraction, A 29
—, halogen derivatives, A 38

Cyclopropane, nitration, A 45
—, organometallic compounds, A 40
—, structure, A 28
Cyclopropanecarbaldehyde, A 23, 47, 48, 57
Cyclopropanecarboxamide, A 45, 47, 53
Cyclopropanecarboxylic acids, A 19, 22, 39, 47, 52, 58, 92
—, terpenoid, B 214
Cyclopropanecarboxylic esters, A 26, 55, 63
Cyclopropane-*cis*-1,2-diacetic ester, C 37
Cyclopropane diazo compounds, A 46
Cyclopropanecarboxylic acids, A 58
Cyclopropane-1,1-dicarboxylic acid, A 58
—, ester, A 1, 58
Cyclopropane-1,2-dicarboxylic acid, A 41, 59
Cyclopropane methyl ketone, A 23
Cyclopropane monocarboxylic acids, A 52
Cyclopropane polycarboxylic acids, A 58
Cyclopropane polymers, A 31
Cyclopropanetetracarboxylic acid, A 62, 63
Cyclopropane-1,1,2-tricarboxylic acid, A 59, 61; C 195
Cyclopropane-1,2,3-tricarboxylic acid, A 61, 63; C 195, 230
—, ester, 46
Cyclopropanols, A 42, 52
Cyclopropanones, A 48
Cyclopropenes, A 23, 26, 33, 46
Cyclopropene, A 46
—, polymerisation, A 32
—, reaction with iodine, A 39
Cyclopropenecarboxylic acids, A 55
Cyclopropenecarboxylic esters, A 55
Cyclopropenium perchlorate, A 55
Cyclopropenone, A 49
Cyclopropenonemetalcarbonyl complex, A 49
Cyclopropenylidene cyclopentadienes, A 37
Cyclopropenylium cations, A 49
Cyclopropenylium compounds, A 36
Cyclopropyl acetate, A 42, 51
α-Cyclopropylacetophenone, A 47
Cyclopropylalkane carboxylic acids, A 43
Cyclopropylalkylamines, A 46
Cyclopropylaluminium compounds, A 42
Cyclopropylamines, A 45
Cyclopropyl bromide, Grignard reagents, A 40
Cyclopropyl *tert*-butyl ketone, A 50
Cyclopropylcarbaldehyde, A 64
Cyclopropylcarbanion, A 63
Cyclopropylcarbinols, A 35, 43
—, physical properties, A 43
Cyclopropylcarbinol, A 47, 64, 77
Cyclopropylcarbinylamines, A 46
Cyclopropylcarbinyl derivatives, A 23
Cyclopropylcarbinyl iodide, A 41
Cyclopropylcarbinyl-lithium, A 41
Cyclopropylcarbinyl radicals, A 64
Cyclopropylcarbonyl chloride, A 47
Cyclopropyl chloride, A 29, 31; C 6
—, Grignard reagents, A 40
—, reaction with lithium A 41
Cyclopropyl cyanide, A 46, 47
Cyclopropylcyclohexane compounds, C 13
Cyclopropylcyclopentane, C 3
Cyclopropylcyclopentane compounds, C 13
Cyclopropyldiazomethane, A 47
Cyclopropyldimethylcarbinol, A 43
Cyclopropyl esters, A 28
Cyclopropyl ethers, A 43
1-Cyclopropyl ethyl acetate, A 114
Cyclopropylethylene, A 27
Cyclopropyl ethyl ether, A 43
Cyclopropyl ethyl ketone, A 50, 52
Cyclopropylethylmethylcarbinol, A 43
Cyclopropyl groups, detection in organic compounds, A 30
Cyclopropyl isopropyl ketone, A 50, 52
Cyclopropyl ketones, A 28, 31
Cyclopropyl-lithium, A 41; C 9
Cyclopropylmagnesium bromide, A 40
Cyclopropylmagnesium chloride, A 42
Cyclopropyl methanol, A 47
Cyclopropylmethylamine, A 46, 77, 91
2-Cyclopropyl-1-methyl bicyclopropyl, C 7
Cyclopropylmethylcarbinol, A 43
Cyclopropylmethylcarbinyltrimethylammonium hydroxide, A 35
Cyclopropyl methyl ether, A 43
Cyclopropyl methyl ketone, A 21, 27, 42, 50, 51, 52, 57
Cyclopropyl methyl ketoxime, A 46
Cyclopropylmethylvinylcarbinol, A 43
N-Cyclopropyl-*N*-nitrosourea, A 26

β-Cyclopropyl-β-oxopropionic ester, A 57
Cyclopropylphenylcarbinol, A 64
Cyclopropyl phenyl ketones, A 50, 51, 52
3-Cyclopropyl-5-phenyl-2-pyrazoline, C 7
2-Cyclopropylpropene, C 7, 8
2-Cyclopropylpyridine, A 27
Cyclopropyl radicals, A 64
Cyclopropyl sodium, A 41
Cyclopropyl styryl ketone, C 7
Cyclopropyl thioethers, A 44
Cyclopropyltrimethylammonium bromide, A 33
Cyclopropyl vinyl ether, A 43
Cyclopseudoisogenin, D *435*
Cyclopseudoneogenin, D *435*
Cyclopseudosapogenins, E 9, 11, 12, 14, 15, 19
Cyclopseudosarsasapogenin, E 17
Cyclopseudotigogenin, E 46
Cyclopseudotigogenin acetate, E 15
Cyclopyrethrosin C 363
3α,5-Cyclo-6,7-seco-5α-cholestane-6,7-dioic acid, D 81
Cyclosteroids D 6, **334**
3,5-Cyclosteroids, rearrangement, D 77
—, 6β-substituted, D 42
3,5-Cyclo-Δ^6-steroids, D 80
5,7α-Cyclosteroids, D 81
5β,19-Cyclosteroids, androstane series, D 334
3α,5-Cyclo-5α-stigmastan-6β-ol, D 78
3α-5-Cyclo-5α-stigmast-22-en-6β-yl methyl ether, D 132
Cyclotetracontadecayne, B 389
Cyclotetracontane, B 381
Cyclotetracontaoctayne, B 389
Cyclotetracosahexayne, B 389
Cyclotetracosane, B 381
Cyclotetracosane-1,13-dione, B 413
Cyclotetracosanone, B 409
Cyclotetracosatetrayne, B 389
Cyclotetradeca-1,3-diene, B 386
Cyclotetradeca-1,7-diene, B 386
Cyclotetradecadiynes, B 389
Cyclotetradeca-1,8-diyne, B 388
Cyclotetradecane, B 381
Cyclotetradecane-1,2-dione, B 413
Cyclotetradecane-1,8-dione, B 413
Cyclotetradecanol, B 404
Cyclotetradecanone, B 406, 409
Cyclotetradecan-2-onecarboxylic methyl ester, B 417
Cyclotetradecatetrayne, B 389
Cyclotetradec-2-en-1-one, B 410, 416
Cyclotetrapentacontane, B 381
Cyclotetratriacontane, B 381, 423
Cyclotetratriacontane-1,18-dione, B 413
Cyclotetratriacontanone, B 409
Cyclotriacontahexayne, B 389
Cyclotriacontane, B 381
Cyclotriacontane-1,16-dione, B 413
Cyclotricontanone, B 409
Cyclotricosane, B 381
Cyclotricosanone, B 409
Cyclotrideca-1,3-diene, B 386
Cyclotrideca-1,7-diene, B 386
Cyclotrideca-1,7-diyne, B 389
Cyclotridecane, B 381
Cyclotridecanol, B 402, 404
Cyclotridecanone, B 409
Cyclotritriacontanone, B 409
Cyclo-undeca-1,3-diene, B 386
Cyclo-undeca-1,5-diyne, B 389
Cyclo-undecane, B 381
Cyclo-undecanecarboxylic acid, B 410, 417
Cyclo-undecane-1,2-dione, B 413
Cyclo-undecane-1,6-dione, B 414
Cyclo-undecanol, B 404
Cyclo-undecanone, B 409
Cyclo-undecene, B 386
Cyclo-undecyne, B 386
Cylindrin, C 457
Cymantrene, A 140
Cymarin, D 361, 362, 384, 398
Cymarin acetate, D 384
Cymarol, D 362, 373, **384**, 398
Cymarose, D 361, 363, 365, 385
Cymarylic acid, D 362, **384**
Cymbopogon densiflorus, B 179, 192
Cymbopogon martini, B 162
o-Cymene, C 152, 295
m-Cymene, C **152**
p-Cymene, B 173, 178, 182, 217; C 152, 198, 200; E 77
p-Cymene derivatives, B 173
Cynanchum caudatum, D 285
Cyperene, C 278, **279**, 289
α-Cyperone, C **301**, 334
β-Cyperone, C 301, 342
Cyperus rotundus, C 279, 298, 301, 329
Cypreth, A 43
Cyprethylene, A 43
5α-Cyprinol, D 195
5β-Cyprinol, D 195
Cyprome, A 43
Cyprus origanum, C 138

Dacrydium biforme, C 269
Dambonitol, B 61
Dammadienol, C 425
Dammaranes, C 478
Dammarane group, triterpenoids, C 409, 425
5α-Dammarane, D *429*
Dammarenediols, C 409, 410, 425; E 115, 116
Dammarenolic acid, C 410, 411, 425
Damsin, C **359**, 360, 368
Daniellic acid, C 375, 376, 384
Darutigenol, C 388
Darzens reaction, B 87, 353, 406; D 277
Daucol, C **323**, 325, 336
Daucus carota, C 323; E 81
Davallic acid, C 455
16-Deacetyl-16-anhydroacoschimperoside P, D 387
ψ-Deacetylbufotalin, D 413, 414
Deacetyl-14α-digacetigenin, D 404
Deacetyl-14β-digacetigenin, D 404
Deacetyldigacetigenin 3-acetate, D 404
Deacetyldigilanide A, D 362, 386
Deacetyldigilanide B, D 362, **391**
Deacetylnomilin, C 430
Deacetyltanghinin, D **406**
Deacetylcynanchogenin, D 285
Debromoaplysin, C 333
Decachloro-bi(cyclopenta-2,4-dienyl), C 12
Decafluorocyclohexanes, B 38, 39
Decafluorocyclohexene, B 42
Decafluorocyclopentane, A 151
Decahydrodimethanonaphthalene, C 105
Decahydro-ethanonaphthalene, C 116
Decalins, C **60**, 61, **62**
Decalin, C 4, 5
Decalin-2-carboxamide, C 67
Decalin-1-carboxylic acid, C 67
Decalin-2-carboxylic acid, C 67
Decalin-9-carboxylic acid, C 68
Decalin-9,10-diamine, C 29
Decalin-1,2-dicarboxylic acids, C 68
Decalin-1,4-diols, C **64**, 65
Decalin-2,3-diols, C **64** 65
Decalin-2,7-diol, C 64
Decalin-9,10-diol, C 29, 64, 69
Decalindiones, C 67
Decalin-1,4-dione, C 67
Decalin-1,5-dione, C 67; D 205
Decalin-1,8-dione, C 67
Decalin-2,7-dione, C 64
Decalinhydroperoxide, B 411; C 63
Decalin-β-spirocyclopropane-1,2-dicarboxylic acids, C 25
Decalols, C 62, **63**, 66, 68, 69
1-Decalol, B 405
α-Decalol xanthate, C 68
Decalones, C 61, 62, **66**, 307
—, oxidation, B 149
—, reduction, C 63
α-Decalones, D 11, 90
β-Decalones, B 150
Decalylamines, C 63, **65**, 66, 68
α-Decalylamine, C 69
β-Decalylamines, C 62
β-Decalylmagnesium bromide, C 67
Decalyl perbenzoate, B 411
Decapreno-β-carotene, B 281
Decaprenoxanthin, E 132
Dec-1-ene, B 19
n-Decyclcyclopentane, A 113
1-*n*-Decylcyclopentan-1-ol, A 157
Decylcyclopent-1-enes, A 116
Deglucocheirotoxin, D 362, **382**, 384
Deglucodigitalinum verum, D 362, 391
Deglucohellebrin, D 366, **411**
Deglucolhellebrol, D 366
Degras, D 96
Dehydroabietic acid, C **469**
Dehydroandrosterone, D 257
Dehydroangustione, B 130
Dehydroannulenes, B 396, 397, 400
Dehydro-[14]-annulenes, B 398
15,15′-Dehydro-β-apo-12′-carotenal, B 272
9(1)-Dehydroarenobufagin, D 421
Dehydroaspidospermine, E 75, 76
15,15′-Dehydroastacene, B 322
Dehydrobufotenine, D 412
15,15′-Dehydrocanthaxanthin, B 319, 322
3,4-Dehydro-α-carotene, B 261
Dehydro-β-carotenes, B 263
15,15′-Dehydro-β-carotene, B 249, 320
15,15′-*all trans*-Dehydro-β-carotene, B 236
retro-Dehydro-β-carotene, B 253, 263, 271
7-Dehydrocholesterol, D 43, 50, 54, 67, 69, 74, 79, 85, 105, **106**, 131, 137, 145, 146, 158, 164
—, ultraviolet absorption spectra, D 68
22-Dehydrocholesterol, D **107**, 131, 154, 158, 164; E 110
24-Dehydrocholesterol, D **158**
7-Dehydro-[4-^{14}C]cholesterol, D 107

7-Dehydro-[19-$^3H_1$]cholesterol, D 107
7-Dehydrocholesteryl acetate, D 107
7-Dehydrocholesteryl benzoate, D 107
7-Dehydrocholesteryl tosylate, D 80
Dehydrocholic acid, D 20, 180, 181, 182, 190
ψ-Dehydrocholic acid, D 190
Dehydrocinerin A 183
11-Dehydrocorticosterone, D 287, 289, 291, **296**, 297, 298, 299, **321**, 324, 329
11-Dehydrocorticosterone acetate, D 321, 322, 323, 324
1,2-Dehydrocortisol, D **333**, 354
6,7-Dehydrocortisone, D **333**
1,2-Dehydrocortisone acetate, D **332**
Dehydrocostus lactone, C **351**, 367
Dehydrodeoxycholic acid, D 19
14-Dehydrodigitoxigenin, D 390
6-Dehydro-7,17α-dimethyltestosterone, D 272
25(27)-Dehydrodiosgenin, E 49
Dehydroeburicoic acid, C 421
15,15′-Dehydroechinenone, B 317
Dehydroelsholtzione, B 170, 171
Dehydroepianderosterone, D 38, 42, 200, 203, 233, 240, 254, 256, **257**, 258, 259, 263, 264, 268, 269, 270, 275, 278, 279, 337, 357, 358
Dehydroepiandrosterone acetate, D 83, 232, 257, 277, 284, 286
Dehydroepiandrosterone oxime, D 241
Dehydroergosterol, D 136 137, 140, 141, 318
9(11)-Dehydroergosterol, D 328
14-Dehydroergosterol, D 151, 165
—, ultraviolet absorption spectra, D 68
Dehydroergosteryl acetate, D 149, 318
6-Dehydroestradiol diacetate, D 221
Δ^8-Dehydroestradiol 3-methyl ether, D 218
Δ^6-Dehydroestrone, D 217
Dehydroeudesmane, C 299
Dehydroferruginol, C 392
Dehydrogenase, E 74, 101
9-Dehydrohecogenin, E 26, **35**
9-Dehydrohecogenin acetate, E 35
Dehydrohomocamphoric acid, C 201, 203
Dehydroisoandrosterone, D 257
Δ^8-Dehydro-14-isoestrone, D 217
15,15′-Dehydroisozeaxanthin diacetate, B 319
Dehydrolinalyl acetoacetate, B 168
24-Dehydrolophenol, E 104
Dehydrolycopene, B 281, 336
Dehydrolumisterol, D 140, 141
9-Dehydromanogenin, E 27, **35**, **52**
1-Dehydro-17α-methyltestosterone, D 274
6-Dehydro-17α-methyltestosterone, D 272
Dehydroneoergosterol, D 135, 136
Dehydronorcamphor, C 93, 95
Dehydronorbornium ions, C 80
Dehydronorcholene, D 24, 25
9(11)-Dehydro-18-nor-D-homo-estrone D 209
Dehydroperilla ketone, B 171
Dehydroperillic acid, B 215
Dehydropregnenolone, E 23
16-Dehydroprogesterone, D 234, 349
3,4-Dehydroretinal, B **288**
3,4-Dehydroretinol, B **287**
3,4-Dehydrorhodopin, B 257, 307, 336
Dehydroroyleanone, C 392
Dehydroselinene, C 274
Dehydrosterols, D 85, 86
7-Dehydrostigmasterol, D 69, 156, 166
16-Dehydrostrophanthidin, D 384
Dehydrotestosterone, D 329
Dehydrotetrahydroivalin, C 338, 339
Dehydrotropolone, B 377
Dehydro-vitamin A_1, B 287
Delphinium spp., E 90
Demethylated spirilloxanthin, B 337
Demjanov rearrangement, A 56; C 83, 112, 118, 122, 133
Dendrolasin, C **290**, 334
Dendropanoxide, C 446
17-Deoxoestrone, D 217
Deoxoestrone methyl ether, D 203
Deoxotetrahydrosantonin, C 346
6-Deoxy-δ-allose, D 385
Deoxybilianic acid, D 19, 20, 22
Deoxycholic acid, D 11, 12, 19, 20, 24, 28, 54, 57, 100, 108, 169, 170, 172, 175, 180, 181, 182, **184**, 185, 186, 187, 251, 304, 312, 323, 324, 327
Deoxycholic methyl ester, D 37
Deoxycholic acid methyl ester 3-acetate, D 24
Deoxycorticosterone, D 7, 57, 232, 235, 238, 301, 350
—, *nomenclature*, D *439*
—, total synthesis, D 102
11-Deoxycorticosterone, D 287, 289, **300**, **305**, 306, 311, 312, 384
11-Deoxycorticosterone acetate, D 299, 300, 305, 354
[5-^{14}C]-Deoxydigitoxigenin, E 111
3-Deoxyequilenin, D 216

Deoxyflexixanthin, B **318**, 322
2,6-Deoxyhexose, D 363
Deoxyinositols, B **57**
Deoxyinososes, B 58
Deoxylimonin, C 430
6-Deoxy-3-*O*-methyl-L-altrose, D 387
6-Deoxy-6-nitro-D-glucose, B 60
Deoxynitroinositols, B 60, 85
Deoxyrosenonolactone, C 389
Deoxyscymnol, D **194**
3-Deoxysmilagenin, E 31
2-Deoxystreptamine, B **83**, 84
Dephanthanic acid, D 380
Dephanthic acid, D 380
7-Desacetoxyhelvolic acid, C 420
Desacetoxymatricarin, C 353, 354, 367
Desacetylglobicin, C 355
Desacetylmatricarin, C 353
α-Desdimethylgranatenine, B 418, 420
Desglucodigitonin, E 2
Desisopropylidenepetasol, C 307, 308
N-Desmethylcona-4,20(*N*)-dien-3-one, D 335
Desmethyl-lanostane derivatives, C 408
Desmosterol, D 74, **107**, 131, 149, 154, **158**, 164, E 104
[3-$^3H_1$]-Desmosterol, E 113
Desmotroposantonin, C 340, 341, 342
Desmotroposantoninic acid, C 340
11-Desoxoglabrolide, C 441
Desoxotetrahydrosantonin, C 337
Desoxycholic acid, E 34
Desoxy-ψ-santonin, C 346
Desoxy-C-santonin, C 367
(25*R*)-Des-A-spirostan-5-one, E 41
Deuterochloroform, A 84
Deuterocyclopentadiene, A 122
3-Deutero-*p*-menthane, B 217
Devadarene, C 389
Dewar benzene, C 43
Dexamethasone pivalate, D 333
Diacetate A, C 429
3β,21-Diacetoxyallopregn-17(20)-ene, D 309
3β,16α-Diacetoxy-5α-androstan-17-one, E 21
3β-16α-Diacetoxybisnor-5α,20-isocholanoic ester, E 21
3,6-Diacetoxycholest-4-enes, D 49
3,5-Diacetoxycyclopent-1-enes, A 159
1,2-Diacetoxycyclo-octane, B 403
1,4-Diacetoxycyclo-octane, B 403
1β,3β-Diacetoxy-5β-etianic methyl ester, E 30
3,17α-Diacetoxy-5α,5α-fluoro-6-oxopregnane, D 348
3α-,12β-Diacetoxy-11β-hydroxy-5β-cholanoic methyl ester, D 57
3β,6α-Diacetoxy-5-hydroxy-5α-cholestane, D 49
3β,6β-Diacetoxy-5-hydroxy-5α-cholestane, D 49
3β,11α-Diacetoxy-14-hydroxy-5β,14β-etianic acid, D 13
3β,16β-Diacetoxy-14-hydroxy-5β,14β-etianic acid, D 13
3β,11α-Diacetoxy-14β-hydroxy-14-isoetianic acid, D 399
3β,11α-Diacetoxy-14β-hydroxy-14-isoetianic methyl ester, D 420
3β,16β-Diacetoxy-14β-hydroxy-14-isoetianic methyl ester, D 418
2,2-Diacetoxy-4-methyl-cyclohexa-3,5-dien-1-one, B 122
2,6-Diacetoxy-3-methylcyclohexan-1-one, B 53
3α,17α-Diacetoxy-17β-methyl-D-homoetiocholane-11,17a-dione, D 281
3α,21-Diacetoxypregnane-11,20-dione, D 325
3α,11α-Diacetoxy-5β-pregnane-20β-ol, D 303
3β,21-Diacetoxypregn-5-en-20-one, D 397
3α,21-Diacetoxypregn-17(20)-en-11-one, D 324
N,N'-Diacetylactinamine, B 84
Diacetyladipic ester, A 175
3,16-Diacetyl-Δ^{14}-anhydrogitoxigenin, D 395
5,7-Diacetylbicylco[2.2.2]oct-2-ene, C 109
1,1'Diacetylcobalticinium perchlorate, A 139
1,3-Diacetylcyclopentane, C 188
αω-Diacetylenes, B 380, 388
1,1'-Diacetylferrocene, A 147
Diadinoxanthin, B 332
1,1-Dialkoxycyclopropanes, A 49
1,1-Dialkoxy-2,2-dichlorocyclopropanes, A 49
17-[*N*-(Dialkylaminoalkyl)imino]-androstan-3-ols, D 357
Dialkylaminocyclobutanes, A 76
2-Dialkylaminomethylcyclohexan-1-ones, B 108
Dialkylanilines, hydrogenation, B 92
1,3-Dialkylbicyclobutane-2,4-dicarboxylic esters, C 33

5,7-Dialkylcyclohepta-1,3-dienes, B 349
Dialkylcyclopropylboranes, A 41
1,1′-Dialkylferrocenes, A 133
Dialkylfulvenes, A 120, 124
2,6-Dialkylphenoxides, B 120
Diallyl, B **36**
1,3-Diaminoadamantane, C 131
3β,17β-Diaminoandrost-5-ene, D 357
2β,16β-Diamino-5α-androstane-3α,17β-diol dimetholalides, D 359
1,2-Diaminocyclobutane, A 91
1,3-Diaminocyclobutane, A 91
1,6-Diaminocyclodecane, A 8
1,6-Diaminocyclodecane dihydrochloride, B 421
1,2-Diaminocyclohexanes, B 79, **81**, 82
1,3-Diaminocyclohexanes, B **81**
1,4-Diaminocyclohexanes, B **82**
Diaminocyclohexane phosphates, B 25
1,5-Diaminocyclohexane-2,3,4-triol, B 83
Diaminocyclopentanes, A 153
1,2-Diaminocyclopropane, A 46
9,10-Diaminodecalin, C 66
1,1′-Diaminoferrocene, A 135
1,3-Diamino-*scyllo*-inositol, B 84
3,16β-Diaminopregna-3,5-dienes, D 357
3β,20α-Diaminopregnane, D 2
3,20-Diaminopregnane derivatives, D 357
Diaminotetrahydroxybenzene, A 188
2,5-Diaminotroponeimine, B 378
Dianabol, D 274
Dianhydroantiarigenin, D 400, 401
Dianhydrogitoxigenin, D 391, 394, 395, 418
Dianhydrogitoxigenin acetate, D 389
Dianhydrogitoxigenone, D 394
Dianhydro-17-isostrophanthidin ethylals, D 385
Dianhydrodrostrophanthidin, D 378
Dianhydrostrophanthidin ethylal, D 378, 385
Dianhydrostrophanthidin methylal, D 382
Dianhydrostrophanthidonic ester, D 377
2,2-Diarylcyclopropanecarboxylic acids, A 54
1,2-Diarylcyclopropenones, A 38
Diatoxanthin, B 332
Diazepines, A 46
Diazoacetic ester, A 59; C 8
—, cyclopropene adduct, A 33
Diazoacetic ester, reaction with acetylenes, A 55
—, — with allyl acetate, A 56
—, — with allylbenzene, A 54
—, — with 1,3-butadiene, A 53
—, — with ethyl vinyl ether, A 56
—, — with olefins, A 26, 52
—, — with phenyl vinyl ether, A 56
—, — with stilbene, A 54
3-Diazoacetyl-1,2-dimethylcyclopropene, C 34
Diazoalkanes, B 368
16-Diazoandrost-5-en-3β-ol-17-one, D 286
1-Diazobut-3-ene, C 32
Diazocamphor, C 85, **233**
Diazo compounds, cyclopropanes, A 46
Diazocyclopentadiene, A 122, 123, **128**
Diazocyclopentane, A 114
2-Diazo-4,4-dimethylandrost-5-ene-3,17-dione, D 286
2-Diazo-17β-hydroxy-4,4-dimethylandrost-5-en-3-one, A 286
Diazomalonic ester, A 59
Diazomethane, A 59; D 276
—, photolysis, A 24, 25
—, reaction with fumaric ester, A 59
—, — with keten, A 48, 78, 92
α-Diazonopinone, C 173
Diazonorbornanone, C 86
16-Diazo-17-oxosteroids, D 286
21-Diazoprogesterone, D 305, 306
Diazo-steroids, D **286**
1-Diazo-2,3,4-triphenylcyclopentadiene, C 12
Dibenzometallocenes, A 128
1,2-Dibenzoylcyclobutanes, A 94
1,2-Dibenzoylcyclobutane, A 95
1,2-Dibenzoylcyclobut-1-ene, A 95
1,4-Dibenzoyl-1,4-dibromobutane, A 95
1,2-Dibenzoyl-3,4-diphenylcyclobutane, A 94
1,3-Dibenzoyl-2,4-diphenyl-cyclobutane, A 94
1,1′-Dibenzoylferrocene, A 132, 145
3,4-5,6-Dibenztropolone, B 370
1,4-Dibenzylcyclohexane-1,4-diols, B 51
2,6-Dibenzoylcyclohexanones, B 95
2,6-Dibenzyl-2,6-dibromocyclohexanone, B 104
2,6-Dibenzylidenecyclohexanones, B 100
2,5-Dibenzylidenecylopentanone, A 164, 170

2,5-Dibenzylidenecyclopent-3-en-1-one, A 170
Dibenzyl ketone, reaction with oxalic ester, A 188
3,7-Dibenzyltropolone, B 369
Diborane, A 20, 154; B 19, 45, 80; C 173, 175, 178, 182
—, reaction with cyclohexenes, B 20
Dibromoacetic methyl ester, C 132
1,3-Dibromoadamantane, C 130
αα′-Dibromoadipic ester, A 192
1,3-Dibromoalkanes, A 19
αω-Dibromoalkanes, B 388
—, reaction with acetoacetic ester, A 9
7,7-Dibromobicylco[4.1.0]heptane, C 34
7,7-Dibromobicyclo[4.1.0.]hept-3-ene, C 38
1,1′-Dibromobicycloheptyl, C 18
6,6-Dibromobicyclo-[3.1.0.]hexane, B 41; C 36
7,8-Dibromobicyclo[4.2.0]octa-2,4-diene, B 396
2,4-Dibromobicyclo[3.2.1]octan-3-ol, C 120
2,3-Dibromobicyclo[**2.2.2**]oct-5-ene, C 109
3,4-Dibromobicyclo[3.2.1]oct-2-ene, C 120
1,1′-Dibromobicyclopentyl, C 10
2,6-Dibromobornane, C 192
1,4-Dibromobutane, A 105, 190, 191; B 354; C 27, 29
1,4-Dibromobut-2-ene, A 22 58, 162, 192
αγ-Dibromobutyric acid, A 48
Dibromocamphors, physical properties, C 211
αα′-Dibromocamphor, C 203, 212, 223
απ-Dibromocamphor, C 209
Dibromocarbene, A 25, 49, 55; C 23, 36, 38, 120, 121; D 276
1,2-Dibromo-1-chlorocycloheptane, B 351
4α,5-Dibromo-5α-cholestane, D 115
5,6α-Dibromo-5β-cholestane, D 115
5,6β-Dibromo-5α-cholestane, D 115
2,2-Dibromo-5α-cholestane-1,3-dione, D 124
5,6α-Dibromo-5β-cholestan-3β-ol, D 97
5,6β-Dibromo-5α-cholestan-3β-ol, D 97
Dibromocholestan-3-ones, D 122
2,2-Dibromo-5α-cholestan-3-one, D 120, 121
2α,4α-Dibromo-5α-cholestan-3-one, D 120, 130
2β,4β-Dibromo-5β-cholestan-3-one, D 121
3β,5-Dibromo-5α-cholestan-6-one, D 80
5,6α-Dibromo-5α-cholestan-3-one, D 17
5,6α-Dibromo-5β-cholestan-3-one, D 128
5,6β-Dibromo-5α-cholestan-3-one, D 123, 128, 130, 131
2,2-Dibromocholest-4-en-3-one, D 120
2α,6β-Dibromocholest-4-en-3-one, D 128, 130
1,1-Dibromocyclobutane, A 87, 88
1,2-Dibromocyclobutane, A 82, 87, 88
3,4-Dibromocyclobutane-1,2-dicarboxylic acids, A 96
1,2-Dibromocyclobutane-1,2-dicarboxylic methyl ester, A 96
1,2-Dibromocyclobutene, A 87, 88
1,6-Dibromocyclodecane, B 401, 402
2,2′-Dibromocyclodecanone, B 410
2,12-Dibromocyclododecanone, B 411
3,7-Dibromocyclohepta-1,4-diene, B 351
1,2-Dibromocycloheptane, B **350**, 352, 358
2,7-Dibromocycloheptanone, B 354
1,2-Dibromocycloheptene, B 349
1,4-Dibromocyclohept-2-ene, B 352
1,1-Dibromocyclohexane, B 33, 34
1,2-Dibromocyclohexane, B 22, 27, 33, 34, 35, 41
1,3-Dibromocyclohexane, B 42
1,4-Dibromocyclohexane, B 73
Dibromocyclohexane-1,3-dicarboxylic acids, B 147, 148
Dibromocyclohexanetetrols, B 57
Dibromocyclohexanones, B 104
1,2-Dibromocyclohex-1-ene, B 31
1,4-Dibromocyclohex-2-ene, B 27
4,5-Dibromocyclohex-1-ene, B 28
2,2′-Dibromocyclononanone, B 410
Dibromocyclo-octadiene, B 395
1,2-Dibromocyclo-octane, B 401, **402**, 403
2,8-Dibromocyclo-octanone, B 356; C 39
1,1-Dibromocyclopentane, A 149
1,2-Dibromocyclopentane, A 119, 148, 158
Dibromocyclopentanone, A 166, 168
1,2-Dibromocyclopent-1-ene, A 118
3,4-Dibromocyclopent-1-ene, A 149
3,5-Dibromocyclopent-1-ene, A 149, 159, 160, 194

2,2-Dibromocyclopent-4-ene-1,3-dione, A 186
Dibromocyclopropane, A 39, 40, 41
2,2-Dibromocyclopropane-carboxylic acid, A 55
Dibromodeoxyinositols, B 85
2,6-Dibromo-2,6-dibenzylcyclohexanone, B 104
$\alpha\alpha'$-Dibromo-$\beta\beta$-diethylglutaric acid, A 61
Dibromodimedone, B 112
1,3-Dibromo-1,3-dimethylcyclobutane, C 32
N,*N*-Dibromo-5,5-dimethylhydantoin, D 107
1,4-Dibromo-1,4-dinitrosocyclohexane, B 79
1,2-Dibromoethane, A 58; C 24, 113
1,1′-Dibromoferrocene, A 135
2,5-Dibromohexane, A 190
2,6-Dibromohexanone, B 110
Dibromomethylcyclohexanones, B 104
Dibromomethane, C 132
1,1-Dibromo-2-methylpropene, A 34
$\alpha\omega$-Dibromo-α'-nitrocamphor, C 213
$\alpha\pi$-Dibromo-α'-nitrocamphor, C 213
Dibromopalladium, A 86
1,3-Dibromopentane, B 151
1,5-Dibromopentane, C 31
$\alpha\alpha'$-Dibromopimelic ester, A 191
1,2-Dibromopropionic ester, A 61, 97
$\alpha\beta$-Dibromopropionic ester, B 357
1,1-Dibromospirohexane, C 22, 23
1,2-Dibromosuccinic methyl ester, A 63
Dibromotetraphenylcyclobutadienenickel, A 85
3,5-Dibromotricyclo[2.2.2.$0^{2,6}$]-octane, C 109
Dibromotropolones, B 368, 373
2,7-Dibromotropone, B 363
$\alpha\alpha'$-Dibromo-*m*-xylene, B 28
Dibutoxytitanium dichloride, A 129
2,6-Di-*tert*-butylcyclohexa-2,5-dien-1-ones, B 124
1,3-Di-*tert*-butylcyclohexane, B 8
4,6-Di-*tert*-butylcyclohexane,1,3-dione, B 109
2,5-Di-*tert*-butylcyclohexane-1,4-dione, B 109
6-Di-*tert*-butyl-4-methylcyclohex-2-enone, B 117
2,6-Di-*tert*-butylphenoxide, B 120
Dicarboxylic acids, exocyclic hydroaromatic B 149
$\alpha\omega$-Dicarboxylic acids, B 411
$\alpha\omega$-Dicarboxylic acid chlorides, cyclisation, B 407
$\alpha\omega$-Dicarboxylic esters, conversion to cyclanones, B 406
—, cyclization, B 407
Dichloroacetic ester, A 25
Dichloroacetyl chloride, C 46
9α,11β-Dichloroandrosta-1,4-diene-3,17-dione, D 218
p-Dichlorobenzene, B 112
7,7-Dichlorobicyclo-[4.1.0]-heptane, B 349
1,4-Dichlorobicyclo[2.2.1]heptane-7-one, C 42
6,6-Dichlorobicyclo[3.2.0]heptan-7-one, C 46
5,6-Dichlorobicyclo[2.2.1]hept-2-ene, C 90
6,6-Dichlorobicyclo[3.2.0]hept-2-en-7-one, C 46
2,3-Dichlorobicyclo[3.2.1]octa-2,4-diene, C 118
7,8-Dichlorobicyclo[4.2.0]octane, C 47
3,4-Dichlorobicyclo[3.2.1]oct-2-ene, C 118
9,9-Dichlorobicyclo[4.2.1]nona-2,4-diene, C 128
9,9-Dichlorobicyclo[4.2.1]nonane, C 128
9,9-Dichlorobicyclo[4.2.1]non-2-ene, C 128
Dichlorobiscyclopentadienylmolybdenum, A 126
Dichlorobiscyclopentadienyltitanium, A 126
Dichlorobiscyclopentadienylvanadium A 126
2,2-Dichlorobornane, C 166
1,3-Dichlorobut-2-ene, D 209, 214, 230, 246, 247
1,4-Dichlorobut-2-ene, A 192
1,4-Dichlorobut-2-yne, D 215
Dichlorocarbene, A 25, 42, 45, 49, 55; B 349; C 36, 102, 118, 128; D 344, 351
Dichlorocamphors, physical properties, C 211
$\alpha\pi$-Dichlorocamphor, C 209
3α,6β-Dichloro-5α-cholestane, D 51
3β,6α-Dichloro-5α-cholestane, D 51
5,6α-Dichloro-5α-cholestan-3β-ol, D 97
5,6β-Dichloro-5α-cholestan-3β-ol, D 97
Dichlorocholestan-3-ones, D 122
2,2-Dichloro-5α-cholestan-3-one, D 120
3,4-Dichlorocyclobutane-1,2-dicarboxylic acids, A 96

3,4-Dichlorocyclobutene, A 85
5,9(10)-Dichlorocyclododec-1-ene, B 402
1,3-Dichlorocyclohexa-1,3-diene, B 111
2,4-Dichlorocyclohexa-1,3-diene, B 41
1,1-Dichlorocyclohexane, B 33, 34, 94
1,2-Dichlorocyclohexane, B 33 35
3,3-Dichlorocyclohexane-1,2-dione, B 110
2,2-Dichlorocyclohexanone, B 104
Dichlorocyclohex-1-enes, B 41
2,3-Dichlorocyclohex-1-ene, B 40
1,1-Dichloro-1-cyclohexylethane, B 31
1,6-Dichlorocyclo-octa-1,5-diene, B 383, 402
1,2-Dichlorocyclo-octane, B 402
2,3-Dichlorocyclopentanone, A 171
Dichlorocyclopropane, A 39, 40
2,2-Dichlorocyclopropane-carboxylic acid, A 55
2,2-Dichlorocyclopropanone acetals, A 49
2,2-Dichlorocyclopropyltrimethylsilene, A 42
2,3-Dichloro-5,6-dicyanobenzoquinone B, 4, 265, 322; D 84, 94, 130, 218, 341
1-(4′,4′-Dichloro-3′,3′-difluorocyclobutenyl)cyclohexene, C 13
Dichlorodifluoroethene, C 13
4,4-Dichloro-3,3-difluoro-1-phenylcyclobutene, A 88
3,3-Dichloro-1,2-diphenyl-cyclopropene, C 13
—, derivatives, A 38
1,1-Dichloroethane, C 41
1,7-Dichloroheptan-4-one, A 51; C 8
Dichlorohexafluorocyclopent-1-ene, A 150, 151
(Dichloroiodo)benzene, D 97
Dichloroketene, C 46
1,1′-Dichloromercuriferrocenes, A 135
18,18-Dichloro-20α-methylamino-pregn-4-en-3-one trifluoroacetate, D 356
3,5-Dichloro-2-methylbenzoic acid, B 65
2,3-Dichloro-3-methylcyclohex-1-ene, B 41
1,3-Dichloro-2-methylenepropane, A 34
2,6-Dichloro-4-methyl-4-trichloromethylcyclohexa-2,5-dien-1-ol, B 65
α,α′-Dichloromuconic ester, C 8
Dichloro-octafluorocyclopentane, A 150, 151
Dichloro-octanedione, C 472
Dichloropalladium, A 86
3,5-Dichloropentan-2-one, C 472
1,1′-Di-(*p*-chlorophenyl)ferrocene, A 147
9α,11β-Dichloroprogesterone, D 349
1,3-Dichloropropan-2-one, A 42
αα-Dichloropropionic ester, D 277
18,18-Dichlorosteroids, D 355
2α,4-Dichlorotestosterone acetate, D 340
4,4-Dichlorotetracylco[6.2.1.0^{2,7},0^{3,5}]-undec-9-ene, C 102
Dichloro(tetramethylcyclobutadiene)-nickel, A 84, 86; B 392
Dichlorotetramethylcyclobutene, A 77, 84, 86; C 43
8,8-Dichlorotricyclo[4.3.0.0^{7,9}]nona-2,4-diene, C 36
3,3-Dichloro-*exo*-tricyclo[3.2.1.0^{2,4}] octane, C 118
1,1-Dichloro-2-vinylcyclopropane, A 110
Dicobalt octacarbonyl, A 130
Dictylostelium discoideum, E 109
Dicyanoacetylene, C 110
αω-Dicyanoalkanes, A 2
7,7-Dicyanocyclohepta-1,3,5-triene, C 38
1,3-Dicyanocyclohexane-1,3-diol, B 111
Dicyanodiazomethane, C 38
7,7-Dicyano-1,4-dimethyl norcaradiene, C 38
7,7-Dicyano-2,5-dimethylnorcaradiene C 38
7,7-Dicyanonorcaradiene, C 38
Dicyclobutyl ketone, C 9
Dicyclobutylmethane, C 9
Dicyclohexadiene, B 27; C 115
Dicyclohexane, B 20
Dicyclohex-2-enyl ether, B 42
Dicyclohex-1-enylethyne, C 15
Dicyclohexylalkanes, C 15
Dicyclohexylamine, B 81
Dicyclohexylcarbodiimide, D 83
Dicyclohexyl carbonates, B 33
2,5-Dicyclohexylcyclohexane, 1,4-diol, B 51
Dicyclohexyl disulphide, B 76
1,2-Dicyclohexylethane, C 15
1,1-Dicyclohexylethene, C 15
Dicyclohexyl ether, B 49

Dicyclohexylidenecyclohexanone, B 100
Dicyclopentadienes, A 119
Dicyclopentadiene, C 5, **99**, 101, 102, 103
Dicyclopentadienedicarboxylic acid, C 103
Dicyclopentadienol, A 170; C 12
Dicyclopentadienyl-lead, A 147
Dicyclopentadienylmolybdenum, A 126
Dicyclopentadienylniobium A 126
Dicyclopentadienylrhodium, A 131
Dicyclopentadienyltechnetium, A 125
Dicyclopentadienyltin, A 147
Dicyclopentylalkanes, C 12
Dicyclopentylamine, A 153
Dicyclopentylcarbinol, C 12
1,2-Dicyclopentylethane, C 12
2,5-Dicyclopentylidenecyclopentanone, C 11
Dicyclopentylmethane, C 12
Dicyclopropyl, A 41
1,1-Dicyclopropyl-1,3-butadiene, C 9
Dicyclopropylcarbinol, C 8
Dicyclopropyl ethers, A 43
Dicyclopropylglycollic acid, C 9
Dicyclopropyl ketone, A 51; C 8, 9
Dicyclopropylmercury, A 40
Dicyclopropylmethane, C 8
Dicyclopropylmethyl chloride, C 9
Dicyclopropylpropylcarbinol, C 9
1,4-Dideoxy-1,4-dinitro-inositols, B 80
αα-Dideuteriocarvonecamphor, B 201
2,3-Dideutero-*p*-menthane, B 217
Didymella lycopersici, D 332
Dieckmann reaction, A 105, 106; C 11, 37, 53, 54, 112, 121 224,; D 101
Dieldrin, C **105**
Diels-Alder reaction, B 1, 5, 6, 25, 78, 136, 140; C 10, 17, 30, 49, 58, 78, 89, 92, 93, 96, 108, 111, 121, 123; D 328
—, vitamin A_1, B 292
Diels' hydrocarbon, D 370; E 7
Dienes, condensation with acetylenes and olefins, B 5
—, Diels-Alder addition to olefins, B 16
—, Diels-Alder synthesis, B 5
—, dimerisation, A 75
—, reaction with αβ-unsaturated acids, B **136**
—, — with αβ-unsaturated aldehydes, B 86
Dienones, cyclisation, A 108
1,4-Dien-3-ones, D 93, 94
1,5-Dien-3-ones, D 93
4,7-Dien-3-ones, D 92
5,6-Dien-3-ones, D 94
Dienone-phenol rearrangements, B 124
Dienynes, cyclisation, A 108
2,3,6,7-Diepoxyallo-ocimene, B 160
16,22:22,26-Diepoxycholestane, D 432
Diepoxycyclohexane, B 28, 57, 72
1,7:4,8-Diepoxy-1,4-dimethylcyclohexane, B 112
5,6,5',6'-Diepoxyzeaxanthin, E 127
17α,17'α-Diestradiol, D 337
Diethoxyaluminohydride, B 91
3,3-Diethoxycarbonylbicyclo-[3.3.1] nonane, C 126
1,3-Diethoxycarbonylbicyclo [3.3.1]-nonan-2-one, C 125
3-(3,3-Diethoxycarbonylcyclohexyl)-propionic ester, C 125, 126
2,5-Diethoxycarbonyl-1,3,4-cyclopentanetrione, A 187
1,4-Diethoxycarbonyl-2,5-dioxobicyclo [2.2.2]octane, C 113
2,4-Diethoxycarbonyl-1,3-diphenylbicyclo[1.1.0] butane, C 33
1,6-Di(ethoxycarbonyl)hexan-2-one, A 106
3,4-Diethoxycyclopent-2-en-1-one, A 186
1,1-Diethoxycyclopropane, A 49
3,3-Diethoxy-4,4-difluorocyclobutene, A 89
1,2-Diethoxytetrafluorocyclobutene, A 93
1-Diethylaminobutadiene, B 91
1-Diethylaminobutan-3-one, B 144
1-Diethylaminobutan-3-one methiodide, D 100
2-Diethylamino-3-formylcyclohex-1-ene, B 90
3-Diethylamino-4-formylcyclohex-1-ene, B 91
Diethyl-(2-chloro-1,1,2-trifluoroethyl)amine, D 334
4,6-Diethylcyclohexane-1,3-dione, B 109
Diethylcyclohexan-6-one-1,3-dicarboxylate, B 150
1,1-Diethylcyclopropane, A 32
3,3,Diethylcyclopropane-1,2-dicarboxylic acids, A 61
αα'-Diethyl-4,4'-dihydroxystilbene, C 16
13β,17α-Diethylgon-4-en-17β-ol, D 274
Difenchlolylurea, C 235

Difluorocarbene, D 344, 351
Difluorocyclobutenes, A 89
1,1-Difluorocyclohexane, B 33, 34
Difluorohexachlorocyclopent-1-ene, A 150, 151
6α,9α-Difluoro-16α-methyl-1,2-dehydrocortisol, D 351
5β,6β-Difluoromethylenepregna-16-en-3β-ol-17-one, D 351
6α,9α-Difluoro-16α-methylhydrocortisone acetate, D 351
9α,16α-Difluoroprednisolone, D 354
9α,16β-Difluoroprednisolone, D 354
9α,15β-Difluoropregna-1,4-diene-11β,17α,21-triol-3,20-dione 21-acetate, D 354
9α,15β-Difluoropregn-4-ene-11β,17α,21-triol-3,20-dione 21-acetate, D 354
21,21-Difluoroprogesterone, D 350
1,2-Diformylcyclo-octatetraene, C 73
1,1′-Diformylferrocene, A 132
Digacetigenin, D **404**
Digacetinin, D 404
Digalogenin, E 27, **36**, 37
Digalonin, E 2
Digilanidase, D 360, 386, 391
Digilanides A, B and C, D 360
Digilanide A, D 362, **386**
Digilanide B, D 362, **391**
Digilanide C, D 362, **385**
Diginane, D 2
Diginigenin, D 2 362, **403**
Diginin, D 29, 362, **403**
Diginose, D 363, 365, 396
D-Diginose, D 361, 391, 395, 403
L-Diginose, D 402
γ-Digiprogenin, D 404
14α-Digiprogenin, D **404**
γ-Digipronin, D 404
14α-Digipronin, D **404**
Digiproside, D 387
Digitaligenin, D 394
Digitalinium verum, D 362, 391
—, monoacetate, D **362**
Digitalis canariensis, D 388
Digitalis germanicum, E 2
Digitalis lanata, D 360, 387, 388, 390, 391; E 2, 5, 25, 27, 111, 112
Digitalis orientalis, D 391
Digitalis purpurea, D 360, 387, 390, 391, 403, 404; E 2, 5, 25, 27, 111
Digitalis species, D 360, 385, 386; E 5, 25, 111, 113
Digitaloid glycosides, D 361, 362
Digitaloid lactones, D 7
Digitalose, D 361, 363, 365, 368, 384, 385, 403
Digitenolides, D 29
Digitogenic acid, E 36, 37
Digitogenin, E 2, 27, **36**, 37
Digitonides, D 40
Digitonide formation, D 39
Digitonin, D 39, 40, 77, 102, 104, 252, 253, 288, 323, 396; E 2, 4, 5
Digitoxan dioic acid, D 370
Digitoxanol dioic acid, D 388, 394
γ-Digitoxanol dioic acid, D 370, 387
Digitoxigenin, D 2, 12, 13, 39, 59, 362, 369, 370, 372, 381, 382, 384, **386**, 387, **388**, 389, 390, 393, 399, 405, 416; E 111
—, *nomenclature*, D *431*
Digitoxigenin acetate, D 389, 390
Digitoxigenin-acetyldigitaloside-[glucoside]$_2$, D 386
Digitoxigenin-α-L-acofrio-pyranoside, D 386
Digitoxigenin-allomethyloside, D 387
Digitoxigenin-cymaroside, D 386
Digitoxigenin-cymaroside-[glucoside]$_2$, D 386
Digitoxigenin-digilanidobioside, D 388
Digitoxigenin-diginoside, D 386
Digitoxigenin-digitaloside, D 386
Digitoxigenin-digitaloside-glucoside, D 386
Digitoxigenin-[digitoxoside]$_3$, D 386
Digitoxigenin-[digitoxoside]$_3$-glucoside, D 386
Digitoxigenin-D-fucoside, D 387
Digitoxigenin glycosides, D 386
Digitoxigenin-thevetoside, D 386
Digitoxigenin-thevetoside-[glucoside]$_2$ D 386
Digitoxigenin-α-L-vallaropyranoside, D 386
Digitoxigenone, D 388
Digitoxin, D 360, 362, 369, 386
Digitoxose, D 361, 363, 365, 384, 404
3-*O*-Digitoxosyl-16-dehydrostrophanthidin, D 384
Diglyme, D 344
Digoxigenin, D 362, **385**, 386, 399; E 111
Digoxigenin diacetate, D 385
Digoxigenin-[digitoxoside]$_3$ acetates, D 385
Digoxigenin glycosides, D **385**
Digoxigenone, D 385, 399
Digoxin, D 360, 362, 369, **385**
Diguanidine streptidine, B 84

Dihalogenoalkanes, B 151
1,3-Dihalogenoalkanes, A 19, 31
$\alpha\omega$-Dihalogenoalkanes, A 9
7,8-Dihalogeno-[4.2.0]-bicyclo-octanes A 96
Dihalogenocamphors, C **210**
Dihalogenocarbenes, A 26, 34; B 362; D 344, 351
3,4-Dihalogenocyclobutane-*cis*-1,2-dicarboxylic acids, B 395
Dihalogenocyclohexanes, B 18, 25, 27, 33, **34**, 36, 40
7,8-Dihalogenocyclo-octa-1,3,5-trienes, B 395
1,2-Dihalogenocyclopentanes, A 149
Dihalogenocyclopentenes, A 149
4,5-Dihalogenocyclopent-4-ene-1,3-diones, A 187
1,2-Dihalogenocyclopropanes, A 33
1,3-Dihalogenopropanes, A 21
9,11-Dihalogenosteroids, D 349
D-*Dihomo-5α-androstane*, D *444*
Dihydroagarofuran, C 303, **304**, **335**
Dihydroaglycons, D 371
Dihydroagnosterol, C 407, **421**; D 75
Dihydroalantolactone, C 340
Dihydro-allo-aromadendrene, C 321
Dihydroambrosin, C 357, 360
2,3-Dihydroanisole, B 64
2,4-Dihydroanisole, B 3
2,5-Dihydroanisole, B 64, 362
3,4-Dihydroanisole, B 3
Dihydroaristolone, C 308, 309
2,5-Dihydroarylcyclopropanes, A 36
Dihydrobenzenes, B 25
Dihydrobenzoic acids, B 138
3,5-Dihydrobenzoic acid, B **362**
2,5-Dihydrobenzo[b]thiophen-1,1-dioxide, B 29
1,4-Dihydrobenzyl alcohol, B 349
Dihydro-α-betulenol, C 331
Dihydrobrassicasterol, D 155, 156; E 103, 105
Dihydrobulnesol, C 315
Dihydrocalotropagenin, D 406
Dihydrocarotol, C 324, 325
Dihydrocarquejols, B 193
Dihydrocarquejone, B 193
Dihydrocarveol, B **191**, 192, 200, 201, 222
Dihydrocarveol hydrate, B **221**
Dihydrocarveol hydroiodide, B 187
Dihydrocarveol methylxanthate, B 180
Dihydrocarvone, B **200**; E 74
Dihydrocarvone hydrate, B 222
Dihydrocarvone hydrazone, B 219
Dihydrocarvone hydrobromide, C 154, 252
Dihydrochamic acid, C 157, 158
Dihydrochaulmoogric acid, A 197, 198
Dihydrocinerolone, A 179
Dihydrocinerone, A 179
Dihydrocitronellol, C 468
Dihydrocivetone oxime, B 414
Dihydrocopaene, C 298
Dihydrocostunolide, C 337, **364**
Dihydrocostus lactone, C **351**, 367
Dihydro-ψ-cumene, C 145
Dihydrocumin alcohol, C 140
Dihydrocuparene, C 268
Dihydrocyclocitral, B 88
2,5-Dihydrocyclohexanes, B 4
16,17-Dihydro-15*H*-cyclopenta-[α]phenanthrene, D 1
2α,3α-Dihydro-2′H-cyclopropa [2,3]-5α-androstane, D *452*
2α,3-Dihydro-2′H-cyclopropa [2,3]-5α-androstan-3α-ol, D 452
Dihydro-*p*-cymenes, B 25
Dihydrocyperene, C 279
Dihydrodesoxytodomatuic acid, C 293
Dihydrodicyclopentadiene, C 97, 100
Dihydrodigacetigenin, D 404
Dihydrodihydroxyzierone, C 319
Dihydroenmein, C 396, 397
Dihydroepitestosterone, D 262, 263
Dihydroequilenins, D 223, 224
17α-Dihydroequilenin, D 201, 223
17β-Dihydroequilenin, D 201, 223
Dihydroequilenin diacetate, D 221
Dihydroequilins, D 210, **223**
Dihydroequilin 3-methyl ether, D 218
Dihydroeremophilenolide, C 349
5,6-Dihydroergosterol, D 86, 134, 148, 151, 165
Dihydroergosterol, D 134, 165; E 108
1,4-Dihydroestron-3-enol methyl ethers, D 228
3β,11α-Dihydroetianic acid, D 399
Dihydroeucarveol, C 155
Dihydroeucarvone, C 155
Dihydroeudesmol, C 299
Dihydrofulvulene, C 12
Dihydrofurans, thermal ring contraction, A 23
2,3-Dihydrofuran, A 47
Dihydrogedunin, C 429
Dihydrogenin, D *435*
α-Dihydrogitoxigenin, D 392, 393
β-Dihydrogitoxigenin, D 393
Dihydrogitoxigenone, D 392

Dihydroguaiol, C 314, 315
Dihydroheptalene, C 73
Dihydrohydnocarpic acid, A 197
Dihydro-icterogenin, C 441
4,7-Dihydroindane, C 58
8,9-Dihydroindanes, C 58
8,9-Dihydroindene, C 58
8,9-Dihydroindenone, A 172, 173
Dihydro-γ-ionone, B 176
$\alpha\beta$-Dihydro-ψ-ionone, C 289, 290
Dihydroiresone, C 349, 367
Dihydroisoalantolactone, C 337, 366
Dihydroisoparthenin, C 356
Dihydroisophorone, B 95
Dihydroisophotosantonic lactone, C 352
Dihydroisophotosantonic lactone acetate, C 355
Dihydrojasmone, A 180
Dihydrojunenol, C 300
Dihydrokryptogenin diacetate, E 43, 44, 45
Dihydrolanosterol, C 407, 421; D 75, E 102
Dihydrolaserol, C 325
Dihydromalvalic acid, A 70
Dihydro-12-methoxycostunolide, C 363
1′,3′-Dihydro-1′-methyl-5α-androstano-[5,6-b]azirine, D *453*
Dihydromexicanin, C 361, 362
Dihydromuconic ester, A 202
Dihydromustakone, C 298
Dihydromyrcene, B 159
1,4-Dihydronaphthalene, C 71
9,10-Dihydronaphthalene, B 399; C 71
Dihydroneohelenalin, C 361
Dihydronepetalactone, B 210
4,5-Dihydro-octalene, C 73
Dihydropalitantin, B 106
Dihydroparthenin, C 357
Dihydropentalene, A 128; C 52
Dihydroperiplogenin, D 374, 384
Dihydrophellandrols, B 196
9,10-Dihydrophenanthrene, D 213
Dihydrophenols, B 64
Dihydrophytyl bromide, B 266
Dihydropinnatifidin, C 339
Dihydropinol, B 223
Dihydropseudogenin, D *435*
Dihydropseudotigogenin, D *435, 436*
Dihydropyran, A 9, 164; D 215
Dihydropyrazoles, C 79
Dihydropyrethrosin, C 363
Dihydropyrogallol, B 112
Dihydroresorcinol, B 111
Dihydroresorcinol dioxime, B 82
Dihydrorhemmanic acid, C 440
Dihydrosantonins, C 343
Dihydrosapogenins, E 8, 12, 19
Dihydrosarsasapogenin, E 12
—, *nomenclature*, D *435*
Dihydroshikimic acid, B 139
Dihydrosmilagenin, E 12
Dihydrosobrerols, B 194, 221
Dihydrosterculic acid, A 69
5α,11α-Dihydrosteroids, D 318
5,6-Dihydrosterols, D 73
Dihydrostrophanthidin, D 373, 374, 378, 384
Dihydrostrophanthidinic acid, D 373, 374
Dihydrotachysterol, D 142, 143
Dihydrotauremisin A, C 347
Dihydrotelekin, C 339
Dihydroterephthalic acids, B 148
Dihydro-α-terpineol, B 189, 220
Dihydro-β-terpineol, B 220
Dihydrotestosterone, D 213, 261, 264, 271, 280
Dihydrotestosterone acetate cyanohydrin, D 276
Dihydrotigogenin, E 44
Dihydrotigogenin 3-acetate, E 14
Dihydrotigogenin diacetate, E 43, 44
Dihydrotriazoles, C 79
β-Dihydroumbellulol, C 148, 150
Dihydroumbellulone, C 150, 254
20,22-Dihydrouzarigenin, D 398
Dihydrovitamin D_2, D 143, 144
1,3-Dihydroxyadamantane, C 127
3β,17α-Dihydroxyalloetianic acid, D 291, 292
3β,17α-Dihydroxyallopregnane-11,20-dione, D 326
3α,17α-Dihydroxyallopregnan-20-one, D 309
3β,19-Dihydroxyandrost-5-en-17-one 3-acetate-19-tosylate, D 335
2,4-Dihydroxybenzaldehyde, B 54
2,6-Di-α-hydroxybenzylcyclohexanones, B 100
1,1′-Dihydroxybicyclobutyl, C 9
1,1′-Dihydroxybicycloheptyl, C 18
1,1′-Dihydroxybicycloheptylethyne, C 18
1,1′-Dihydroxybicyclopentyl, C 10
7,8-Dihydroxybicyclo[4.2.0]octane, C 47
4,4′-Dihydroxybiphenyl, C 16
2,3-Dihydroxybornane, C 220

3α,17α-Dihydroxy-21-bromopregnan-20-one, D 310
3β,14-Dihydroxy-5β,14β-bufa-20,22-dienolide, D *432*
3β,14-Dihydroxy-14β-bufa-4,20,22-trienolide, D *432*
2,5-Dihydroxycamphane, C 223
3β,14-Dihydroxy-5β,14β-card-20(22)-enolide, D 431
3,14-Dihydroxycard-1,3,5(10),20(22)-tetraenolide, D 381
3,3′-Dihydroxy-α-carotene, E 127
Dihydroxy-β-carotene, B 260, 263, 266, **299**
Dihydroxycarotenoids, B 295
3α,11α-Dihydroxycholanic acid, D 313, 315
Dihydroxy-5α-cholanoic acids, D 171, 172
Dihydroxy-5β-cholanoic acids, D 171, 172
3,6-Dihydroxy-5α-cholanoic acid, D 185
3,6-Dihydroxy-5β-cholanoic acids, D 10, 49, 180, 185
3α,7α-Dihydroxy-5β-cholanoic acid, D 55
3α,11-Dihydroxy-5β-cholanoic acids, D 56, 180
3α,11β-Dihydroxy-5β-cholanoic acid, D 55
3α,12α-Dihydroxy-5β,17β- cholanoic acid, D 28
7α,12α-Dihydroxy-5β-cholanoic acid, D 182
3β,7β-Dihydroxychol-5-enic acid, D 53
3α,12α-Dihydroxy-5β-chol-8(14)-enic acid, D 187
3,12-Dihydroxychol-14-enic acid, D 411
3α,12-Dihydroxychol-9(11)-enic methyl ester, D 316
3α,12α-Dihydroxy-5β-chol-7-enoic acid, D 186, 188
3α,12α-Dihydroxy-5β-chol-14-enoic acid, D 186
3β,5-Dihydroxy-5β-cholestan-6-one, D 45
4,5-Dihydroxycholestan-3-ones, D 125, 127
3β,9α-Dihydroxy-5α-cholest-7-en-6-one, D **152**
20α,22-Dihydroxycholesterol, E 110
1,3-Di(1′-hydroxycyclohexyl)-prop-1-yne, B 66
4′,4″-Dihydroxydicyclohexylethanes, C 16
3,4-Dihydroxydihydroagarofuran, C 304, 335
1,4-Dihydroxy-5,8-dihydronaphthalene, C 71
11β,21-Dihydroxy-3,20-dioxo-4-pregnen-18-al, D *439*
20,26-Dihydroxyecdysone, D 161
3β,5β-Dihydroxy-14β,15β-epoxy-5-bufa-20,22-dienolide, D 421
3β,16β-Dihydroxy-14β,15β-epoxy-5β-bufa-20,22-dienolide 16-monoacetate, D 420
3α,5-Dihydroxy-5α-ergosta-7,22-diene, D 136
3β,5-Dihydroxy-5α-ergosta-7,22-diene, D 136
Dihydroxyestrapentaenes, D 201
Dihydroxyestratetraenes, D 201
Dihydroxyestratrienes, D 201
3α,11α-Dihydroxyetianic acid, D 56, 399
3β,11α-Dihydroxyetianic acid, D 399
3,12-Dihydroxyetianic acids, D 28, 385
3β,19-Dihydroxyetianic acid, D 374, 412
2β,17β-Dihydroxy-2α-ethynyl-A-nor-5α-androstane, D 281
Dihydroxyferrocene, A 135
5,6-Dihydroxyindane, C 57
3β,14β-Dihydroxy-14-isoallocholanic acid, D 409
3β,14β-Dihydroxy-14-isoalloetianic acid, D 396, 397
3β,12β-Dihydroxy-22-isoallospirostan 11-one, D 327
3β,14β-Dihydroxy-14-isoetianic acid, D 388, 416
14β,19-Dihydroxy-14-isoetianic acid, D 374
α′α′-Dihydroxy-α-isopropyl-α′-methyladipic acid, B 207
αδ-Dihydroxy-δ-isopropyl-α-methyladipic acid, B 181
Dihydroxykauranoic acid, C 395
Dihydroxykaurene, C 395
Dihydroxykaurenolide, C 396
3β,14β-Dihydroxylactone, D 369
1,1-Di(hydroxymethyl)cyclobutane, A 90
1,6-Dihydroxy-1-methylcyclodecane, B 403
Di(hydroxymethyl)cyclohexanes, B 69, **70**

1,1-Di(hydroxymethyl)cyclohexane, B 89
1,4-Di(hydroxymethyl)cyclohexane, B 70
1,2-Dihydroxymethylcyclohex-4-ene, B 29
1,1′-Dihydroxymethylferrocene, A 132
3β,20β-Dihydroxy-17a methyl-C-nor-D-homopregn-17(17a)-ene, D 285
11β,17α-Dihydroxy-6α-methyl-pregna-1,4-diene-3,20-dione, D 352
17α,21-Dihydroxy-19-methylpregn-4-ene-3,20-dione, D 331
3β,17α-Dihydroxy-6-methylpregn-5-en-20-one, D 240
1,5-Dihydroxynapthalene, C 67
2,7-Dihydroxynaphthalene, C 64
3β,14β-Dihydroxy-19-oxo-5α-bufa-dienolide, D 412
3α,11α-Dihydroxy-12-oxocholanic acid, D 315
3α,11β-Dihydroxy-12-oxocholanic acid, D 315
3α,12α-Dihydroxy-11-oxocholanoic acid, D 315
3α,12β-Dihydroxy-11-oxocholanic acid, D 315
3α,11β-Dihydroxy-12-oxo-5β-cholanoic acid, D 56
3α,12α-Dihydroxy-7-oxo-5β-cholanoic acid, D 181
3α,12β-Dihydroxy-11-oxo-5β-cholanoic acid, D 57
3α,7α-Dihydroxy-12-oxo-5β-cholanoic acid, D 181, 182
7α,12α-Dihydroxy-3-oxo-5-β cholanoic acid, D 186
3α,11-Dihydroxy-12-oxo-5β-etianic acids, D 56
11β,17α-Dihydroxy-3-oxoeti-4-enic acid, D 295
11α,15-Dihydroxy-9-oxoprosta-5,13-dienoic acid, A 203
11α,15-Dihydroxy-9-oxoprosta-5,13,17-trienoic acid, A 203
11α,15-Dihydroxy-9-oxoprost-13-enoic acid, A 202
Dihydroxyoxotricyclohexyl, C 17
Dihydroxyperuenoic acid, C 383
3β,17β-Dihydroxy-5α,17α-pregnane, D 39
3α,17α-Dihydroxypregnane-11,20-dione, D 327
17α,21-Dihydroxypregnane-3,11,20-trione, D 326
3β,20β-Dihydroxy-5α-pregnan-12-one, D 285
11β,21-Dihydroxy-4-pregnene-3,20-dione, D *439*
3β,17β-Dihydroxy-14α,17α-pregn-5-ene-11,15,20-trione, D 405
17β,21-Dihydroxy-4-pregnene-3,11,20-trione, D *439*
3β,16α-Dihydroxypregn-5-en-20-one, D 60
11,12-Dihydroprogesterones, D 235
Dihydroxysemi-β-carotenone, D 260
Dihydroxyspirostans, E 46
(25*R*)-2,3-Dihydroxy-5α-spirostan, E 32
Dihydroxysteroids, D 87
Dihydroxytetrahydrolycopene, E 130
1,4-Dihydroxy-5,6,7,8-tetrahydro-naphthalene, C 71
Dihydroxythymoquinone, E 77
p,p′-Dihydroxy-α-truxillic acid, A 100
Dihydrozingiberene, C 260, 261
Di-imide, C 173, 175
1,4-Di-iodobutane, C 129
Di-iodocamphors, physical properties, C 211
αα′-Di-iodocamphor, C 223
Di-iodocyclopentadienylchromium, A 126
1,2-Di-iodocyclopropane, A 39,40
Di-iron-nonacarbonyl, B 363
Di-isoamylborane, B 45, 62, 63, 71, 159
Di-isoamylcadmium, D 99, 153
Di(isobutyl)aluminium hydride, B 405; C 117
Di-isohexylcadmium, D 108, 153
Di-isopinocamphenylborane, C 173
1,1′-Di-isopropenylcobalticinium perchlorate, A 139
Di-isopropylcadmium, D 152
Di-*O*-isopropylidene-inositol, B 61
αω-Diketens, conversion to cyclanones, B 407
Diketones, aldol condensation, B 6
—, cyclisation, B 113
—, macrocyclic, B 412
1,2-Diketones, B 411
—, conversion to cyclobutanones, A 78
—, macrocyclic, B 413
—, ring, A 184
1,3-Diketones, ring, A 185
1,5-Diketones, cyclisation, B 114
2,5-Diketones, A 107
—, cyclisation, A 181
αω-Diketones, macrocyclic, B 413

Dimedone, B 15, 49, 54, 109, **111**, 112; C 31
Dimeric steroid hormones, D **336**
Dimerisation, alkenes, A 74
Dimesylsodium, A 127
1,3-Dimethoxycarbonylcyclobutane, A 84
Dimethoxycycloheptatrienes, B 367
3,3-Dimethoxy-1,2-diphenylcyclopropene, A 49
1,1-Dimethoxyethene, C 49
2,2-Dimethoxypropane, D 92
ω-Dimethylacetonylacetone, C 145
3,4-Dimethylacetophenone, C 237, 238
2,2-Dimethyl-3-acetylcyclobutane-1-acetic acid, A 97
5,5-Dimethyl-3-acetylcyclopent-3-ene-1-carboxylic acid, C 185
Dimethylacetylene, A 77; B 392
ββ-Dimethylacrylic acid, B 199
2,2-Dimethyladipic acid, C 275, 282
αα-Dimethyladipic acid, B 340
ββ-Dimethyladipic acid, B 216
γγ-Dimethylallyl bromide, B 280
γγ-Dimethylallyl epoxides, E 64
Dimethylallyl pyrophosphate, B 155, 341; C 371, 466; E 55, 60, 73
Dimethylamine, B 108; D 357
3β-Dimethylamino-5α-cholestane, D 119
N,*N*-Dimethylaminocyclodecane, B 420
N,*N*-Dimethylaminocyclododecane methiodide, B 420
N,*N*-Dimethylaminocyclododec-2-ene, B 420
N,*N*-Dimethylaminocyclohepta-2,4-diene, B 358
3-Dimethylaminocyclohept-1-ene, B 351
N,*N*-Dimethylaminocyclohept-3-ene, B 358
N,*N*-Dimethylaminocyclohept-4-ene, B 358
N,*N*-Dimethylaminocyclononane, B 420
4-Dimethylaminocyclo-octa-1,3-diene, B 418
5-Dimethylaminocyclo-octa-1,3-diene, B 418
N,*N*-Dimethylaminocyclo-octa-1,3-diene, B 420
N,*N*-Dimethylaminocyclo-octa-2,4-diene, B 420
N,*N*-Dimethylaminocyclo-octane, B 420
5-Dimethylaminocyclo-oct-1-ene, B 418
Dimethylaminocyclo-oct-2-ene, B 420
N,*N*-Dimethylaminocyclo-oct-4-ene, B 420
6-Dimethylamino-3,5-dien-3-yl ethers, D 95
2-Dimethylaminoethanol, D 337
β-Dimethylaminoethyl chloride, B 108
2-(β-Dimethylaminoethyl)cyclohexanone, B 108
17-*N*-(2-Dimethylaminoethyl)imino-androst-5-en-3β-ol acetate, D 357
6-Dimethylaminofulvene, A 123
2-Dimethylaminomethylcyclohexanone, B 108, 126; C 30
Dimethylaminomethylcyclopentadiene, A 123
2-Dimethylaminomethylestrone, D 228
N,*N*-Dimethylaminomethylferrocene, A 133, 134
2-Dimethylaminomethyl-1-formylcyclohexane, B **91**
21-Dimethylaminopregna-5,16-dien-3β-ol-20-one acetate, D 356
3β-(Dimethylamino)-5α-pregnan-20α-ol, D 438
N,*N*-Dimethylaminotropone, B 379
Dimethylanhydroacetonebenzil, A 184
1,4-Dimethylbenzene, C 341
2,4-Dimethylbenzoic acid, C 185
1,3-Dimethylbicyclo[1.1.0]butane, C 32
7,7-Dimethylbicyclo[4.2.0]octan-2-ones, C 49, 286
2,2′-Dimethylbicyclopentyl, C 9
3,3′-Dimethylbicyclopentyl, C 9
2,3-Dimethylbutanal, D 155
2,3-Dimethylbutyric acid, D 155
Dimethylcadmium, C 43
Dimethylcamphor, E 77
2,2′-Dimethyl-β-carotene, B 283
4,4-Dimethylcholesta-5,7-dien-3β-ol, D 103
4,4-Dimethyl-5α-cholesta-8,14-dien-3β-ol, E 102
4,4-Dimethylcholesta-8,24-dien-3β-ol, E 99
2,2-Dimethyl-5α-cholestan-3-one, D 123
4,4-Dimethyl-5α-cholestan-3-one, D 123
4,4-Dimethylcholest-4-en-3β-ol, E 100
4,4-Dimethylcholest-5-en-3-one, D 129
4,4-Dimethylcholesterol, D 103, 129

4β,6-Dimethylcholesterol, D 103
7,7-Dimethylcholesterol, D 103
4,4-Dimethylcholesteryl mesylate, D 79
4,4-Dimethylcholesteryl tosylate, D 79
2,2-Dimethylchromane, E 64
2,2-Dimethylchromenes, E 63
4,6-Dimethylcoumalin, B 392
Dimethylcyclobutadienoquinone, A 93
1,2-Dimethylcyclobutane, A 80
2,2-Dimethylcyclobutane-3-carboxy-1-acetic acid, A 97
2,2-Dimethylcyclobutane-1,3-dicarboxylic acid, A 97
3,3-Dimethylcyclobutane-1,2-dicarboxylic acid, A 97
N,*N*-Dimethylcyclobutylamine, A 91
Dimethylcyclobutylmethylamine oxide, A 82
4-Dimethylcycloheptadiene, carboxylic acids, B 216, 357
4,4-Dimethylcycloheptanecarboxylic acid, B 357
2,2-Dimethylcycloheptanol, B 351
Dimethylcycloheptanones, B 353
2,5-Dimethylcycloheptanone, C 324
4,4-Dimethylcyclohepta-1,3,6-trienecarboxylic acid, B 357
4,4-Dimethylcyclohept-1-enecarbaldehyde, B 353
Dimethylcyclohexadienes, B 26
2,2-Dimethylcyclohexa-3,5-dien-1-one, B 122
6,6-Dimethylcyclohexa-2,4-dien-1-one, B 29
Dimethylcyclohexanes, B **15**, 16
1,1-Dimethylcyclohexane, B 13, 112
1,2-Dimethylcyclohexane, A 17; B 13
1,3-Dimethylcyclohexanes, B 13
1,4-Dimethylcyclohexanes, B 13
2,6-Dimethylcyclohexanecarbaldehyde, B 88
Dimethylcyclohexane-1^c-carboxylic acids, B 133
Dimethylcyclohexane-1,2-diols, B 51
1,2-Dimethylcyclohexane-1,2-diol, B 45, **53**
1,3-Dimethylcyclohexane-1,3-diol, B 54
2,5-Dimethylcyclohexane-1,4-diol, B 51
5,5-Dimethylcyclohexane-1,3-diol, B 54, 112
Dimethylcyclohexane-1,3-diones, B 109
2,5-Dimethylcyclohexane-1,4-dione, B 109
5,5-Dimethylcyclohexane-1,3-dione, B **111**
2,2-Dimethylcyclohexane-1,3,5-trione, B **113**
Dimethylcyclohexan-1-ols, B 47
1,1-Dimethylcyclohexan-2-ol, C 164
3,3-Dimethylcyclohexanol, B 15, **49**, 112
Dimethylcyclohexanones, B 95, 101
1,1-Dimethylcyclohexan-2-one, C 159
2,4-Dimethylcyclohexanone, B **100**
3,3-Dimethylcyclohexanone, C 156
Dimethylcyclohex-1-enes, physical properties, B 21
1,2-Dimethylcyclohexene, B 16, 17, 19, 33, 45, 50, 53, 75; C 159, 164
1,6-Dimethylcyclohex-1-ene, B 17
3,3-Dimethylcyclohex-1-ene, B 45
Dimethylcyclohex-3-enecarbaldehydes B 88
4,4-Dimethylcyclohex-5-ene-1,3-dione, B 130
Dimethylcyclohex-2-enones, B 117, 119
1,2-Dimethylcyclohexyl acetate, B 17
1,2-Dimethylcyclohexyl bromide, B 33
1,2-Dimethylcyclopentanes, A 113
1,3-Dimethylcyclopentanes, A 111, 113
2,5-Dimethylcyclopentane-1-carboxylic acid, A 190
1,1-Dimethylcyclopentane-2,3-dicarboxylic acid, C 246
4,5-Dimethylcyclopentane-1,3-dicarboxylic acids, A 192
Dimethylcyclopentane-1,2-diols, A 157, 160
Dimethylcyclopentan-1-ols, A 114, 155, 157
Dimethylcyclopentan-1-ones, A 167
Dimethylcyclopentenes, A 116; B 348
1,2-Dimethylcyclopentene, A 114, 148, 156, 159
2,3-Dimethylcyclopent-2-enone, B 116
3,4-Dimethylcyclopent-2-en-1-one, A 170
1,2-Dimethylcyclopentyl bromides, A 148
1,2-Dimethylcyclopentyl chloride, A 148
1,1-Dimethylcyclopropane, A 32, 73; C 21
1,2-Dimethylcyclopropane, A 32
3,3-Dimethylcyclopropane-1,2-dicarboxylic acid, A 60
3,3-Dimethylcyclopropanetetracarboxylic acid, A 62

1,3-Dimethylcyclopropene, A 33
3,3-Dimethylcyclopropene, A 30, 33
N,*N*-Dimethylcyclopropylamine, A 46
4,9-Dimethyldecalin, C 63
6,16α-Dimethyl-6-dehydroprogesterone, D 240, **241**
Dimethyldiazomethane, A 68
3,17-Dimethyl-16,17-dihydrocyclopentaphenanthrene, D 7
1,2-Dimethylenecyclohexanes, B 24
4,5-Dimethyl-1,2-dimethylenecyclohex-1-ene, B 29
2,2-Dimethyl-1,3-dioxane-4,6-dione, A 105, 191
2,5-Dimethyl-3,4-diphenylcyclopentan-1-one, A 167
5,5-Dimethyl-3,4-diphenylcyclopent-2-en-4-ol-1-one, A 184
N-Dimethyl-*N'*-(2,3-diphenylcycloprop-2-enyl)urea, A 47
Dimethyleneaminomethylcyclopropane *N*-oxide, A 34
2,3-Dimethylenebicyclo[2.2.1]-heptane, C 91
1,2-Dimethylenecyclobutane, A 81, 83
1,3-Dimethylenecyclobutane, A 81
3,4-Dimethylenecyclobutene, A 84
1,2-Dimethylenecyclohexadiene, B 29, 30
1,4-Dimethylenecyclohexadiene, B 30, 31
1,2-Dimethylenecyclohexane, B **24**
1,3-Dimethylenecyclohexane, B 25
3,5-Dimethylenecyclohex-1-ene, B 28
3,6-Dimethylenecyclohex-1-ene, B 28
4,5-Dimethylenecyclohex-1-ene, B 29
7,8-Dimethylenecyclo-octa-1,3,5-triene, C 50
1,2-Dimethylenecyclopentane, A 118
Dimethylenecyclopropanes, A 34
Dimethylene-β-decal-6-en-2-ol, C 104
Dimethylenedecalins, C **104**
Dimethylene-β-decalol, C 104
Dimethylene-β-decalone, C 104
1,2-Dimethylene-3,4-diphenylcyclobutene, A 81
1,4:5,8-Dimethyleneoctalin, C 104
1,2-Dimethylene-4-*n*-octylcyclohexane, B 24
Dimethylenesteroids, D 336
Dimethyl-1,2-epoxycyclohexanes, B 72
4,4-Dimethylergosta-8,24(28)-dien-3β-ol, E 108
1,4-Dimethylestrones, D356
Dimethylethylenediamine, D 357
N,*N*-Dimethylformamide, B 353
Dimethylfulvene, A 124
αα-Dimethylglutaric acid, B 244, 295, 316, 340; C 282
2,6-Dimethylheptanoic acid, C 292
5,6-Dimethylheptan-2-one, D 13, 14, 27, **156**
2,5-Dimethylhexa-2,4-diene, A 65
1,2-Dimethyl-1-hydroxy-3-isopropenylcyclopentane, B 208
1,1-Dimethyl-2-isobutenylcyclopropane, A 36
2,3-Dimethyl-4-isopropylcyclopent-2-enone, C 146
Dimethylisopropyldecalin, C 324
1,1-Dimethyl-3-isopropylcyclopentane, A **113**
2,2-Dimethyl-3-isopropylidenecyclopropyl propionate, A 71
Dimethylketen, A **76**
—, trimerisation, B 113
Dimethylketen dimer, A 90
ω-Dimethyllaevulic acid, C 145
Dimethylmagnesium, B 73
Dimethylmaleic acid, A 109, 187
Dimethylmalonic acid, B 130, 165, 244, 295, 320; C 151, 282
1,2-Dimethyl-5-methoxycarbonylcycloheptatriene, B 216
N,*N'*-Dimethyl-1-methylcyclopentylamine oxide, A 115, 117
3,3-Dimethyl-1-methylenecyclobutane, A 81
6,6-Dimethyl-5-methylenecyclohexa-1,3-diene, B **29**
Dimethylmethylenenorbornanes, C 234
2,2-Dimethyl-3-methylenenorbornane, C 137
1,4-Dimethylnaphthalene, C **341**
2,6-Dimethylnaphthalene, from carotenoids, B 258
2,2-Dimethylnorbornane, C 196, 199
2,3-Dimethylnorbornane, C 188
2,3-Dimethyl-5-norbornenes, A 192
2,3-Dimethylnorborn-2-ene, C 187
Dimethylnorcampholide, C 194, 195
7,7-Dimethylnorcar-3-ene-3-carboxylic acid, C 158
7,7-Dimethylnorcar-2-ene-4-carboxylic acid, C 157
2,6-Dimethylocta-2,7-diene, C **261**
3,7-Dimethylocta-3,7-dienol, D **234**
3,7-Dimethyloctan-6-one-1,3-dicarboxylic acid, C **310**
Dimethyl-1,3,5-octatriene, B 182
2,7-Dimethyloctatrienedial, B **274**

2,6-Dimethyloct-7-en-4-one, B 168
2-Dimethyl-1-oxo-*n*-butane-1,3-dicarboxylic acid, C 203
1,1-Dimethyl-2(3′-oxo-*n*-butyl)-cyclopropane-3-carboxylic acid, C 154
2,2-Dimethyl-3-(3′-oxobutyl)-cyclopropanecarboxylic acid, A 58
β-2-(3,5-Dimethyl-2-oxocyclohexyl)-2-hydroxyethylglutarimide, B 101
Dimethyloxosulphonium methylide, A 25
2,3-Dimethylpentane, D 133
1,2-Dimethylphenanthrene, D 22, 23, 202
1,2-Dimethylphenanthr-7-ol, D 202
3,4-Dimethylphenol, B 121
5,5-Dimethyl-3-phenylcyclohexanone, B 95
5,5-Dimethyl-3-phenylcyclohex-2-enone, B 117
ββ-Dimethylpimelic acid, B 216; C 155
6α,16α-Dimethylprogesterone, D 240
2,3-Dimethylnaphthalene, D 144
4,8-Dimethylnon-2-en-1-al, D 25
7α,17α-Dimethyl-19-nortestosterone, D 273
Dimethyl sciadinonate, C 384, 385
2,5-Dimethylsorbic methyl ester, A 67
Dimethylsuccinic acid, B 244, 295, 320, 340; C 151, 156, 282, 284
Dimethylsuccinic dimethyl ester, D 220
Dimethylsulphonium methylide, A 25; B 70, 94
Dimethyl sulphoxide, D 83, 258, 337
Dimethylsulphoxonium methylide, D 336
6α,17α-Dimethyltestosterone, D 266
7α,17α-Dimethyltestosterone, D 266, 272, 273
7β,17α-Dimethyltestosterone, D 266, 272
3,3-Dimethyltetracyanocyclopropane, A 62
1,1-Dimethyl-1,2,3,4-tetrahydronaphthalene, C 275
1,7-Dimethyltricycloheptane-7-carboxylic acid, C 176
2,4-Dimethyltricyclo[1.1.1.0^{2,4}]-pentan-5-one, C 34, 84
2,3-Dimethyltriphenylphosphonium, iodide, D 155
Dimethyltropolones, B 369
1,3-Dimethyl-2-vinylcyclohex-1-ene, B 28
4,4-Dimethylzymosterol, D 166
αω-Dinitriles, B 406
1,3-Dinitroadamantane, C 131
3,5-Dinitrobenzoic acid, C 319
1,2-Dinitrocyclohexane, B **79**, 80, 81
1,3-Dinitrocyclohexanes, B 79, 80 82
1,4-Dinitrocyclohexanes, B 79, 80, 82
Dinophyceae, B 238, 331
18,19-Dinorandrostane, D *443*
18,19-Dinor-5α-pregnane-20α-carboxylic acid, D *443*
18,19-Dinorspirostan, D *443*
17α,17′α-Di-19-nortestosterone, D 337
Dinoxanthin, B 332
Dioscin, E 2, 6
Dioscoreaceae, E 5
Dioscorea chiapasensis, E 26
Dioscorea gracillima, E 3
Dioscorea macrostachya, E 46
Dioscorea mexicana, E 27
Dioscorea sativa, E 3
Dioscorea species, E 25, 43
Dioscorea septemloba, E 3
Dioscorea spiculiflora, E 26, 113
Dioscorea tenuipes, E 27
Dioscorea tokoro, E 2, 3, 25, 27
Diosgenin, D 232, 233, 254, 326, 328; E 1, 2, 3, 9, 11, 19, 21, 23, 24, 25, **30**, 32, 33, 34, 43, 44, 49, 112
Diosgenin acetate, E 19, 44, 47, 48
ψ-Diosgenin diacetate, D 232
Diosgenone, E 30
Diosphenol, B 108, **202**, 203; C 399
Diosphenol A, D **124**, E 125
Diosphenol B, D **124**, 125
Diosphenolene, B 204
Diosterol I, D 131
Diosterol II, D 128, 131
Diotigenin, E 27
Dioximinodicyclopentadiene, C 103
3,11-Dioxoalloetianic acid, D 291, 296, 297
3,3′-Dioxobicyclohexyls, C 17
2,2′-Dioxobicyclopentyl, C 11
3,16-Dioxobisnor-5α-cholanoic methyl ester, E 21, 22
3,16-Dioxobisnor-5α,17α-cholanoic methyl ester, E 21, 22
3,16-Dioxobisnor-5α,20-isocholanoic methyl ester, E 21
2,3-Dioxobornane, C 218, 219
2,5-Dioxobornane, C 222
2,6-Dioxobornane, C 212, 223
2,5-Dioxocamphane, C 217

4,4′-Di-oxo-β-carotene, B 318
3,11-Dioxocholanic acid, D 313, 314
Dioxocholanoic acids, D 189, 190
3,6-Dioxo-5β-cholanoic acid, D 15, 16, 189
3,7-Dioxo-5β-cholanoic acid, D 182
3,12-Dioxo-5β-cholanoic acid, D 19, 21, 182
7,12-Dioxo-5β-cholanoic acid, D 11, 182
6,7-Dioxocholestanol, D 36
6,7-Dioxocholestanyl acetate, D 37
16,22-Dioxocholesterol, D 99
3,11-Dioxo-20-cyanopregn-17(20)-en-21-ol, D 329
Dioxocyclobutenediol, A 93, 94
5,20-Dioxodes-A-19-norpregn-9-ene, D 246
3,11-Dioxoetianic acid, D399
3,11-Dioxoetianic methyl ester, D 322
3,11-Dioxoeti-4-enic acid, D 296, 297
3,21-Dioxoeti-4-enic acid, D 296
9,11-Dioxononadecanoic acid, A 69
2,2′-Dioxospirilloxanthin, B 312, 345
Dioxospiro[5.5]undecanes,, C 31
11,12-Dioxotigogenin, D 284
Dipentene, B 178, 179, 218; C 172
Dipentene dihydrochloride, B 180; C 151
Dipentene nitrosochloride, B 179
Diphenylacetylene, A 49, 55, 59, 85, 109, 126, 139
Diphenyladipic methyl ester, A 166
1,1-Diphenylallene, A 46
Diphenylamidines, B 89
Diphenylamine, B 334; E 124, 126
1,1′-Diphenylazoferrocene, A 135
2,5-Diphenyl-3,4-benzofuran, A 118
5,6-Diphenylbicyclo[3.1.0]hexan-2-one, C 37
1,4-Diphenylbutane-1,3-dione, A 50
2,3-Diphenylbutanetetracarboxylic ester, A 99
Diphenylcarbene, A 25
Diphenylcholenes, D 26, 233
1,3-Diphenylcyclobutadiene, A 87
Diphenylcyclobutadienoquinone, A 93
Diphenylcyclobutane-1-carboxylic acids, A 95
Diphenylcyclobutanedicarboxylic acids, A 72, 98
1,2-Diphenylcyclohexane-1,2-diols, B 51, 53
1,2-Diphenylcyclohexene, B 53
4,4-Diphenylcyclohex-2-en-1-one, C 37
1,4-Diphenylcyclopentadiene, A 121
1,2-Diphenylcyclopentane-3,4-dione, A 109
2,4-Diphenylcyclopentane-1,3-dione, A 185
3,4-Diphenylcyclopentane-1,2-dione, A 184
2,5-Diphenylcyclopentane-1,3,4-trione A 188
Diphenylcyclopentanones, A 166, 167
3,4-Diphenylcyclopent-2-en-4-ol-1-ones, A 183
3,4-Diphenylcyclopent-2-en-1-one, A 170
1,2-Diphenylcyclopropanes, A 36
2,2-Diphenylcyclopropanecarboxylic ester, A 54
2,3-Diphenylcyclopropanecarboxylic esters, A 54
2,3-Diphenylcyclopropenecarboxylic ester, A 55
1,2-Diphenylcyclopropene-3,3 dicarboxylic methyl ester, A 59
Diphenylcyclopropenone, A 37, 49; B 363; C 13
Diphenylcyclopropenylium salts, A 37
2,3-Diphenylcycloprop-2-enyl nitrite, A 47
2,2-Diphenylcyclopropyl cyanide, A 63
2,2-Diphenylcyclopropylidene, A 46
2,2-Diphenylcyclopropylidene carbene A 26
Diphenyldiazocyclopropane, A 46
Diphenyldiazomethane, cyclopropene adduct, A 33
Diphenyldibromoquinol- cyclopropene adduct, A 37
1,2-Diphenyl-3 (dicyanomethylene)-cyclopropene, A 37
Diphenyl ether, B 49
1,1-Diphenylethylene, A 54
1,1′-Diphenylferrocene, A 134
Diphenylfulvene, A 124
2,5-Diphenyl-3,4-isobenzofuran, B 348
Diphenylketen, A 76, 92
Diphenyl-laevulic esters, A 185
Diphenylmethane, D 204
—, reduction, C 6
2,2-Diphenyl-1-methylcyclopropane-carboxylic acid, A 64
1,2-Diphenyl-3-nitrocyclopropene, A47
2,6-Diphenylpentalenes, C 54
1,2-Diphenyltetrafluorocyclobutene, A 93
Diphenyltropones, B 363
2,7-Diphenyltropone, B 364
Diphosphoranes, C 73

$2\beta,16\beta$-Dipiperidino-5α-androstane-$3\alpha,17\beta$-diol diacetate, dimethobromide, D 359
Dipiperitones, B 198
Diploptene, C 417, 452
Diplopterol, C 452
Dipropenyl ketone, A 170
1,4-Dipropylcyclohexanes, B 14
Dipropylcyclopropenone, A 37, 49
1,2-Dipropyl-3,(dicyanomethylene)-cyclopropene, A 37
Dipterocarpol, C 425
Dipterocarpus species, C 278
Disodio-bismalonates, B 151
Disodiopropanetetracarboxylic ester, A 62
Disodiumphenanthrene, B 396
Dispiro[3.1.5.1]dodecane-2 carboxylic acid, C 27
Dissociation constants, structure determination, C 480
Distyranic acid, A 95
Distyrinic acid, A 95
Diterpenes, C 60
—, bicyclic, E 86
—, biosynthesis, E 85 ff
—, monocyclic, E 86
—, *nomenclature*, D *423*
—, tetracyclic, E 87
—, tricyclic, E 86
Diterpene alkaloids, C 380, 470
Diterpenoids, C 369 ff, 373 ff, 478
—, abietanes, physical properties, C 392
—, atisirane, C 400
—, bicyclic, C 375, 479
—, biosynthesis, C 460
—, cembrane group, C 401
—, gibberellic acid group, C 398
—, kauranes, physical constants, C 395
—, labdanes, physical properties, C 382
—, phyllocladanes, physical properties, C 393
—, stachane—hibane group, C 399
—, synthesis, C 468
—, tetracyclic C 379
—, totarols, physical properties, C 393
—, trachylobane, C 400
—, tricyclic, C 377
—, verticillol and taxicin group, C 402
$17\alpha,17'\alpha$-Ditestosterone, D **337**
1,2-Dithiocyanatocyclohexane, B 77
Ditropyl, B **360**
Ditropyl ether, B **360, 363**
Divinylacetylene, C 9
1,2-Divinylcyclobutane, B 383
1,2-Divinylcyclopentane, A 118
1,2-Divinylcyclopropane, A 35; B 349
Divinyl ethers, A 43
Diynes, B 388
—, cyclic, B 390
1,3-Diynes, condensation with acetylenes and olefins, B
DMAPP (= Dimethylallyl pyrophosphate), E 55, 60, 61, 62, 65, 66, 67, 118
DOCA (= 11-Deoxycorticosterone 21-acetate), D 300
Dodecacarbonyltri-iron, B 292
Dodecahydrotriphenylene, B 31, 100
1,12-Dodecanedicarboxylic acid, A196
1,12-Dodecanedicarboxylic acid, B387
1,12-Dodecanedioic acid, B 416
ω-Dodecanolactam, B 387
Dodecapreno-β-carotene, B 281
n-Dodecyclcyclohexane, B 14
Doisynolic acid, D 202, 212, 223, 224, 226
Dolabradiene, C 378, 389, 470
β-Dolabrin, B 365
Dolichoderus species, B 212
Dolichodial, B **211,** 212
Dolichols, E 134
Drimanic acid, C 311
Drimene derivatives, C 310
Drimenin, C **347**, 348, 367
Drimenol, C **310**, 311, 335, 347
Drimic acid, C 311, 348
Drimone, C 311
Drimys confertigolia, C 347
Drimys lanceolata, C 311
Drimys winteri, C 310, 347, 349
Dryobalanops aromatica, C 226
Dumortierigenin, C 441
Duodephanthon diacid, D 380
Durene, C 110
Dustanin, C 452
Duvatrienediols, C 401
Duvatrienol oxide, C 401
Dysoxylon frazeranum, C 263

Ebelin lactone, C 412, 413, 435
Eburicoic acid, C 407, 408, 421, 460, **467**
Ecdysones, E 113
Ecdysone, D 3, 33, 36, 69, 71, **159,** 160; E 113
—, ultraviolet absorption spectra, D 68
β-Ecdysone, D **161**

Ecdysterone, D **161**; E 113, 114
Ecgonine, B 356
Echimidinic acid, E 66
Echinenone, B 238, 239, 249, 262, **314**, **316**, **317**, 346; E 120, 121, 128
Echinocystic acid, C 440
Echinulin, E 62, 63, 65
Echubioside, D 362, **386**
Echujin, D 362, **386**
Egomaketone, B **171**
Ekasantalal, C 264
Ekasantalic acid, C 264, 265
Elemane, C 263, 295, 329
Elemane group, C 262
Eleman-2,3-11-triol, C 295
Elemenes, C **262**, 288
α-Elemene, C 263, 295
β-Elemene. C 262, 263
δ-Elemene, C 263
Elemol, C 263, **295**, 334
α-Elemolic acid, C 410, 411, 426
β-Elemonic acid, C 426
Elettaria cardomomum, B 181
Elleryone, **C** 318,
Elliotinol, C 383
Elsholtzia cristata, B 170
Elsholtzia oldhami, B 170
Elsholtzione B **170**, 171; E 67
Elymocalvine, E 65
Emicymarin, D 362, 384, 398
Eminii mirabilis, D 398
Emmolactone, C 414, 436
Emulsin, B 91
Enamines, A 76, 165; B 93, 97
Endrin, C **105**
4-Ene-3,6-diones, D 94
Enmein, C 374, 380, 396, 397
Enmein acetate, C 396
5-En-3β-ols, D 95
Enolide, D 7
4-En-3-ones, dehydrogenation, D 94
—, enamines, D 93
—, enol ethers, D 92
—, reactions, D 92, 94
5-En-3-ones, D 93
Entagenic acid, C 440
1,3-Enynes, condensation with acetylenes and olefins, B 5
Eperuanediol, C 382
Eperuanetriol, C 382
Eperuenedioic acid, C 384
Eperuic acid, C 383; E 86
Epiandrosterone, D 9, 36, 38, 39, 41, 101, 255, 256, 263, 264, 291, 292, 294, **328**
5β-Epiandrosterone, D 38, 39
Epiandrosterone acetate, D 308, 309
$\Delta^{9(11)}$-Epiandrosterone acetate, D 290
Epiborneol, C **231**, **232**
Epicalciferol, D 147
Epicamphor, C **199**, 205, 206, 219, 221, **224**, 225, 226, 231, 232
Epicamphor-2-carboxylic acid, C 225, 226
Epicholestanol, B 32; D 38, 39, 40, 43, 74, **110**, 255, 256
Epicholestanyl acetate, D 185, 255
Epicholestanyl bromide, D 43
Epicholesterol, D 43, 45, 48, 66, 74, 85, **103**
Epi-ψ-cholesterol, D 52
Epicholesterol β-oxide, D 48
Epicholesterol 5,6-oxides, D 104
Epicholesteryl acetate, C 45, 46, 104
Epicholesteryl tosylate, D 43, 44
Epichlorhydrin, A 42
Epicoprostanol, D 38, 39, 41, 75, 110, 180
Epicoprostanyl acetate, D 110, **255**
11-Epicorticosterone, D **312**
11-Epicortisol, D 328
—, *nomenclature*, D *439*
2-Epicucurbitacin B, C 424
Epi-α-cyperone, C 295, 305
12-Epideoxycholic acid, D 172
3α,9-Epidioxy-5α-androstan-17-one, D *451*
Epiergostanol, D 75
3-Epiergosterol peroxide, D **136**
16-Epiestriol, D **211**
17-Epiestriol, D **211**
12-Epietiodeoxycholic acid, D 58
Epigermanicol, C 438
16-Epigitoxigenin, D 394
Epihydrotestosterone, D 264
Epi-isoborneol, C **231**, 232
Epi-isofenchone, C 242, 243
Epilithochloic acid, D **185**
Epilumisterol, D **142** 147
3-Epilupeol, C 435
3-Epimanool, C **382**
13-Epi-*enantio*-manoyl oxide, C 382
Epineoergosterol, D 135, 136
3-Epi-oleanolic acid, C 439
Epiphasic carotenoids, B 235, 240
13-Epimaric acid, C 377
3-Episamogenin, E 31
Epi-β-santalene, C 265, 266
Episarsasapogenin, D 251, 252; E 46
21-Episerratenediol, C 458
Epishellolic acid, C 326
3-Epismilagenin, E 46

Episterol, D **150**, 165
Epistigmastanol, D 75, 256
Episulphides, steroids, D **114**
ψ-Epitaraxastanediol, C 449
Epitestosterone, D 264, 271
17-Epitestosterone, D 61, 260
Epithiocholestanes, D **113**, 114
1,2-Epithiocyclohexane, B **76**
Epitigogenin, D 251, 253
Epitorulosol, C 382
Epitruxillic acid, A 100
3-Epiuzarigenin, D 364, **396**, 398
3-Epiuzarigenin-[glucoside]$_2$, D 398
Epivitamin D_2, D 147
5-Epixanthoperol, C 392
Epoxides, biosynthesis, B 295
—, macrocyclic alcohols, B **403**
—, reaction with phosphoranes, A 21
—, steroids, D 113
Epoxycardanolides, D *431*
Epoxycardanolide, D 7
16β,21ξ-Epoxy-14β,20ξ-cardanolide, D *431*
16β,22-Epoxycholestane, D 434
16β,22-Epoxycholestan-3β-ol, E 14
1,2-Epoxycyclodecane, B 405
1,2-Epoxycyclododecane, B 405
1,2-Epoxycycloheptane, B 352
1,2-Epoxycyclohexadecane, B 405
Epoxycyclohexanes, B 50, 52, **71**, 72, 73, 94
1,2-Epoxycyclohexane, B 67, 76, 77, 82
3,6-Epoxycyclohexane-1,2-dicarboxylic acid, B 150
Epoxycyclohexanols, B **73**
2,3-Epoxycyclohexan-1-ol, B 62
Epoxycyclohexenes, B 72, 73
1,2-Epoxycyclohex-3-ene, B 27
1,2-Epoxycyclohex-4-ene, B 28, 63
3,6-Epoxycyclohex-4-ene-1,2-dicarboxylic anhydrides, B 150
α:1-Epoxycyclohexylacetic ester, B 142
1,2-Epoxycyclo-octane, B **405**
1,2-Epoxycyclotridecane, B 405
1,2-Epoxydecane, B 405
Epoxygrindelic acid, C 384
8,19-Epoxy-19-hydroxycortexone, D 377
16α,17α-Epoxy-3β-hydroxy-6-methylpregn-5-en-20-one, D 240
8,19-Epoxy-19-hydroxy-17α-progesterone, D 377
Epoxyketones, terpenoid, B 202
2,3-Epoxyketones, B 408
Epoxylinalool, B **169**
1,2-Epoxy-*p*-menthan-3-one, B 203
1,2-Epoxy-*p*-menth-4(8)-en-3-one, B 204
1,2-Epoxy methylcyclohexanes, B 72
17β,20-Epoxy-21-norpregn-5-en-3β-ol, D 268
1,2-Epoxy-1-phenylcyclohexene, B 83
(20S)-18,20-Epoxypregnane, D *426*
11β,12β-Epoxypregnane-3,20-dione-bisethylene acetal, D 285
5,6-Epoxyzeaxanthin, E 127
Equilenin, D 9, 198, 199, 201, 205, **216**, 217, 219, 220, 222, 224, 225, 227, 261, 262, 280
—, synthesis, D 218
Equilenin acetate, D 227, 262
Equilenin methyl ether, D 220, 221
Equilin, D 198, 199, 201, **216**, 217, 227
Eremophilane lactones, C 349
Eremophilene, C **273**, 288
Eremophilene derivatives, C 306
Eremophilenolide, C **349**, 350, 367
Eremophilla mitchelli, C 306
Eremiphilone, C 273, **306**, 307, 308, 335, 349, 371
Ergocalciferol, D 138, **144**
—, *nomenclature*, D *439*
Ergometrine, E 65
Ergopinacol, D 135
Ergosine, E 65
Ergostadienetriol, D 137
Ergostadienetriol I, D 136
Ergostadienetriol II, D 134, 135, 136
Ergosta-5,7-dien-3β-ol, E 108
5α-Ergosta-7,11-dien-3β-ol, D 86
Ergosta-7,23-dien-3β-ol, E 108
Ergosta-8,24(28)-dien-3β-ol, E 108
5α-Ergosta-7,22-dien-3-one, D 68, 159
Ergosta-7-en-3β-ol, E 108
Ergosta-8-en-3β-ol, E 108
Ergostanes, D 321
—, *nomenclature*, D *428*, *429*
Ergostane, D 3, 4
—, molecular rotation, D 63
5α-Ergostane-3,6-dione, D 134
5α-Ergostane-3β,5,6α-triol, D 134
Ergostanol, D 6, 75, 155, 156
—, *nomenclature*, D *427*
5α-Ergostanol, D 13, 27, 28
5α-Ergostan-3β-ol, D 73, 134, 135
9β,10α-Ergostanol, D 141
Ergostanyl acetate, D 156
5α-Ergosta-6,8(14),22-triene 3β,5-diol, D 137

5α-Ergosta-6,8(14),22-trien-3β-ol, D 148
5α-Ergosta-7,9(11),22-trien-3β-ol, D 148
5α-Ergosta-7,14,22-trien-3β-ol, D 148
5,7,22-Ergostatrien,3β-ol, D *439*
Ergost-4-ene-3,6-dione, D 134
5α-Ergost-7-en-3β-ol, D 86
5α-Ergost-8(14)-en-3β-ol, D 86, 150
Ergosterol, C 467; D 3, 12, 13, 27, 28, 29, 36, 43, 67, 69, 75, 107, **134**, 135, 136, 139, 141, 142, 143, 144, 145, 155, 165, 326, 413; E 107, 108, 109
—, biosynthesis, E 108
—, — in yeast, E 108
—, irradiation, D 148
—, irradiation products, D 137
—, *nomenclature*, D *439*
—, photo and thermo isomers, D 138
—, structure, D 9
—, ultraviolet absorption spectra, D 68
9α,10α-Ergosterol, D 138
9β,10α-Ergosterol, D 138
9β-Ergosterol, D 138, 140
γ-Ergosterol, D 86
i-Ergosterol, D 155
Ergosterol B, D **148**, 149
Ergosterol D, D **148**, 149, 319
Ergosterol peroxide, D 136, 137
i-Ergosterone, D 155
Ergosterone peroxide, D 136
Ergosteryl acetate, D 134, 137, 321
Ergot, yeast, D 75
Ergotamine, E 65
Erysimoside, D **384**
Erysimum diffusum, D 384
Erythrodiol, C 438
Erythrodiol-3-acetate, E 118
Erythrophlamic acid, C 391
Erythrophleadienolic acid, C 391
Erythroxydiols, C 389, 399
Erythroxylol, C 399
Erythroxylon coca, A 98
Erythroxytriols, C 389
Escherichia coli, A 69
Escholtzia californica, B 305
Eschscholtzxanthin, B 249, **305**, 314, 315; E 120, 122, 129
—, biosynthesis, E 131
—, structure, B 233
Eschscholtzxanthone, B **314**
Escobedia scabrifolia, B 328
Estafiatin, C **354**, 367
Esters, αβ-unsaturated, A 20
Estra-4,9-diene-3,17-dione, D 209
Estradiols, D **200**
Estradiol, D 2, 61, 102, 199, 200, **211**, 212
—, vinyl homologues, D 214
Estradiol-17α, D 7, 201, 211, 212, **214**
—, *nomenclature*, D *439*
Estradiol-17β, D 7, 61, 197, 199, 201, 211, **212**, 213, 214, 215, 223, 227, 230, 262, 342
—, *nomenclature*, D *439*
Estradiol benzoate, D 210, 212
Estradiol-17β dibenzoate, D 211
Estradiol-17β digitonide, D 212
Estradiol ethers, D 215
Estradiol 3-methyl ether, D 207, 227, 242, 262, 301
Estradiol-17β 17-monoacetate, D *437*
Estradiol-17β monobenzoate, D 213
Δ^6-Estradiol-17β-ol acetate, D 262
17β-Estradiol 3-tetrahydropyranyl ether, D 215
17β-Estradiol 17-tetrahydropyranyl ether, D **215**
Estranes, 5,10-disubstituted, D 229
—, *nomenclature*, D *427*, *443*, *446*
Estrane, D 2, 5, 6
Estrane-3α,17β-diol-10β,5α bromohydrin diacetate, D 229
Estrane-3α,17β-diol-5α,10α-epoxide diacetate, D 229
Estrane series, steroids, D 356
Estrane-3α,5β,10β,17β-tetrol- 3α,17β-diacetate, D 230
1,3,5(10)-Estratriene, D 200
—, *nomenclature*, D *428*
1,3,5(10)-Estratriene-3,17α-diol, D *439*
1,3,5(10)-Estratriene-3,16α,17β-triol, D *439*
Estr-5(10)-enes, D 284
Estr-5(10)-ene-3α,17β-diol diacetate, D 230
Estr-5(10)-ene-3,17-dione, D 229
5(10)-Estrenolone, D 350
Estrenones, D 228
Estriol, D 7, 60, 61, 198, 199, **200**, 201, 202, **210**, 223
—, isomerides, D 211
—, *nomenclature*, D *439*
Estriol-3-methyl ether, D 223
Estrogens see Oestrogens
Estrogenic lactones, D **226**
Estrogenolic acids, D **223**, 226
Estrolic acid, D 226
Estrololactone, D 227
Estrones, ring A substituted, D 228

Estrones, stereoisomeric racemic, D 205
—, troponoid analogues, D 228
Estrone, D 2, 7, 9, 198, 199, **200**, 201, 202, 204, 209, 211, 212, 214, 216, 217, 223, 224, 227, 228, 230, 248, 262, 280, 289, 302, 339, 358, 381
—, from cholesterol, D 204
—, isomerides, D 209
—, *nomenclature*, D *439*
—, total synthesis, D 204, 206
Estrone a, D 204, 205
Estrone b, D 205
Estrone d, D 205
Estrone e, D 205
Estrone f, D 205
Estrone g, D 205
Estrone h, D 205
Δ^6-Estrone, D 262
Δ^7-Estrone, D 216
Estrone acetate, D 209, 226
Estrone acetate cyanohydrin, D 280
Estrone methyl ether, D 202, 203, 204, 207, 208, 210, 215, 222, 245, 336, 339
Estrone 3-methyl ether 17-enol acetate, D 339
Estrone-quinoline complex, D 199
Estrone tetrahydropyranyl ether, D 215
Ethane, conformation, A 6
Ethane-1,2-diol, D 92
Ethanedithiol B 110; D 92
Ethanetetracarboxylic ester, A 22, 97; B 144
17α-Ethinyl-3,17β-dihydroxyestra-1,3,5(10)-triene, D 201
17α-Ethinylestradiol, D 201, 214, 241
17α-Ethinylestradiol-3-methyl ether, D 241
17α-Ethinylestr-4-en-17β-ol, D 241
17α-Ethinyltestosterone, D 239, 240, 271
Ethoxalylcyclopentadienylsodium, A 120
2-Ethoxalyltestosterone, D 340
Ethoxyacetylene, A 76; B 91; D 258
3-Ethoxyanhydrovitamin A_1, B 292
3-Ethoxyandrosta-2,5-dien-17-one, D 259
1-Ethoxybicyclo[3.2.0]hepta-3,6-diene, C 45
2-Ethoxybuta-1,3-diene, B 151
Ethoxycarbonylcarbene, A 26, 33
2-Ethoxycarbonylcyclohexanone, C 124
2-Ethoxycarbonylcyclopropylidene acetic ester, A 60
3-Ethoxycarbonyl-1,2-diphenylcyclopropene, C 33
1-Ethoxycarbonyl-2-hydrocyclopentyl butyric ester, C 28
2-Ethoxycarbonyl-1-methyl-bicyclopropyl, C 8
4-Ethoxycarbonylmethylene-2,2-diphenylcyclopropene, A 37
10-Ethoxycarbonyl-$\Delta^{1(9)}$-octal-2-one, C 64
5-Ethoxycarbonylspiro[5.5]undecane-2,4-diene, C 31
3β-Ethoxy-5α-cholan-24-oic acid, D *438*
6-Ethoxy-3β,5-cyclo-5β-cholestane, D 117
3-Ethoxycyclohept-1-ene, B 352
3-Ethoxycyclohex-1-ene, B 27
Ethoxycyclo-octatetraene, B 396
5-Ethoxycyclopent-2-en-1-one, A 186
2-Ethoxycyclopropanecarboxylic ester, A 56
1-Ethoxy-4,4-dimethylcyclobuten-3-one, A 76
β-Ethoxyethanol, C 206
Ethoxyethynylmagnesium bromide, D 329
3-Ethoxy-1-hydroxyadamantane, C 127
3-Ethoxypenta-1,3-diene, D 328
3-Ethoxy-Δ^3-5α-steroids, D 88
3-Ethoxy-3,4,4-trifluorocyclobutene, A 89
Ethylamine, B 3
24α-Ethylcholesta-5,22-dien-3β-ol, D 75
24α-Ethylcholestane, D 3
24β-Ethylcholestane, D 3
24α-Ethyl-5α-cholestan-3α-ol, D 75
24α-Ethyl-5α-cholestan-3β-ol, D 75
24β-Ethyl-5α-cholestan-3β-ol, D 75, 158
24α-Ethylcholest-5-en-3β-ol, D 13, 75, 154
Ethylcyclobutane, A 80
Ethylcyclohexa-1,3-dienes, B 26
Ethylcyclohexane, B 14
2-Ethylcyclohexane-1,3-dione, B 109
Ethylcyclohexane-2,5-dione-1,4-dicarboxylate, B 112
Ethylcyclohexan-1-ols, B 47
2-Ethylcyclohexanone, B 12
3-Ethylcyclohexanone, B 95
Ethylcyclohex-1-enes, B 3, 17, 21
3-Ethylcyclohex-2-enone, B 117

Ethylcyclopentane, A 113
1-Ethylcyclopentan-1-ol, A 157
Ethylcyclopent-1-enes, A 116
1-Ethylcyclopentyl acetate, A 117
Ethylcyclopropane, A 26, 32
3-Ethyl-2,5-dimethylnaphthalene, C 338
1-Ethyl-3,5-dimethylcyclohexane, B 14
Ethyl 1,1-dimethylpentane-2,4-dione-1,3,5-tricarboxylate, B 113
Ethylene, C 105; E 96
—, conversion to cyclohexene, B 16
—, reaction with cyclohexadiene, C 117
—, — with cyclopentadiene, C 89
—, — with furan, B 73
Ethylenedecalin, C 116
Ethylene dibromide, A 57, 58; C 23
3,3-Ethylenedioxy-4,6-dienes, D 92
3,3-Ethylenedioxy-5,7-dienes, D 92
8-Ethylenedioxy-1-methoxy-10a-methyldecahydrochrysene, D 259
24-Ethylenelophenol, D 166
Ethylenic ketones, D 127
13-Ethyl-17-ethynyl-3-methoxygona-1,3,5(10),8-tetra-en-17-ol, D 274
Ethylferrocene, A 143
2-Ethylglutaric acid, A 197
17β-Ethyl-D-homo-5α-androstane, D *446*
1-Ethylidenecyclohexane, B 17
Ethylidenecyclopentane, A 117
Ethylidenecyclopentane nitrosochloride, A 175
2,4-Ethylidenelophenol, C 408; E 107
Ethylidenetriphenylphosphorane, D 245
Ethylisopropylacetaldehyde, D 27, 132, 158
Ethylmagnesium bromide, A 42; D 207
—, reaction with imines, A 165
13-Ethyl-3-methoxygona-1,3,5(10), 8-tetra-en-17-one, D 274
Ethylmethylacetonitrile, A 63
2-Ethyl-3-methylbutanal, D 133, 158
4-Ethyl-4-methylcholesteryl mesylates, D 79
3-Ethyl-1-methylcyclobutane, A 80
1-Ethyl-3-methylcyclohexanes, B 14
Ethyl-4-methylcyclohexane-1,2-diols, B 51
Ethylmethylcyclohex-1-enes, B 21
Ethylmethylcyclohex-2-enones, B 117
2-Ethyl-1-methylcyclopentanes, A 113
1-Ethyl-1-methylcyclopropane, A 32
Ethylmethylfulvene, A 124
5-Ethyl-6-methylheptan-2-one, D 13, 27, 154
Ethyl methyl ketone, A 43
1-Ethyl-2-methylphenanthrene, D 223
1-Ethyl-2-methylphenanthr-7-ol, D 223
17α-Ethylnortestosterone, D 273
21-Ethylprogesterone, D 240
Ethyl propenyl ether, B 246, 247
17α-Ethyltestosterone, D 265
1-Ethyl-2,2,3,3-tetrafluorocyclobutane A 87
5-Ethyltropolone, B 369
Ethyl vinyl ether, A 56; B 246, 247
Ethyl vinyl ketone, B 317
Ethynyl carbenes, A 25, 34
1-Ethynylcycloheptanol, B 351
1-Ethynylcyclohept-1-ene, B 348
1-Ethynylcyclohexan-1-ols, B 32, **65**, 127, 129
1-Ethynylcyclohex-1-enes, B 32, 66
1-Ethynycyclohexene, C 13
1-Ethynylcyclopentanol, A 157, 175; C 13
Ethynylcyclopentene, A 157
Ethynylcyclopropane, A 35
17α-Ethynyl-6-dehydro-19-nortestosterone, D 346
1-Ethynyl-6,6-dimethylcyclohex-1-ene, B 32
17α-Ethynylestradiol, D 215
17γ-Ethynylestradiol, D 228
17α-Ethynylestradiol 3,17-bis tetrahydropropyranyl ether, D 215
Ethynylestradiol 3-cyclopentyl ether, D 215
17α-Ethynylestradiol 3-methyl-17-tetrahydropyranyl bis-ether, D 215
17α-Ethynylestr-5(10)-en-17-ol-3-one, D 240, 241, 242, 243, 244
17α-Ethynyl-6α-fluorotestosterone, D 346
17α-Ethynyl-6α-fluorotestosterone acetate, D 346
17α-Ethynyl-3-methoxyestra-2,5(10)-dien-17β-ol, D 242
1-Ethynyl-2-methylcyclohexan-1-ol, B 129
1-Ethynyl-2-methylcyclohex-1-ene, B 32
Ethynyl methyl ketone, B 130
17α-Ethynyl-7α-methyl-19-nortestosterone, D 243

17α-Ethynyl-18-methyl-19-nortestosterone, D 243, 347
17α-Ethynyl-18-norandrosta-3,5-dien-17β-ol acetate, D 344
17α-Ethynylnorethisterone, D 240
Ethynylnortestosterone, D 273, 345, 346
17α-Ethynyl-19-nortestosterone, D 241, 242, 243, 283, 346, 347, 350
17α-Ethynyl-19-nortestosterone acetate, D 242
Ethynyltestosterone, D 247, 308, 345, 346
1-Ethynyl-2,6,6-trimethylcyclohexan-1-ol, B 32
Etianic acids, D 191, 416
—, *nomenclature*, D *439*
Etianic acid, D 191, 360, 372, 388
5α-Etianic acid, D 179
5β-Etianic acid, D 169, 179
5α,17α-Etianic acid, D 7, 179
5β,17α-Etianic acid, D 7
5α,17(αH)Etianic acid, D *439*
5α,17(βH)-Etianic acid, D *440*
5β,17(αH)-Etianic acid, D *439*
5β,17(βH)-Etianic acid, D *440*
Eti-5-enic acid chloride, D 108
Etioallobilianic acid, E 7
Etioallocholanic acid, D 179
Etiobilianic acids, D 22, 23, **170**; E 7, 8
Etiocholane, D 6
—, derivatives, D 254, 263
Etiocholane-3α,17α-diol, D 264
Etiocholane-3α,17β-diol, D 264
Etiocholane-3β,17β-diol, D 264
Etiocholane-3,17-dione, D 265
Etiocholanic acid, D 169, 179
Etiocholanoic acid, D 191
Etiocholan-3α-ol, D 263
Etiocholan-3β-ol, D 263
Etiocholanolones, D 264, 269
Etiocholan-3-one, D 263
Etiocholan-17-one, D 263
Etiodeoxycholic acids, D 28
Eucalyptus camfieldi, B 130
Eucalyptus coneorifolia, B 183; E 77
Eucalyptus dives, C 138
Eucalyptus eudesmioides, C 319
Eucalyptus globulus, C 180, 186, 277, 319; E 67, 69
Eucalyptus globulus oil, B 208
Eucalyptus oil, B 166, 182, 204
Eucalyptus piperita, C 298
Eucalyptus species, B 198
Eucarvone, B **215**, 350, 355; C 46, **155**, 156, 157; E 69
Eudalene, C 256, 273, 274, 298, 299, 303, 304, 330
Eudesmol, C 256, **298**, 299, 300, 302, 304, 322, 334; E 85
α-Eudesmol, E 80
Eugenis caryophyllata, C 284
Euglena, B 332; E 127
Euglena gracilis, B 332, 340
Euglenineae, B 238
Eupatoriopicrin, C **365**, 366, 368
Euphanes, C 478
5α-Euphane, D *429*
Euphanetirucallane group, triterpenoids, C 410, 426
Euphols, C 410, 411, 426; E 115, 116
Euphorbia latex, E 105
Euphorbia peplus, E 106
Euphorbol, C 426
Eupteleogenin, C 442
Eurycomalactone, C 434
Evobioside, D 405
Evodia elleryana, C 318
Evodone, B **171**
Evomonoside, D 405
Evonogenin-[glucoside]$_2$-rhamnoside, D 405
Evonoside, D **405**
Evonymus europaea, D 405
Evoxine, E 63
Exaltone, B 409
Exocyclic cyclohexane alcohols, B 68
Exocyclic ketones, A 94

False cubebs, C 298
Faradiol, C 449
Farnesal, C 290
Farnesenes, C 258
α-Farnesene, C **259**
β-Farnesene, C **258**, 288
Farnesenic acid, C 290
Farnesiferol a, E 84
Farnesiferol b, E 84
Farnesol, C 259, **289**, 290, 334; E 54, 80, 81, 84, 85
Farnesol pyrophosphate, C 371, 372; E 81
Farnesyl bromide, B 277
Farnesyl pyrophosphate, B 341, 342, 343; C 460; E 66, 79, 80, 83, 84, 85, 93, 136, 137
Farnesyl pyrophosphate[1-^{3}H], C 467
Farnesyl pyrophosphate synthetase, E 66, 79, 85
Favorskii reaction, A 92; B 103, 134; C 42
Fecosterol, D **150**, 165

Feist's acid, A 59
Fenchanes, C 137, **243**
α-Fenchane, C 245
Fenchenes, C 198, 243
α-Fenchene, C 172, 235, 240, 244, 245, 248
β-Fenchene, C 245, 246, 248
γ-Fenchene, C 246, 249
δ-Fenchene, C 246, 249
ε-Fenchene, C 247, 249
ξ-Fenchene, C 247, 249
δ-Fenchene-3-carboxylic acid, C 242
γ-Fenchene chloride, C 245
α-Fenchenylanic acid, C 245
α-Fenchocamphorone, C 235, 245, 248
β-Fenchocamphorone, C 246, 249
Fencholamide, C 235
α-Fencholenamide, C 237
α-Fencholenic acid, C 236, 237
β-Fencholenic acid, C 237
γ-Fencholenic acid, C 236
Fencholenonitriles, C 236, 237
Fencholic acid, C 235, 250
Fenchone, C **233, 234,** 235, 236, 237, 238, 239, 243, 249, 250, 252
Fenchone hydrazone, C 243, 248
Fenchone oxime, C 236, 238, 244
Fenchyl alcohols, C 239, 240, 241
—, physical properties, C 240
Fenchyl alcohol, C 89, 172, 188, **238,** 247, 252
α-Fenchyl alcohol, C 163, 235, 244, 245, 246
β-Fenchyl alcohol, C 247
Fenchylamines, C 244
α-Fenchylamine, C 247
Fenchyl chloride, C 171, 244, 245
Fenchylene, C 246
α-Fenchyl methyl carbonate, C 248
Fennel oil, C 234
Fernane group, triterpenoids, C 455
Fernenes, C 417, 418, 455; E 119
Fernenol, C 455
Fern tannins, B 113
Ferricinium ions, A 125
Ferricinium salts, A 131
Ferrocene, A 120, 125, 127, 131, **132,** 135, 140, 145, 147, 148
—, halogenation, A 135
—, polymers, A 133
—, reactivities, A 135
—, sulphonation, A 134
—, synthesis, A 129
Ferrocene carborane analogue, A 144
Ferrocenecarboxylic acids, A 134, 136
Ferrocene-1,1′-disulphonic acid, A 134
Ferrocene-1,1′-disulphonyl chloride, A 145
Ferrocenyl acetate, A 135
Ferrocenylacetylene, A 133
1-Ferrocenylalkynes, A 133
Ferrocenylallenes, A 134
Ferrocenylamine, A **136**
Ferrocenylboronic acid, A 135
Ferrocenyl-1,1′-diboronic acid, A 135
Ferrocenyl-lithium, A 135
Ferrocenylmethylenetriphenylphosphorane, A 133
Ferruginol, C 378, 379, 392
Ferula pyramidata, C 314
Fesogenin, E **44,** 45
Festuclavine, E 66
Ficaprenols, E 133
Filicane group, triterpenoids, C 456
Filicene, C 417, 418, 456
Filicene epoxide, C 456
Filicic acid, B 113, 131
Filicinic acid, B 113
Filixic acid, B **131**
Finitin, C **346,** 367
Fish liver oils, B 282
Flacourtiacea, A 196
Flavaspidic acid, B 113, **131**
Flavobacterium dehydrogenans, E 132
Flavochrome, B **302**
Flavoglaucin, E 62, 63
Flavenoids, B 140
Flavoxanthin, B 296
Flavoxanthin diacetate, B 297
Flexibacter, species, B 318, 322
Flexianthin, B 258, 321, **322**
Flindersine, E 63
Flindissol, C 410, 411, 426
Floridin, D 96
Flumethasone, D 351, 352
Flumethasone pivalate, D 333
Fluorene, C 74
Fluorenone, C 92
6β-Fluoro-17α-acetoxy-9β,10α-pregn-4-ene-3,20-dione, D 348
Fluoroalkanedicarboxylic acids, B 38
Fluoroalkenes, A 76
Fluoroalkylpyrazolines, A 38
2α-Fluoroandrostan-17-ol-3-one, D 340
9α-Fluoroandrost-4-ene-3,11,17-trione, D 342
Fluoroanthene, C 76
1,2-Fluorobromocyclohexane, B 35, 40,
Fluorocarboxylic acids, B 42
Fluorocholestan-3-ones, D 122
Fluorocyclohexadienes, B 38

Fluorocyclohexanes, B 33, 38
2-Fluorocyclohexan-1-ol, B 35, 66
2-Fluorocyclohexanone, B 102, 104
Fluorocyclohexenes, B 38, **40**, 41
Fluorocyclopropanes, A 38
9α-Fluoro-1,2-dehydrocortisol, D 352
2α-Fluorodihydroallotestosterone, D 340, 341
2-Fluoro-3,5-dimethylcyclohexanone, B 104
5α-Fluoroestrane-10β,17β-diol-3-one, D 229
Fluoroferrocenes, A 135
9α-Fluorohydrocortisone, D 342
15β-Fluorohydrocortisone acetate, D 354
6α-Fluoro-17β-hydroxy-9β,10α-androst-4-en-3-one acetate, D 341
9α-Fluoro-11β-hydroxy-17α-methyltestosterone, D 342
9α-Fluoro-11β-hydrocytestosterone, D **342**
1-Fluoro-2-iodocyclohexane, B 33, 35
5α-Fluoro-10β-iodoestrane-3α,17β-diol, D 229
9α-Fluoro-17α-methylandrost-4-en-3,11-dione, D **342**
16α-Fluoro-16β-methylandrost-4-en-3,17-dione, D 343
2-Fluoro-4-methylcyclohexanone, B 104
9α-Fluoro-16β-methyl-1,2-dehydrocortisol, D **352**
6α-Fluoro-16α-methyl-1,2-dehydrocortisol acetate, D 351, 352
1-Fluoro-4-methylestra-1,3,5(10)-trien-17-one, D 356
9α-Fluoro-6α-methylpregna-1,4-diene-11β,17α-diol-3,20-dione-16α-acetic acid, γ-lactone, D 352
21-Fluoro-17α-methylpregna-4,6-diene-3,20-dione, D 350
2α-Fluoro-17α-methyltestosterone, D 340
16α-Fluoro-16β-methyltestosterone, D 343
10-Fluoronorandrosta-1,4-dien-17β-ol-3-one, D 343
10-Fluoro-19-norsteroids, D **342**
Fluoro-olefins, dimerisation, A 75
2-Fluoro-oxosteroids, B 102
9α-Fluoro-11-oxotestosterone, D 342
9α-Fluoroprednisolone, D 352, 354
16α-Fluoroprednisolone, D 353, 354
15β-Fluoropregna-1,4-diene-11β,17α,21-triol-3,20-dione 21-acetate, D 354
16α-Fluoropregna-1,4-diene-11β,17α,21-triol-3,20-dione 21-acetate, D 353
16β-Fluoropregna-1,4,17(20)-triene-11β,21-diol-3-one 21 acetates, D 353
6α-Fluoro-9β,10α-pregn-4-ene-3,20-dione, D 347
3β-Fluoropregn-5-en-17β-ol-20-one acetate, D 345
6α-Fluoro-9β,10α-progesterone, D 347
21-Fluoroprogesterone, D 350
2α-Fluorotestosterone, D 340
2-Fluorotropone, B 376
Fluoxymesterone, D 272, 273, **342**
Foeniculin, E 62, 63
Fokiena hadginsii, C 301
Foliachrome, B 329, 330
Foliaxanthin, B 258, **330**
Fomes fomentarius, D 159
Formaldehyde, B 157
—, reaction with acetoacetic ester, B 144
Formic acid, A 91; B 75, 157; D 416
3α-Formoxy-17α-hydroxypregnan-20-one, D 310
1-Formylaminoadamantane, C 131
17β-Formalandrostane-3β,17α-diol, D 307
6-Formylbicyclo[3.1.0]hex-2-ene, C 117
6-Formylbicyclo[3.2.2]non-8-ene, C 122
N-Formylcarvomenthylamine, B 229
β-Formylcinnamic ester, A 99
Formylcyclobutane, A 91
1-Formylcycloheptatriene dimethyl acetal, B 352, 395
1-Formylcyclohexa-1,3-diene, B 91
1-Formylcyclohexa-1,4-diene, B **90**
Formylcyclohexane, B **87**
2-Formylcyclohexanone, C 70
1-Formylcyclohexene, B 86, **89**
4-Formylcyclohex-1-ene, B 87, **90**
1-Formylcyclopent-1-ene, B 22, 75
Formylcyclopropane, A 47
(25*R*)-10-Formyl-des-A-spirostan-5-one, E **42**
1-Formylferrocene, A 132
3-Formyl-2-hydroxymethyl-1,1-dimethylcyclohex-3-ene, B 173
1-Formyl-4-hydroxy-2,6,6-trimethylcyclohex-1-ene, B 91
N-Formylmenthylamine, B 228

4-Formyl-1-methoxycyclohexene, B 91
6β-Formyl-B-nor-5β-cholestane-3β,5-diol 3-acetate, D 284
3-Formyltetracyclododec-8-ene, C 104
2-Formyl-2,4,4-trimethylcyclopentanone, A 166
Formyltropolones, B 373
Franseria dumosa, C 357
Fraxinellone, C 410, 412, 434
Free hemiterpenes, E 61
Frequentin, B **105**, 106
Freund reaction, A 80
Friedalin, E 119
Friedelan-y-al, C 447
Friedelane, C 447
Friedelane-diones, C 447
Friedelane group, triterpenoids, C 447
Friedelane-3,x,y-trione, C 447
Friedelanols, C 413, 414, 415, 447, 476, 477
Friedelan-3β-ol acetate, C 447
Friedel-Crafts reaction, D 208
Frugoside, D 364, **398**
D-Fucose, D 365
Fucostadienone, D 152
Fucosterol, D **152**, 153, 166; E 103, 105, 109
i-Fucosteryl methyl ether, D 152
Fucoxanthin, B 238, 241, 243, 258, 266, **328**, 330; E 128
Fucoxanthin acetate, B 329
Fucoxanthin monobenzoate, B 329
Fucus vesiculosus, B 328
Fuegin, C **349**, 367
Fuerstion, C 392
Fugenal, C 397
Fugenoic acid, C 397
Fujenal, C 396
Fujenoic acid, C 380, 396
Fulvalenes, C **11**
Fulvenes, A 84, 120, 123, 124, 128
Fulvulene, C 12
Fumaric acid, dinitrile, A 97
—, ester, A 61, 76; C 115
—, methyl ester, A 59; B 150
—, reaction with butadiene, B 146
—, — with cyclopentadiene, A 192
Fumaronitrile, B 146; C 123
Fumaroyl chloride, B 146, 151; C 115, 123
Fungal metabolites, C 460
Furans, B 31
Furan, reaction with ethylene, B 73
—, — with maleic anhydride, B 150
Furanodecalins, C 304
Furanoeremophilene, C **308**, 335
Furanoeremophilone, C **308**, 335
Furanoid oxides, B **232**
Furanoid sesquiterpenes, C 302
Furanopetasin, C **308**, 335
Furan rings, monoterpenes, E 78
Furostans, E 18
—, *nomenclature*, D *434*, *435*, *436*, *446*
Furostan, D 5, 7
5β,22ξ,25δ-Furostan-3β,26-diol D *435*
(20S,22ξ,25R)-5α-Furostan-3β,26-diol, D *436*
(20S,22ξ,25S)-5α-Furostan-26-ol, D *435*
Furostantriol, E 46
Furost-20(22)-en, D 327
—, *nomenclature*, D *435*
Furost-22(23)-en, E 45
(25R)-5α-Furost-20(22)-en-3β,26-diol, D *436*
(25R)-5β-Furost-20(22)-en-3β,26-diol, D *435*, *436*
Fusarim lini, D 419, **420**
Fusarium moniliforme, E 90
Fusarium solani, E 19
Fuscin, E 63, 64
Fused cyclic systems, C 32 ff
Fusicoccin, E 87, 92
Fusicossum amygdali, E 87, 92
Fusidane, D 163
Fusidane group, triterpenoids, C 420
Fusidic acid, C 406, 420; D 60, **162**, 163
Fusidium coccineum, D 162
Futronolide, C **349**, 367

Gafrinin, C **365**, 368
Gaillardia pulchella, C 360
D-Galactose, E 2
Gamabufogenin, D 414, **419**, 420
Gamabufotalin, D 13, 413, **419**
Gamabufotoxin, D 414, 416
Gammaoleananes, E 116
Gammexane, B 36
Garryfoline, C 380
Garryine, C 470, 471
Gascardic acid, C 403, 404, 405
Gascardic madagascariensis, C 403
Gas-liquid chromatography, bile acids, D 178
—, steroids, D 72
Gazania rigens, B 304
Gazaniaxanthin, B **304**
Gedunin, C 411, 429
Gedunin group, triterpenoids, C 429
Geigera africana, C 365

Geigera aspera, C 350, 360
Geigerin, C **350**, 351, 367
Geigerinin, C **360**, 368
Geijerene, C 263
Genins, D 360 ff
—, *nomenclature*, D *435*
Genipa americana, B 213
Genipic acid, B **213**
Genipin, B **213**
Genipic acid, B **213**
Gentrogenin, E 26, **34**, 36
Gentrogenin acetate, E 35
Geranic acid, B 167
Geraniol, B 154, 155, 156, **162**, 163, 167, 170; C 266; E 54, 67, 71, 75, 78
Geraniol pyrophosphate, C 371, 372
Geranium macrorhizum, C 329
Geranium oil, B 161
Geranyl acetate, B 163, 172
Geranyl acetone, E 96, 97
Geranylamine hydrochloride, B 157
Geranyl bromide, B 267, 270, 273, 279
Geranyl chloride, E 96, 97
Geranylfarnesol pyrophosphate, C 371, 372
Geranylfarnesyl pyrophosphate, C 403, 404; E 92
Geranylgeraniol, C 373, 374, 381, 461; E 87
Geranylgeraniol pyrophosphate, C 371, 372
Geranylgeranyl pyrophosphate, B 341, 342, 343; C 463; E 66, 85, 89, 90, 91, 92, 122, 123, 136, 137
Geranyl-linalool, C 373, 375
Geranyl pyrophosphate, B 155, 341; E 66, 79, 83, 85, 136, 137
Geranyl resorcinol, E 70
Geranyl transferase, E 66
Germacrane, C 329
Germacranolides, C 362
Germacrol, C 329
Germacrone, B 414; C **329**, 330, 336, 363
Germanicol, C 413, 414, 415, 438
Germanidiol, C 438
Geronic acid, B 205, 271, 340; C 282
Gestogens, D **230**
—, halogenation, D **345**
—, trienic, D 247
Gibberella fujikuroi, C 380
Gibberellic acid, C 380, 398, 460, 462; E 87, 88, 89, 90, 119
—, biosynthesis, C 461
Gibberellic acid, group, physical properties, C 398
Gibberellins, C 398, 470; E 87, 90
—, biosynthesis, E 89
Gibberellin A_{12}, C 461
Ginger oil, C 258
Girard reagent T, D 231
Girard test, D 199
Gitaloxigenin, D 388
Gitogenic acid, E 32, 36, 37
Gitonin, E 2, 5, 7, 25, **32**, 33, 34, 36, 37, 52, **112**
F-Gitonin, E 2
Gitorin, D 362, **391**
Gitoside, D **391**
Gitoxigenin, D 39, 60, 362, 388, **391**, 392, 393, 394, 418; E 111
Gitoxigenin 16-acetate, D 364, 391
Gitoxigenin-acetyldigitaloside-glucoside, D 391
Δ^{16}-Gitoxigenin-cymaroside-glucoside, D 391
Gitoxigenin diacetate, D 392, 383
Gitoxigenin-digitaloside, D 391
Gitoxigenin-digitaloside-glucoside, D 391
Gitoxigenin-D-digitoxoside, D 391
Gitoxigenin-[digitoxoside]$_3$, D 391
Gitoxigenin-[digitoxoside]$_3$-glycoside acetate, D 391
Gitoxigenin glucoside, D 391
α-Gitoxigenone, D 392
β-Gitoxigenone, D 392, 393
Gitoxin, D 360, 362, **391**
Glabrolide, C 442
Glaucarubin, C 433
Glaucarubinone, C 412, 433
Glaucarubolone, C 412, 433; E 120
Globicin, C **354**, 367
Globulol, C **319**, 320, 321, 336
Glochidiol, C 435
Glochidone, C 436
Glomerella fusarioides, D 210
Glucoconvallasaponin-B, E 3, 5
Glucodigifucoside, D **387**
Glucose, D 360, 363, 365, 367, 368, 383, 384, 385, 386, 395, 408, 411, 412
D-Glucose, D 361; E 2, 3
δ-Glucose, D 411
Glucuronic acid, C 217
Glutamic acid, A 70
Glutaraldehyde, reaction with nitromethane, B 80
Glutaric acid, B 75, 98
Glutaric dimethyl ester, C 11
Glutaric esters, A 197, 201
—, reaction with methylene iodide, A 73

Glutaric methyl ester, A 107
Glutinane group, triterpenoids, C 446
Glutinol, C 413, 414, 415, 446
Glutinone, C 446
Glyceraldehyde, A 66, 197; C 249
Glycine, D 169
Glyco-acids, D 169, 170
Glycochenodeoxycholic acid, D 177
Glycocholanic acid, D 177
Glycocholic acid, D 177
Glycodehydrocholic acid, D 177
Glycodehydrodeoxycholic acid, D 177
Glycodeoxycholic acid, D 177
1,2-Glycols, B 411
Glycolithocholic acid, D 177
Glycollic acid, C 348
Glycose, D 384
Glycosides, D 361
—, cardiac, D 360 ff
—, cardiotonic, D 360 ff
—, digitalised, D 360 ff
—, digitaloid, D 361, 362
—, squill, D 408
—, strophanthidin, D 382
β-Glycosides, D 361
Glycoursodeoxycholic acid, D 177
Glycoverodoxin, D 388
β-Glycuronidase, D 236
Glycyrrhetic acid, C 441
Glyoxal, B 246
Glyoxalates, A 188
Glyoxylic acid, D 416
Gofruside, D 364, **398**
Gomphocarpus fructicosus, D 398
Gonadotrophin, D 241
Gonanes, D 5, 6, 274
—, *nomenclature*, D *427*, *443*, *446*
Gonocaryum pyriforme, B 274
Gorlic acid, A 161, 196, 198
Gossypol, E 80, 83, 84
Gracillin, E 3
Grapeseed oil, D 165
Gratiogenin, C 409, 424
Grayanotoxins, C 380, 395, 397
Grignard reactions, D 25, 232
Grignard reagents, from cyclopropyl halides, A 40
—, preparation of cyclopropanes, A 21
Grindelic acid, C 383
DL-Griseofulvin, B 115
Guaia-3,7-diene, C **278**, 288
Guaianolides, C 258, 350
—, abnormal, C 355
Guaiazulene, C 277, 297, 298, 314, 316, 317, 327, 329, 350
Guiacolsulphonic acid, D 199
Guaiol, B 347; C 256, 314, 315, 316, 335
Guareschi imides, A 73; C 21, 23, 25
Guayule, C 316
Gummosgenin, C 439
Gurjun balsam, C 263, 274, 278
α-Gurjunene, C **278**, 289
β-Gurjunene, C **274**, 288
Gutta percha, E 135
Gypsogenin, C 411
Gypsogenin lactone, C 440

Haematoporphyrin, D 84
Hagemann's ester, B 144
Half-cage hydrocarbons, C **106**
"Half-sandwich" complexes, A 129
Halogenated steroids, D **338**
ω-Halogenoalkylmalonic ester, A 9
2-Halogenoallyl alcohols, A 39
6-Halogeno-9β,10α-androsta-1,4,6-trien-3-ones, D 341
1-Halogenobutadienes, B 147
Halogenocamphors, physical properties, C 211
3-Halogenocamphors, C 208
α-Halogenocamphors, C 29, 212, 224
π-Halogenocamphors, C 209
α-Halogenocamphor-π-sulphonic acids, C 210, 215
Halogenocarbenes, A 26, 55
Halogenocholestan-3-ones, physical properties, D 122
1-Halogenocycloalkenes, B 388, 401
Halogenocyclobutanes, A 87
α-Halogenocycloheptanones, B 354
6-Halogenocyclohexanepentols, B 67
Halogenocyclohexanes, B 32, 34, 46, 80, 149
—, physical constants, B 33
Halogenocyclohexanols, B **66**, 67
2-Halogenocyclohexan-1-ol, B 82
2-Halogenocyclohexan-1-ol-acetates, B 67
Halogenocyclohexanones, B 67, 94, 100, 102, 103, 104, 113
Halogenocyclohexenes, B 27, 40, 41, 62
Halogenocyclo-octatetraenes, B 395
Halogenocyclopentanones, A 112, 166, 168
Halogenocyclopentenes, A 121, 149
Halogenocyclopropanecarboxylic acids, A 55
α-Halogeno-esters, A 20; B 93
α-Halogeno-ethers, B 93
Halogenoferrocenes, A 135
2-Halogeno-1-hydroxymethylcyclohexanes, B 71

Halogenoketones, macrocyclic, B 410
α-Halogenoketones, B 36, 93, 417
Halogenomethanes, A **26**
2-Halogenomethylcyclopropane-carboxylic acids, A 56
α-Halogenonitriles, B 93
α-Halogeno-α′-nitrocamphors, C 212
2-Halogeno-7-oxabicyclo[2.2.1]-heptanes, B 73
α-Halogeno-oxosteroids, D 33
6-Halogeno-9β,10α-pregna-4,6-dien-3-ones, D 348
6-Halogeno-9β,10α-pregna-1,4,6-trien-3-ones, D 348
6-Halogenopregnenes, D 347
Halogenoprogesterones, D 346, 347
3-Halogenopropyl ethers, A 21
Halogenoquercitols, B 67, 68
2-Halogenotropones, B 368, 375, **376**, 377
Hammarsten test, D 178
Haplopappus heterophyllus, E 111
Hardwickia pinnata, C 269
Hardwickiic acid, C 387
Hechtia texensis, E 26
Hecogenic acid, E 34, 36
Hecogenin, D 285, 312, 315, 327; E 1, 26, **33**, 34, 35, 43
Hecogenone, E 34, 35
Hederagenin, C 440
Helenalin, C 357, **358**, 359, 361, 368
Helenien, B 296
Helenium amarum, C 360
Helenium aromaticum, C 360
Hellenium autumnale, C 358
Helenium bigelovii, C 360
Helenium flexuosum, C 359
Helenium linifolium, C 359
Helenium mexicanum, C 359, 361
Helenium microcephalum, C 358
Helenium pinnatifidum, C 339
Helenium species, C 358
Hellebrigenin, D 366, **411**; E 113
Hellebrigenin acetate, D 412
Hellebrigenin-19-acid, D 412
Hellebrigenin-rhammoside-glucoside, D 411
Hellibrigenol, B 366, 412
Hellebrin, D 366, 369, **411**; E 113
Helloborus atrorubens, E 113
Hell-Volhard-Zelinsky reaction, C 97
Helminthosporal, C **333**, 336; E 81
Helminthosporium sativium, C 333; E 81
Helminthosporium species, C 403
Heloniogenin, E 26, **35**, 36
Heloniopsis orientalis, E 26
Helvolic acid, C 406, 420; D **162**, 163
Helvolinic acid, C 420
Hemiterpenes, E **61** ff
Hemiterpenoids, biosynthesis, E 65
Heptachlor, C **102**
Heptacyclo-octadecane, C 135
1,17-Heptadecanedioic acid, B 414
Heptadecapeptide, D 287
1,17-Heptadec-8-enedioic ester, B 415
Heptadec-8-ynoic acid, A 70
1,3-Heptadiene, A 35
Hepta-1,6-diene-3-one, C 85
Heptafluorocyclohexadienes, B 43
Heptafluorocyclohexane, B 39
Heptafluorocyclohexenes, B 42, 43
Heptafluorocyclopentanes, A 150, 151
Heptafluorocyclopentenes, A 151
Heptafulvalene, B 360
Heptafulvene, B **350**, 364
Heptalene, C 73
Heptalenium fluoroborate, C 73
Heptatetracarboxylic acid, C 28
n-Heptane-2,3,6-tricarboxylic acid, C 341
Heptan-2-one, B 74
Hepta-1,2,6-triene, C 22
n-Heptylcyclohexane, B 14
1-*n*-Heptylcyclopent-1-ene, A 116
Heterocholestenone, D 77
Hevea (rubber) latex, E 87
Hevea species, E 135
Hexa-alkylcyclohexa-2,4-dien-1-ones, B 121
Hexabromocyclobutane, A 87, 88
Hexabromocyclododecane, B 402
Hexabromocyclohexanes, B **37**
Hexa-(α-bromoethyl)-benzene, B 25
Hexachlorobicyclo[4.2.0]octane, C 47
Hexachlorobirdcage hydrocarbon, C 107
Hexachlorocyclohexanes, B 18, 36
Hexachlorocyclopentadiene, A 119, 150, 152
Hexachlorocyclopentan-1-ol-1-carboxylic acid, A 198
Hexachlorocyclopentenone, A 152, 171
Hexachlorodeca-*endo*-5-ol, C 107
Hexachlorohexacyclododecane, C 106
Hexachloro-1-hydroxycyclopentane-1-carboxylic acids, A 171
Hexachloropentacyclododecan-5-one, C 107
Hexacyclododecane, C 106
Hexacyclopentadienes, C 103
Hexadecahydrochrysene, C 76

Hexadecahydropyrene, C 76
Hexadecane-2,15-dione, B 416
Hexa-1,5-diene, C 42, 84, 372
Hexa-1,5-diyne, A 84
Hexaethylbenzenes, B 25, 121
Hexaethylidenecyclohexane, B 25
Hexafluorobenzene, C 43
Hexafluorobicyclo[2.2.0]hexa-2,5-diene, C 43
Hexafluorobutadiene, C 44
Hexafluorobut-2-yne, C 110
Hexafluorocyclobutene, A 87, 88
Hexafluorocyclohexadienes, B 43
Hexafluorocyclopentanes, A 151
Hexafluorocyclopentenes, A 151
Hexafluorocyclopropane, A 40
Hexahydroaniline, B 80
Hexahydroanisole, B 48
Hexahydrobenzaldehyde, B 87
Hexahydrobenz[2,3]-5α-androstane, B *452*
Hexahydrobenzoyl chloride, B 15
Hexahydrobenzyl alcohol, B 68
Hexahydrobenzylideneacetophenone, B 89
Hexahydrocuminic acid, B 195
Hexahydro-*p*-cymene, B 216
Hexahydrodimethanonaphthalene, C 105, 107
Hexahydro-1,4-ethanonaphthalene, C 115
Hexahydrogeranyl geraniol, E 87
Hexahydrohumulene, C 283
Hexahydroisonaphthalic acid, C 88
Hexahydroisophthalic acids, B 143, 145
Hexahydroisophthaloyl chloride, C 89
Hexahydroisophthalyl diacetate, B 25
Hexahydronaphthalenes, C **70**
Hexahydrophthalic acids, B 142, 144, 395
Hexahydrophthalyl diacetate, B 24
Hexahydropyrethrone, A 179, 180
Hexahydrosalicylic acids, B 138
Hexahydroscillaren A, 409
Hexahydroterephthalic acid, B 145
Hexahydrothujic acid, B **216**
Hexahydroxybenzene, B 60
Hexahydroxy-5β-cholestane, D 194, 195
Hexahydroxycyclohexanes, B 44
Hexahydrozingiberene, C 260
Hexalindione, C 71
Hexalins, C **70**
Hexamethylbenzene, B 121
Hexamethylcyclohexa-3,5-dien-1-one, B **123**
Hexamethylcyclohexane, B 14
Hexamethylcyclohexane-1,3,5-trione, B 113
Hexamethylradialene, B **25**
Hexane-1,2,5,6-tetracarboxylic acid, C 54, **112**
Hexan-2-one, A 112
Hexan-5-one-1-carboxylic acid, A 189
Hexaphenylbenzene, A 38, C 8
Hexaphenyldihydropentalene, C 52
Hexaphenylfulvalene, C **12**
Hexaphenylpentalene, C 52
Hexaphenyltriapentafulvalene, C **13**
Hexa-1,3,5-triene, B **27**
Hex-5-enoic acid, B 116
n-Hexylcyclohexane, B 14
n-Hexylcyclopentane, A 112, 113
1-*n*-Hexylcyclopent-1-ene, A 116
2-Hexyl-1-cyclopropanedec-9-enoic acid, A 69
Hibaene, C 379, 399, 470
—, synthesis, C **469**
Himachalenes, C **281,** 289
β-Himachalene, C **282**
Himatantandra baceata, C 320
Himbaccol, C **320**
Hinesol, C **322,** 323, 336
Hinokiic acid, C **275,** 288
Hinokiol, C 392
Hinokione, C 392
Hinokitiol, B 365
Hirtin, C 428
Hispidin, E 3
Hispidogenin, E 3, 26
HMG = β-Hydroxy-β-methylglutaryl, E 59
HMG-ACP = β-Hydroxy-β-methylglutaryl aryl-acyl carrier protein, E 57
HMG-CoA = β-Hydroxy-β-methylglutaryl-co-enzyme-A, B 155; E 55, 57, 58, 59
HMG-CoA reductase, E 58
Hofmann degradation, B 80, 86; C 108; D 254, 312
Hog bile, D 169
Holaphyllamine, E 111
Holaphylline, E 110
Holarrhena, E 111
Holarrhena floribunda, E 110
Holarrhimine, D 2
Holothurinogenins, C 421
Holothurinogenin I, C 408
Homoadamantane, C **133**
Homoadamantylcarbonium ions, C 133

D-Homoandrostane, D 279
A-*Homo-5α-androstane*, D 445
D-Homo-5α-androstane, D 444
C(14a)-Homo-5α-androstane, D 445
BC(8a)-Homo-5α-androstane, D 445
D-Homoandrostane derivatives, D 280
B-Homoandrostane-3β,17β-diol, D 277
D-Homoandrostane-3,17-diones, D 279
D-Homoandrosterone, D 279
Homocamphor, C 220
Homocamphoric acid, C 201, 202, 204, 224, 328
Homocamphoronic acid, C 203
Homocamphorquinone, C 220
Homocaronic acid, C 151
Homocaryophyllenic acid, A 97; C 284, 331, 332
D-Homocholane, D 445
A-Homo-5α-cholestan-4-one, D 119
A-Homodihydrotestosterone, D 276
B-Homodihydrotestosterone D 277
D-Homodihydrotestosterone, D 280
D-Homoepiandrosterone, D 279, 280
Homoepicamphor, C 220
D-Homoquilenin, D 280
D-Homoestradiol, D 280
D-Homoestrogens, D 280
D-Homoestrone, D 280
δ-Homoestrone, D 209
(22β)-D(17a)-Homo-5β-furostan, D 446
Homomarrianolic acid methyl ether, D 208
Homomycin, B 84
B(9a)-Homo-19-nor-5α,10a(H)-pregnane, D 445
Homophthalic acid, B 149
D-*Homopregnane*, D 445, 446
D-Homo-sex hormones, D 279
Homosteroids, D **276, 277**
Homoterpenylformic acid, C 169
Homoterpenylic acid, C 153, 169, 175
Homoterpenyl methyl ketone, B 188; C 145, 152, 168, 250, 252
A-Homotestosterone propionate, D 276
Homothujadicarboxylic acid, C 139, 148, 254
Homotropone, C 40
Homotropylium ion, B394; C 40
Hongheloside A, D 364, **391**, 394
Hongheloside B, D **391**
Hongheloside C, D 364, **391**
Hongheloside G, D **386**
Hopanes, C 376
Hopane group, triterpenoids, C 452
Hopenone-I, C 474
Hop resin, B 131, 132
Hops oil, C 258
Hormones, adrenocorticol, D 2, **286**
—, insect moulting, D 3
—, steroid, D 336
Hortincyclanone, C 85
Hosta kiyosumiensis, E 52
Hosta species, E 113
Huang-Minlon reduction, D 52, 108, 188
Huckel's rule, B 391
Hudson-Klyne rule, C 313
Human bile, D 169
Humbertia madagascariensis, C 292
Humbertiol, C **292**, 334
Humulane, C 283, 330
Humulenes, C **282**, 289
Humulene, B 414; C **282**, 283, 284, 287, **330**
α-Humulene, C 283; E 80, 85
β-Humulene, C 283
Humulene epoxides, C **331**, 336
Humulenol, C **331**, 336
Humulone, B **131**; E 65
Humulus lupulus, C 282
Hunsdiecker reaction, A 39; B 32, 35, 407; C 26
Hydantoins, D 358
Hydnocarpic acid, A 104, 161, 193, 196, 197, 198
Hydnocarpus alcalea, A 196
2-Hydrazinotropones, B 376
Hydrindan-5,6-diols, C 57
Hydrindan-1,7-dione, C 59
Hydrindanes, C 4, 5, **55**, 56, 57
Hydrindanols, C 57
Hydrindanones, C 57, 60
Hydrindan-1-one, C 58
Hydrindan-2-one, C 58
α-Hydrindanone, C 56, 58; D 12, 23
β-Hydrindanone, C 56, 57, 58
Hydrindanylamines, C 57, 58
Hydrind-2,5-diene-1-one, C 59
Hydrind-5-ene-1,3-dione, C 59
Hydrind-5-ene-1-one, C 58
Hydrocarbons, biosynthesis, B 333
—, macrocyclic, B 380 ff, 381
Hydrochrysene, D 303
Hydrocortisone, D 331, 352, 354
Hydrogenation, benzene compounds, B 1
Hydromellitic acid, B 152
Hydronaphthalene derivatives, sesquiterpenoids, C 296
Δ^5-7α-Hydroperoxides, D 84

5-Hydroperoxy-5α-cholest-6-en-3β-ol, D 96
6α-Hydroperoxycholest-4-en-3-one, D 128
6β-Hydroperoxycholest-4-en-3-one, D 130
7α-Hydroperoxycholesterol, D 96
3-Hydroperoxycyclohex-1-enes, B 75
3-Hydroperoxy-2,3-dimethylcyclohex-1-ene, B 19
10β-Hydroperoxy-17α-ethynylestr-4-en-17β-ol-3-one, D 244
10-Hydroperoxy-19-nor-Δ^4-3-oxosteroids, D 243
Hydroquinone, D 352
—, hydrogenation, B 54
—, reaction with maleic anhydride, C 115
3β-Hydroxy-11α-acetoxy-5β-etianic acid, D 39
19-Hydroxy-17α-acetoxyprogesterone, D 245
8-Hydroxyachillan, C 354
1-Hydroxyadamantanes, C 127
3-Hydroxyadamantane, C 133
α-Hydroxyadipic acid, B 75
3α-Hydroxyeticholane-11,17-dione, D 324, 325
3β-Hydroxyallocholanic acid, D 250, 256, 409
3β-Hydroxyallochol-14-enic acid, D 409
3β-Hydroxyalloetianic acid, D 308
19-Hydroxyalloetianic acid, D 374
3β-Hydroxyallopregnane-11,20-dione, D 326, 327
17β-Hydroxyandrosta-4,9(11)-dien-3-one, D 302
3α-Hydroxy-5α-androstan-17-one, nomenclature, D *438*, *439*
3α-Hydroxy-5α-androstan-17-one, D 36, 37
3β-Hydroxy-5α-androstan-17-one, E 21, 22, 23
17β-Hydroxy-5α-androstan-3-one, D 185
17β-Hydroxyandrosta-1,4,6-triene-3-one, D 282
19-Hydroxyandrost-4-ene-3,17-dione, D 204 ,334, 335
3β-Hydroxyandrost-5-en-17-one, D 38
17β-Hydroxy-4-androsten-3-one, D *438*, *439*, *440*, *442*
17β-Hydroxy-8α,9β-androst-4-en-3-one, D *442*
17α-Hydroxy-13α,14β-androst-4-en-3-one, D *441*
N-(3β-Hydroxyandrost-5-en-17-ylidene)pyrrolidinium p-toluenesulphonate, D 357
3-Hydroxy-β-apo-8′-carotenal, B 322; E 132
Hydroxybalchanolide, C 365, 368
7β-Hydroxy-A-barrigenol, C 439
Hydroxybenzsuberane, C 72
α-Hydrobenzylcyclohexanones, B 100
3-α-Hydroxybenzyl-2-phenylcyclopropanecarboxylic acids, A 56
1-Hydroxybicyclo[2.2.1]heptane, C 91
2-Hydroxybicycloheptyl, C 18
2-Hydroxybicyclohexyl, C 14, **15**
4-Hydroxybicyclohexyl, C 15
2-Hydroxybiphenyl, C 15
3β-Hydroxybisnorallocholanic acid, D 250
3β-Hydroxybisnor-5α-cholanic acid, D 158
Hydroxybisnorcholanoic acids, D 191
3β-Hydroxybisnorchol-5-enic acid, D 158, 231
3β-Hydroxybisnorchol-5-enoic acid, D 132, 152, 155, 191, 192, 232
5-Hydroxyborneol, C 223
Hydroxybovogenin A, D 366
16β-Hydroxybovoside A, D 412
3β-Hydroxy-5α-bromo-6,19-oxide-androstane-17-one, D 242
3-Hydroxybuta-1,3-diene-1-carboxylic acid lactone, A 103
2-Hydroxybutyraldehyde, A 48
Hydroxycampholic acid, C 201
Hydroxycamphors, C **216**, 218, 220, 223
—, physical properties, C 217
3-Hydroxycamphor, C 216, 217; E 77
5-Hydroxycamphor, C 217, 222; E 77
10-Hydroxycamphor, C 218
α-Hydroxycamphor, C 216, 217
π-Hydroxycamphor, C 217
α-Hydroxycamphor-α-carboxylic acid, C 216
N-Hydroxycamphorimide, C 213
Hydroxycarbaldehydes, B **90**
14β-Hydroxycardenolides, D 389
3-Hydroxy-α-carotene, B 331
4-Hydroxy-α-carotene, B 263
4-Hydroxy-β-carotene, B 263, 302, 316, 317
Hydroxycarotenoids, B 306
1-Hydroxycarvomenthol, B 221
Hydroxycarvomenthone, B 222

1-Hydroxy-3-chloroadamantane, C 127
3α-Hydroxy-5β-cholanic acid, D 9
12α-Hydroxycholanic acid, D 312
Hydroxycholanoic acids, D 171, 180 ff
3α-Hydroxy-5α-cholanoic acid, D 17, 185
3α-Hydroxy-5β-cholanoic acid, D 181, 185
3α-Hydroxy-5β-cholan-24-oicacid, D*439*
7α-Hydroxy-5β-cholanoic acid, D 21, 181
12α-Hydroxy-5β-cholanoic acid, D 182, 184
12β-Hydroxy-5β-cholan-24-oic acid, D *448*
3β-Hydroxy-5α-cholano-24,17α-lactone, D *437*
25-Hydroxycholecalciferol, D 146
3α-Hydroxychol-11-enic acid, D 315
3β-Hydroxychol-5-enic acid, D 180, 233, 257
3α-Hydroxy-5β-chol-9-enoic acid, D 188
3α-Hydroxy-5β-chol-11-enoic acid, D 56, 188
3β-Hydroxychol-5-enoic acid, D 183, 185, 189
3-Hydroxycholesta-3,5-dien-7-one, D 132
Hydroxycholestanes, D 31
5α-Hydroxycholestane-3,6-dione, D 45
3β-Hydroxycholestane-6:7-seco-dioic acid, D 37
2α-Hydroxy-5α-cholestan-3-one, D 124
3β-Hydroxy-5α-cholestan-2-one, D 124
3β-Hydroxy-5α-cholestan-6-one, D 17, 18
3β-Hydroxy-5α-cholestan-15-one, D 90
5-Hydroxy-5α-cholestan-2-one, D 117
7α-Hydroxy-5β-cholestan-3-one, D 51, 52
3β-Hydroxycholest-5-enoic acid, D 189
Hydroxycholestenone, D 36
2-Hydroxy-5α-cholest-1-en-3-one, D 124
3-Hydroxy-5α-cholest-3-en-2-one, D 124
3β-Hydroxycholest-5-en-7-one, D **109**
3β-Hydroxycholest-5-en-22-one, D **109**
3β-Hydroxycholest-5-en-23-one, D 109
3β-Hydroxycholest-7-en-6-one, D 164
4-Hydroxycholest-4-en-3-one, D 124
5-Hydroxy-5α-cholest-3-en-2-one, D 117
6α-Hydroxycholest-4-en-3-one, D 49
6β-Hydroxycholest-4-en-3-one, D 128
7α-Hydroxycholest-4-en-3-one, D 51, 52
3β-Hydroxy-5α-cholest-7-en-6-one, ultraviolet absorption spectra, D 68
Hydroxycholesterols, D 85, **107**; E 113
7-Hydroxycholesterols, D 51, 76, 96, 108, 146, 184
22-Hydroxycholesterol, D 164
25-Hydroxycholesterol, D 96, 158
25-Hydroxycholesteryl acetate, D 158
7α-Hydroxycholesteryl benzoate, D 51, 52
6′-Hydroxyconvallatoxin, D **383**
11-Hydroxycorticosteroids, D 334
17α-Hydroxycorticosterone, D 287, 290, 294
17α-Hydroxycorticosterone cortisol, D 289
Hydroxycostunolide, C 364, 368
2-Hydroxycycloalkyltrimethyl-ammonium hydroxides, A 16
2-Hydroxycyclobutanones, A 78
6β-Hydroxy-3α,5-cyclo-5α-cholestane, D 43
4-Hydroxy-β-cyclocitral, B 91
2-Hydroxycyclodecanone, B 412
6-Hydroxycyclodecanone, B 411
2-Hydroxycyclododecanone, B 412
2-Hydroxycyclo-eicosanone, 412
2-Hydroxycycloheptadecanone, B 412
1-Hydroxycycloheptanecarboxylic acid, B **357**
Hydroxycycloheptanones, B 355, 367
2-Hydroxycyclohepta-2,4,6-trienone, B 359
2-Hydroxycyclohexadecanone, B 412
4-Hydroxycyclohexane, B 355
Hydroxycyclohexanecarboxylic acids, B 49, 83, 138
Hydroxycyclohexanedicarboxylic acids, B 150
Hydroxycyclohexanones, B 103
2-Hydroxycyclohexanone, B 53, 110
Hydroxycyclohexylalkanecarboxylic acids, B 142
3-(1′-Hydroxycyclohexyl)-prop-1-yne, B 66
1-(1′-Hydroxycyclohexyl)-prop-1-yn-3-ol, B 66
2-Hydroxycyclononanone, B 412
2-Hydroxycyclo-octadecanone, B 412
2-Hydroxycyclo-octanone, B 412
2-Hydroxycyclopentadecanone, B 412
1-Hydroxycyclopentane-1-carboxylic acid, A 191

1(or α)-Hydroxycyclopentanone, A 166
2-Hydroxycyclopentanone, A 184; C 11
Hydroxycyclopentenones, A 176
1-(2′-Hydroxycyclopentylidene)-
2-hydroxycyclopentan-3-one, C 11
Hydroxycyclopropanecarboxylic
acids, A 56
20α-Hydroxycyclopseudosarsasa-
pogenin, E 17
6-Hydroxy-3α,5-cyclosteroids, D 79
2-Hydroxycyclotetradecanone, B 412
2-Hydroxycyclotridecanone, B 412
2-Hydroxycyclo-undecanone, B 412
17α-Hydroxy-11-dehydrocortico-
sterone, D 287, 289, 294
17α-Hydroxy-11-deoxycorticosterone,
D 287, 289, 310
2-Hydroxydicyclopentyl, C 69
4-Hydroxy-4,6-dien-3-ones, D 94
15α-Hydroxydigitoxigenin, D 373
17α-Hydroxydigitoxigenin 3-acetate,
D 389
*3-Hydroxy-D-dihomo-1,3,5(10)-
estratrien-17b-one*, D 445
4-Hydroxydihydroagarofuran, C 304,
335
3-α-Hydroxydihydroalantolactone,
C 340
1′-Hydroxy-1′,2′-dihydro-γ-carotene,
B 304
Hydroxydihydro-eremophilone, C 306,
307
Hydroxydihydrospheroidene, E 130
Hydroxydihydrospheroidenone, E 130
17β-Hydroxy-4,4-dimethylalandrost-
5-en-3-one, D 286
1-Hydroxy-4,4-dimethylcyclobuten-
3-one, A 92
17β-Hydroxy-4,4-dimethyl-2-
oximinoandrost-5-en-3-one, D 286
17α-Hydroxy-3,11-dioxoalloetianic
acid, D 290, 291
3α-Hydroxy-7,12-dioxo-5β-cholanoic
acid, D 181
17α-Hydroxy-3,11-dioxoeti-4-enic
acid, D 295
3β-Hydroxy-5,6-dioxo-5,6-secocholes-
tane, D 96
4-Hydroxy-3,4-diphenylcyclopent-
2-en-1-one, A 170
20-Hydroxyecdysone, D 161
Hydroxyechinenone, B 332
6β-Hydroxy-4-en-3-ones, D 94
15-Hydroxyeperuenoic acid, C 383
Hydroxyepicamphors, physical
properties, C 217
Hydroxyepicamphor, C 220
2-Hydroxyepicamphor, C 216, 217,
225
6-Hydroxyepicamphor, C 222
12α-Hydroxy-13-epimanoyl oxide,
C 382
*3β-Hydroxy-14,21ξ-epoxy-5β,14β,20ξ-
cardanolide*, D *431*
3β-Hydroxy-12α-20α,epoxy-14-iso-
17-isopregn-5-ene-11,15-dione,
D 403
Hydroxyeremophilane, C 349
Hydroxyeremophilone, C 306, 307
5-Hydroxy-5α-ergosta-7,22-diene-
3,6-dione, D 136
5-Hydroxy-5α-ergosta-7,22-dien-3-
one, D 136
5-Hydroxy-5α-ergostane-3,6-dione,
D 134
3β-Hydroxyergosta-5,8,22-trien-7-
one, D 137
7α-Hydroxyestradiol, D 210
15α-Hydroxyestradiol, D 210, 214
3-Hydroxyestra-1,3,5(10)-triene-
17β-carboxylic methyl ester, D 301
3-Hydroxy-1,3,5(10)-estratrien-17-one,
D *439*
17β-Hydroxyestr-5(10)-en-3-one,
D 229
7-Hydroxyestrone, D 210, 217
15α-Hydroxyestrone, D 210
16α-Hydroxyestrone, D 210
18-Hydroxyestrone, D 227
4-Hydroxyestrone 3-methyl ether,
D 216
1-(β-Hydroxyethoxy)-4-methyl-
estrones, D 356
Hydroxyetianic acids, D 191
3β-Hydroxyetianic acid, D 379, 388,
393, 416
3-Hydroxyeti-5-enoic acid, D 49, 192,
306
3β-Hydroxyetiobilianic acid, E 8, 29
3α-Hydroxyetiocholan-17-one, D 180,
255
3β-Hydroxyetiocholan-17-one, D 255,
263
3α-Hydroxyetiocholane-11,17-dione,
D 324, 325
Hydroxy-α-fenchenic acid, C 235
Hydroxy-β-fenchenic acid, C 246
Hydroxyferrocene, A 135, 136
Hydroxyfriedelanediones, C 447
Hydroxyfriedelan-3-ones, C 447
6-Hydroxyfulvenes, A 124
25-Hydroxygitogenin, E 113

19-Hydroxygitoxigenin, D 396
Hydroxygratiogenin, C 424
2-Hydroxyheptanoic acid, A 202
7-Hydroxyheptan-2-one, B 74
6-Hydroxyhexan-2-one, A 112
4-Hydroxyhex-5-enoic acid lactone, A 58
3-Hydroxyhomoadamantane, C 133
Hydroxyhopane, C 452
Hydroxyhopanone, C 452
3β-Hydroxy-Δ^6-5α-hydroperoxides, D 84
3β-Hydroxy-5α-hydroperoxycholest-6-ene, D 109
2-Hydroxy-1-2′-hydroxy-1′-cyclopentyl cyclopent-1-en-3-one, A 107
17α-Hydroxy-17β-hydroxymethylandrosta-1,4-dien-3-one, D 276
17α-Hydroxy-17β-hydroxymethylandrost-4-en-3-one, D 275
17β-Hydroxy-2-hydroxymethylene-5α-androstan-3-one, D 337
3β-Hydroxy-22-isoallospirostan-11-one, D 327
3β-Hydroxy-20-isobisnorchol-5-enoic acid, D 153
β-Hydroxyisobutyric acid, C 363
1-Hydroxyisocarvomenthol, B 221
14-Hydroxy-14-iso-11-deoxycorticosterone acetate, D 300
3β-Hydroxy-14-isoetianic acid, D 385
14β-Hydroxy-14-iso-etio-acids, D 372
5-Hydroxyisofenchol, C 242
Hydroxyisofenchones, C 242
3β-Hydroxyl-14-iso-17-isoallo-etianic acid, D 403
3β-Hydroxy-14-iso-17-isoetianic acid, D 388, 393
2-Hydroxyisopinocamphone, C 155
17β-Hydroxy-17-isoprogesterone, D 308
Hydroxyisopropyldihydrobenzofuran, E 64
9α-Hydroxyjurubidine, E 53
Hydroxykauranoic acid, C 395
Hydroxykaurenoic acid, C 396
7β-Hydroxykauren-19-oic acid, E 90
Hydroxykaurenolide, C 380, 396
Hydroxyketones, macrocyclic B **411**
—, terpenoid, B 202
20-Hydroxy-23,21-lactones, D 407
6′-Hydroxylutein-5,6-epoxide, B 332
2α-Hydroxymanoyl oxide, C 382
2-Hydroxy-*p*-menth-1-en-3-one, B 202
1-Hydroxy-1-methoxycyclopropane, A 48
6α-Hydroxy-3β-methoxy-5-methyl-19-nor-5β-cholest-9(10)-ene, D 85
1-(Hydroxymethyl)adamantane, C 133
17β-Hydroxy-17α-methylandrosta-4,9(11)-dien-3-one, D 342
17β-Hydroxy-17α-methylandrostano-[3.2.c] pyrazole, D 275
3-Hydroxy-4-methylbenzoic acid, B 191
2-(Hydroxymethyl)bicyclo[2.2.1]-heptane *p*-toluenesulphonate, C 119
2-Hydroxymethylbicyclo[2.2.1]-hept-5-ene, C 117
9-Hydroxy-9-methylbicyclo[4.2.1]-nonatriene, C 128
4-Hydroxy-3-methylbut-2-enadenosine, E 64
3β-Hydroxy-4α-methylcholest-7-en-24-one, D 157
1-(Hydroxymethyl)cycloheptanol, B 352
Hydroxymethylcyclohexane, B 89
Hydrocymethylcyclohexan-1-ols, B 69, **70**
1-Hydroxymethylcyclohexan-1-ol, B 24
2-Hydroxymethylcyclohexan-1-ol, B 22
Hydroxymethylcyclohexenes, B **71**
1-(Hydroxymethyl)cyclopentan-1-ol, A 161
2-(Hydroxymethyl)cyclopentan-1-ol, A 162
Hydroxymethylcyclopentanones, A 165
2-Hydroxymethylcyclopropanecarboxylic acid, A 56
9-Hydroxymethyldecalin, C 68
10-Hydroxymethyldecal-2-ol, C 64
17β-Hydroxy-2-methylene-5α-androstan-3-one, D 337
Hydroxymethylenecamphor, C 224, 232
2-Hydroxymethylene-5α-cholestan-3-one, D 123
2-Hydroxymethylenecholest-4-en-3-one, D 129
4-Hydroxymethylene[^{14}C-2-$^3H_2$]-cholest-7-en-3-one, E 101
2-Hydroxymethylenecyclohexanones, B 64, 97, 102
2-Hydroxymethylene-17α-methylandrostan-17β-ol-3-one, D273
2-Hydroxymethylene-17α-methyltestosterone, D 340

2-Hydroxymethylene-19-norandrostan-17-ol-3-one, D 282
1-Hydroxy-4-methylestra-1,3,5(10)-trien-17-one, D 356
1-Hydroxy-4-methylestrone, D 203, 356
β-Hydroxy-β-methylglutaric acid-[3-^{14}C], D 390
β-Hydroxy-β-methylglutaryl coenzyme A, B 155; E 55, 57, 58, 59
4α-Hydroxymethyl-17α-methylcholestan-3-ol, E 100
3β-Hydroxymethyl-A-norcholest-5-ene, D 80
16α-Hydroxymethylprogesterone, D 235
17α-Hydroxymethyltestosterone, D 275
4-Hydroxy-17α-methyltestosterone, D 273
3-Hydroxymethyltropolones, B 373
5-Hydroxy-4-methyltropolone, B 372
1-Hydroxyneocarvomenthol, B 221
7-Hydroxyneoergosterol, D 137
1-Hydroxyneoisocarvomenthol, B 221
17β-Hydroxy-A-norandrost-3(5)-en-1,2-dione, D 281
3β-Hydroxy-19-norandrost-5(10)-ene, D 85
Hydroxynorcholanoic acids, D 191
3β-Hydroxynor-5α-cholanoic acid, D 27, 28, 132, 134, 156
3-Hydroxynorchol-5-enoic acid, D 192
2-Hydroxy-A-nor-5α-cholestane-2-carboxylic acid, D 124
3β-Hydroxy-A-nor-5α-cholestane-3α-carboxylic acid, D 126
3β-Hydroxy-19-norergosta-5,7,9-triene, D 136
17a-β-Hydroxy-19-nor-D-homotestosterones, D 281
3-Hydroxynortricyclene, C 92, 93
17β-Hydroxy-4-oxa-5-androsten-3-one, D 448
19-Hydroxy-17α-oxa-D-homoandrost-4-ene-3,17-dione, D 227
3β-Hydroxy-4β,5-oxido-5β-cholestane, D 47
3β-Hydroxy-5,6α-oxido-5α-ergost-22-en-7-one, D 137
3β-Hydroxy-11-oxoalloetianic acid, D 296
3β-Hydroxy-16-oxobisnorcholanic acid, E 20
Hydroxyoxocarotenoids, B 312, 313
3-Hydroxy-6-oxo-5β-cholanoic acids, D 49
3α-Hydroxy-5,6-oxocholanoic acids, D 185
3α-Hydroxy-6-oxo-5α-cholanoic acid, D 185
3α-Hydroxy-6-oxo-5β-cholanoic acid, D 183, 185
3α-Hydroxy-7-oxo-5β-cholanoic acid, D 182, 184, 189
3α-Hydroxy-11-oxocholanoic acid, D 314, 315, 316
3β-Hydroxy-11-oxocholanoic acid, D 318
3α-Hydroxy-12-oxo-5β-cholanoic acid, D 182, 189
6α-Hydroxy-3-oxo-5β-cholanoic acid, D 183
12α-Hydroxy-7-oxo-5β-cholanoic acid, D 182
3α-Hydroxy-12-oxo-5β-cholanoic methyl ester, D 184
12α-Hydroxy-3-oxo-5β-cholanoic methyl ester, D 184
12α-Hydroxy-3-oxo-5β-cholenoic acid, D 11
1′-Hydroxy-4-oxo-1′,2′-dihydro-1′-γ-carotene, B 317
3-Hydroxy-17-oxo-estra-1,3,5(10),6,8-pentaene, D 201
3-Hydroxy-17-oxo-estra-1,3,5(10),7-tetraene, D 201
3-Hydroxy-17-oxo-estra-1,3,5(10)-triene, D 201
3α-Hydroxy-11-oxoetianic acid, D 317, 318
17α-Hydroxy-3-oxoeti-4-enic acid, D 275, 299
3-Hydroxy-17-oxo-14-iso-estra-1,3,5,(10),6,8-pentaene, D 201
19-Hydroxy-2-oxomanoyl oxide, C 383
3α-Hydroxy-11-oxonorcholanic methyl ester, D 317
12α-Hydroxy-11-oxoprogesterone, D 235
3β-Hydroxy-5-oxo-5,6-secocholestan-6-ol-acetate, D 284
17β-Hydroxy-5-oxo-3,5-seco-4-norestr-9-en-3-oic acid, D 302
5α-Hydroxy-6-oxosteroids, D 33
1-Hydroxy-4-pentanone, A 50
4-*p*-Hydroxyphenyl-1-butyl *p*-bromobenzenesulphonate, C 30
2-Hydroxy-1-phenylcyclobutene, dione, A 93

2-Hydroxy-2-phenylcyclohexan-1-one, B 53
2-*p*-Hydroxyphenylethyl bromide, C 24
ω-(*p*-Hydroxyphenyl)pentyl bromide, C 31
Hydroxyphthalic acids, B 150
Hydropinic acid, C 168
3β-Hydroxypregna-5,16-diene-21,20-dione, E 46
3β-Hydroxypregna-5,16-dien-20-one, D 232; E 10, 30
3β-Hydroxypregna-5,17(2)-dien-16-one, D 99
11β-Hydroxypregnanes, D 254
20-Hydroxypregnanes, D 324, 404
(20S)-16β-Hydroxypregnane-20-carboxylic acid lactone, D *426*
3α-Hydroxypregnane-11,20-dione, D 327, 328
3β-Hydroxy-5α-pregnane-6,20-dione, D 185
11α-Hydroxypregnane-3,20-dione, D 328
11-Hydroxy-5α-pregnane-3,20-diones, D 31
11α-Hydroxy-5β-pregnane-3,20-dione, D 38
20α-Hydroxy-5β-pregnane-3,11-dionoic acid, D 303
17α-Hydroxypregnanolone, D 251
16α-Hydroxypregn-17(20)-enes, E 23
(20R)-3β-Hydroxy-5-pregnene-20,18-carbolactone, D *437*
(20S)-3α-Hydroxy-5-pregnene-20-carboxylic acid, D *437*
21-Hydroxy-4-pregnene-3,20-dione, D *439*
19-Hydroxypregn-5-ene-3,20-dione, D 244
21-Hydroxypregn-17(20)-ene-3,11-dione, D 324
16α-Hydroxypregnenolone, D 60
17α-Hydroxypregnenolone 3,17-diacetate, D 244
3α-Hydroxy-5β-pregn-16-en-20-one, E 46
3β-Hydroxypregn-5-en-20-one, D 61, 231
3β-Hydroxy-17α-pregn-5-en-20-one, D 39
11-Hydroxyprogesterone, D 235
11α-Hydroxyprogesterone, D 38, 312, 327, 328
11β-Hydroxyprogesterone, D 235, 298
14α-Hydroxyprogesterone, E 111
15β-Hydroxyprogesterone, E 36
17α-Hydroxyprogesterone, D 235, 238, 239, 278, 289, **300**, **301**, 305, 308
19-Hydroxyprogesterone, D 244
21-Hydroxyprogesterone, D 238
2-(2-Hydroxy-2-propyl)-5-hydroxy-5-methyltetrahydrofuran, B 170
3-(2-Hydroxy-2-propyl)-A-norcholest-5-ene, D 79
12β-Hydroxyresibufogenin, D 419
3-Hydroxyretinene, B 260
16α-Hydroxysapogenin, E 44
27-Hydroxysapogenin, E 23
3-Hydroxy-16,17-seco-1,3,5(10)-estratriene-16,17-dioic acid, D *447*
11α-Hydroxyspirostans, E 43
25-Hydroxyspirostans, E 52
23-Hydroxyspirostan, E 50
(25*R*)-25-Hydroxyspirostan, E 46
(25*R*)-3β-Hydroxy-5α-spirostan-6-one, E **33**
(25*R*)-11α-Hydroxy-5β-spirostan-3-one, E **33**
(25*R*)-3β-Hydroxyspirost-4-en, E 42
Hydroxysteroids, D 34, 369
—, chromatography, D **32**
—, separation, D 40
3-Hydroxysteroids, 33, D 73
3-Hydroxy-Δ^4-steroids, D 69
3α-Hydroxysteroids, D 39
3α-Hydroxy-Δ^5-steroids, D 33
3β-Hydroxysteroids, D 39
3β-Hydroxy-Δ^5-steroids, D 42, 43, 46, 77, 84, 99
5β-Hydroxysteroids, D 47
7α-Hydroxysteroids, D 11
11α-Hydroxysteroids, D 11, 55, 312, 319, 321
11β-Hydroxysteroids, D 11, 55, 312
12-Hydroxysteroids, D 312
14β-Hydroxysteroids, D 58
16α-Hydroxysteroids, D 60
17-Hydroxysteroids, D 61, 304
20-Hydroxysteroids, D 6
7-Hydroxysterols, D 85
19-Hydroxy-Δ^5-sterols, D 86
12β-Hydroxystrophanthidin, D 400
16β-Hydroxystrophanthidin, D 384
2-Hydroxy-sugars, D 368
Hydroxyterephthalic acid, B 150
16β-Hydroxytestosterone, D 271
4-Hydroxytestosterone-17-acetate, D 283
19-Hydroxytestosterone diacetate, D 212
8-Hydroxy-α-tetralone, C 67

3-Hydroxytorulene, B 304
18-Hydroxytotarol, C 393
Hydroxytrachylobanic acid, C 400
3-Hydroxytricyclo[4.3.1.1^{4,8}]-undecane, C 133
4-Hydroxy-1,2,2-trimethylcyclopentane-1-carboxylic acid, B 323
5-Hydroxy-2,6,6-trimethyl-2-vinyltetrahydropyran, B 169, 170
6-Hydroxytropinone, B 362
Hydroxytropolones, B 372, 373
Hydroxytropones, B **365**, 367, 369
—, oxidation, B 371
—, properties, B 368
—, synthesis, B 367
2-Hydroxytropone, B 368
3-Hydroxytropone, B 371
Hydroxytropylium bromide, B 363
Hydroxytropylium ion, B 377
Hydroxyvalerenic acid, C 333
14α-Hydroxyviperidone, D 152
Hygromycin, B **84**
Hyocholic acid, D 172
Hyodeoxycholic acid, D 10, 16, 49, 169, 171, 183, **185**
Hyodeoxycholic methyl ester, D 188
Hypnea japonica, D 154
Hypoglycin A, A 70
Hypoglycin B, A 70
Hypophasic carotenoids, B 235, 240
Hyposantonin, C 340, 341, 342
Hyssop oil, C 176
Hyssopus officinalis, C 176

Ichangin, C 430
Icterogenin, C 441
DL-Idosaccharic acid, B 60
Iffaconic acid, C 450
Illicium religiosum, B 140
Imbricatolal, C 383
Imbricatolal acetate, C 383
Imbricatolic acid, C 383
Imbricatolic acid acetate, C 383
Imines, reaction with ethylmagnesium bromide, A 165; B 93
7-Iminocyclohepta-1,3,5-trienylamine, B 378, 379
Indacene, A 128
Indane, C 58
Indanones, C 57
Indene, hydrogenation, C 57
Indole alkaloids, E 70
Inertogenin, D **403**
Inertoside, D **403**
Infrared absorption spectra, steroids, D 65
Infrared spectra, carotenoids, B 256
Infrared spectrometry, steroids, D 32
Infrared spectroscopy, structure determination, C 479
Inhoffen synthesis, estrone, D 203
Inosadiamine, B 83
Inosamines, B 60, 74, **84**, 85, 86
Inoses, B 57, 58
Inositols, B 44, **58**, 60, **61**, 68, 73, 106
—, conformation, B 59
myo-Inositol, B **59**, 60
Inositol hexa-acetates, B 68
Inososes, B 60, **106**
—, imine derivatives, B 85
—, physical constants, B 107
Inosose oximes, B 58
Intergerrinecic acid, B **169**
Intermedioside, D 364, **398**, **403**
Intermedeol C **301**, 334
Inula helenium, C 337
Iodine monofluoride, B 35
Iodine-nitrobenzene reagent, B 367
N-Iodoacetamide, B 35
ω-Iodobutyric ester, A 164
Iodocamphors, physical properties, C 211
Iodocholestan-3-ones, D 122
2α-Iodo-5α-cholestan-3-one, D 120
2α-Iodocholest-4-en-3-one, D 120
4-Iodocyclobutane-1-carbonitriles, C 33
Iodocyclohexane, B 33
2-Iodocyclohexan-1-ol, A 163; B 66
4-Iodocyclohexan-1-ol, B 62
3-Iodocyclohex-1-ene, B 41, 42
Iodocyclononane, B 402
Iodocyclo-octane, B **402**
Iodocyclopentadienyltricarbonylmanganese, A 141
Iodocyclopropanes, A 39
2-Iodo-*N*,*N*-dimethylcyclohexylamine A **163**
ω-Iodo-esters, A 108
Iodoferrocenes, A 133, 135
Iodoform B 375
2-Iodomethylcyclopropane-carboxylic acid, A 56
Iodomethylzinc iodide, A 42
21-Iodoprogesterone, D 350
Iodoquercitols, B 68
1-Iodo-2-trifluoromethyldecafluorocyclohexane, B 42
Iodotropolones, B 373
2-Iodotropone, B 376
Ionene, B 258

Ionones, B 127, 174, 175, 271
α-Ionone, B 176, 271, 274
β-Ionone, B 176, 267, 268, 285, 286, 319
—, conversion to vitamin A, B 284
γ-Ionone, B 175, 176
ψ-Ionones, B 167, 168, 174, 175, 273, 274, 276
β-Ionylideneacetaldehyde, B 270, 340
Ipomeamarone, C **291,** 334; E 82, 83
IPP = Isopentenyl pyrophosphate, E 55, 60, 61, 65, 66, 79, 85, 118, 136, 137
Irene, B 177
Iresin, C **348,** 367; E 80, 84
Iresine celosioides, C 348, 349
Irididoids, B **213**
Iridodial, B **211,** 212; E 72
Iridolactones, B **210,** 211
Iridomymecin, B **210,** 211
Iridomyrmex detectus, B 211
Iridomyrmex humilis, B 211
Iridomyrmex nitidus, B 211
Iris oils, B 176
Iron carbonyls, A 85
Irone, B 127, **176,** 177
α-Irone, B 177
β-Irone, B 177, 178
γ-Irone, B 177
Iron pentacarbonyl, A 87, 130; B 394
Isacorone, C 335
Isarborinyl methyl ether, C 457
Isatropic acids, A 98
Isoaborinol, C 457
Isoacorone, C 312, 313, 314
Isoaglycons, D 369, 372, 385
Isoalantolactone, C 337, 339, 366
17-Isoallopregnane-3β,17β-diol 3-monoacetate, D 306
17-Isoallopregnane-3,20-dione, D 236
17-Isoallopregnane-3β,17β,20α-triol, D 307
17-Isoallopregnane-3β,17β,20β-triol, D 307
17-Isoallopregnanolone, D 236
Isoamyl alcohol, E 61
Isoamylmagnesium bromide, D 109
Isoamyl nitrite, C 219, 221, 226
5-Isoandrostane, derivatives, D 254
—, D-homo derivatives, D 279
13-Isoandrostan-3α-ol-17-one, D 264
5-Isoandrost-16-en-3α-ol, D 263
13-Isoandrosterone, D 269
5-Isoandrost-16-en-3β-ol, D 263
Isoandrosterone, D 38, 255, 306
Isoanhydrovitamin A_1, B **291**
Isoartemisia ketone, B 165
Isoatisirene, C 400
Isobalchanolide, C **365,** 368
Isobilianic acids, D 20, **169**
Isobisabolene, C **259,** 288
Isobornane, C 245
Isoborneols, C 231
Isoborneol, C 137, 162, 164, 192, 207, **226,** 227, 228, 233
Isoborneol acetate, C 228
Isoborneol-3-carboxylic acid, C 224
Isoborneol nitrite esters, C 228
Isobornyl acetate, C 207
Isobornylamine, C **232,** 233
Isobornylamine hydrochloride, C 232
Isobornyl chloride, C 160, 162, 164, 229, 231
Isobornyl esters, C 164
Isobornyl ethers, C 164
Isobufocholanic acid, D 413, 418
2-Isobutoxymethylenecyclohexanone, B 89
Isobutylene, C 49, 286
Isobutyl methyl ketone, A 109, 187
2-Isobutynyl-4,4,6-trimethylcyclohex-5-ene-1,3-dione, B 130
Isobutyric acid, B 130; C 202
2-Isobutyryl-5-methoxy-4,4,6-trimethylcyclohex-5-ene-1,3-dione, B 130
Isocamphane, C 197, 199
Isocamphoronic acid, C 169, 170
Isocamphoric acids, C 200
Isocamphorquinone, C 220
Isocaproic acid, D **25**
Isocaproic aldehyde, E 110
Isocaprylaldehyde, E 111
Isocard-22-enolide, D 408
Isocarneagenin, E 48, 49
Isocarotene, B **271**
Isocarvomenthol, B 230
Isocarvomenthone, B 197
Isocarvomenthoxime, B 229
Isocarvomenthylamine, B 229, 230
Isocarvone, camphor, B 201
Isocaryophyllene, C 286, 287
Isochamic acid, C 158, 255
Isochiapagenin, E 26, 35, 36
Isocholegenin, E **46,** 47, 48, 49, 52
20-Isocholesterol, D 153, 154
20-Isocholesteryl acetate, D 154
Isocitral, B **167**
Isoclovene, C 287
Isocupressic acid, C 383
17-Isocymarin, D 385

Isodehydrocholesterol, D 106, **107**, 118, 150, 151
17-Iso-11-deoxycorticosterone, D **300**
Isodicyclopentadiene, C 52
Isodigitoxigenic acid, D 387, 388
Isodigitoxigenin, D 59, 372, 387, 388, 390
—, *nomenclature*, D *431*
Isodigitoxigenin acid, D 388
Isodigitoxigeninic acid, D 387, 388
Isodihydronepetalactone, B 210
Isodrimenin, C **347**, 348, 367
Isodrin, C **105**, 106
17-Isoemicymarin, D 385
14-Isoepiandrosterone, D 264
14-Iso-17-epitestosterone, D 269
14-Isoequilenin, D 201, 202, 209, 217, 220, 222, 223, 224, 225, 280
14-Isoequilenin acetate, D 227
14-Isoequilenin methyl ether, D 220
Isoequilin-A, D 216
8-Isoestradiol-17β, D 210
Isoestriol-A, D 211
8-Isoestrone, D 210
13-Isoestrone, D 209, 223
Δ^{8}-14-Isoestrone, D 216
$\Delta^{8,11}$-14-Isoestrone, D 217
17-Isoetiodeoxycholic acid, D 28
Isoeuphanes, C 478
Isofenchocamphoric acid, C 241, 246
Isofenchol-3-carboxylic acid, C 242
Isofencholic acid, C 241
Isofenchones, C 243
Isofenchone, C 187, **240**, 241
Isofenchone hydrazone, C 243
Isofenchoquinone, C 242, 243
Isofenchyl alcohols, C **240**, 241, 242, 243, 245, 246
Isofenchyl chloride, C 245
Isofucosterol, E 103, 105
20-Isofucosterol, D 153
28-Isofucosterol, D 153
29-Isofucosterol, D 153
Isofucoxanthin, B 330
Isogenin, D *435*
Isogermacrone, C **330**
Isogeronic acid, B 271
Isogitoxigenic acid, D 393, 394
Isogitoxigenin, D 391, 392
Isogitoxigeninic acid, D 393
Isogitoxigenin methyl ester, D 392
Isoglabrolide, C 442
Isoglycosides, D 369, 372, **385**
Isohelenalin, C **359**, 368
Isohellebrinic methyl ester, D 411
Isohexylmagnesium bromide, D 102, 108
Isohexyl methyl ketone, D 25
14-Iso-D-homoequilenin, D 280
Isohumulones, B 132
Isohydromellitic acid, B 152
Isoiresin, C **349**
Isoiresin diacetate, C 367
Isoiridomyrmecin, B **210**, 211
Iso-α-irone, B 178
14-Iso-17-isoalloetianic acid, D 373
14-Iso-17-isoallopregnane, D 403
14-Iso-17-isoallopregnane-3,20-dione, D 403
14-Iso-17-isocholanic acid, D 413, 418
14-Iso-17-iso-11-deoxycorticosterone acetate, D 300
14-Iso-17-isoetianic acid, D 375, 376
14-Iso-17-isopregnane, D 388
14-Iso-17-isopregnenolone, D 237
14-Iso-17-isoprogesterone, D 237
Isojateorin, C 385, 387
Isokaurene, C 395
Isokaurenic acid, C 395
Isoketocamphoronic acid, C 169, 170
Isoketopinic acid, C 192, 216, 218
Isokryptoxanthin, B **302**
Isokryptoxanthin acetate, B 263
Isoleucine, E 61
Isolimonene, B 180
Isolinderalactone, C **366**, 368
Isolithobilianic acid, D 9, 10, 18, 19
Isomasticadienoic acid, C 426
p-Isomenthane-2,8-diol, B 222
Isomenthol, B 185, 186, 218, 224, 226; E 73
Isomenthone, B 185, 187, 217, 218; E 73
Isomenthylamine, B 225, 226
Isomenthyl bromide, B 217
Isomenthyl *p*-toluenesulphonate, B 218
Isomenthyltrimethylammonium hydroxide, B 218
Isonarthogenin, E 23, 24, **49**
Isonitrosocamphor, C 200, 219
Isonitrosoisofenchone, C 242
Isonuatigenin, E **48**
Iso-odoroside, D 396
Iso-ouabain, D 401
20-Iso-24-oxocholesterol, D 109, 153
17-Iso-11-oxoprogesterone, D 329
Isopentane diols, E 63
Isopentenyl adenosine, E 64
N^{6}-(Δ^{2}-Isopentenyl)adenosine, E 66
Isopentenyl epoxide, E 63

Isopentenyl pyrophosphate, B 155, 341, 342, 343; C 371, 466; E 55, 136
—, biosynthesis, E 56
Isopentenyl pyrophosphate isomerase, E 60
17-Isoperiplocymarin, D 385
α-Isoperiplogenic acid, D 379, 384
α-Isoperiplogenic methyl ester, D 387
17-Isoperiplogenin, D 373, **385**
α-Isoperiplogenin methyl ester, D 388
Isopetasol, C 307
Isophorone, B 117, 119
Isophotoartemisin acetate, C 351
Isophotosantonic acid, C 342, **344**
Isophotosantonic lactone, C 322, 344, 354
Isophthalic acid, C 158
Isophthalic dimethyl ester, hydrogenation, B 145
Isophyllocladene, C 393
Isopimaradien-3-ol, C 388
Isopimaric acid, C 388
Isopinocampheol, C 176
Isopinocamphone, C 176, 177, 178, 179
Isopiperitenone, B **202**
17-Isopregnan-3,20-dione, D 236
17-Isopregn-5-ene-3β,17β-diol-20-one, D 278
17-Isopregneninolone, D 240
17-Isopregnenolone, D 236
17-Isopregn-20-one, D 253
Isoprene, A 55; C 291
—, reaction with diazoacetic ester, A 52
Isoprene Rule, E 54
—, biogenetic, C 370
17-Isoprogesterone, D 236
Isopropenyl acetate, B 129; D 93, 129
4-Isopropenylcyclohex-1-ene-1-carboxaldehyde, B 196
Isopropenyl methyl ketone, B 319
2-Isopropenyl-3-methylphenol, B 193
Isopropenylcyclopropane, A 36
β-Isopropyladipic acid, B 195
p-Isopropylbenzoic acid, B 216, 374
2-Isopropylbutanal, D 156
m-Isopropylbenzoic acid, C 158
p-Isopropylbenzoic acid, C 158
Isopropylcyclobutane, A 80
3-Isopropylcycloheptane-1,2-dione, B 355
2-Isopropylcyclohexanone, B 12
3-Isopropylcyclohexanone, B 195
Isopropylcyclohex-2-en-4-one, C 175
4-Isopropylcyclohex-2-en-1-one, C 140
4-Isopropylcyclohexylcarbinol, B 69
Isopropylcyclopentane, A 113
1-Isopropylcyclopentane-1,α-diol, C 159
2-Isopropylcyclopentan-1-ols, A 155
2-Isopropylcyclopentanone, A 154
1-Isopropylcyclopent-1-ene, C 159
3-Isopropylcyclopent-2-en-1-one, A 171
Isopropylcyclopropane, A 32
1-Isopropylcyclopropane-1,2-dicarboxylic acid, A 61; C 150
4-Isopropyl-1,2-dimethylbenzene, C 309
3-Isopropyl-4,5-dimethylcyclopent-2-enone, C 147
4-Isopropyl-1,7-dimethylnaphthalene, C 324, 325
α-Isopropylglutaric acid, C 249, 250
β-Isopropylglutaric acid, C 236
5-Isopropylheptan-2-one, D 133
3-Isopropyl-5-hydroxy-2-methylbenzoic acid, C 308
Isopropylideneacetic acid, E 61
Isopropylidenecyclohexan-4-one, C 175
2-Isopropylidenecyclopentanone, A 170
1,2-*O*-Isopropylidene-*myo*-inositol, B 61
Isopropylidene malonate, A 105
4-Isopropylidene-2-methylcyclohexane-1,3-dione, C 220
Isopropyl iodide, A 19
Isopropyllaevulic acid, C 146
Isopropylmethylacetaldehyde, D 27 134, 144, 158
α-Isopropyl-α′-methyladipic acid, B 203; C 138
1-Isopropyl-4-methylbicyclo[3.1.0]-hexane, C 137
1-Isopropylmethylcyclohexanes, B 14
5-Isopropyl-3-methylcyclohexane-1,2-dione, B 108
5-Isopropyl-2-methylcyclopentanone, B 209
1-Isopropyl-3-methylcyclopent-1-ene, A 116
α-Isopropyl-α-methylglutaric acid, C 253
β-Isopropyl-α-methylglutaric acid, C 146
Isopropyl methyl ketone, C 325
3-Isopropyl-6-methyl-2(2-methylbutyl)phenol, E 83

1-Isopropyl-4-methyl-7-oxabicyclo-[2.2.1]heptane, B **206**
α-Isopropyl-α-methylsuccinic acid, C 149, 253
2-Isopropylnaphthalene, C 299
3-Isopropyl-5-oxohexanoic acid, C 235
p-Isopropylphenol, E 77
Isopropylsuccinic acid, B 182; C 146, 249, 251, 270
4-Isopropyl-1,2,6,7-tetramethylnaphthalene, C 261, 269
4-Isopropyl-1,2,6-trimethylnaphthalene, C 269
4-Isopropyl-1,5,6-trimethylnaphthalene, C 296
4-Isopropyl-1,6,7-trimethylnaphthalene, C 271, 312
Isopropyltriphenylphosphonium bromide, D 158
Isopropyltropolones, B 365, 369, 374
2-Isopropyltropone, B 364
Isopulegol, B **191**, 223, 224
Isopulegol hydrate, B 191
Isopulegone, B 191, 200
Isopulegyl chloride, B 219
Isopyrocalciferol, D 138, 139, 140, 141, 148
Isoreineckiagenin, E **48**, 49
Isorenieral, B 277
Isorenieratene, B 254, **277**, 278, 279, 346; E 120, 122, 130
—, structure, B 233
Isorhodeasapogenin, E 3, 25, **30**, 38, 49, 51
Isorosenoic acid, C 389
Isorubijervine, D 1
Isosantene, C 190
Isosapogenins, E 11, 12, 13, 16
Isosarmentogenin, D 399
Isoscillaridin methyl ester, D 409
α-Isostrophanthic acid, D 380
β-Isostrophanthic acid, D 375
α-Isostrophanthidic acid, D 384
Isostrophanthidin, D 375, 376, **385**
Isostrophanthidin acid, D 375
α-Isostrophanthidinic acid, D 375, 379
Isosylvestrene, B 184
Isotelekin, C **339**, 340, 366
Isotenulin, C 358, 368
13-Isotestosterone, D 259
Isotetralin, C 71
Isothujane hydrazone, C 138
Isothujone, C 141, 146, 147
Isothujyl alcohol, C 138, 141, 142
Isotricyclene, C 192, 193, 231
Iso-uzarigenin, D 396
Iso-uzarin, D 396
Isovaleraldehyde, D 154; E 61
Isovaleric acid, B **131**; C **312**; D 53, 54; E 61
Isovaleryl-CoA, E 57
2-Isovaleryl-tetramethylcyclohexane-1,3,5-trione, B 131
Isoverbanol, C 182, 184
Isoxylitone, B 202
Isozeaxanthin, B **299**
Isozingiberene, C **261**
IUPAC nomenclature, steroids, D *422*
Iva asperifolia, C 340
Iva imbricata, C 338
Ivalin, C **338**, **366**
Iva microcephala, C 338, 340
Ivasperin, C **340**, 366
Iva texensis, C 340

Jacquinic acid, C 440
Jalaric acid, C **326**, 336
Jalari sticklac, C 326
Jasminium grandiflorum, A 180
Jasmone, A 104, 107, 176, 180, 181, 182
Jasus lalandei, D 161
Jatamanshic acid, C **318**, 335
Jatamansone, C 309
5α-Jervane, D *450*
Jervanine, D *449*, *450*
Jervine, D **285**
Junenol, C **300**, **334**
Juniperus communis, C 296, 300
Juniperus sabina, C 139, 142
Juniperus utahensis, C **325**
Juniperus virginiana, C 279
Jurubidine, E **52**, 53
Jurubine, E **53**

Kahweol, C 396, 397
Kammogenin, E 27, **34**
Kanamycins, B 83
Kanokonol, C **310**, 335
Kapok seed oil, A 70
Katonic acid, C 439
Kaurandioic acid, C 396
Kaurandiol, C 395
Kaurane diterpenoids, C 395
Kaurane group, C 380
Kauranol, C 395
Kaurantriol, C 395
Kaurane, C 380, 395, 460, 461, 462, 470; E 89, 90
Kaurenic acid, C 395
Kaurenoic acid, C 395, 470; E 89, 90
Kauren-19-ol, E 89, 90

Kaur-16-en-19-ol, C 470
Kaurine, E 90
Kendall's Compound A, D 296
Kendall's Compound E, D 287, 294, 295, 324
Kendall's compound F, D 294
Kessane, C **317**, 318, 335
Kessanol, C **317**, 318, 335
Kessoglycol, C **317**, 318, 335
α-Kessyl alcohol, C **317**, 318, 335
Keten acetals, A 49
—, reaction with diazomethane, A 49
Ketens, dimerisation, A 75
—, reaction with enamines, A 76
—, — with vinyl ethers, A 76
Keten, photolysis, A 25
—, reaction with diazomethane, A 48, 78, 92
Keto compounds, see Oxo compounds
α-Ketol, C 399
Ketones, acetylenic, B 128
—, condensation with cyclopentadienes, A 120
—, cyclic acetylenic, B 408
—, cyclohexanes, B 92 ff
—, ethylenic, B 127
—, exocyclic, A 50, 94
—, exocyclic hydroaromatic, B 127, 128
—, macrocyclic, B 406
—, —, physical properties, B 409
—, naturally-occurring, B 414
—, photochemical cyclisation, A 78
—, reaction with benzil, A 183
—, — with oxalic ester, A 188
—, terpenoid, B 196
—, αβ-unsaturated photodimerisation, A 94
Ketone 250, D 97
Ketopinic acid, C 214
Khivol, C 411, 412
Khivorin, C 411, 412, 429
Khusinol, C **297**, 334
Khusol, C **297**, 334
Kikuba-saponin, E 3
Kishner-Wolff reduction, C 218
Kitigenin, E 27, **41**, 42
Kitigenin triacetate, E 41
Kitol, B **290**
Klaineanone, C 433
Knoevenagel synthesis, B 7
Kogagenin, E 27, **40**
Kogagenin 2,3-diacetate, E 41
Kolavenic acid, C 387
Kolavenol, C 387
Kolavic acid, C 387
Köster-Logemann ketone, D 101
Kryptocapsin, B 257, **325**, 345
Kryptocapsone, B 325
Kryptochrome, B 302
Kryptoflavin, B **302**
Kryptogenin, D 108; E 23, **43**, 44, 45, 46, 113
Kryptogenin diacetate, E 44
Kryptoxanthins, B 237, 246, 263, 294, **300**
α-Kryptoxanthin, B 302
β-Kryptoxanthin, B **300**, 302
Kryptoxanthin acetate, B 301, 302
β-Kryptoxanthin-5,6:5′,6′-diepoxide, B 302
β-Kryptoxanthin-5′,6′-epoxide, B 301, 302, 345
Kuhn-Roth degradation, E 96, 97
Kusmi seedlac, C 326

Labda-8,13-dien-15-ol pyrophosphate, E 90
Labdanes, rearranged, C 377
—, —, physical properties, C 387
—, unrearranged, C 375
Labdanediol, C 382
Labdanediterpenoids, physical properties, C 382
Labdane group, C 375
Labdanolic acid, C 375, 382
—, synthesis, C 468
Labiatae family, B 199
Labile bixin, B **327**
Laccifera lacca, C **326**
Lactams, B 409
—, cyclanones, B 409
Lactarius deliciosus, C 318
Lactaroviolin, C **318**, 336
Lactobacilli, E 59
Lactobacillic acid, A 68, 70
Lactobacillus acidophilus, E 66
Lactobacillus arabinosus, A 68
Lactobacillus plantarum, E 66
Lactones, B 98, 116
—, macrocarbocylic, B 409; C 362
—, monocarbocylic, C 336
—, *nomenclature*, D *437*
—, sesquiterpenoids, C 336
$\Delta^{\alpha\beta}$-γ-Lactone ring, D 371
Lactone Sector Rule, C 479
Lactuca virosa, C 352
Lactucin, C **352**, 367
Laevulic acid, A 179; B 202; C 259, 261, 282, 284, 468
Laevulic ester, A 180
Lambertianic acid, C 375, 376, 384

Lanatigonin I, E 2
Lanceol, C **291,** 334
Lannolene, A 116
Lanosta-8,24-diene, C 407, 466
Lanosta-7,24-dien-3β-ol, E 99
Lanosta-8,24-dienol, C 473
8,24-Lanostadien-3β-ol, D *439*
Lanostane, D 3, 4
—, molecular rotation, D 63
5α-Lanostane, D *429*
Lanostane group, triterpenoids, C 421
Lanostanol, D 75
Lanostenol, C 473
5α-Lanost-8-en-3β-ol, D 103
5α-Lanost-8-en-3β-yl acetate, D 87
Lanosterol, C 164, 187, 371, 372, 373, 406, 407, 408, 410, 421, 460, 465, 466, 473, 480; D 3, 36, 71, 74, 75, 96, 151, 162; E 95, 96, 98, 99, 100, 101, 104, 106, 108, 116, 119
—, biosynthesis, C **464**
—, from squalene oxide, E 96
—, *nomenclature,* D *439*
—, structure, E 54
—, ultraviolet absorption spectra, D 68
[14C]Lanosterol, E 95
Lanost-7,9(11)-dienol, C 473
Lanost-7,9,(11),24-trienol, C 473
Lansium annamalayanum, C 278
Lantadene B, 441
Larixol, C 382
Larixyl acetate, C 382
Laserol, C **325**
Laserpitin, C **325,** 336
Laserpitium lactifolium, C 325
Lasius (Dendrolasius) fulginosus, C 290
Lathosterol, D 96, 164; E 104
Latimerol, D 19
Lavandula dictata, C 234
Lavandulol, B 158, **164,** 173
Lavender oil, B 164, 170
Lavendulic acid, B 164
Laxogenin, E 26, **33**
Leaf xanthophyll, B 295
Ledglycol, C 321
Ledium palustre, C 319
Ledol, C **319,** 320, 321, 336
Legal test, D 371, 385, 393, 400, 403, 408, 410
Lemon grass oil, B 286
Leprotene, B 279; E 130
Leptogenin, D 403
Leptoside, D **403**
Leptospermone, B **131**
Leptospermum flavescens, B 131
Leptospermum liversidgei, B 191
Leptospermum scoparium, B 131
Leucine, E 61
—, from HMG-CoA, E 57
Leuckart reaction, B 80, 81
Leucomysin, C **353**
Leucotylic acid, C 452
Leucotylin, C 452
α-Levantenolide, C 375, 376, 383, 384
β-Levantenolide, C 384
Levopimaric acid, C 392
Lewis acids, B 408; C 103, 160
Libiacaea, D 360
Libocedrus bidwilli, C 262
Libocedrus formosana, B 216
Liebermann-Burchard reaction, D 76
Lignin, B 140
Lilagenin, E **25**
Liliaceae species, B 366; E 5
Lilium candidum, B 299
Lilium species, E 25
Lilium tigrinum, B 299, 325
Limonene, B **178,** 179, 180, 196, 209, 218; C 251; D 133; E 74, 77
—, autoxidation, B 192, 221
—, conversion to carvone, B 201
—, hydration, B 189
Limonene-1,2-epoxide, B 179
Limonene nitrosochloride, B 179
Limonin, C 411, 412, 430, 480
Limonin diosphenol, C 430
Limonoids, C 478
Linaloe oil, B 169
Linalool, B 154, 159, **163,** 164, 169, 170, 267; E 67
—, cyclization, B 208
Linalool-2,3-epoxide, B 169
Linalool oxide, E 67
Linalool pyrophosphate, C 371, 372
Lindane, B **36**
Linderalactone, C **366,** 368
Linderane, C **365,** 368
Lindera strychnifolia, C 272, 282, 302, 365
Linderazulene, C 353
Linderene, C **302,** 335
Linderene acetate, C 303, 335
Linderone, A 187
Lindestrene, C 303, 335
Lindlar catalyst, B 247
Linifolins, C **359,** 368
Lippia asperifolia, B 168
Lippiaphenol, B 204
Lippia turbinata oil, B 203
Lippione, B 203
Liquiritic acid, C 441

Liquoric acid, C 442
Liriodendritol, B **61**
Lithioferrocenes, A 134
Lithium acetylide, B 301
Lithium acetylide-ethylenediamine, D 283
Lithium alkyls, A 26
Lithium-amine reduction, B 3, 4, 16
Lithium aryls, B 73
Lithium cyclohexenyl, B 40
Lithium d_5-cyclopentadienide, A 120
Lithium diethoxyaluminohydride, B 89
Lithium diethylamide, B 353, 395, 405
Lithium ethylenediamide, C 72
Lithium phenoxide, B 3, 361
Lithium tri-*tert*-butoxyaluminium-hydride, B 89; D 89
Lithiobilianic acid, D 9, 10, 18, 19
Lithocholic acid, D 9, 17, 18, 19, 36, 38, 56, 169, 170, 171, 180, 181, 182, 184, **185**, 186, **250**, 253, 306, 315
—, *nomenclature*, D *439*
Loganin, B **214**; E 75, 76
Loliolide, B **178**
Lolium perenne, B 178
Longicyclene, C **281**, 289
Longifolene, C **279**, 280, 289; E 80, 82, 84
α-Longipinene, C **281**, 289
Longispinogenin, C 438
Lonicera japonica, B 275
Lopanes, E 116
Lophanone, D 151
Lophengol, C 408
Lophenol, C 422; D 3, **151**, 157, 165; E 100, 104
Lophocereus schottii, D 151
Lumiandrosterone, D 264, 269, 271
Lumicholestenone, D 128
Lumicolchicines, B 376
Lumidoisynolic acid, D 223, 225, 226
Lumiepiandrosterone, D 264
Lumiestrone, D 205, 209, 223, 225
Lumimarrianolic acid, D 223, 225
Lumisantonin, C **344**
Lumista-4,22-dien-3-one, D 237
Lumistanols, D 141, 142
5α-Lumistan-3β-ol, D 141
5α,14β-Lumistan-3β-ol, D 141
5β-Lumistan-3β-ol, D 141
Lumistanol A, D 141
Lumista-4,6,22-trien-3-one, D 237
Lumista-4,7,22-trien-3-one, D 237
Lumisterol, D 39, 138, 139, 140, 141, 142, 143, 144, 237
Lumisteryl acetate, D 139
Lupanes, C 478; E 116
Lupane group, triterpenoids, C 414, 435
Lupenone, C 435
Lupeol, C 406, 413, 435, 461, **466**, 467; E 116, 117, 119
Lupenyl acetate, C 435
Lupinus gibberellin, C 398
Lupulone, B **132**
Lutein, B 237, 238, 239, 259, 293, 294, **295**, 297, 344; E 120, 121, 127, 128
—, structure, B 235
Lutein diacetate, B 296
Lutein dipalmitate, B 296
Lutein epoxide, B 296
Luteochrome, B 300, **303**
Luteoxanthins, B 300
Luvanga scandens, C 238
Luvigenin, E **42**, 43
Lycoctonine, C 381
Lycopenes, natural, B 267
Lycopene, B 238, 239, 244, 249, 250, 253, **266**, 267, 270, 280, 306, 327, 334, 335, 336, 338, 341, 342; E 120 121, 124, 126
—, labelled, B 339
—, oxidation, B 259
—, pyrolysis, B 258
—, structure, B 231, 232, 234
all trans-Lycopene, B 252
cis-Lycopenes, B 252
Lycopene 5,6-epoxide, B 293
Lycopersane, B 335
Lycopersene, B 342, 343; E 124
Lycopersicum esculentum, B 266
Lycopersylsulphonium ion, E 124
Lycophyll, B **306**
Lycopodium alkaloids, C 127
Lycoxanthin, B **306**, 307; E 120, 121
D-Lyxose, D **363**

β-Maaliene, C **274**, 288
Maaliol, C 303, **304**, 305, 335
Macdougallin, C 408, 422; D 3, **152**, 165; E 99
—, ultraviolet absorption spectra, D 68
Macdougallin diacetate, D 152
MacFadyen-Stevens reaction, A 47; B **362**
Machaeric acid, C441
Machaerinic acid, C 440
Machaerium incorruptible, E 118
Macrocyclic acyloins, B 411, 412
Macrocyclic alcohols, B 402, 404

Macrocyclic aldehydes, B 406
Macrocyclic amines, B **418**
Macrocyclic-2-aminoalcohols, B 418
Macrocyclic compounds, B 347 ff, 380 ff
Macrocyclic 1,2-diketones, B 413
Macrocyclic $\alpha\omega$-diketones, B 413
Macrocyclic halogenoketones, B 410
Macrocyclic hydrocarbons, unsaturated, B 386
Macrocyclic hydroxyketones, B 411
Macrocyclic ketones, A 106; B 406
—, physical properties, B 409
Macrocyclic lactones, B 410; C 362
Macrocyclic polyacetylenes, B 389
Macrocyclic rings, conformations, B 420
Macrolides, B 414
Magnesium methyl carbonate, B 151
Maguey cacaya, E 27
Maleic acid, from cyclohexane, B 15
—, reaction with butadiene, B 6
—, — with cyclopentadiene, A 192
Maleic anhydride, A 76, 173; B 16, 141, 147; C 50, 79, 90, 98, 115, 123
—, adduct with allene, A 96
—, Diels-Alder adducts, A 118
—, reaction with butadiene, B 146
—, — with furan, B 150
—, — with tropone, C 124
Maleic ester, C 49, 115
Malic acid, A 179
Malonic acid, A 196
Malonic ester, A 201, 202; C 31, 117, 124
—, condensation with 1,4-dihalides, A 104
—, reaction with dihalides, A 22
Malonic methyl ester, C 55
Malonitrile, A 63; B 350
Malonyl-CoA, E 57
Malvaceae, A 69
Malvalic acid, A 69, 70
Malva seed oils, A 70
Mandelic acid, B 135
Manduca sexta, D 161
Manfreda maculosa, E 27
Manganesebiscyclopentadienide, A 125, 140
Manganese biscyclopentadienyls, A 131
Mangiferolic acid, C 423
Maniladiol, C 438
Mannich reaction, B 108; D 228
Manogenin, E 27, **34**, 35, **36**, 52
Manool, C 60, 372, 375, 377, 382; E 86, 91
—, synthesis, C 468
Manoyl oxide, C 382
Marasmic acid, E 83
Marine sterols, D 3, 73
Marinobufagin, D 413, 419, **420**
Marinobufogenin, D 414
Marinobufotoxin, D 414
Marker-Lawson acid, D 315
Marker's acid, D 43
Markogenin, E 25
Marmin, E 70
Marrianolic acid, D 202, 210, 223, **224**
Marrianolic methyl ester, D 208
Marrubiin, C 384, 385
Maslinic acid, C 440
Mass spectrometry, carotenoids, B **257**
—, steroids, D 70
—, structure determination, C 477, 478
Masticadienoic acid, C 426
Matatabilactone, B **212**
Matricaria chamomilla, C 258, 353
Matricaria globifera, C 354, 355
Matricaria inodorata, C 258
Matricia parthenium, C 199
Matricarin, C **353**, 354, 367
Matricin, C **353**, 367
Maytenone, C 392, 394
Medicagenic acid, C 442
Meerwein-Ponndorf reduction, B 44; C 53, 181, 186; D 104, 108, 259, 384, 411
Megestrol, D 239, 240
Melaleuca viridiflora, C 320
Melaleucic acid, C 414, 436
Meldrum's acid, A 105
Melianone, C 410, 411, 426
Mellitic acid, B 152
Mentha arvensis, B 186
Mentha crispa, E 74
m-Menthadienes, B 183
p-Mentha-2,8(9)-diene, B **180**
p-Mentha-1(7),8-dien-2-ol, B 192, 193
p-Mentha-2,8-dien-1-ol, B **192**, 193
p-Mentha-4(8),6-dien-2-one, C 155
Menthanes, B 216 ff
m-Menthane, B 171
o-Menthane, B 172; C 183
p-Menthanes, B 166, 173, 175, 178
p-Menthane, B 158, 171, **216**, 217; E 68
Menthanediols, B **220**
p-Menthane-1,2-diol, B **221**
p-Menthane-1,8-diol, B **194**
p-Menthane-2,4-diol, B **221**

p-Menthane-2,5-diol, B **221**
p-Menthane-2,8-diol, B **221**, 222, 223
p-Menthane-3,4-diol, B **223**
p-Menthane-3,8-diol, B **223**
p-Menthane-1,2,8-triol, B 188
Menthanols, B 220
p-Menthan-1-ol, B 218, **220**
p-Menthan-2-ol, B 187, 220
p-Menthan-3-ol, B 220
p-Menthan-4-ol, B **220**
p-Menthan-8-ol, B **220**
p-Menthan-2-one, B 197, 221
p-Menthan-9-yl acetate, B 219
Mentha pulegium, B 202
Mentha pulegium oil, B 202
Mentha rotundifolia oil, B 204
Mentha sylvestris oil, B 203
Mentha viridis oil, B 203
Menthenes, B 230
p-Menthene, B 197, 217, **218**, 220, 221, 230
p-Menth-2-ene, B 18, 186, 217, **218**, 219, 225, 226, 230
p-Menth-3-ene, B 18, 186, 218, **219**, 220, 223, 224, 226
p-Menth-4(8)-ene, B **219**
p-Menth-8-ene, B **219**
Δ^2-Menthene, C 251
p-Menthene-1,2-diols, B 179
p-Menth-1-ene-3,6-diols, B 182, 221
p-Menth-1-ene-6,7-diol, B 194
p-Menth-2-ene-1,4-diol, B 207
p-Menthene-6-ene-2,4-diol, B 221
p-Menth-1-en-3-ol, B 190
p-Menth-1-en-4-ol, B 189, 194
p-Menth-1-en-8-ol, B 188
p-Menth-3-en-1-ol, B **190**
p-Menth-4(8)-en-2-ol, B 223
p-Menth-8-en-1-ol, B 188
p-Menth-1-en-3-one, B 198
$\Delta^{1,2}$-*o*-Menthen-3-one, C 183
Menthhydrazide, B 227
Menthofuran, B 208; E 78
Menthols, B 224, 225
—, infra red spectra, B 186
—, physical constants, B 186
Menthol, B 44, **184**, 185, 186, 199, 218, 219, 220, 226; E 73, 74, 75
—, oxidation, B 196
Menthomenthene, B **219**
Menthones, B 224
—, physical constants, B 187
Menthone, B 185, 187, 196, 197, 198, 199, 217, 218; C 251; E 74
Menthoximes, B 224
Menthoxyacetyl chloride, D 259
Menthyl acetate, B 219
Menthylamines, B **224**, 225, 227
—, derivatives, B 228
Menthylamine, B 226, 230
o-Menth-2-ylamine, B **227**
p-Menth-3-ylamines, B 224
Menthylaminoacetic acid, B 227
N-Menthylaminoacetic ester, B 227
Menthyl *N*-aminocarbamate, B 227
Menthyl bromide, B 217
Menthyl carbonate, B 227
Menthyl chloride, B 18, 218, 226
Menthyl isocyanate, B 227
Menthyl-*p*-menthylxanthate, B 219
N-Menthylphthalamic acid, B 227
N-Menthylphthalamoyl chloride, B 227
Menthyl *p*-toluenesulphonate, B 186, 217, 218
Menthyltrimethylammonium hydroxide, B 218
2β-Mercapto-5α-cholestan-3α-ol, D 113
3α-Mercapto-5α-cholestan-2β-ol, D 113
β-Mercaptoethanol, D 92; 110
Mercaptotropones, B **378**, 379
Mesembryanthemoidigenic acid, C 440
Mesitylene, A 143
Mesityl oxide, A 57
—, reaction with methyl vinyl ketone, B 202
—, reaction with sodiomalonic ester, B 111
—, self condensation, A 176
Metagenin, E 27, **33**
Metahexahydrobenzaldehyde, B 89
Metal acetylides, B 65
Metal cyclopentadienyls, A 124
Metallicinium ions, A 125
Metallocenes, A 124
—, chemical properties, A 131
—, iron group, reactivities, A 135
—, reactivities, A 135
—, synthesis, A 127
α-Metallocenylethyl acetates, A 136
Metal tripropyls, A **32**
Metanarthecium luteoviride, E 26, 27, 42, 49
Metandrolone, D 273
Metasaccharinic acid, B 58
Metasantonins, C **345**
Meteogenin, E **42**, 43
Methandrostenolone, D 275
Methane, C 136
Methanethiol, A 34
Methione, B 344
Methionine[methyl-^{14}C], C 467
Methostenol, D 151, 165

Methosterol, E 104
5-Methoxy-17β-acetoxy-17α-acetyl-des-A-estra-5,7,9-triene, D 246
5-Methoxy-17β-acetoxy-17α-ethynyl-des-A-estra-5,7,9-triene, D 246
Methoxyacetylene, C 468
3-Methoxy-17-acetylestra-1,3,5(10)-triene, D 244
17β-Methoxy-4-androsten-3-one, D *438*
1-Methoxybicyclo[3.2.0]hepta-3,6-diene, C 45
7-Methoxybicyclo[3.2.0]hepta-3,6-dien-2-one, B 376
1-Methoxybicyclo[3.2.0]hepta-2,6-dien-4-one, C 46
7-Methoxycarbonylbicyclo[2.2.1]-hept-2-ene, C 97
3β-Methoxy-23,24-bisnorchol-5-en-22-al, D 155
3β-Methoxybisnorchol-5-enoyl chloride, D 99
16-Methoxycarbonylequilenin, D 219
Methoxycarbonylmethylenetriphenyl-phosphorane, D 408
2α-Methoxycarbonyl-A-nor-5α-cholestane, D 119
α-Methoxycarbonyl sulpho-acetic acid, C 172
4-Methoxy-α-carotene, B 263
4-Methoxy-β-carotene, B 263
Methoxycarotenoids, B 306, 344
3α-Methoxy-5α-cholestane, D 44
3β-Methoxycholest-4-ene, D 105
3β-Methylcholest-5-ene, D 42
2-Methoxy-5β-cholest-1-en-3-one, D 124, 125
3-Methoxy-5α-cholest-2-en-4-one, D 125
4-Methoxycholest-4-en-3-one, D 125
5α-Methoxy-6β,19-cycloandrostane-3,17-dione, D 335
6-Methoxy-3,5-cyclo-5α-cholestane, D 42, 77
Methoxycycloheptadienones, B 375
7-Methoxycyclohepta-1,3,5-triene, C 45
1-Methoxycyclohexa-1,3-diene, B 64
1-Methoxycyclohexa-1,4-diene, B 64
4-Methoxycyclohexane-1-carboxylic acid, B 135
2-Methoxycyclohexanol, B 53
4-Methoxycyclohexyl tosylate, A 14; B 55
7-Methoxy-1,2-cyclopentenophen-anthrene, D 202, 203
3-Methoxy-6-deoxyhexose thevetose, D 386
12-Methoxydihydrocostunolide, C 368
7-Methoxy-3′,3′-dimethyl-1,2-cyclopentenophenanthrene, D 202, 203, 216
3-Methoxyestra-2,5(10)-dien-17-ethylene acetal, D 222
3-Methoxyestra-2,5(10)-dien-17β-ol, D 242
2-Methoxyestrone, D 216
16α-(β-Methoxyethyl)-6α-methyl-16b,20-oxidopregna-tetraene-3-one, D 352
2-Methoxymethylbenzyl chloride, B 29
4-Methoxymethylbenzyl chloride, B 30
3-Methoxymethylcyclobutane-1,1-dicarboxylic ester, C 88
3-Methoxy-17-methyl-16,17-dihydro-cyclopentaphenanthrene, D 212
2-Methoxymethylestrone, D 228
7-Methoxy-2-methyl-1-oxo-1,2,3,4-tetrahydrophenanthrene, D 221
2-Methoxymethylpropane-1,3-diol, C 88
5-Methoxy-1-methyl-2-tetralone, D 100
6-Methoxy-2-naphthyl methyl ketone, D 222
3-Methoxy*cis*-19-norpregna-1,3,5(10), 17(20)-tetraene, D 245
3-Methoxy-19-norpregna-1,3,5(10)-trien-20α-ol, D 245
Methoxyoxocarotenoids, B 306
5-Methoxy-17-oxodes-A-estra-5,7,9-triene, D 246
7-Methoxy-1-oxo-1,2,3,4,9,10-hexahydrophenanthrene, D 247
8-Methoxy-1-oxo-isodecahydro-chrysene, D 205
12α-Methoxy-11-oxoprogesterone, D 235
7-Methoxy-1-oxo-1,2,3,4-tetra-hydrophenanthrene, D 218, 220
m-Methoxyphenethyl toluene-*p*-sulphonate, D 206
p-Methoxyphenol, B 55
m-Methoxyphenylacetylene, D 205
6-(*m*-Methoxyphenyl)-hex-2-yn-1-ol, D 207
5-(*m*-Methoxyphenyl)-pent-1-yne, D 207
16α-Methoxysapogenin, E 44
α-Methoxysuccinic acid, A 179

2-(6-Methoxy-1-tetrahydronaphthylidene)ethylcyclopentane-1,3-dione, D 207
5-Methoxy-2-tetralone, D 100
6-Methoxy-1-tetralone, C 471; D 213, 329
4-Methoxy-2,5-toluquinone, D 101, 102
Methoxytriphenylcyclopropenylium salts, A 37
Methoxytropolones, B 369, 373
2-Methoxytropone, B 375
3-Methoxytropone, B 376
4-Methoxytropone, B 375, 376
6-Methoxy-1-vinyl-1-tetralol, D 206, 209
p-Methylacetophenone, C 197
Methyl 3β-acetoxy-20-oxo-21-norchola-5,22-dienate, D 408
Methylacetylene, A 33
Methyladipic acids, B 100
β-Methyladipic acid, B 199
16α-Methylaldosterone-21-acetate, D 303
Methylamine, B 3, 340
Methylamine hydrochloride, E 10
α-Methylaminocamphor, C 221
17β-Methylaminoandrost-5-en-3-ol, D 357
N-Methylaminocyclododecane, B 420
7-Methylaminocyclohepta-2,4,6-trienethione, B 380
3-Methylaminocyclopentane-1-acetic-2,2-dicarboxylic acid, A 194
3-Methylaminocyclopent-1-ene, A 153
N-Methylaminocyclo-undecane, B 420
17α-Methylandrosta-1,5-diene-3β,17β-diol, D 274
17α-Methylandrosta-1,4-dien-17β-ol-3-one, D 273
17-Methylandrostane, D *443*
17α-Methylandrostane-3α,17β-diol, D 266
17α-Methylandrostane-3β,17β-diol, D 266
6β-Methylandrostan-17β-ol-3-one, D 265
17α-Methylandrostan-17β-ol-3-one, D 266
1'-Methyl-1'H-5α-androstano[3.2-c]-pyrazole, D *453*
17α-Methylandrost-5-ene-3β,17β-diol, D 268
1-Methylandrost-1-en-17β-ol-3-one, D 273
2'-Methyl-2'H-5α-androst-2-eno-[3,2-c]pyrazole, D *452*
Methyl angolensate, C 431
Methyl benzyl ketone, reaction with oxalic ester, A 188
Methylbetulinate acetate, C 436
Methylbicyclo[1.1.0]butane-1,2,4-tricarboxylic acids, C 33
2-Methylbicyclo[2.2.1]heptane, C 89, 117
1-Methylbicyclo[3.1.1]heptan-6-one, C 88
Methylbicyclo[3.3.1]nonane-2,6-dione 1,3,5,7-tetracarboxylate, C 124
9-Methylbicyclo[4.2.1]nona-9-ol, C 128
9-Methylbicyclo[6.1.0]nona-2,4,6-triene, C 41
7-β-Methylbicyclo[3.3.1]nonene, C 127
Methylbicyclo[2.2.2]octanes, C 121
1-Methylbicyclo[1.1.1]pentane, C 84
1-Methylbicyclopropyl, C 7
1-Methyl-1,1-'bicyclopropyl-2-carboxylic acid, C 8
p-Methylbiphenyl, B 65
7-Methylbisdehydrodoisynolic acids, D 226
7-Methylbisdehydromarrianolic acids, D 219, 224
Methylbixins, B 327, 328
all trans-Methylbixin, B 327
Methylborneol, C 165
1-Methylbutadiene, B 16
2-Methylbut-1-ene, A 73
3-Methylbut-2-ene, C 49
2-Methylbutyric acid, E 61
γ-Methylbutyrolactone, C 316
α-Methylcamphene, C 165
β-Methylcamphene, C 165
Methylcamphenilol, C 164, 195
20-Methylcholanthrene, D 24, 25
Methylcholate 3-acetate, D 188
4-Methylcholesta-3,5-diene, D 78
4-Methylcholesta-5,24-dien-3β-ol, E 104
4α-Methylcholesta-8,24-dien-3β-ol, E 103
24β-Methylcholesta-5,22-dien-3β-ol, D 155
14β-Methyl-5α-cholesta-7,9(11)-dien-3β-yl-acetate, D 152
4α-Methylcholestane, D 3
14α-Methylcholestane, D 3
24α-Methylcholestane, D 3
24β-Methylcholestane, D 3

6α-Methyl-5α-cholestane-3β,6β-diol, D 103
6β-Methyl-5α-cholestane-3β,5-diol, D 103
24α-Methyl-5α-cholestan-3β-ol, D 75
24β-Methyl-5α-cholestan-3α-ol, D 75
24β-Methyl-5α-cholestan-3β-ol, D 75
24-Methyl-5α-cholestan-3β-ols, D *427*
Methyl-5α-cholestan-3-ones, D 123
4α-Methylcholestanone, D 151
4β-Methyl-5β-cholestan-3-one, D 124
24β-Methylcholesta-5,7,22-trien-3β-ol, D 75
14α-Methyl-5α-cholest-8(9)-ene-3β,6α-diol, D 152
3-Methylcholest-4-en-3β-ol, D 85
4-Methylcholest-4-en-3α-ol, D 85
4α-Methyl-5α-cholest-7-en-3β-ol, D 151
14α-Methylcholest-7-en-3β-ol, E 100
24α-Methylcholest-5-en-3β-ol, D 75, 156
(24*R*)-24-Methylcholest-5-en-3β-ol, D 156
24α-Methylcholest-22-en-3β-ol, D 157
2α-Methylcholest-4-en-3-one, D 129
2β-Methyl-5α-cholest-6-en-3-one, D 123
4-Methylcholestenone, D 123
4-Methylcholest-4-en-3-one, D 95, 103, **129**
6α-Methylcholest-4-en-3-one, D 129
6β-Methylcholest-4-en-3-one, D 129
14α-Methyl-5α-cholest-7-en-3β-yl acetate, D 152
Methylcholesterols, D 103
6-Methylcholesterol, D 103, 129
4α-Methylcholesteryl tosylate, D 78
1-Methylchrysene, D 278
β-Methylcinnamic acid, B 136
4β-Methylcoprostanone, D 123
α-Methylcrotonaldehyde, B 299
β-Methylcrotonic acid, E 61
β-Methylcrotonyl-CoA, E 57
Methylcyclobutane, A 80
2-Methylcyclobutane-1-carboxylic acid, A 96
1-Methylcyclobutenes, A 81, 83
3-Methylcyclobut-2-ene-1,2-dicarboxylic acid, A 96
6β-Methyl-3α,5-cyclo-5α-cholestan-3β-ol, D 103
6-Methylcyclodecanone, B 403
Methylcyclogeraniolene, B 172
Methylcycloheptane, B 348, 350
2-Methylcycloheptanecarbaldehyde, B 405
2-Methylcycloheptanediols, B 405
4-Methylcycloheptane-1,2-dione, B 335
Methylcycloheptanols, B 351
1-Methylcycloheptanol, B 358
Methylcycloheptanones, B 353, 354
Methylcycloheptene, B 348
2-Methylcyclohept-2-enone, B 353
1-Methylcycloheptylamine, B **358**
Methylcyclohexadienes, B 26
Methylcyclohexadienecarbaldehydes, B 88
Methylcyclohexane, B 129, 134, 351
—, physical properties, B 13
—, reaction with carbon monoxide, B 134
Methylcyclohexanecarbaldehydes, B 88
1-Methylcyclohexane-1-carboxamide, B 81
1-Methylcyclohexanecarboxylic acid, B 133, 134
2-Methylcyclohexane-1-carboxylic acid, B 81
4-Methylcyclohexanecarboxylic acids, B 134, 216
4-Methylcyclohexane-1-carboxylic ester, B 220
Methylcyclohexane-1,2-diols, B 51, **53**
4-Methylcyclohexane-1,3-diols, B 54
3-Methylcyclohexane-1,2-dione, B 109
Methylcyclohexane-1,3-diones, B 109
2-Methylcyclohexane-1,3-dione, D 209
Methylcyclohexanols, B 23, 32, 47
1-Methylcyclohexanol, B 74
2-Methylcyclohexanol, B 18, 44, 45, **49**, 100
3-Methylcyclohexanol, B **49**, 62
4-Methylcyclohexanol, B **49**, 100
2-Methylcyclohexan-2-ol-1-one, B **105**
Methylcyclohexanones, B 95
2-Methylcyclohexanone, B 12, 49, 64, **100**
3-Methylcyclohexanone, B 49, 62, **100**, 199, 354
4-Methylcyclohexanone, B 49, **100**, 219, 220; C 259
1-Methylcyclohexan-2-one-1-carboxylic esters, B 143
2-Methylcyclohexanoneoxime, B 81
Methylcyclohexenes, B **23**, 348
—, physical properties, B 21
1-Methylcyclohex-1-ene, B 18, 24, 33, 45, 68, 75, 81, 129

1-Methylcyclohex-1-ene, autoxidation, B 19
—, hydroxylation, B 53
3-Methylcyclohex-1-ene, B 18, 45
Methylcyclohex-1-enecarbaldehydes, B 88
Methylcyclohexenecarboxylic acids, physical constants, B 136
1-Methylcyclohex-1-en-3-ol, B 23
5-Methylcyclohex-2-en-1-ol, B **62**
Methylcyclohex-2-enones, B 117
1-Methylcyclohex-1-en-3-one, C 181
2-Methylcyclohex-2-en-1-one, B 23
3-Methylcyclohex-2-enone, B 116, 119, 202; C 24, 29, 31
3-Methyl-Δ^3-cyclohexen-1-one, C 185
2-Methylcyclohex-2-en-4-one-1-carboxylic ester, B 144
1-Methylcyclohexylamine, B 358
4-Methylcyclohexyl bromide, B 134
Methylcyclohexylcarbinols, B 69, **70**
1-Methylcyclohexyl hydroperoxide, B **74**
4-Methylcyclohexylideneacetic acid, A 17
N-Methylcyclo-oct-4-ene, B 420
Methylcyclopentadecane, B 416
3-Methylcyclopentadecanone, B 414, 416
Methylcyclopentadienes, A 121
Methylcyclopentadienyltricarbonylmanganese, A 141
Methylcyclopentane, A 112, 113, 152
—, acetylation, A 176
—, from cyclohexane, B 15
—, reaction with carbon monoxide, B 134
2-Methylcyclopentaneacetic acid, A 193
Methylcyclopentane-1-carboxylic acids, A 190
2-Methylcyclopentane-1,1-diacetic acid, A 194
3-Methylcyclopentane-1,1-diacetic acid, A 194
1-Methylcyclopentane-1,2-dicarboxylic acids, A 191
Methylcyclopentane-1,2-diols, A 160
2-Methylcyclopentane-1,3-dione, D 207, 209
3-Methylcyclopentane-1,2-dione, A 184
1-Methylcyclopentanol, A 112, 115, 157
2-Methylcyclopentanol, A 114, 154, 155
3-Methylcyclopentan-1-ols, A 155
1-Methylcyclopentanones, A 167
2-Methylcyclopentanone, A 157; C 14
3-Methylcyclopentanones, A 166
2-Methylcyclopentan-1-one oxime, A 153
1-Methylcyclopentene, A 114, 115, 116, 153, 154, 155, 157, 159, 176
1-Methylcyclopent-2-ene-1,2-dicarboxylic acid, A 191
2-Methylcyclopent-2-ene-1,5-dione, A 184
2-Methylcyclopent-2-en-1-ol, A 121
2-Methylcyclopent-2-en-1-one, A 170, 171, 184
3-Methylcyclopent-2-en-1-one, A 128; B 416
Methylcyclopentenophenanthrene, D 370, 413
7-Methyl-1,2-cyclopentenophenanthrene, E 7
3-Methylcyclopentenylacetaldehyde, A 163
2-Methylcyclopent-1-enylacetone, A 176
1-Methylcyclopentyl acetate, A 117
1-Methylcyclopentylamine, A 152, 153
2-Methylcyclopentylamine, A 153
1-Methylcyclopentylcarbinol, A 161
2-Methylcyclopentylcyclohexane, C 14
2-Methylcyclopentylideneacetone, A 176
Methylcyclopentylmalonic acid, A 193
Methylcyclopropane, A 32, 34, 39, 48, 64
2-Methylcyclopropanecarboxylic acid, A 39
1-Methylcyclopropane-1,2-dicarboxylic acid, A 59
3-Methylcyclopropane-1,2-dicarboxylic acid, A 59
3-Methylcyclopropanetetracarboxylic acid, A 59
3-Methylcyclopropane-1,1,2-tricarboxylic acid, A 59
1-Methylcyclopropanol, A 43
3-Methylcyclopropene, A 33
1-Methylcyclopropene-2,3-dicarboxylic acid, A 60
2-Methylcyclopropyl acetate, A 43
2-Methylcyclopropylglyoxylic acid, A 58
9-Methyldecalins, C 63
10-Methyldecalin-2,9-diol, C 64
9-Methyl-α-decalone, C 66
1-Methyl-Δ^6-dehydroestrone, D 228

6α-Methyl-16-dehydroprogesterone, D 349
1-Methyl-17-dihydroequilenin-17β, D 223
2-Methyl-4,5-dihydrofuran, A 23
17-Methyl-16,17-dihydrophenanthrene D 17
1-Methyl-2,4-dimethylenecyclobutene, A 81
3-Methyl-1,2-dimethylenecyclohexane, B 24
Methyl-3,11-dioxoeti-4-enate, D 322
1-Methyl-2,2-diphenylcyclopropane, A 36
1-Methyl-2,2-diphenylcyclopropyl-magnesiumbromide, A 41
7-Methyldoisynolic acid A, D 226
Methylene, A 24, 25
2-Methylenebicyclo[2.1.1.]hexane, C 85
3-Methylenebicyclo[3.3.1]nonane, C 127
3-Methylenebicyclo[3.3.1]nonan-7-one, C 124, 126, 127
3-Methylenebicyclo[3.3.1]nonene, C 127
3-Methylenebicyclo[3.3.1]non-7-yl *p*-toluenesulphonate, C 127
2α,2′α-Methylenebis-17β-hydroxy-5-androstan-3-one, D 337
Methylenebis(3,3′-perhydrobenzo-oxazole), B 83
24-Methylene-5α-cholest-7-enol, D 165
7-Methylenecholesterol, D 103
23-Methylenecholesterol, D 157
24-Methylenecholesterol, D 159, 165; E 107
24-Methylenecycloartanol, E 107
24-Methylenecycloartenol, E 105
Methylenecyclobutanes, A 76, 82, 83
Methylenecyclobutane, A 73, 81, 82, 92; C 21, 22, 23
3-Methylenecyclobutane-1,2-dicarboxylic anhydride, A 96
2-Methylenecyclobutanol, C 22
3-Methylenecyclobutanol, C 22
Methylenecyclobutenes, A 81, 82
Methylenecycloheptane, B 348, 352; C 5
2-Methylenecycloheptanone, B 353
7-Methylenecyclohepta-1,3,5-triene, B 350
5-Methylenecyclohexa-1,3-diene, B 29
Methylenecyclohexane, B 17, 23, 24, 69, 70
2-Methylenecyclohexan-1-ol, B 64
2-Methylenecyclohexanone, B 108, 126
Methylenecyclohexene nitroso-chloride, B 89
2-Methylenecyclohexyl acetate, B 24, 28
Methylenecyclo-octane oxide, B 406
Methylenecyclopentane, A 117, 161, 194
2-Methylenecyclopentan-3-one-1-carboxylic acid, A 201
1-Methylenecyclopent-2-ene, A 118, 162
Methylenecyclopropane, A 34; C 23
3-Methylenecyclopropane-1,2-dicarboxylic acid, A 59
(2-Methylenecyclopropyl)acetic acid, A 70
L-β-(Methylenecyclopropyl)-α-aminopropionic acid, A 70
Methylenedecalin, C 103
2′-Methylenedicyclohexylidene-ethane, C 15
24-Methylenedihydrolanosterol, E 110
2α,3α-Methylenedioxy-5-pregnene, D 451
24-Methylene-elophenol, E 107
α-Methyleneglutaric acid, E 51
8,9-Methyleneheptadecanoic acid, A 70
3-Methylenehexa-1,5-diene, C 85
9,10-Methylenehexadecanoic acid, A 69
Methylene iodide, A 42; C 7
24-Methylenelanosterol, E 106, 107, 108, 109
24-Methylenelophenol, D 166
Methylenemalonic ester, C 99, 124
Methylenemalonic methyl esters, C 132
9,10-Methyleneoctadecanoic acid, A 69
11,12-Methyleneoctadecanoic acid, A 69
10-Methylenespiro[4.5]decan-5-one, C 29
8-Methylenespiro[2.5]octan-4-one, C 24
Methylenespiropentane, C 21
5-Methylenespiro[5.5]undecan-1-one C 31
2-Methylenetetrahydrofuran, A 23
Methylenetriphenylphosphorane, A 117; D 331
1-Methylene-2-vinylcyclopentane, A 109, 117
4-Methyl-4-en-3-ones, D 95
6α-Methyl-4-en-3-ones, D 95

3β-Methylepicholesterol, D 103
Methyl-1,2-epoxycyclohexanes, B 53, 72
1-Methylestradiol-17β, D 213
1-Methylestrone, D 203, **228**
1-Methyl-2-ethylcyclopentanes, A 113
Methylethylcyclopentan-1-ones, A 167
Methylfenchyl alcohol, C 165
Methylfulvene, A 123, 124
2-Methylfuran, A 50
Methyl geranates, B 163
β-Methylgutaconyl-CoA, E 57
α-Methylglutaric acid, C 315, 323, 356; E 12
α-Methylglutaric-α-carboxylic acid, D 22
10-Methylgonane, D *428*
N-Methylgranatonine, B 383
3-Methylheptane-2,6-dione, C 247
6-Methylheptan-2-one, C 259
Methylheptenone, B 54, 163, 167, 205, 267, 309
x-Methylheteroestradiol-17β, D 213
x-Methylheteroestrone, D 203
3-Methylhexadecane-1,16-dioic acids, B 417
γ-Methyl-δ-hexanolactone, B 116
3-Methylhexan-5-one-1-carboxylic ester, A 189
17α-Methyl-D-homoandrostane, D 281
17α-Methyl-D-homoandrostane-3,17a-dione, D 280
17α-Methyl-D-homoandrostane-3β-ol-17a-one acetate, D 280
17a-Methyl-D-homoandrost-5-ene-3β,17aβ-17-one, D 280
17a-Methyl-D-homoandrost-5-en-3β-ol-17-one, D 277
7 Methylhomomarrianolic acid, D 204, 206
8-Methylhydrindanones, C 59
19-Methylhydrocortisone, D 331
β-Methyl hydrogen camphorate, C 201, 203
3-Methylhydroxycyclopentadiene, A 128
7-Methyliminocyclohepta-1,3,5-trienylmethylamine, B 379
3-O-Methylinositol, B 60
Methyl-ψ-ionone, B 178
4-Methylisoborneol, C 165
Methyllinderone, A 187
Methyllithium, B 41, 361; C 8, 34, 86, 321, 327
Methyllithocholate tosylate, D 188
Methylmagnesium bromide, D 108, 272, 307, 308
Methylmagnesium iodide, B 105; D 108
Methylmalondialdehyde, B 247
7-Methylmarrianolic methyl ester, D 204
Methyl methacrylate, B 134
3-Methyl-1-methylenecyclobutane, A 81
1-Methyl-3-methylenecyclobut-1-ene, A 81, 83
Methylmethylenecyclohexene, B 17 ,28
7-Methylmonodehydrodoisynolic acids, D 226
N-Methylmorpholine peroxide, D 353
2-Methylmuconic methyl ester, A 68
7-Methyl-2′,1′-naphtha-1,2-fluorene, D 370
1-Methylnaphthalene, C 275
5-Methyl-2-naphthol, C 308
Methyl β-naphthyl ether, D 247
Methyl natural bixin, B 327
1-Methyl-1-nitrocyclohexane, B 81
1-Methyl-2-nitrocyclohexane, B 78
1-Methyl-1-nitrocyclopentane, A 152, 153
1-Methyl-2-nitrocyclopentane, A 152,
4-Methyl-1-nitromethylpentane, D 99
2-Methyl-5-nitrophenyl acetate, E 23
4-Methyl-5-nitrotropolone, B 374
5-Methyl-3-nitrotropolone, B 372
2-Methylnona-3,8-dione, B 193
Methylnorbornanes, C 51
β-(Methylnorcampholidyl)propionic acid, C 265
5-Methyl-19-nor-5β-cholest-9(10)-ene-3β,6β-diol diacetate, D 82
5-Methyl-19-nor-5β-cholest-9(10)-ene-4β,7β-diol diacetate, D 82
(25R)-4-Methyl-19-norspirosta-1,3,5(10)-triene, E 42
(25R)-1-Methyl-19-norspirosta-1,3,5(10)-trien-11α-ol, E 43
7α-Methyl-19-nortestosterone, D 243
17α-Methylnortestosterone, D 273
10-Methyl-$\Delta 1^{(9)}$-octal-2-one, C 64
Methyl oleanolate, C 480
4β-Methyl-4α,5-oxido-5α-cholestan-3β-yl acetate, D 103
6α-Methyl-9,11β-oxidopregna-1,4-dien-17α-ol-3,20-dione-16-acetic acid γ-lactone, D 352
4-Methyl-1-(4-oxo-2-phenyl-3(4H)-quinazolinylestra-1,3,5(10)-triene-17-one, D 356

2-Methyl-4-oxotetrahydrofuran, D 406
α-Methylparaconic acid, B 205
3-Methylpentadecanedioic acid, B 416
3-Methyl-2-penta-2′,4′-dienylcyclopent-2-en-4-ol-1-one, A 178
3-Methylpentane-1,5-dioic acid, B 417
2-Methylpent-2-enal, A 42
2-Methylpent-4-enoic acid, E 23
3-Methylpentenolone, A 178
3-Methyl-2-pent-2′-enylcyclopent-2-en-1-one, A 180
2-(4-Methylpent-3-enyl)furan, B 171
3-Methylpent-4-enylmagnesium bromide, E 23
2-Methylpent-1-en-3-yne, A 55
Methylpentenynol, B 285
1-Methylphenanthrene, D 23
2-Methyl-4-phenylbenzoic acid, B 65
1-Methyl-4-phenylbuta-1,3-diene-1-carboxylic acid, A 99
1-Methyl-1-phenylcyclobutane, A 80
3-Methyl-1-phenylcyclopentane-1-carboxylic acid, A 190
4-Methyl-1(2-phenyl-4-quinazolinyloxy)estra-1,3,5(10)-trien-17-one, D 356
4-Methyl-1-phenyl-4-trichloromethylcyclohexa-2,5-dien-1-ol, B 65
4-Methylphthalic dimethyl ester, C 261
16-Methylpregna-5,16-dien-3β-ol-20-one acetate, D 343
6α-Methylpregna-1,4,9(11),16-tetraene-3,20-dione, D 352
17α-Methylpregn-5-en-3β-ol-20-one, D 350
16α-Methylpregn-5-en-3β-ol-20-one acetate, D 351
6α-Methylprogesterone, D 185
17α-Methylprogesterone, D 240
21-Methylprogesterone, D 240
2-(2-Methyl-1-propenyl)-4-methyltetrahydropyran, B 170
1-Methyl-2-propionylcyclopropane, A 58
1-Methyl-3-*n*-propylcyclohexanes, B 14
3-Methyl-1-*n*-propylcyclohexan-1-ol, B 47
Methylpropylcyclohex-2-enone, B 117
Methyl *n*-propyl ketone, reaction with oxalic ester, A 188
4-Methyl-2-(2-methyl-1-propyl)-tetrahydropyran, B 170
5-Methylpyridine-3,4-di-carboxylic acid, B 212
2-Methylpyrrolidine, A 153
4-Methylsalicylaldehyde, B 375
Methyl salicylate, B 70
1-Methylsantene, C 247
Methylsantenyl alcohol, C 247
γ-Methylsantonin, C **345**
4′R-Methyl-(R)-spiro [5α-androstane-3,2′-(1′,3′-oxathiolane)], D *453*
2-Methylspiro[4.4]nonane-6,8-dione, C 28
6-Methylspiro[2.5]octane-1,2-dicarboxylic acids, C 25
8-Methylspiro[2.5]oct-7-en-4-one, C 24
Methyl stable bixin, B 327
14α-Methyl-Δ^8-stenols, D 87
Methylsteroids, C 408; D 6
4α-Methylsterols, E 100
4α-Methyl-5α-stigmasta-7,24(28) dien-3β-ol, D 156
4β-Methyl-5α-stigmasta-7,24(28)-dien-3β-ol, D 166
4α-Methylstigmastane, D 3
4α-Methylstigmastan-3-one, D 157
Methylsuccinic acid, B 210; C 249, 292; E 12
Methylsuccinyl chloride, A 193
Methyltestosterone, D 273, 340
2α-Methyltestosterone, D 265
4-Methyltestosterone, D 265
6α-Methyltestosterone, D 265
6β-Methyltestosterone, D 265
7α-Methyltestosterone, D 265
16β-Methyltestosterone, D 265
17-Methyltestosterone, D 241
17α-Methyltestosterone, D 265, **268**, 271, 272, 273, 274, 275, 342
17-Methyl-Δ^{11}-testosterone, D 265
8-Methyl-5,6,7,8-tetrahydroindane-1,5-dione, D 209
3-Methylthiotropolones, B 373
2-Methyl-6-(*p*-tolyl)heptane, C 281
4-Methyl-4-tribromomethyl-cyclohexa-2,5-dien-1-one, B 126
3-Methyl-4-trichloromethylbiphenyl, B 65
4-Methyl-4-trichlorormethylcyclohexa-2,5-dien-1-one, B 121
α-Methyltropidine, B 358,
γ-Methyltropolone, B 372
4-Methyltropolone, B 372, 374
Methyltropones, B 364
2-Methyltropone, B 361
4-Methyltropone, B 362

Methyl tropyl ether, B 361
Methylvinylcyclohexanes, B 24
2-Methyl-1-vinylcyclohexan-1-ol, B 28
Methylvinylcyclohex-1-enes, B 21, 28
1-Methyl-1-vinylcyclopropane, A 36
2-Methyl-2-vinylcyclopropane-1-carboxylic ester, A 53
Methyl vinyl ether, A 77; D 352
Methyl vinyl ketone, B 129, 285; C 291, 469; D 100, 102, 328
—, reaction with acetylacetone, B 116
—, — with mesityl oxide, B 202
Methylzinc iodide, A 176
4α-Methylzymosterol, D 165
Metrosiderene, C **269,** 288
Metrosideros scandens, C 278
Metrosideros umbellata, C 269
Mevaldic acid, E 58, 59
Mevaldic dehydrogenase, E 59
Mevalonate[2-^{14}C], C 466
Mevalonic acid, B 155, 341, 342; C 174, 465; E 55, 56, 59, 61, 65, 66, 71 84, 89, 91, 98, 101, 104, 105, 114, 119, 122, 123, 126, 128, 129, 135, 137
—, enzyme complex, E 58
Mevalonic acid[2-^{14}C], D 390; E 72, 73, 75, 82, 88, 113, 119, 120, 133, 135
Mevalonic acid,[3-^{14}C], D 390
Mevalonic acid kinases, E 59, 137
Mevalonic lactone[2-^{14}C], C 463; E 113
Mexicanins, C 359, **361,** 368
Mexicanolide, C 411, 412, 431
Mexicanolide group, triterpenoids, C 431
Mexogenin, E 27, **35**
Michael reaction, C 31; D 99, 141; E 23
Microcephalin, C **340,** 366
Micrococcus lysodeikticus, E 66, 86, 134
Miescher-Wettstein degradation, D 26, 233, 253, 323
Miliacin, C 438
Milloside, D 364, **398**
Mint glyoxal, B 203
Mokko lactone, C **351,** 367
Molecular complexes, saponins, E 4
Molecular rotation, steroid hydrocarbons, D 62, 63
Molybdenumhexacarbonyl, A 126, 130; C 91
Momordica charantia, D 166
Monardo fistulosa, E 77
Monoacetylenes, B 387
Mono-*O*-acetylneriifolin, D 406
Monoanhydrostrophanthidin, D 378, 379
Monocyclic monoterpenoids, B 153 ff
Monohydrolycopene, B **281**
Monogynol, B 435, 470
Monohalogenocyclohexanes, physical constants, B 33
Monohydrodoisynolic acids, D 224
Monohydroxysteroids, physical properties, D 35
Mono-oxo steroids, physical properties, D 90
Monoterpenes, E 66 ff
—, absolute configuration, C 249
—, bicyclic, E 79
—, biological oxidation, E 76
—, biosynthesis, E 71
—, cycloheptanes, E 78
—, furan rings, E 78
—, oxidation, E 76
—, peppermint, E 73
Monoterpenoids, B 153 ff
—, absolute configuration, B 154
—, acyclic, B 156
—, aldehydes, B 195
—, bicyclic, C 136 ff
—, biogenesis, B 154
—, cyclic ethers, B 204
—, diols, B 193
—, epoxyketones, B 202
—, heterocyclic, B 169
—, hydroxyketones, B 202
—, ketones, B 196
—, monocyclic B 171 ff
—, phenolic, E 70
Monoterpenoid diosphenol, B 202
Monotropein, B 214
Moretenol, C 417, 418, 454
Morolic acid, C 439
Morpholine, B 105; D 357
Moschus moschiferus, B 416
Motidiol, C 455
Motiol, C 455
Mould metabolites, C 256
Muconic acid, B 371
Mucor hiemalis, B 340
Multiflorenol, C 413, 414, 415, 445, 475, 476, 477
Multiflorenyl acetate, C 476
Munetone, E 64
Muricholic acid, D 172
Musaroside, D **403**
Muscone, A 1; B 380, 409, 414, **416**
Musk, B 414, 416
Mustakone, C **298,** 334
Mutatochrome B **303**

Mutatoxanthin, B **299**
Mutilin, biosynthesis, C **463**
MVA = Mevalonic acid, *q.v.*
Mycelinamide, E 72
Mycobacteriaceae, B 279
Mycobacterium phlei, B 305, 317, 331
Mycosterols, D 2, 3, 134
Myoporum crassifolium, C 292
Myoporum lactum, C 290
Myrcene, B 154, **158**, 159, 160; C 176; E 67, 71, 76
Myrcenol, B 158, 159
Myrcia acris, B 158
Myricadiol, C 445
Myricolal, C 445
Myrothecium species, C 333
Myrtenal, C **180**, 187; E 77
Myrtenal epoxide, C 187
Myrtenic acid, C **180**, 181
Myrtenol, C 175, **180**, 186; E 77
Myrtenyl chloride, C 180
Myrtillogenic acid, C 440
Myrtus communis, C 180
Myxinol, D 195
Myxoxanthin, B **314**
Myxoxanthophyll, B 238, 239, **331**

NAD = Nicotinamine adenine dinucleotide, E 56
NADPH = Nicotinamide adenine dinucleotide phosphate, E 56, 58, 93, 98, 101, 102, 103, 122
Naginata ketone, B 170
Nametkin rearrangement, C 137, 159, **164**, 165, 166, 210, 216, 238
Naphtha, B 13
Naphtha[2′,3′:2,3]-5α-androst-2-ene, D *452*
Naphthalene, D 144
—, hydrogenation, C 62
—, reduction, C 69
N-Naphthalene-2-sulphonylmenthylamines, B 228
Naphthaquinones, C 116
Naphthenates, A 195
Naphthenes, B 13
Naphthenic acids, A 195
Naphthoic acids, hydrogenation, C 67
β-Naphthoic acid, D 144
Naphthols, hydrogenation, C 63
β-Naphthol methyl ether, D 247
Naramycin A, B 101
Narcissus majalis, B 237
Nardostachys jatamansis, B 274, 304, 309, 318
Narthecium ossifragum, D 164
Narthogenin, E **49**
Natural products, structure and configuration, C 477
Navicula torquatum, B 274
Nef reaction, C 213; D 99
Neoabietic acid, C 392
Neoacorone, C **312**
Neoallo-ocimene, B 160, 161
Neobotogenin, E 26, **34**
Neo-β-carotenes, B 251
Neocarvomenthol, B 230
Neocarvomenthylamine, B 229, 230
Neochlorogenin, E 26, 50
Neochlorogenin diacetate, E 50
Neodigalogenin, E 27
Neodigitogenin, E 27
Neoergosterol, D 135
Neogenin, D *435*
Neoglucodigifucoside, D **387**
Neohelenalin, C 359, 361, 368
Neointermedeol, C **301**, 334
Neo-α-irone, B 178
Neo-β-irone, B 178
Neo-γ-irone, B 178
Neoiso-α-irone, B 178
p-Neoisomenthane-2,8-diol, B 222
Neoisomenthol, B 185, 186, 219, 224
Neoisomenthylamine, B 219, 226
Neoisomenthyltrimethylammonium hydroxide, B 219
Neoisopinocampheol, C 178
Neoisopulegol, B 191
Neoisothujyl alcohol, C 141, 142, 253
Neoisoverbanol, C 184
Neoisoverbenol, C 182
Neokammogenin, E 27
Neo-β-kryptoxanthin A, B 302
Neomanogenin, E 27
p-Neomenthane-2,8-diol, B 222
p-Neomenth-8-ene-1,2-diol, B 221
Neomenthol, B 185, 186, 219, 224
Neomenthylamine, B 219, 226, 230
Neomenthyl chloride, B 18
Neomenthyltrimethylammonium hydroxide, B 219
Neomexogenin, E 27
Neomotiol, C 417, 418, 455
Neomycins, B 83
Neonepetalactone, B 210
Neonogiragenin, E 26
Neopinocampheol, C 178
Neopregnenolone, D 278
Neoquassin, C 433
Neoruscogenin, E 51
Neosapogenins, E 11
Neospongosterol, D 3

Neothujyl alcohol, C 138, 141, 142, 143, 253
Neotigogenin, E 11, 12, 14, 21, 22, 23, 25, **30**, 52, 112
Neotigogenone, E 52
Neoverbanol, C 182
Neoxanthin, B 237, 238, 241, **330**, 344; E 120, 122, 130
—, biosynthesis, E 131
Neozeaxanthin A, B 252
Nepata cataria, B 209
Nepetalactone, B **209**, 210; E 72
Nepetalic acid, A 166; B 210
Nepetalinic acids, B 210, 211; C 315
Neriantin, D 364, **394**, **395**
Neriantogenin, D 364, **394**, 395
Neriifolin, D 362, **386**, 388, **406**
Nerium indicum, D 391
Nerium odorum, D 386, 405
Nerium oleander, D 391
Nerol, B **162**, 163, 167, 170; E 68
Nerolidol, B 276; C 259, **289**, 290, 334
Neroli oil, B **162**
Nerolidyl pyrophosphate, B 343
Neryl pyrophosphate, E 68, 74, 83
Neurospora crassa, B 275, 276, 326
Neurospora sitophila, B 326
Nerosporaxanthin, B **326**
Neurosporene, B 231, 239, 257, **276**, 310, 334, 335, 336, 338, 344; E 124, 125, 126, 130
Ngai-camphor, C 226
Ngaione, C **290**, 291, 334
Nickel-acetoacetic ester, B 393
Nickel acetylacetonate, B 28
Nickelcarbonyl, C 99
Nickel(II)2-ethylhexanonate, B 2
Nickelocene, A 130, 131, 139, 140, 147
Nickeltetracarbonyl, A 130
Nicotiana tabacum, E 105, 106
Nicotinamide adenine dinucleotide, E 56
Nicotinamide adenine dinucleotide phosphate, E 56, 58, 93, 98, 101, 102, 103, 122
Nimbin, C 411, 412, 432
Nimbin group, triterpenoids, C 432
Nimbiol, C 378, 379, 392
Nitriles, $\alpha\beta$-unsaturated, A 20
p-Nitroaniline, D 199
—, reduction, B 82
p-Nitrobenzoic acid, E 101
ω-Nitrocamphene, C 196
Nitrocamphors, C 212
—, physical properties, C **213**
3-Nitrocamphor, C 213
α-Nitrocamphor, C 233
α'-Nitrocamphor, C 212
6-Nitrocholest-4-en-3-ones, D 128
6-Nitrocholesteryl acetate, D 17, 37
2-Nitrocyclohexa-2,4-diene, B 80
Nitrocyclohexanes, B **77**
—, reduction, B 80
Nitrocyclohexane, B **78**
2-Nitrocyclohexane-1,3-diols, B 80
1-Nitrocyclohex-1-ene, B **80**
1-Nitrocyclohexylmethanol, B 78
Nitrocyclopentanes, A 152
Nitrocyclopropanes, A 45
9-Nitrodecalins, C 66
p-Nitrodimethylaniline, A 174
3-Nitro-2,3-diphenylcyclopropane-1,1'-dicarboxylic methyl ester, A 59
Nitroethylene, C 111
Nitroferrocene, A 134
Nitro-inositols, B 85
Nitromethane, reaction with glutaraldehyde, B 80
1-Nitromethylcycloheptane, B 353
1-Nitronorbornane, C 89
22-Nitro-16-oxocholesterol, D 99
Nitropyrazolines, A 45
Nitrosates, cyclohexenes, B 19
Nitrosites, cyclohexenes, B 19
Nitrosochlorides, cyclohexenes, B 19
N-Nitrosocyclopropyl methylurethane, A 47
N-Nitroso-*N*-(2,2-diphenylcyclopropyl)urea, A 46
Nitrosolimonene, B 201
Nitrosopinene, C 177, 178, 186
Nitrosotropolones, B 370, 373, 375, 377
γ-Nitrosotropolone dioxime, B 378
β-Nitrostyrenes, B 79
—, reaction with butadiene, B 78
p-Nitrotoluene, E 101
4-Nitro-*m*-toluic acid, B 374
Nitropolones, B 373, 377
Nocardia restrictus, D 204, 381
Nogiragenin, E 26, **33**
Nologenin, E **44**, 45
Nolonin, E 44
Nomenclature, steroids, D *422*
Nomilin, C 430
Nonacarbonyldiiron, A 85
Nona-1,8-diene, B **382**
Nonadecapeptide, D 287
Nonafluorocyclohexanes,B 39
Nonafluorocyclohexenes, B 42, 43

Nonafluorocyclopentane, A 150, 151
Nonane-1,3,9-tricarboxylic acid, A 198
Nonaprenoxanthin, E 132
Nootkatene, C **273**, 288
Nootkatin, B 365
Nopinene, C **174**
Nopinic acid, C 174, 175, 176
Nopinone, C 175, 176
19-Nor-17α-acetoxyprogesterone, D **245**
Norallocholanic acid, D 179
19-Norandrosta-4,7-dien-3,17-dione, D 218
A-Nor-5α-androstane, D *444*
18-Norandrostane, D *428*
19-Norandrostane, D 443
A-Norandrostane derivatives, D 283
A-Norandrostane-3,17β-diol-3α-carboxylic acid, D **283**
D-Norandrostane-3,16-dione, D 286
19-Norandrosta-1,4,6-trien-17β-ol-3-one acetate, D 221
19-Norandrost-4-ene-3,17-dione, D **301**
19-Norandrost-5(10)-en-17β-ol-3-one, D 229
19-Norandrosterone, D 256
Norascaridole, B **75**
Norbixins, B 327
Norbolethone, D 274, 275
Norbornadiene, C 89, 91, 93, 94, 110
7-Norbornadienol, C 93
7-Norbornadienol acetate, C 94
Norbornanes, C **89**, 233
Norbornane, C 89, 136
Norbornane-1-carbaldehyde, C 111
Norbornane group, C **193**, **233**
Norbornanols, C 91
2-Norbornanol, C 83
7-Norbornanol, C 93
Norbornanyl *p*-toluenesulphonate, C92
Norbornenes, C 233
Norbornene, C 80, 89, 92
Norbornene epoxide, C 91, 93, 94
Norbornenols, C 92, 93
7-Norbornenone, C 123
7-Norbornenyl cations, C 80
7-Norbornenyl *p*-toluenesulphate, C 93
7-*anti*-Norbornenyl tosylate, C 46
π-Norborneol, C 189
exo-Norborneol, A **162**
2-Norbornyl acetate, C 82
exo-Norbornyl acetate, A 162
Norbornylane, C 137
*endo*Norbornyl arenesulphonates, C 162
2-Norbornyl brosylate, C 80, 81, 82, 83
1-Norbornylcarbinol, C 111
Norbornylcarbonium ion, C 92
Norbornylene C 97
2-Norbornyl *p*-toluenesulphonates, C 82
Norcamphane, C 137
Norcamphor, C 95, 265
Norcamphoric acid, A 192
Norcaradienes, C 38
Norcarane, 38, C 136
Norcardienecarboxylic acid, B 356
Norcaryophyllenic acid, A 67; C 284
Norcedrenedicarboxylic anhydride, C 279
24-Nor-5β-cholane, D *443*
Norcholanic acids, D 191
5α-Norcholanoic acid, D 179
5β-Norcholanoic acid, D 24, 25, 179
5β-Norcholan-24-oic acid, D 169
24-Norcholanoic acid, D 191
24-Nor-5β-cholan-23-oic acid, D *443*
Norchol-4-ene-3,22-dione, D 240
A-Nor-5α-cholestane-2α-carboxylic methyl ester, D 119
A-Nor-5α-cholestan-2-one, D 124
A-Norcholest-3(5)-ene-1,2-dione, D 281
B-Norcholesterol, D 82
B-Norcholesteryl acetate, D 284
31-Norcycloartanol, E 100
19-Nordeoxycorticosterone, D 301, 376
19-Nordeoxycorticosterone acetate, D 301
Norechinocystadienol, C 442
19-Norergosta-5,7,9-triene, D 132
19-Norergosta-5,7,9-trien-3-one, D 136
18-Norestradiol methyl ether, D 262
18-Norestrone, D **227**
—, *nomenclature*, D *443*
18-Norestrone methyl ether, D 227, 262
Norethisterone, D 240, 262, 347
19-Nor-10-ethylhydrocortisone, D 331
18-Nor-17α-ethynyl-13β-propyl-estradiol, D 228
19-Nor-17α-ethynyltestosterone, D 301
Norethynodrel, D 240, 244
A-Nor-B-homosteroids, D 284,
C-Nor-D-homosteroids, D 285
C-Nor-D-homotestosterone acetate, D 285
18-Nor-14-isoequilenin, D 223
18-Nor-14-isoequilenin 3-methyl ether, D 222

A-Nor-19-nortestosterone, D 282
Noroxo-agarofuran, C 304, 335
19-Nor-11-oxocard-20(22)-enolides, D 402
21-Nor-20-oxocholesteryl acetate, D 153
19-Nor-Δ^4-3-oxosteroids, D 256
Norparthenone, C 356
Norpinanane, C 136
Norpinic acids, A 73, 97; C 168, 169
Norpinic anhydride, C 170, 171
C-Norprednisone acetate, D 284, 285
19-Norpregnanes, D 284
19-Norpregnane, D *443*
21-Norpregnane, D *443*
C-Norpregnane derivatives, D 284
C-Norpregnane-3,11,20-trione, D 285
18-Nor-4-pregnane-3,20-dione, D *443*
19-Norpregn-4-en-20α-ol-3-one, D 246
A-Norpregn-3(5)-en-20β-ol-2-one, D 282
D-Norpregn-5-en-3β-ol-20-one, D 286
18-Norprogesterone, D 241
19-Norprogesterone, D 240, 241, 244, 245, 246, 376
21-Norprogesterone, D 240
A-Norprogesterone, D 282, 283
B-Norprogesterone, D 284
D-Norprogesterone, D 286
Norsantonins, C 343
Norsolannellic acid, D 20
19-Norsteroids, D 87, 88, 204, 229, 274
—, gestonic, D 241
A-Norsteroids, D 281
B-Norsteroids, D **284**
C-Norsteroids, D **284**
D-Norsteroids, D **286**
19-Norsterols, D 86
18-Nortestosterone, D 259
19-Nortestosterone, D 243, 256, 301
A-Nortestosterone, D 282, 283
19-Nortestosterone acetate, D 217, 221
A-Nortestosterone acetate, D 282
19-Nortestosterone adamantoic ester, D 273
19-Nortestosterone 17-(cyclopent-1-enyl) ether, D 273
Nortricyclanone, C 93
Nortricyclene, C 80
3-Nortricyclenecarboxylic acid, C 92
Nortricycloekasantalic acid, C 263
Nortricyclonium ions, C 80
19-Nor-17α-vinyltestosterone, D 301
Nuatigenin, E **47**, 48
Nuciferal, C **293**, 334
Nuciferol, C **293**, 334
Nuclear magnetic resonance spectrometry, D 69
—, steroids, D 33, 69
—, structure determination, C 477
Nutriacholic acid, D 189
Nyctanthic acid, C 410, 414, 439
Nylon, B 387

Obacunone, C 411, 430
Obtusifoldienol, C 421
Obitusifoliol, E 107
Occidentalol, C **306**, 335
Occidol, C **306**, 335
Ochromonas malhamensis, E 102, 105, 106, 109, 110
Ocimene, B **159**, 160
Ocimum basilicum, B 159
Octacholorocyclopentene, A 150, 152
Octachlorodicyclopentadiene, C 102
Octachlorofulvalene, C 12
Octachloropenta-1,3-diene, A 152
1,18-Octadec-9-enedioic acid, B 415
n-Octadecyclopentane, A 113
Octadecyclopent-1-enes, A 116
Octa-1,7-diene-3,5-diyne, B 384
Octafluorocyclobutane, A 87, 88
Octafluorocyclohexa-1,4-diene, B 43
Octafluorocyclohexanes, B 39
Octafluorocyclohexenes, B 42,43
Octafluorocyclopentanes, A 150, 151
Octahydroanthracene, C 76
Octahydroanthraquinone, C 76
Octahydrodehydroxylinderene, C 303
Octahydro-1,4:5,8-dimethanonaphthalene, C 104, 105
Octahydrodimethanonaphthyl cations C 80
Octahydro-1,4-ethanonaphthalene, C 116
Octahydroestrone, D 255
Octahydro-4,7-methano-indene-*cis*-1,3-dicarboxylic acid, C 105
Octahydronaphthalenes, A 76; C **68**
Octahydrophenanthrene, C 75
Octahydrotrianhydroperiplogenin, D 397
Octahydroxycyclobutane, A 94
Octalene, C 73
Octalins, C 62, **68**, 69
Δ^4-Octalin, C 10
$\Delta^{9(10)}$-Octalin, C 64
Octalin-1,5-dione, C 69
α-Octalol, C 69
Octalones, C 307
α-Octalone, C 69

Octamethylcyclo-octatetraene, B 392
Octant Rule, C 479
Octamethyltricyclo[$4.2.0.0^{2,5}$]-octa-3,7-diene, C 43
Octaphenylcubane, C 44
Octaphenylcyclo-octatetraene, B 393; C 45
Octaphenylfulvalene, C 12
Octaphenyltetramethylenecyclobutane, A 81, 84
6-Octen-1-yne, A 109, 117
Octotillol, C 425
1-*n*-Octylcyclopent-1-ene, A 116
Oct-4-yne, A 49
Odoratin, C 412, 434
Odorigenin B, D **396**
Odorobioside G, D 362, 386, 388
Odorosides, D **405**
Odoroside A, D 362, **386**, 388
Odoroside B, D 364, **396**
Odoroside G, D **386**
Odoroside H, D 362, **386**
Odoroside G acetate, D 386
Odorotrioside G mono-acetate, D 362
Oestrogens, D 2, 197, 198 ff, 199
—, assay, D 199
—, carcinogenesis, D 227
—, chromatographic assay, D 200
—, halogenation, D **338**
—, properties, D 201
—, structure, D 200
Oestrones, biosynthesis, E 112
Oestrone; see Estrone
OH-Chlorobactene, B **305**
OH-Rhodopin, B 307
OH-Spheroidene, B 311, 338, 345
OH-Spheroidenone, B 311, 337, 345
OH-Spirilloxanthin, B **309**, 336, 337
Oil, agarwood, C 303
—, bergamot, C 259, 267
—, calycanthus, D 154
—, camomile, C 292
—, cedar wood, C 279, 325
—, chloranthus, C 271
—, citronella, C 269, 296
—, cloves, C 284
—, copaiba, C 271, 284
—, costus root, C 336, 337, 363
—, cotton seed, D 154
—, cubebs, C 268
—, deodar, C 281
—, elemi, C 295
—, eucalyptus, C 277
—, false cubebs, C 296
—, fusel, C 290
—, galbanum, C 296
Oil, ginger, C 258
—, grape seed, D 165
—, hops, C 258
—, juniper, C 296
—, lavender, C 284
—, lemon grass, C 269
—, neroli, C 290
—, olibanum, E 76
—, orange, C 272
—, *Picea excelsea*, C 118
—, pine needle, B 183; C 188
—, pine tar, B 172
—, rapeseed, D 155, 165
—, rose, B 161
—, sandalwood, C 188, 189, 192, 263, 293
—, soya, D 156
—, sweet-flag, C 262, 274, 292, 312
—, tall, D 154
—, tansy, C 145
—, thuja, C 234
—, turpentine, B 172, 193, 194; C 167 199, 227, 229, 281
—, vetiver, C 269, 279, 296, 300, 323
—, wheat germ, D 154, 156, 165
—, wild bergamot, E 77
—, wormseed, B 204
—, wormwood, C 141; E 86
—, ylang-ylang, C 258, 270
Okenone, B 322; E 128
Oleananes, E 116
Oleanane group, triterpenoids, C 414, 438
Olean-11,13(18)-diene, C 475, 476
Oleandrigenin, D 364, **391**, **394**, 418
Oleandrigenin-cymaroside-glucoside, D 391
Oleandrigenin-sarmentoside-glucoside, D 391
Oleandrigenin-α-L-vallaro-pyranoside, D 387, 391
Oleandrigenone, D 394
Oleandrin, D 364, **391**, 394
Oleandrose, D 361, 365, 402
Oleanolaldehyde, C 439
Oleanolic acid, C 439
Oleanolic acid acetate, C 439
Oleanonic acid, C 441
Olefins, A 31
—, cyclic, B **382**
—, Diels-Alder addition to dienes, B 16
—, reaction with diazoacetic ester, A 26, 52
—, — with dihalogenocarbenes, A 38
—, — with enamines, A 76

Olefins, reaction with ethynylcarbenes, A 34
cis-Olefins, conversion to optically active alcohols, C 173
Oleum bergamottae verum, C 267
α-Onicerin, C 458
Onoceradiene, C 474
Onocerin, C 419, **473**
α-Onocerin, C 418, 419; E 116
—, synthesis, C 473
γ-Onocerin, C 419
α-Onocerin acetate, C 474
Onocerin diacetates, C 474
Ononitol, B 61
Ophiobolin, C 403, 404, 405, **463**, 464, 480; E 92
Ophiobolus species, C 403
Oppenauer oxidation, D 83, 131, 140, 143, 152, 153, 154, 158, 184, 232, 233, 234, 237, 238, 240, 256, 258, 259, 279, 305, 310, 322, 323, 337, 350; E 30
Opsin, B 289
Optical rotary dispersion, steroids, D 64
—, structure determination C 477, 479
Optical rotation, steroids, D 33, 62
Oreoselone, E 64
Organometallic compounds, from cyclopropanes, A 40
Origanene, C 138
Origidin, D 362, **391**
Oscillatoria rubescens, B 331
Oscillaxanthin, B 331
Osmane, B **208**
Osmanthus fragrans, B 208
Osmocene, A 131, 135, 145
Ostreasterol, D **157**
Ostruthin, E 70
Ouabagenin, D 364, 368, **401**
Ouabagenin tetra-acetate A, D 402
Ouabain, D 364, 368, **401**
Ouabain 19-methyl ether heptahydrate, D 401
Ovatodiolide, C 401
Ovo-ester, B 320
Ovoverdin, B 320
7-Oxabicyclo[2.2.1]heptane, B 73
7-Oxabicyclo[4.2.0]octane, B 70
3-Oxacyclohept-4-enyl acetate, B 19
2-Oxacyclohex-3-enyl acetate, A 114
2-Oxahydrindanes, C 56
Oxalic acid, A 114; C 327; D 416
Oxalic ester, A 58, 120, 201
—, condensation to give 5-membered rings, A 106
Oxalic ester, reaction with ketones, A 187
Oxalic methyl ester, D 219
Oxalyl chloride, D 93, 338
1-Oxaspiro[5.3]nonane, B 71
1-Oxaspiro-[2.5]-octane, B 24, 94
1-Oxa-2,5,5,9-tetramethylhexahydronaphthalene, B 176
Oxazolines, B 418
Ox bile, D 169
α-Oxidase, E 74
5α,6α-Oxides, D 95
16,17-Oxido-3β-acetoxypregn-5-en-20-one, D 310
16α,17α-Oxido-21-acetoxyprogesterone, D 310
6,19-Oxidoandrost-4-ene-3,17-dione, D 204
3α,9α-Oxido-5β-chol-11-enoic dibromides, D 57
Oxidocholestanes, D **113**, 114
2α,3α-Oxido-5α-cholestane, D 120
3β,4β-Oxido-5α-cholestane, D 44
3α,5-Oxido-5α-cholestane, D 45, 104
5,6α-Oxido-5α-cholestane, D 46, 47
5,6-Oxido-5β-cholestane, D 46
1α,2α-Oxido-5α-cholestan-3-one, D 34, 124, 129
3β,5-Oxido-5β-cholestan-6-one, D 45
4α,5-Oxido-5α-cholestan-3-one, D 47, **105**, 125
4β,5-Oxido-5β-cholestan-3-one, D 47, 105, 125, 128
4α,5-Oxido-5α-cholestan-3β-yl acetate D 103
6,19-Oxidocholest-4-en-3-one, D 204
7,8-Oxido-digitoxigenin, D 406
6,19-Oxido-9α-hydroxyandrost-4-ene-3,17-dione, D 204
4β,5β-Oxido-3-ones, D 94
16α,17α-Oxido-5α-pregnane-3,20-dione, D 60
20β,21-Oxidopregna-1,4,16-triene-11β-ol-3-one, D 353
16α,17α-Oxidoprogesterone, D 357
2,3-Oxidosqualene, C 407, 466
5,6-Oxidosterols, D 83
11α,12α-Oxidotaraxerol, C 476
4β,5β-Oxidotestan-17β-ol-3-one, D 340
24,26-Oxido-3α,7α,12α,27-tetrahydroxy-5β-cholestane, D 195
24,26-Oxido-3α,7α,12α-trihydroxy-5β-cholestane, D 195
Oximinocamphor, C 212, 219, 221, 233
Oximinoepicamphor, C 219, 226, 233
2-Oxoadamantane, C 126

β-Oxoadipic acid ethyl methyl ester ethylene acetal, A 185
11-Oxo-α-amyrin, C 450
16-Oxoandrostanes, E 19
7-Oxoandrostane-3β,17β-diol diacetate, D **277**
2-Oxobicyclo[3.2.2]nonane-6,7-dicarboxylic acid, C 124
2-Oxobicyclopentyl, C **11**
11-Oxo-bile acids, D 184,
3-Oxobisnorchol-11-enic esters, D 313
2-Oxobornane, C 199
3-Oxobornane, C **224**
5-Oxoborneol, C **222**
4-(3-Oxobutyl)androst-4-ene-3,17-dione, D **230**
Oxocamphors, C **218, 223**
5-Oxocamphor, C **222,** 223
6-Oxocamphor, C **223**
10-Oxocamphor, C **223**
Oxo-β-carotene, B 316, 332
4-Oxo-γ-carotene, B 317
Oxocarotenoids, B 312, 313
6-Oxocativic acid, C 383
Oxocholanic acids, D 189
Oxocholanoic acids, D 190
3-Oxo-5β-cholanoic acid, D 185
11-Oxo-5α-cholan-24-oic acid, D *437, 438*
12-Oxo-5β-cholanoic acid, D 21, 22, 24, 25, 54
3-Oxochol-11-enic esters, D 313
3-Oxochol-4-enoic acid, D 185
3-Oxochol-4-enoic methyl ester, D 188
11-Oxo-5α-cholestan-3β-ol, D 106
6-Oxocholestanyl acetate, D 37, 48
7-Oxocholestanyl acetate, D 37, 106
1-[2-(3-Oxo-5α-cholestanyl)]pyridinium bromide, D 120
Oxocholesterols, D **109**
7-Oxocholesterol, D 96, **109,** 146
22-Oxocholesterol, D 108, **109,** E 113
24-Oxocholesterol, D 108, **109,** 152, 157
24-Oxocholesterol tetrahydropyranyl ether, D 153
7-Oxocholesteryl acetate, D 37, 53, 85, 107, 109
7-Oxocholesteryl chloride, D 42, 104
22-Oxocholesteryl methyl ether, D 99
2-Oxocycloheptane-1-carboxylic ester, B 357
2-Oxocycloheptane-1-glyoxylic ester, B **357**
Oxocyclohexanecarboxylic acids, B 143
Oxocyclohexanedicarboxylic acids, B 150
Oxocyclohexenecarboxylic acids, B 144
(2-Oxocyclohexylmethyl)malonic ester, B 108
3-Oxocyclopent-2-enylacetic acid, B 376
3-Oxocyclopent-4-enylacetic acid, B 376
3-Oxocyclopentylacetic ester, C 54
2-Oxo-6,7-dicarboxybicyclo[3.2.2]-nona-3,8-diene anhydride, C 124
Oxodicyclopentadiene, A 172
7-Oxoepicholesteryl acetate, D 104
γ-Oxo-esters, condensation with succinic ester, A 169
7-Oxoestrone, D **217**
16-Oxoestrone, D **211**
19-Oxo-5α,17(αH)-etianic acid, D *438*
3-Oxo-5α-etianic methyl ester, D 102
3-Oxoetia-4,9(11),16-trienic methyl ester, D **328**
3-Oxoeti-4-enic acid, D 298, **300**
3-Oxoeti-11-enoic esters, D **313**
3-Oxoetioallocholanic methyl ester, D **234**
11-Oxoetiolithocholic acid, D **317**
3-Oxoeti-4,9(11),16-trienic methyl ester, D 101, **102**
7-Oxogitogenic acid, E 33
19-Oxogitoxigenin, D 396
19-Oxogitoxigenin rhamnoside, D 396
6-Oxogrindelic acid, C 384
9-Oxo-1,17-heptadecanedioic acid, B 415
5-Oxohexanoic acid, A 112
22-Oxo-20-isocholesterol, D **153**
Oxoisodrimenin, C 348
Oxoketenes, dimerisation, A 75
Oxomalonic ester, A 63
2-Oxomanoyl oxide, C 383
3-Oxomanoyl oxide, C 383
19-Oxomarinobufagin, D 419
12-Oxonor-5β-cholanoic acid, D 24, **25**
20-Oxo-21-norcholesterol, D 108
25-Oxo-26-norcholesterol, D 108
4-Oxopentane-1-carboxylic acid, C 181
3-Oxoplatanic acid, C **436**
20-Oxopregnanes, E 1
16-Oxopregnan-20-ol, E 19
11-Oxoprogesterone, D **235,** 329
2-Oxorhodovibrin, B **312,** 337; E 130
12-Oxosapogenins, E 35
δ-Oxosebacic acid, C 66
6-Oxo-2,3-seco-5α-cholestane-2,3-dioic acid, D 17
5-Oxo-3,5-seco-A-norcholestan-3-oic acid, D **126, 127**

2-Oxospirilloxanthin, E 130
6-Oxospiro[5.4]decane, C 66
12-Oxospirostans, E 35
1-Oxospiro[5.5]undec-3-ene, B 108
Oxosteroids, physical properties, D 90
Oxo-5β-steroids, D 90
3-Oxosteroids, D 88, 110, 340, 345
—, assay, D 270
11-Oxosteroids, D 11, 55, 88, 89, 319, 320, 321
12-Oxosteroids, D 54, 312
15-Oxosteroids, D 90, 268
17-Oxosteroids, D 60, 76, 200, 268
—, assay, D 269, 270
—, enamines, D 357
20-Oxosteroids, D 61, 407
6-Oxotestosterone, D 271
16-Oxotestosterone, D 271
3-Oxototarol, C 393
18-Oxototarol, C 393
19-Oxouzarigenin, D 364, 398
Δ^2-Oxyallobetulene, C 437
Oxyallobetulin, C 437
Oxyallobetulone, C 437
Oxy-Cope rearrangement, C 122
Oxymethalone, D 273, 275
Oxistigma manii, C 271
Ozic acid, C 384

Pi = Phosphate ion, E 56
Pachymic acid, C 421
Palitantin, B **105**, 106
Palitantol, B 106
Palmarin, C 385, 387
Palmitoyl chloride, B 296, 297
Palustric acid, C 392
Palustrol, C **320**, 321, 336
Pancuronium bromide, D 359
Paniculidine, E **53**
Paniculogenin, E **50**
Panstroside, D 364, **398**, **403**
Paper chromatography, bile acids, D 178
—, steroids, D 71
Paracentrotus lividus, B 314
Paramycin, B 83
Parasantonide, C **345**
Parchoulenol, C 328
Parillin, E 3
Parkeol, C 407, 421; E 106
Parquetina nigrescens, D 384
Parthenin, C **355**, 356, 358, 367
Partheniol C **316**, 317, 335
Parthenium argentatum, C 316
Parthenium hysterophorus, C 355
Parthenium species, C 357
Parthenolide, C **363**, 368
Pasteurella pestis, A 69
Patchoulenes, C 327, 328, 329
Patchoulenone, C 279, **329**, 336
Patchouli acetate, C 328
Patchouli alcohol, C 279, **327**, 328, 329, 336
Patchouli diol, C 328
Paulioside, D 364, **398**
Pectenaxanthin, B 331
Pencil-wood, C 275
Penicillium frequentans, B 105
Penicillium palitans, B 105
Penillium puberulum, B 366
Penicillium species, metabolites, B 366
Penicillium stipitatum, B 366
Penicereus fosterianus, D 151, 152
Peniocereus macdougalli, D 152
Peniocerol, D **151**, 152, 164
—, ultraviolet absorption spectra, D 68
Pennogenin, E **44**, 45
Pennyroyal oil, B 100
Pentabromocyclobutane, A 87, 88
Pentachloro-4-pentachlorophenoxy-cyclohexa-2,5-dien-1-one, B 121
Pentachloropentafluorocyclopentanes, A 150
Pentachlorophenol, B 121
Pentacyclic triterpenes, C 478, 479
Pentacyclododecan-4-ol, C 106
Pentacyclopentadeca-3,11-diene, C 102
Pentacyclopentadienes, C **103**
Pentacyclo [$4.2.0.0^{2,5}.0^{3,8}.0^{4,7}$] octane 1,4-dicarboxylic acid, C 44
Pentacyclosqualene, C 418
Pentacyclotetradecane, C 134
Pentadecan-3-one-1,15-dicarboxylic acid, A 196
Pentadecan-1,3,15-tricarboxylic acid, A 196
Penta-2,4-dienoic acid, B **146**
Penta-3,4-dienoic methyl ester, C 22
Pentadienolides, D **408**
Pentadienylcobalt, A 86
Penta-2,4-dienylrethrin, A 180
Pentaerythrityl tetrabromide, A 73, 82; C 21, 26
Pentafluorocyclopentenes, A 151
Pentahydroxy-5α-bishomocholane, D 195
Pentahydroxy-5β-bishomocholane, D 195
Pentahydroxycholestanes, D 195
Pentahydroxy-5β-cholestane, D 194

Pentahydroxy-5β-cholest-7-en-6-one, D 160
Pentahydroxycyclohexanes, B 44
Pentahydroxycyclohexanones, B 57, **106**
Pentalanes, C **51**
Pentalane, C 108
Pentalane-1,3-dicarboxylic acid, C 100
Pentalene, C 51
Pentamethylcyclohexane, B 14
Pentamethylcyclohexane-1,3,5-trione, B 113
Pentamethylcyclopentadienylcarbinol, A 163
Pentamethylene dibromide, B 144
Pentamethylenefulvene, C 14
n-Pentane, A 26
Pentanetetracarboxylic ester, A 191; B 150, 357
Pentanetricarboxylic ester, B 150
Pentan-4-one-1-carboxylic acid, A 176
Pentaphenylcyclopentadiene, A 157
Pentaphenylcyclopentadienol, A 122, 157
Pentaphenylcyclopentadienyl bromide A 122
Pentaphenylcyclopent-3-en-1-one, A 157
Pentaxanthin, B 331
Pent-4-ene-1-al, A 164
Pentologenin, E 27, **42**
Penupogenin, D 285
Peppermint monoterpenes, E 73
Peppermint oil, B 100, 184, 196, 208
Peptides, B 414
—, from adrenocortical extracts, D 287
Peptidoglycan, biosynthesis, E 134
Perbenzoic acid, C 38; D 410
Perchloroallene, A 83
Perchlorocyclobutenone, A 94
Perchlorocyclopentenes, A 150
Perchloro-1,2-dimethylenecyclobutane, A 83
Perchloroindene, A 198
Perchloryl fluoride, B 102; D 341, 342, 343, 350
Perezone, C **292,** 334
Perfluoroalkyl iodides, A 139
Perfluorocyclobutene, A 93
Perfluorocyclohexanes, B 37, 38, 39
Perfluorocyclohexene, B 42, 43
Perfluorocyclopentane, A 150
Perfluorocyclopentene, A 150, 151
Perfluoro-(1,2-dimethylcyclohexane), B 42
Perfluorotricyclo[4.2.0.0^{2,5}]octane, C 44
Perhydroacenaphthene, C 74
Perhydroanthracene, B 8; C 76
Perhydroboraphenalene, B 385
Perhydrochrysene, C 74; D 9
Perhydrocoronenes, C 76
Perhydrocrocetin, B 328
Perhydrocyclopenta[α]phenanthrene, D 9
Perhydro-1,2-cyclopentenophenanthrene, D 9
Perhydro-9,10-dihydroxy-phenanthrenes, C 75
Perhydrodimethanonaphthalenes, C **104**
Perhydrodiphenic acids, C 17, 75
—, physical properties, C 19
Perhydrofluorene, C 74
Perhydrofluorenone, B 22; C 74
Perhydrofluoroanthene, C 76
Perhydrolutein, B **295**
Perhydrolycopene, B 266, 335
Perhydronorbixin, B 326, 328
Perhydrophenanthrenes, C **74,** 75
Perhydrophenanthrene system, D 22
Perhydro-9-phenanthrol, C 75
Perhydro-9-phenanthrone, C 74, 75
Perhydropicene, C 76
Perhydropyrocalciferol, D 141
Perhydrostilboestrols, C 16
Perhydrotriphenylene, C 76
Perhydrozeaxanthin, B 299
Peridinin, B 238, 331
Peridinium species, B 331
Perilla citriodora, B 171; E 78
Perilla frutescens, B 170, 171; E 78
Perilla ketone, B 171; E 78
Perillaldehyde, B 86, **196**
Perillen, B **171**; E 78
Perillyl alcohol, B **196**
Periplaneta americana, A 71
Periploca nigrescens, D 384
Periplocin, D 362, 384
Periplocymarin, D 362, 384, 398
Periplogenin, D 362, 374, 379, 384, 385, 388, 397, 417
Peritruxillic acid, A 100
Peritruxillic anhydride, A 97
Pernitrosocamphor, C 232
Perovskia scrophulariaefolia, C 277
Peroxides, cyclohexane series, B 74
2α,5-Peroxido-5α-cholest-3-ene, D 117
5,8α-Peroxido-5α-ergosta-6,22-dien-3-one, D 136
Peroxyadipic acid, B 98

Peroxytrifluoroacetic acid, B 78, 79; C 86
Peroxytrifluoroacetic acid-boron trifluoride, B 121
Perphthalic acid, B 299
Persicaxanthin, B 333
Petasalbin, C **308**, 335
Petasin, C **307**, 335
Petasitolides, C **350**, 367
Petasol, C 307
Petasytes albus, C 273
Petasytes officinalis, C 273, 307, 308, 349, 350
Pettenkofer reaction, D 178
Pettenkofer reagent, D 270
Phaeophyceae, B 238, 328
Phebalium aquameus, C 320
Phellandral, B 86, **195**; E 75
α-Phellandrene, B 25, **182**, 183, 207, 218; C 251; E 75
—, autoxidation, B 221
β-Phellandrene, B **183**, 200
α-Phellandrene nitrosite, B 182, 229
Phellandric acid, B 195; E 75
Phellandrol, B 195
Phellodendron, D 166
9,10-Phenanthraquinone, C 75
Phenanthrene, C 18, 74
Phenethylcyclohexanols, C 469
Phenols, B 98
—, catalytic hydrogenation, B 2
—, from cyclohexadienones, B 124
—, hydrogenation, B 44, 46, 92
Phenolic hemiterpenes, E 62
2-Phenoxycyclopropane carboxylic acids, A 56
Phenylacetaldehyde, B 395
Phenylacetic acid, hydrogenation, B 135
Phenylacetonitrile, A 95, 190
—, condensation with 1,4-dihalides, A 104
Phenylalanine, B 140
Phenylalkylamines, hydrogenation, B 86
α-Phenylaminocamphor, C **221**
Phenyl azide, A 114
2-Phenylbicyclopropyl, C 7
4-Phenylbuta-1,3-dien-1,1-dicarboxylic acid, A 99
N-Phenylcarbamoylmethylamine, B **228**
Phenylcyclobutadienoquinone, A **93**
Phenylcyclobutane, A 80
1-Phenylcyclobutane-1-carboxylic acid, A 95
Phenylcyclobutenedione, A **93**
1-Phenylcyclobutyl acetate, A 77
1-Phenylcyclohexa-1,3-diene, B **26**
Phenylcyclohexane, B 14, **16**
Phenylcyclohexane-1-carboxylic acids, B **133**, **136**
4-Phenylcyclohexanecarboxylic acid, C 17
1-Phenylcyclohexane-1,2-diols, B 51, 53
3-Phenylcyclohexane-1,2-dione, B 110
4-Phenylcyclohexane-1,3-dione, B 109
2-Phenylcyclohexan-1-ol, B 18
4-Phenylcyclohexan-1-ols, B 48
3-Phenylcyclohexan-2-ol-1-one, B 110
4-Phenylcyclohexan-4-ol-1-one, B **105**
Phenylcyclohex-1-enes, B 21
1-Phenylcyclohex-1-ene, B 18, 53
3-Phenylcyclohex-1-ene, B 18
3-Phenylcyclohex-2-enone, B 117, 119
2-Phenylcyclohex-1-enylacetic acid, B 136
2-Phenylcyclohexylacetic acid, B 136
2-Phenylcyclohexyltrimethyl-ammonium cations, B 18
Phenylcyclopentadienes, A 121
1-Phenylcyclopentane-1-carbonitrile, A 190
1-Phenylcyclopentane-1-carboxylic acid, A 190
2-Phenylcyclopentanol, A 115
Phenylcyclopentenes, A 109, 115
1-Phenylcyclopentyl cyanide, A 175
Phenylcyclopropane, A 27, 36, 48
2-Phenylcyclopropanecarboxylic acids, A 28, 54
2-Phenylcyclopropanecarboxylic ester, A 53
Phenylcyclopropylamines, A 46
1-Phenylcyclopropylcarbinol *p*-toluenesulphonate, A 77
2-Phenyldicyclopropyl, A 27
Phenylene-1-acetic-2-propionic acid, B **150**
o-Phenylenediacetic acid, B 149
o-Phenylenediamine, B **320**
1-Phenyl-1,2-epoxycyclohexane, B 53
α-Phenylethylalcohol, B 70, 71
β-Phenylethylamine, hydrogenation, B 86
Phenylferrocenes, A 127
Phenylhydroxycyclopropenone, A 50
2-Phenylimidazoline, B 90
Phenylketen dimethyl acetal, A 49
Phenyllithium, A 93, 118; B **53**, 108, 362, 374, 395, 405, 419

Phenylmagnesium bromide, A 84, 157; B 53, 83, 105, 108, 110
N-Phenylmaleimide, B 30, 363
Phenyl 2-phenylcyclopropyl ketone, A 51
Phenylpotassium B 383
3-Phenylpropyltrimethylammonium halides, A 21
Phenylpropyne, A 33
Phenylsodium A 134
Phenyl styryl ketone, A 51
2,Phenyl-1,1,3,3-tetrachloropropene, A 50
2-Phenyltetrahydro-1,2-oxazines, A 74
N-Phenylthiocarbamoyl-methylamine, B 228
Phenyl transferase, E 85
Phenyl(tribromomethyl)mercury, C 7
Phenyltrimethylammonium bromide, D 268
Phenyltropolones, B 369
2-Phenyltropone, B 362, 364, 374
Phenyl vinyl ether, A 57
Phleixanthophyll, B 331
Phloroglucinol, methylation, B 113
—, reaction with halogens, B 113
Phloroisovalerophenone, B 131
Phocaecholic acids, D 173
Phosphate ion, E 56
Phospholipids, B 60
Phosphomevalonic kinase, E 59
Phosphoranes, reaction with epoxides, A 21
3-Phosphorylated MVA, E 60
Photoisopyrocalciferol, D 138
Photopyrocalciferol, D 138
Photosantonic acid, C **344**
Photosantoninic acid, C **344**
Phthalic acid, reduction, B 146, 147
Phthalic esters, hydrogenation, B 144
Phthalic methyl esters, hydrogenation B 144
N,N-Phthaloylmenthylamine, B 228
Phycomyces blakesleeanus, B 342; E 59, 107, 108, 122
Phyllanthol, C 413, 415, 416, 451; E 119
Phyllirigenin, C 449
Phyllocladane, C **393**
Phyllocladanol, C **393**
Phyllocladene, C 379, 380, 393, 470; E 87, 90
Phylloquinone, E 87, 134, 135, 137
Physalien, B 297
Physalis alkekengi, B 297, 302
Physalis franchetti, B 297
Physalis species, B 300
Physostigma venenosum, D 132
Physoxanthin, B **302**
Phytoene, B 249, 257, 263, 333, 334, 335, 336, 338, 339, 342, 343; E 122, 123, 124, 125, 126, 129, 136, 137
Phytoene synthase, E 123
Phytofluene, B 243, 257, 333, 334, 335, 336, 339; E 124, 125
Phytol, E 87, 91, 136, 137
—, synthesis, C **468**
Phytolaccogenin, C 442, 481
Phytol pyrophosphate, E 137
Phytosterols, D 3, 73, 133, **154**, 159; E 103, 106, 107, 109
Picene, C 76; D 17
Picrocrocin, B 91
Picorcrocin aglycone, B 174
Pig bile, D 169
Pigment OH-P481, B 307
Pigment OH-R, B 311
Pigment OH-Y, B 311
Pigment P481, B 307
Pigment P518, B 311
Pigment R, B 310
Pigment Y, B 310
Pilgerodendron uviferum, C 296
Pilgerol, C 296
Pimaradienes, synthesis, C 469
Pimaradiene, C 377, 378
Pimaranes, rearranged, C 378, 389, 391
—, unrearranged, C 377, 388
Pimarane group, C 377
Pimaric acid, C 377, 388, 469
Pimarinal, C 388
Pimelic acids, B 93
—, cyclisation, B 6
Pimelic acid, B 143, 354
Pimelic esters, B 93, 143, 355
Pinacols, conversion to polycarboxylic compounds, C 6
Pinacol-pinacolone rearrangement, C 164
Pinane, C 136, 137, 173, 174, 175, 178; E 69
Pinane group, C **166**
Pinane-3-xanthate, C 176
Pinan-3-ol, C **176**
Pinan-3-one, C 170, 171, **176**
Pinan-4-one, C 170
Pinenes, C 167, 249
α-Pinene, A 72, 97; B 160, 172; C 86, 112, 152, 161, **166**, 167, 168, 169, 170, 171, 172, 173, 174, 176, 178,

180, 181, 183, 186, 192, 193, 227, 239, 250, 251, 252; E 69, 76, 79
—, hydration, C 163
β-Pinene, C **174**, 175, 176, 180, 186, 187, 198, 239; E 76
δ-Pinene, C 170, 171, 176, 183
Pin-2,3-ene, C **166**
Pin-2,10-ene, C **174**
Pin-3,4-ene, C 170, 176
Pine needle oil, B 183; C 188
Pinene glycol, C 168
Pinene hydrochloride, C 171, 193, 207
α-Pinene nitrosochloride, C 167, 174, 177
Pin-2,3-en-4-ol, C 181
Pin-2,3-en-10-ol, C 180
Pin-2(10)en-3-ol, C 186
Pin-2,3-en-4-yl acetate, C 183
Pin-3,4-en-2-yl acetate, C 183
Pine-tar oil, B 172
Pinic acid, A 97; C 168, 169, 186
Pinicolic acid A, C 421
Pinifolic acid, C 384
Pinitol, B 60, **61**
Pinnatifidin, C **339**, 366
Pinocampheols, C **176**, 178, 180
—, physical properties, C 179
Pinocampheol tosylate, C 176
Pinocamphones, physical properties, C 179
Pinocamphone, C **176**, 177, 180
Pinocamphoric acid, C 177
Pinocarveol, C 180, **186**, 187, 241
Pinocarveol acetate, C 186
Pinocarvone, C 186, 187
Pinocarvone epoxides, C 187
Pinol, B 194, **206**; C 167, 168
Pinol dibromide, C 167
Pinol glycol, C 167
Pinolol, B 209
Pinonic acid, A 97; C 168, 169, 170, 171, 174, 177
Pinonic ester, C 177
Pinononic acid, C 181
Pinoylformic acid, C 168, 169
Pinus formosan, C 297
Pinus karacensis, C 226
Pinus longifolia, C 150, 281
Pinus palustris, C 238
Pinus parviflora, C 297
Pinus sibirica, C 297
Pinus species, B 158, 178, C 279
Pinus sylvestris, C 281
Piper cubeba, B 206; C 268, 298
Piperidine, C 237, 357
2-Piperidinomethylcyclohexan-1-one, B **126**
16-Piperidinopregn-5-en-3β-ol-20-one, D 357
22-(1-Piperidyl)-9β,10α-bisnorchola-4,20(22)-dien-3-one, D 237
3-(2-Piperidyl)-1,4,5,6-tetrahydropyridine, A 101
Piperinic acid, B **208**
Piperitenone, B 164, **202**; E 73, 74
Piperitenone hydrazone, C 154
Piperitenone oxide, B **203**, 204
Piperitol, B **190**, 191
Piperitone, B 190, **198**; C 251; E 73, 75
Piperitone oxide, B **203**, 204
Pittosporum tenuifolium, B 181
Pitzer strain, A 5
Pivalic acid, D 333
Plastochromanol, E 134
Plastoquinones, E 134, 135, 136, 137
Platanic acid, C 414, 436
Pleuromutilin, C 374, 377, 387, 463; E 91
Pleurotus mutilis, E 91
Plinol, B **208**
Plumiera species, B 214
Plumieride, B **214**; E 72
Podocarpic acid, C 378, 379, 392, 474
Podocarpus nakaii, D 161
Podototarin, C 393
Pogostemon patchouli, C 327
Pollinastanol, D 165
Polyacetylenes, B 387
—, macrocyclic, B 389
Polyalthic acid, C 375, 376, 384
Polybromocyclopentanones, A 168
Polycarbocyclic compounds, C 1 ff
—, classification, C 1
—, from cyclopentadiene, C **99**
—, nomenclature, C 2
—, synthesis, C 6
Polycyclobutyl compounds, C 9
Polycyclohexyl compounds, C 14
Polycyclohydrocarbons, C **129**
Polycyclopentyl compounds, C 9
Polycyclopropyl compounds, C 6
Polyene aldehydes, B 245, 294
Polyene ketones, B 245
Polyeneynes, B 396
Polyferrocenyls, A 135
Polyfluorobutadiene, C 34
Polyfluorocyclobutanes, A 90
Polyfluorocyclohexadienes, B 43
Polyfluorocyclohexanes, B 13, 37, 38, 40, 42

—, physical constants, B 39
Polyfluorocyclohexenes, B 42, 43
Polyfluorohalogenocyclohexanes, B 38, 39
Polygalacic acid, C 441
Polygodial, C **311**, 335
Polygonum hydropiper, C 311
Polyhalogenocyclohexanes, B 36
Polyhydric steroids alcohols, D 193
Polyhydroxy carotenoids, B 294
Polyhydroxycyclohexanes, B 55
Polyketones, A 189
—, cyclic, B 390
—, ring, A 184
Poly*cis*-lycopenes, B 252, 267
Polymannan, biosynthesis, E 134
Polymethylcyclopentadienes, A 122
1,2-Polymethylenebenzenes, B 390
Polymethyleneimines, B 409
cis-Polyolefins, B 390
Polyoxocyclohexanes, B 50
Polypodine A, D 161
Polypodine B, E 114
Polypodium vulgare, D 161; E 100
Polyporenic acid, C 407, 408, 421, 422
Polyporus anthrocophilus, C 467
Polyporus species, C 407
Polyporus sulphureus, C 467
Polyprenols, E 133
Polyterpenoids, E 133 ff
Poly-ynes, B 396
—, higher cyclic B 388
Ponasterones A, B, C, E 114
Ponasterone A, D 161
Ponndorf reduction, C 53, 181, 186
Populus nigra, C 282
Populus tremuloides, D 159
Poriferastanol, D 75, 158
Poriferasterol, D 13, **158**, 166; E 102, 105, 109, 110
Potamogeton natans, B 313
Potassamide, B 419
Potassium *tert*-amyl oxide, C 23
Precalciferol, D 138
Prednisolone, D **333**, 352, 354
Prednisolone apocamphene-1-carboxylate, D 333
Prednisolone pivalate, D 333
Prednisone, D 283, 333
Prednisone acetate, D **332**
Pre-ergocalciferol, D 138
Pregna-4,6-dienes, D 254
9β,10α-Pregna-4,6-diene-3,20-dione, D 238, 348
Pregna-5,16-dien-3β-ol-20-one, D 254, 350
Pregna-3,5-dien-17α-ol-20-one acetate, D 351
Pregna-5,16-dien-3β-ol-17-one acetate, D 351
Pregna-5,16-dien-3-ol-20-one-butyrate, D 233
Pregnanes, D 402
—, *nomenclature*, D *425*, *426*, *428*
—, stereoisomers, D 237
—, steroids, D 356
Pregnane, D 2,5, 232, 237, 248
—, derivatives, D 248
—, molecular rotation, D 63
5α-Pregnane, D 9, 35
—, *nomenclature*, D *453*
5β-Pregnane, D 9
5β,9β,17α-Pregnane, D 440
Pregnane-20-carboxylic acid, D 191
Pregnanediol, D 236, 241, 249
Pregnane-3α,20α-diol, D 232, 235, 249, 250, **252**, 253
Pregnane-3α,20β-diol, D 249, 253
Pregnane-3β,20α-diol, D 249, 252
Pregnane-3β,20β-diol, D 249, 252, 253
Pregnane-3α,20α-diol diacetate, D 250
Pregnane-3β,20β-diol diacetate, D 250
Pregnane-11β,21-diol-3,20-dione 21-acetate, D 324
Pregnane-3α,12α-diol-20-one, D 322, 323
5β-Pregnane-3β,14β-diol-20-one, E 111
5β-Pregnane-3β,14β-diol-20-one-3-acetate, D 389
Pregnane-3,20-dione, D 40, 232, 236, 249, 250, 252, 253, 298
5β,9β,10α-Pregnane-3,20-dione, D *440*
Pregnane-3α,11β,17α,20β,21-pentol, D 251
Pregnane-3α,11α,17α,20β-tetrol, D 251
Pregnane-3α,12α,17α,20β-tetrol, D 251
Pregnane-3β,17α,20β,21-tetrol, D 251
Pregnane-3α,17α,20β,21-tetrol-11-one-20,21-acetonide, D 251
Pregnane-3α,11α,20β-triol, D 251
Pregnane-3α,11β,20α-triol, D 251
Pregnane-3α,11β,20β-triol, D 251
Pregnane-3,16,20-triols, D 253
Pregnane-3α,16β,20α-triol, D 251
Pregnane-3α,16β,20β-triol, D 251
Pregnane-3α,17α, 20α-triol, D 251
Pregnane-3α,17α,20β-triol, D 251
Pregnane-3β,16β,20α-triol, D 251
5α-Pregnane-3β,17α,20α-triol 3,17-dimethyl ether, D *438*
5α-Pregnane-3β,17α,20α-triol tri-methyl ether, D *438*

Pregnane-3,11,20-trione, D 323, 324, 328
5α-Pregnane-3,12,20-trione, E 34
5α-Pregnane-3,15,20-trione, E 36, 37
Pregnane-3,11,20-trione-3,20-bis (ethylenecyclodithioacetal), D 249
4′βH-5α-Pregnano[16.17-*d*]-*isoxazole*, D 453
Pregnan-3α-ol, D 248, 249
Pregnan-3β-ol, D 248
Pregnan-7α-ol, D 248, 249
Pregnan-12α-ol, D 248, 249
Pregnan-20α-ol, D 248, 249
Pregnan-20β-ol, D 248
5α-Pregnan-20α-ol, D *426*
Pregnan-3-ol-11,20-diones, D 251, 322
Pregnan-3α-ol-20-one, D 248, 249, 250, **253**
Pregnan-3β-ol-20-one, D 250, **253**
Pregnan-20α-ol-3-one, D 248, 249
Pregnan-20β-ol-3-one, D 250
5α-Pregnan-3β-ol-20-one, E 112
Pregnan-3-one, D 248, 249
Pregnane-7-one, D 249
Pregnane-11-one, D 249
Pregnane-12-one, D 249
Pregnan-20-one, D 248, 249, 253, 254
5α-Pregnanones, D 91
5α-Pregnan-20-one, D 61
5β-Pregnanones, D 90
5α,14β-Pregnanone, D 91
5α,17α-Pregnanones, D 91
Pregna-1,4,17(20)-triene-11β,16α,21-triol-3-one 21-acetate, D 353
5β,13ξ,14ξ-Pregna-6,8,11-trien-20-yn-3α-ol, D *428*
Pregn-5-en-4,20-dione, D 254
Pregn-4-enes, D 254
Pregn-5-enes, D 254
Pregn-9-enes, D 254
Pregn-11-enes, D 254
Pregn-17-enes, D 306
Pregn-16-ene, D 306
$\Delta^{(17)20}$-Pregnene, D 324
Δ^{20}-Pregnene, D 324
Δ^{4}-Pregnene compounds, D 294
Pregnene derivatives, D **253**
Pregnene-1,2-diols, D 281
Pregn-17-ene-3β,21-diol diacetate, D 251
Pregn-4-ene-11β,21-diol-3,20-dione, D 289
Pregn-4-ene-17α,21-diol-3,20-dione, D 289
Pregn-4-ene-20β,21-diol-3,11-dione, D 289
Pregn-5-ene-3β,21-diol-20-one, D 308
Pregn-4-ene-17α,21-diol-3,11,20-trione, D 289
Pregn-4-ene-3,20-dione, D 253, 289
Pregn-5-ene-3,20-dione, D 241
Pregn-11-ene-3,20-dione, D 313
Pregn-16-ene-3,20-dione, D 249
4-Pregnene-3,20-dione, D *439*
5-Pregnene-3,20-dione, D *438*
Pregn-4-ene-11β,17α,20β,21-tetrol-3-one, D 289
Pregn-5-ene-3α,16α,20-triol, D 60
Pregn-4-ene-11β,17α,21-triol-3,20-dione, D 289
Pregn-4-ene-17α,20β,21-triol-3,11-dione, D 289
Pregn-4-ene-17α,20β,21-triol-3-one diacetate, D 300
5α-Pregn-17(20)-en-16α-ol, D 60
Pregn-4-en-17α-ol-3,20-dione, D 289
Pregn-4-en-21-ol-3,20-dione, D 289
Pregnenolone, D 36, 108, 231, 232, 233, 235, 236, 237, 238, 239, 248, 254; E 110, 111, 113
[4-^{14}C]Pregnenolone, E 111
Pregn-4-en-3β-ol-20-one, D 263
Pregn-4-en-20α-ol-3-one, D 241
Pregn-4-en-20β-ol-3-one, D 282
Pregn-5-en-3β-ol-20-one, D 271, 277, 306, 350
Pregn-16-en-3-ol-20-ones, D 252, 253
Pregnenolone acetate, D 237, 244
Pregn-5-en-3β-ol-20-one acetate, D 284, 397, 407
Pregn-4-en-15α-ol-3,11,20-trione, D 354
Pregn-4-en-21-ol-3,11,20-trione, D 289
Pregn-16-en-20-ones, D 254, 304
Pregn-5-en-3β,17α,19-triol-20-one 3,17-diacetate, D 245
Prehnitic acid, D 136
Prenol diphosphogalactose, E 134
Prenol monophosphomannose, E 134
Prenyl isomerase, E 60
Prenyl transferase, E 79
Presenegenin, C 415, 416, 442
Previtamin D_2, D 138, 139, 140
Prevost reagent, D 160
Primulagenin, C 415, 416, 438
Prins reaction, B 22, 70, 166; C 102
Prismane, C **45**
Pristimerin, C 415, 446
Pristnicerin, C 416
Pro-γ-carotene, B **273**
Pro-λ-carotene, B 252
Prochamazulenogen, C 352

Progesterone, D 2, 7, 132, 153, 185, 191, 197, 230, **231**, 232, 233, 234, 236, 237, 238, 240, 241, 244, 247, 248, 250, 253, 258, 271, 289, 298, 300, 302, 312, 328, 345, 350, 377, 384; E 1, 10, 110
—, halogenated, D 347
—, metabolism, D **235**
—, *nomenclature*, D *439*
—, oxygenated, D 235
—, steroisomerides, D 236
—, total synthesis, D 102
9β,10α-Progesterone, D 237, 238
9β,10β-Progesterone, D 347
Δ^6-Progesterone, D 238, 239
$\Delta^{(9)11}$-Progesterone, D 238, 298
Δ^{11}-Progesterone, D 238
Δ^{16}-Progesterone, D **238**
Progestogens, D 2, 197
Prolycopene, B 252, 267
Proneurosporene, B **276**
Propanetetracarboxylic esters, B 144
Propane-1,1,3,3-tetracarboxylic ester, A 22
Propargyl alcohol, B 66
Propargylaldehyde dimethyl acetal, C 286
Propargyl bromide, B 66
Propargyl chloride, A 41
Propargyl halides, B 93
Propargylic aldehyde, B 90
Propene, A 19, 25
—, from cyclopropane, A 32
Propiolic acid, B 138; C 114
Propiolic methyl ester, A 86
Propionaldehyde, A 42
Propionic acid, A 48; C 312
—, complexes with deoxycholic acid, D 176
1-Propionyl-3-methylcyclohexane, B 128
1-Propionyl-5-methylcyclohex-1-ene, B 128
17β-Propionyloxy-A-norandrost-3(5)-ene-1,2-dione, D 281
*iso*Propyl, see under Isopropyl
Propylamine, B 3
n-Propylbenzene, A 36
Propylcyclohexadienes, B 26
Propylcyclohexanes, B 14
Propylcyclohexan-1-ols, B 47
Propylcyclohexanones, B 95
Propylcyclohex-1-enes, B **21**
Propylcyclohex-2-enones, B 117
Propylcyclopentane, A 113
1-*n*-Propylcyclopentan-1-ol, A 157
3-*n*-Propylcyclopentan-1-one, A 167
Propylcyclopropane, A 32
Propylene, reaction with cyclopentadiene, C 89
Propylidenecyclohexanes, B 21
n-Propyl iodide, A 19
Propynylnortestosterone, D 345, 346
Proscillaridin A, D 408
Prosolannellic acid, D 20
Prostaglandins, A 104, 202, 203
Protein, ACTH, D 287
—, metabolism, D 287
Protetrahydrolycopene, B 276
Protoaescigenin, C 439
Protoanemonin, A 102
Proton magnetic resonance, carotenoids, B **256**
Protopanaxadiol, C **425**
Protopanaxatriol, C **425**
Protoquercitol, B 58
Provitamins A, B 283
Provitamin D, D 143
Provitamin D_3, D 106
Prunella vulgaris, C 234
Psallista campestris, E 107
Pseudobotogenin diacetate, E 46
Pseudo-α-carotene, B 250
Pseudocholesteryl tosylate, D 81
Pseudodiosgenin diacetate, E 9, 44
Pseudogenin, D *435*
Pseudokryptogenin diacetate, E 44
Pseudomonas strain, C 205, 217
Pseudopelletierine, B 383, 418
Pseudoneoruscogenin, E 51
Pseudoricogenin, E 46
Pseudosapogenins, E 9, 10, 11, 12, 16, 19
Pseudosapogenin diacetate, E 12
Pseudosapogenin 3,26-di-esters, E 11
Pseudosapogenin esters, E 21
Pseudosarsasapogenin, E 12, 16
—, *nomenclature*, D *435*, *436*
Pseudosmilagenin, E 12
—, *nomenclature*, D *435*
Pseudotigogenin, D *435*, *436*
Pseudotigogenin diacetate, E 45
Pseudowintera colorata, C 271, 322
Pterosterone, E 114
Puberulic acid, B 366, 377
Puberulonic acid, B 366
Pulchellins, C **360**, 368
Pulegene, A 116
Pulegols, B **200**
Pulegone, B 100, 184, **199**, 200; E 73, 74, 75, 78
Pulegone hydrazone, B 219; C 153

Pulenone, B 95
Pulvinolactone, A 188
Pulvinone, A 188
Purpurogallin, B 375
Purpurogallone, B 375
Purpurogenin 3-acetate, D 404
Pyracantha angustifolia, B 273, 276
Pyrazolenines, A 23
Pyrazolines, A 26, 57, 74; C 35
—, conversion to cyclopropanes, A 22
Pyrene, C 76
Pyrethric acid, A 65, 68
Pyrethrins, A 1, 19, 65, 176, 178, 183; B **214**
Pyrethrin insecticides, A 104
Pyrethrin semicarbazones, A 68
Pyrethrolones, A 176
Pyrethrolone, A 65, 103, 178, 182
Pyrethrolone acetate, A 178
Pyrethrone, A 181
Pyrethrosin, C **362**, 368
Pyrethrum cinerariaefolium, B 214
Pyridine, D 312
Pyridine hydrochloride, E 10
Pyridiniumcyclopentadienylide, A **122**
Pyroalloisolithobilianic acid, D 10
Pyrocholoidanic acid, D 19, 20
Pyrocalciferol, D 138, 139, 140, 141, **148**, **238**
Pyrocins, A **66**
Pyroclavine, E 65, 66
Pyrodeoxybilianic acid, D 22, 23
Pyrogallitol, B 55
Pyrogallol bromination, A 184
—, hydrogenation, B 55, 112
Pyrogallol triacetate, B 112
Pyroisolithobilianic acid, D 10, 19
Pyrolithobilianic acid, D 10, 19
Pyromellitic acid, hydrogenation, B 152
Pyronenes, B **172**
α-Pyronene, B 172, 207
β-Pyronene, B 172
5-Pyrophospho-MVA, E 59
5-Pyrophosphomevalonic anhydrocarboxylase, E 59
Pyrosolannellic acid, D 19, 20
Pyrrolidine, A 165; D 357
Pyrrolizidine alkaloids, B 169
Pyrrolylpotassium A 138
Pyruvaldehyde, A 182
Pyruvic acid, A 67; C 198, 332; D 326, 331
Pythocholic acid, D 173
Pyxinic acid, C 452
Quadricyclanol, C 94
Quadricyclanone, C 46, 94
Quadricyclene, C 91
Quadricyclenedicarboxylic acid, C 91
Quadricyclo[9.1.0.0^{3,5}.0^{7,8}]dodecane, C 42
Quassia amara, D 159
Quassin, C 410, 411, 412, 433
Quassin group, triterpenoids, C 433
Quebrachitol, B 60, 61
Quercitols, B 44, **57**, 58, 60
Queretaroic acid, C 440
Quillaic acid, C 441
Quinic acid, B **139**, 140
Quinine, B 139
Quinitol, B 54
Quinones, terpenoid, E 134
Quinone, D 254
Quinonetetrol, B 58
Quinovic acid, C 414, 450

Radix hellebore niger, D 411
Radix sarsaparilla, E 3, 25
Raney-nickel, B 2
Ranunculaceae, D 360
Ranunculus species, A 102
Rapeseed oil, D 155, 165
Reductic acid, A 185
Reformatsky reaction, B 66, 94, 142; C 170, 175; D 101, 204, 208, 407; E 21
Regularobufogenin, D 414
Regularobufotoxin, D 414
Rehydrovitamin A_1, B 291
Rehydrovitamin A_2, B **292**
Reich diketone, D 100
Reichstein's substance S, D 328
Reineckia carnea, E 27, 51
Reineckiagenin, E **48**, 49
Reniera japonica, B 277, 331
Renieral, B 277
Renierapurpurin, B 245, **277**, 278, 279
Renieratene, B 254, **277**, 278, 279; E 130, 132
Renieraxanthin, B 331
Reppe synthesis, B 392
Reserpine, E 75, 76
Resibufogenin, D 414, **419**, 420
Resibufogenone, D 419
Resinic acids, E 86
Resorcinool, hydrogenation, B 53
—, reaction with halogens, B 113
—, reduction, B 1, 111
Resorcitol, B 53
Rethrins, A 180
—, synthetic A 183

Reticulataxanthin, B 332
Retinal, B 282, **288**
Retinene, B 260, 274, 282, 284, 287, 340
Retinene$_1$, B **288**, 289, 290, 292
Retinene$_2$, B **288**, 290
Retinoic acid, B 282, 286
Retinol, B **282**
Retro-Prins reaction, C 415
Reychler's acid, C 214
Rhamnose, D 361, 363, 365, 367, 368, 382, 396, 401, 405, 411, 412; E 2, 3
3-*O*-Rhamnosyl-16-acetoxy-strophanthidin, D 384
Rhemmanic acid, C 441
Rhenium(VII) oxide-tetrahydropyran complex, B 2
Rhodea japonica, E 25
Rhodeasapogenin, E 3, 5, 25
Rhodinol, B 161
Rhodizonic acid, A 188
Rhodomicrobium vannielli, B 307
Rhodophyceae, B 238
Rhodopin, B 306, 307, 332, 336; E 120, 121
Rhodopseudomonas gelatinosa, B 309, 310, 311, 337, 345
Rhodopseudomonas palustris, B 307
Rhodopseudomonas species, B 307, 311
Rhodopseudomonas spheroides, B 310, 344; E 127
Rhodopsin, B 289
Rhodospirillum, B 307
Rhodospirillum rubrum, B 308, 309, 334, 336; E 129
Rhodotorula species, E 126
Rhodovibrin, B 252, 308, 336, 337; E 130
Rhodovibriobacteria, B 308
Rhodoviolascin, B **308**, 309
Rhodoxanthin, B 238, 297, 313, 314, 315
—, structure, B 233
Rhodymenia palmata, D 154
Rhizopus nigricans, D 312
Rhizopus species, D 235
Ribonucleic acid, E 64
Ricogenin, E **46**
Rimuene, C 378, 389, 470
Ring closure, cycloalkanes, A 9
Ring-nor-steroids, D 281
Ring polyketones, A 184
Ring-systems, bridged, C 2
—, fused, C 2
—, separate, C 1
Ritter reaction, B 80, 81; C 127, 131
RNA = Ribonucleic acid, E 64
Robinson annellation, C 68
Roburic acid, C 410, 414, 450
Rockogenin, E 26, **35**
Rolyeanone, C 378, **379**
Rosenheim test, D 76
Rosenonolactone, C 378, 389, 461; E 88, 119
—, biosynthesis, C **460**
Rose oil, B 161
Rosololactone, C 389
Rotundifolone, B 204
Royleanone, C 392
Rubber, E 135
—, biosynthesis, E 135
Rubiaceae species, B 213
Rubichrome, B 304
Rubijervine, D 2, 357
Rubixanthin, B **303**
Rubixanthin acetate, B 304
Rubixanthin-5,6-epoxide, B 304
Rules, Nomenclature of Steroids, D *423*
Ruscogenin, D 404; E 25, **30**, 31, 51
Ruscus aculeatus, E 25
Ruthenocene, A 127, 135, 145
Ruthenocenecarboxylic acid, A 136
Ruzika's method, cyclanones, B 407
Ruzicka's biogenetic isoprene rule, B 234

Sabina chloride, C 139
Sabina glycol, C 143
Sabina hydrate, C **144**, 255
Sabina ketone, C 140, 141, 144, 255
Sabinane, C **137**
Sabinene, C 137, 138, **139**, 140, 141, 144, 252, 253, 255
—, hydration, B 189
Sabinenic acid, C 140, 141
Sabinol, B 221; C **142**, 143, 253
Sabinol acetate, C 253
Sabinone, C 143
Sabinyl acetate, C 253
Saccharic acids, B 60, 61
Saffron, B 174, 327
Safranal, B 88, 91, **174** E 68
Sagittol, C 298
Saikogenin, C 439, 440
Salannin, C 432
Salicylaldehyde, B 375
Salicylic acids, B 375
Salicylic acid, hydrogenation, B 138
Salicylidenecarvomenthylamine, B 229
Salicylidenementhylamine, B 228
Salmonella newingtonii, E 133

Salmonella typhimurium, A 69
Salvin, C 378, 379, **392**
Samaderine B, C 434
Samaderine C, C 434
Samogenic acid, E 38
Samogenin, E 25, **31** 33, 35
Samuela carnerosana, E 25, 27
Sandalwood oil, C 188, 189, 192
Sandaracopimaradiene, C 388
Sandaracopimaradiene-3,18-diol, C 388
Sandaracopimaradiene-3,19-diol, C 388
Sandaracopimaradien-3-ol, C 388
Sandaracopimaradien-3-one, C 388
Sandaracopimaric acid, C 377, 388, 469
Sandmeyer reaction, D 356
α-Santalal, C 293
Santal camphor, C **307**
Santalenes, C 187, 263, 265, 288, 293
α-Santalene, C **263** 264, 294
β-Santalene, C **263**, 264, 265, 266, 294
α-Santalic acid, C **293**
Santalols, C 293, 334
α-Santalol, C **293**, 294
β-Santalol, C **293**, 294
Santalum album, C 188, 263; E 69
Satalum lanceolatum, C 291
Santalum preussianum, C 307
Santane, C 188
Santanolide, C 337
Santenes, C 136
—, alcohols, C 189
—, carboxylic acids, C 189
—, ketones, C 189
Santene, A 192; C 164, **187**, 188, 189, 190, 192
Santene glycol, C 188
Santene group, C **187**
Santene hydrate, C 189
Santene hydrochloride, C 189
Santenic acid, C 189, 190
Santenols, C 190, 191
Santenone, C **189**, 191
α-Santenone, C 190, 191
β-Santenone, C 190, 191, 247
Santenyl formate, C 192
Santiaguine, A 101
Santolina chamaecyparissus, B 165
Santolinelones, B 165
Santonic acid, C **344**, 345
Santonide, C **345**
Santonins, C 295, 340, 343, 367
Santonin, C 60, 306, 337, 344, 364; E 83
α-Santonin, C **340**, 341, 342, 346
β-Santonin, C **341**, 342
ψ-Santonin, C **345**, 346, 363
Santoninic acid, C 340
Santonous acid, C 340, 341
β-Santonous acid, C 341
Sapogenins, D 2, 60, 253, 259; E 1 ff, 5, 6 ff, 8, 9, 11, 13, 14, 15, 16, 20, 21
—, biogenesis, E 23
—, from sea cucumbers, D 69
—, naturally occurring, E 25
—, *nomenclature*, D *435*
—, side-chain reactions, E 19
—, side-chain synthesis, E 21
—, stereo-isomerism, E 11
—, with hydroxy groups in Ring F, E 46
—, with modified side-chains, E 43
—, with normal side-chains, E 29
—, with ring F unsaturated, E 51
ψ-Sapogenins, D 248, 253, 254, 305, 306, 327
Sapogenin B, C 441
Sapogenin F, C 441
Saponins, E 1 ff, 4, 5, 6
Saprospira grandis, B **304**
Saproxanthin, B **304**
Sapucaiuha oil, A 198
Sarcina lutea, B **331**
Sarcinaxanthin, B **331**
Sarcinene, B **331**
Sargassum, D 166
Sargastadienone, D 153
Sargastanol, D 153
Sargasterol, D 13, **153**, 166
Sargenoside, D **398**
Sargenoside diacetate, D 364, 398
Saringosterol, D 166
Sarkomycin, A 201
Sarmentocymarin, D 364, **398**
Sarmentogenin, D 39, 328, 364, **398**, **399**, 403, 420
Sarmentogenin-11α-acetate, D 398
Sarmentogenin-11α-acetate-acetyldigitaloside-glucoside, D 398
Sarmentogenin-digitaloside, D 398
Sarmentogenin-digitaloside-glucoside, D 398
Sarmentogenin-sarmentoside, D 398
Sarmentogenone, D 399
Sarmentose, D 361, 365, 403
Sarmentoside A, D 364, **398**, **400**
Sarmentoside B, D **398**
Sarmentoside E, D **400**
Sarmentoside-A acetate, D 398

Sarmentosigenin, D 398, 400
Sarmentosigenin-A, D **398**, 399, **400**
Sarmentosigenin E, D **400**
Sarmutogenin, D **402**, 403
Sarmutoside, D **402**
Sarnovide, D 364, **398**
Sarsaparilloside, E 3, 5, 6
Sarsasapogenin, D 2, 7, 251; E, 3,7, 11, 12, 25, **29**, 30, 35
—, *nomenclature*, D *433*, *435*
ψ-Sarsasapogenin, D 251, 253
Sarsasapogenin acetate, E 7, 8, 9, 20
Sarsasapogenoic acid, E 8, 9, 44
Sarsasaponin, E 3
Sarverogenin, D 364, **398**, **403**
Sarverogenin-sarmentoside, D 398
Sarverogenone, D 403
Sarveroside, D 364, **398**, **403**
Saussurea lactone, D **336**, 337, 366
Saussurea lappa, C 301, 336, 351, 363
Savin oil, C 139, 142
Sawamilletin, C 445
Scabiolide, C **365**, 366, 368
Schinol, C 426
Schizandra chinensis, C 272
Schmidt reaction, A 51; B 80, 81, 82, 142, 226, 340, 377; C 111
Sciadin, C 375, 376, 384
Sciadinone, C 384, 385
Scilla-20,22-dienolides, D **408**
Scilla maritima, D 408, 410
Scillaren A, D 366, **408**, 410
Scillaren F, D 366, 410
Scillarenase, D 408
Scillarenin, D 2, 366, 411
—, *nomenclature*, D *432*
Scillarenin β-δ-glycoside, D **411**
Scillaridin A, D **408**, 409, 411
Scilliroside, D 366, 410
Scilliroside tetra-acetate, D 410
Sclareol, C 375, 382, **468**
Scrophulariaceae, D 360; E 5
Scyllinosose, B 60
Scymnol, D **194**, 195
Sebacic acid, A 164
Sebacoin, B 412
Seco acid, C 384, 385
9,10-Seco-cholesta-5(10),6-dien-3β-ol, D 97
2,3-Seco-5α-cholestane, D *446*
2,3-Seco-5α-cholestane-2,3-dioic acid, D *446*
2,3-Seco-5α-cholestane-2,3-dioic methyl ester, D 124
3,4-Seco-5α-cholestane-3,4-dioic acid, D 126
2:3,6:7-Seco-5α-cholestane-2,3,6,7-tetroic acid, D 17, 18
9,10-Seco-5,7,10(19)-cholesta-trien-3β-ol, D *439*, *447*
*9,10-Seco-5,7,10(19),22-ergostatetraen-*3β-ol, D *439*
4(5)-Seco-3,5,20-trioxo-19-norpregn-9-ene, D 246
Selin-4,11-diol, C 301
Selinane, C 329
Selinene, C 256, **273**, 303
α-Selinene, C 273
β-Selinene, C 273, 288
δ-Selinene, C 303
Selinene dihydrochloride, C 299
Semibenzenes, B 65
Semi-β-carotenone, B 250, 260
Semitruxinonic acid, A 102
Senecic acid, B **169**
Senecionine, B 169
Senega root, D 156
Senegenic acid, C 415, 416
Sen-mokko, C 351
Senso, D 412, 415
Sequoyitol, B 61
Serini reaction, D 60, 293, 300
Serratenediol, C 419, 458, 475
Seseli sibiricum, C 239
Sesquibenihene, C **270**, 288
Sesquibenihiol, C 301
Sesquifulvalene, C 17
Sesquigoyol, C 297
Sesquiterpenes, C 60; E 79 ff
—, biosynthesis, E 79
—, furanoid derivatives, C 302
—, *nomenclature*, D *423*
Sesquiterpenoid lactones, C 257
—, physical constants, C 366
Sesquiterpenoids, C 256 ff
—, hydrocarbons, C 258 ff
—, lactones, C 336
—, nomenclature, C 258
—, physical constants, C 288, 334
—, spiro compounds, C 312
—, structures, C 257
—, with oxy-groups, C 289
Sesterterpenes, E 92 ff
Sesterterpenoids, C 369 ff, 403 ff
—, biosynthesis, C 463
—, physical properties, C 405
Sex hormones, D 197 ff; E 10
Shairol, C 314
Shellac C 326
Shellolic acid, C **326**, 336
Shikimic acid, A 197; B 139, **140**
Shionone, C 413, 435

Shonanic acid, B **216**, 357
Shonanol, C 393
Siaresinolic acid, C 440
Silvestrene, C 150
Silvestrene dihydrochloride, C 151
Simarolide, C 410, 411, 412, 432
Simaruba glauca, E 120
Simiarenol, C 456
Simmons-Smith reaction, C 22, 36
Sindora wallichii, C 271
Sinensiaxanthin, B 333
Sinogenin D **402**
Sinoside, D **402**
Sinostroside, D **402**
Sintaxanthin, B 332
Siphonales, B 331
Siphonein, B 331
Siphonoxanthin, B 331
Sisalagenin, E 26, 34
Sitostanol, D 75
Sitosterols, D **154**
α-Sitosterol, C 422; D 155, 166
β-Sitosterol, D 3, 13, 75, 154, 155, 158, 166; E 103, 105
—, ultraviolet absorption spectra, D 68
γ-Sitosterol, D 155, 413
β-Sitosteryl acetate, D 133
Sitosteryl acetate dibromide, D 258
β-Sitosteryl-β-D-glucoside, D 154
Skatole, B 414
Skytanthine, B 212; E 76
β-Skytanthine, E 75
Skytanthus acutus, B 213; E 75
Smilagenin, E 11, 12, 25, **30**, 31
—, *nomenclature*, D *435*
Smilagenin acetate, E 7
Smilagenone, E 38
Smilax sieboldi, E 26
Smilax species, E 25
Smithiandienol, C 439
Smokeless powder, C 207
Sobrerol, B 193, **194**, 206, 222
Sodamide, B 31; C 27
Sodioacetoacetic ester, B 100, 144
—, reaction with β-chloroketones, B 114
Sodio-α-bromocamphor, C 219
Sodiocamphor, C 219, 223, 224
Sodiocyanoacetic ester, A 58, 63
Sodiodiethyl malonate, A 162
Sodioepicamphor, C 219
Sodio-21-ethoxalylpregn-5-en-3β-ol-20-one, D 350
Sodiomalonic ester, A 58, 61, 63, 190, 192, 193; B 142, 144, 149; C 8, 26
—, reaction with 1,3-dihalides, A 73
—, — with mesityl oxide, B 111
Sodiomethylmalonic ester, A 193
Sodium acetylide, C 9
Sodium borohydride, C 10
Sodium cylopentadienide, A 86, 137
Sodium propanol, D 321
Sodium tetra-alkylborons, A 41
Sodium triethyleneglycoxide, A 47
Sodium triphenylmethyl, C 32
So-jutsu, C 302
Solanesol, E 133, 134
Solanesol-50, E 137
Solanesyl-45-pyrophosphates, E 136, 137
Solanesyl-50-pyrophosphates, E 136,
Solanidane, D 2, 137
Solanidanine, D *449*, *450*
Solanidine, D 2
Solannellic acid, D 20
Solanoside, D **386**
Solanum alkaloids, E 52
Solanum dulcamara, B 306
Solanum hispidum, E 3, **26**
Solanum paniculatum, E 26, 50, 52, **53**
Solanum sisymbriifolium, E 47
Solanum tuberosum, D 155
Somalin, D 362, **386**
Sophoradiol, C 438
Sorbic ester, A 53
Sorbicortol, C 435
Sorbrerol acetate, C 183
Soya bean, D 75
Soya bean oil, D 156
Soyasapogenol, C 438; E 88
Soyasapogenol A, E 118, **119**
Soyasapogenol-C, C 439
Soyasapogenol D, C 461, **466**, 467
Soyasapogenol-E, C 439
Spergulagenic acid, C 442
Sphaerophysin, E 65
Spheroidene, B **310**, 311, 337, 338, 344; E 127, 128, 129, 130
—, demethylated, E 130
Spheroidenone, B **310**, 337, 345; E 127, 128, 130
Spinacera oleracea, E 106
Spinasterols, C 467; D 156
α-Spinasterol, D, 156, 166; E 104, 105
Spiranes, C 1, **20**
—, nomenclature, C 3
—, preparation, C 20
Spiranones, C 21
Spirilloxanthin, B 239, 252, 257, **308**, 309, 312, 327, 336, 337, 344, 345; E 129, 130

—, demethylated, E 130
(3S)-Spiro[5α-androstane-3,2'-oxiran], D *454*
Spirobicyclohexane, C 4
Spiro compounds, C 1 ff, 3, **20**
Spiro-cyclohexadienones, B 120
Spiro[4.5]deca-6,9-dien-8-one, B 120; C 30
Spiro[4.5]decanes, C **29**; E 80
Spiro[5.4]decane-1,3-dione, B 149
Spiro[4.5]decan-1-ol, C 29
Spiro[4.5]decan-1-one, C 29
Spiro[4.5]decan-6-one, C 29
Spiro[5.4]decan-6-one, C 66
Spirodienes, A 121
Spiro[5.6]dodecane, C 31
Spiro[5.6]dodecan-7-one, C 31
Spiro[2.4]hepta-4,6-diene, C 23
Spiro[2.4]heptane, C 23
Spiro[3.3]heptanes, C 25
Spiro[3.3]heptane-2-carboxylic acid, C 25
Spiro[4.2]heptanecarboxylic acids, C 24
Spiro[2.4]heptane-1,2-dicarboxylic acids, C 23
Spiro[3.3]heptane-2,6-dicarboxylic acid, C 26
Spiro[4.2]heptane-1,2,dicarboxylic-1-malonic ester, C 29
Spiro[2.4]heptane-4,7-dione-5,6-dicarboxylic acid, C 24
Spiro[3.3]heptanetetracarboxylic acid, C 26
Spiro[2,4]heptan-4-ol, C 23
Spiro[2.4]heptan-1-ol-1,2-dicarboxylic acid, C 23
Spiro[2.4]heptan-4-one, C 23, 45
Spiro[2.3]hexanes, C **22**
Spirohexene, C **23**
17-Spirohydantoins, D 358
Spiro[4.4]nona-1,3-diene, C 27
Spiro[4.4]nona-1,6-diene, C 27
Spiro[2.6]nonanes, C **25**
Spiro[3.5]nonanes, C **26**, 27
Spiro[4.4]nonanes, C **27**
Spiro[3.5]nonane-2-carboxylic acid, C 26
Spiro[4.4]nonane-1,6-diamine, C 27
Spiro[2.6]nonane-1,2-dicarboxylic acid, C 25
Spiro[4.4]nonane-1,6-diols, C 28
Spiro[4.4]nonane-1,6-dione, C 28
Spiro[4.4]nonan-1-ol, C 28
Spiro[4.4]nonan-6-ol-1-one, C 28
Spiro[3,5]nonan-1-one, C 27
Spiro[3,5]nonan-2-one, C 27
Spiro[4.4]nonan-1-one, C 27
Spiro[4.4]nonan-3-one-1-carboxylic acid, C 29
Spiro[3.5]nonene, C 3
Spiro[4.4]non-1-ene, C 27
Spiro[3.5]non-7-en-1-one-6-carboxylic acid, C 3
Spiro[4.4]non-1-ylamine, C 27
Spiro[2.5]octadiene, C 24
Spiro[2.5]octa-4,7-dien-6-one, C 24
Spiro[2.5]octane, C 24
Spiro[3.4]octanes, C **26**
Spiro[3.4]octane-2-carboxylic acid, C 26
Spiro[2.5]octane-1,2-dicarboxylic acids, C 25
Spiro[3.4]octanetetracarboxylic acid, C 26
Spiro[3.4]octan-5-ol, C 26
Spiro[3.4]octan-5-one, C 26
Spiro[2.5]oct-5-ene, C 24
Spiropentane, A 73
Spiro[2.2]pentane, C **21**
Spiropentanecarboxylic acid, C 22
Spiropentylamine, C 22
Spiropentyl Grignard reagent, C 22
Spirosesquiterpenoids, C 312
(25*R*)-Spirosta-1,4-dien-3-one, E 42
Spirostans, E 18
—, *nomenclature* D, *425*, *432*, *433*, *435*, *446*
(25*R*)-Spirostans, E 45
Spirostan, D 2, 5, 7
(25*R*)-Spirostandiol, E 33
(25*S*)-Spirostan-3β,12β-diol, E 35
(25*R*)-5β-Spirostandiol, E 31
(25*R*)-5α-Spirostan-2α,3α-diol, E 32
(25*R*)-5α-Spirostan-2β,3α-diol, E 32
(25*R*)-5α-Spirostan-3β,12-diol, E 35
(25*R*)-5α-Spirostan-3β,15β-diol, E 37
(25*R*)-5β-Spirostan-1β,3β-diol, E 38
(25*R*)-5β-Spirostan-3,11α-diol, E 33
5β-Spirostan-3α,27-diol, E 46
5β-Spirostan-3α,x-diols, E 46
5α(25*R*)-Spirostan-3,12-dione, E 34
Spirostanols, biosynthesis, E 113
Spirostan-3β-ols, D *433*, *435*
Spirostanols $C_{(27)}$, in plants, E 112
(25*R*)-5α-Spirostan-1-one, E 42
(25*R*)-Spirostanpenta-ol, E 42
(25*R*)-Spirostantetraol, E 40, 41
(25*R*)-Spirostantriol, E 33, 38
(25*S*)-Spirostantriol, E 39
(25*R*)-5α-Spirostan-2α,3α,5-triol, E 40
(25*R*)-5α-Spirostan-2,3,12-triol, E 36

(25*R*)-5β-Spirostan-1β,3β,25-triol, E 48
(25*R*)-5β-Spirostan-2β,3β,x-triol, E 33
(25*S*)-5β-Spirostan-1β,3β,27-triol, E 49
—, *nomenclature*, D *434*
(25*S*)-5β-Spirostan-1β,3β,25-triol, E 48
(25*R*)-Spirosta-1,4,6-trien-3-one, E 43
5-Spirostene, D *436*
(25*R*)-5β-Spirost-3-en, E 31
5β-Spirost-25(27)-en-1β,3β-diol, E 51
(25*R*)-Spirost-5-en-3β,12-diol, E 35
(25*S*)-Spirost-5-en-3β,27-diol, E 49
Spiro[6.7]tetradecanes, C 31
Spiro[6.7]tetradecan-8-one, C 31
Spiro[5.5]undeca-1,4-dien-3-one, C 31
Spiro[5.5]undecane, C 4, **30**
Spiro[5.5]undecane-2,4-dione, C 31
Spiro[5.5]undecan-1-ol, C 31
Spiro[5.5]undecan-1-one, C 30
Spiro[5.5]undec-2-ene, C 30
Spiro[5.5]undec-7-en-1-one, C 30
Spongesterol, D **157**, 165
Squalene, B 341, 342, 343; C 164, 371, 372, 373, 407, 419, 464, 465, 466, 467, 472; D 74, 162; E 54, 93, 95, 98, 99, 104, 106, 113, 115, 116, 119, 136, 137
—, biosynthesis, E 93, 122
—, cyclisation, C 406, 409, 417, 418; E 96, 98, 119
—, synthesis, C **471**
[^{14}C]Squalene, E 93
Squalene-cholesterol biosynthesis, C 460
Squalene diepoxide, E 116
Squalene 2,3-epoxide, E 93, 96, 117
Squalene oxide, E 96, 104, 116, 119
Squalene 2,3-oxide, D 162; E 98
Squalene 2,3 oxide cyclase, E 96
Squalene sterols, E 136
Squaric acid, A 94
Squill glycosides, D **408**, 416
Stable bixin, B 327
Stachane group, C 381
Stachene, C 379, 380, 399
Stachene-hibaene diterpenoids, physical constants, C 399
Stachenone, C 399
Stanols, D 18, 73 ff, 86
—, oxidation, D 87
—, physical properties, D 74
3β-Stanols, tosylates, D 43
Stanones, D 87 ff
—, epimerisation, D 90
—, Grignard reactions, D 89
—, reduction, D 88
—, ultraviolet absorption chomophores, D 68
Stanozolol, D 274, 275
Stearolic acid, A 70
Stellatogenin, C 436
Stenols, D 73 ff, 76, 85, 86
—, colour reactions, D 76
—, epoxidation, D 85
Stenones, ultraviolet absorption chromophores, D 68
Sterculiaceae, A 69
Sterculia foetida, A 69, 70
Sterculic acid, A 69
Stereochemistry, camphorquinone-oximes, C **221**
—, capsorubin, B 324
—, steroids, D 9 ff
Stereo-isomerism, sapogenins, E 11
Steroids, D 1 ff
—, adrenocortical, D 289, 303
—, allylic oxidation, D 84
—, axial and equatorial substituents, D 30
—, chromatography, D 32
—, —, with a liquid mobile phase, D 71
—, classification, D 2
—, determination of chemical structure, D 15
—, 4-en-3-ones, D 92
—, episulphides, D **114**
—, epoxides, D **113**
—, fusidane skeleton, D 162
—, gas-liquid chromatography, D 72
—, halogenated, D **338**
—, infrared absorption spectrophotometry, D 65
—, infrared spectrometry, D 32
—, mass spectrometry, D 70
—, *nomenclature*, D *422*
—, nuclear magnetic resonance spectrometry, D 33, 69
—, optical rotary dispersicn and circular dichroism, D 64
—, optical rotation, D 62 ff
—, —, and optical rotary dispersion, D 33
—, 1-oxygenated, D 129
—, paper chromatography, D 71
—, ring-nor, D 281
—, stereochemistry, D 9 ff
—, stereochemistry of nuclear substituents, D 29
—, structure, D **8** ff

—, —, in relation to physical properties, D 62 ff
—, —, side chains, D 13
—, substituents at nuclear positions, D 33
—, substituted, D 356
—, 3β-substituted, D 42
—, thin-layer chromatography, D 71
—, ultraviolet absorption spectrophotometry, D 67
—, X-ray crystallography, D 71
14β-Steroids, D 61
Δ^{2}-Steroids, D 43
$\Delta^{7,9(11)}$-Steroids, D 318, 319
$\Delta^{9(11)}$-Steroids, D 55, 312, 315, 316
Δ^{11}-Steroids, D 312, 316
Δ^{16}-Steroids, D 304, 305
$\Delta^{17(20)}$-Steroids, D 304, 305
Steroid aglycones, D 369, 413
Steroidal alkaloids, E 6
Steroid alcohols, D 30, 193 ff; E 4
—, *esters, nomenclature*, D *437*
—, oxidation, D 31
—, *nomenclature*, D *449*
Steroid anthraquinone-β-carboxylates, D 312
Steroid 3-chloro-3,5-dienes, D 340
Steroid cyclopropanes, D 344
Steroid α-diazoketones, D 286
Steroid ethers, D 339
Steroid hormones, D 132, 184, 286, 336; E 30
Steroid hydroperoxides, D 243
Steroid nuclei, conformations, D 14 ff
Steroid sapogenins, D 36; E 1
—, biogenesis, E 23
Steroid saponins, E 1 ff
Steroid spectra, collections, D 66
Steroid-12-trimethylammonium hydroxides, D 312
Sterols, D 1 ff, 73 ff; E 135
—, algae, D 152
—, alkylation mechanism, E 109
—, biogenesis, E 54 ff
—, biosynthetic sequence, E 107
—, bond formation, E 109
—, cactus, D 3, 151
—, chemical reactions, D 77 ff
—, colour reactions, D 76
—, degradation, D 189
—, esters, D 43
—, fluorinated, biological activity, D 354
—, from carbon dioxide, E 57
—, fungal, E 110
—, marine, D 3
—, marine invertebrates, D 157
—, mixed crystals from neutral fats, D 73
—, naturally occurring, physical properties, D 164
—, physical properties, D 74
—, separation, D 73
—, side-chain oxidation, D 83
—, ultraviolet absorption chromophores, D 68
Δ^{5}-Sterols, acetates, D 85
—, esters, D 85
—, hydrogenation, D 86
—, oxidation, D 83, 84
Δ^{22}-Sterols, E 110
$C_{(19)}$Sterols, in plants, E 112
$C_{(21)}$Sterols, in plants, E 110
Sterophenol, D 130
Steviol C 470; E 90
Stevia rebaudiana, E 90
Stigmasta-5,24(28)-dien-3β-ol, D 152
Stigmasta-5,25-dienol, D 166
5α-Stigmasta-8,(14),22-dien-5β-ol, D 166
Stigmasta-3,5-dien-7-one, D 159
Stigmasta-4,22-dien-3-one, D 132
Stigmastanes, nomenclature, D *428, 429*
Stigmastane, D 3, 4
—, molecular rotation, D 63
Stigmastanol, D 27, 28, 75, 152, 153, 156, 157, 166
Stigmastanyl acetate, D 132
Stigmast-7-enol, D 156, 166
Stigmast-22-en-3β-ol, E 109
5α-Stigmast-8(14)-en-3β-ol, D 156
5α-Stigmast-14-en-3β-ol, D 156
5α-Stigmast-22-enol, D 166
Stigmast-4-en-3-one, D 157, 159
Stigmast-4-en-3-one, ultraviolet absorption spectra, D 68
Stigmasterol, D 3, 13, 27, 28, 29, 36, 75, **132**, 133, 154, 155, 158, 166, 231, 326
i-Stigmasterol methyl ether, D 155
Stigmasteryl acetate, D 132
Stigmasteryl acetate dibromide, D 191
i-Stigmasteryl methyl ether, D 132
Stigmasteryl tosylate, D 132
Stilboestrol, C 16
Stipitatic acid, B 359, 366, 367
Stipitatonic acid, B 366
Stobbe condensation, A 169; B 100; D 220, 221
Streblus asper, D 384
Strecker reaction, B 142
Streptamine, B 83, **84**

Streptomyces, D 311, 328
Streptomyces erythochromoglues, A 201
Streptomyces griseus, B 101
Streptomyces spectabilis, B 84
Streptomycins, B 84
Stroboside, D 364, **398**
Strophadogenin, D 384
Strophalloside, D **384**
Strophanthidic acid, cymaroside, D 384
Strophanthidin, D 1, 2, 39, 204, 244, 301, 361, 362, 369, 370, 371, **373**, 374, 375, 376, 377, 379, 382, 383, 384, 385, 412
—, biodegradation, D 381
—, *nomenclature*, D *431*
ψ-Strophanthidin, D 377
Strophanthidin glycosides, toxicity, D 383
Strophanthidinic acid, D 373, 374, 375, 376, 377, 412
Strophanthidinic lactone, D 377
Strophanthidin-δ-lyxoside-glycoside, D 384
Strophanthidin-α-L-mannopyranoside, D 383
Strophanthidin α-δ-mannopyranoside, D 383
Strophanthidin-rhamnoside, D 382, 383
Strophanthidin-rhamnoside-glucoside, D 383
Strophanthidin-3β-tetrahydropyranyl ether, D 383
Strophanthidol, D 1, 362, 373, 374, 398, 412
Strophanthidol-cymaroside, D 384
g-Strophanthidin, D 401
k-Strophanthin, D 361, 362
Strophanthobiase, D 383, 384, 386, 388, 405, 408, 411
Strophanthobiose, D 361
k-Strophanthoside, D 361, 362
Strophanthotriose, D 361
Strophanthus barteri, D 398
Strophanthus courmontii, D 398
Strophanthus digitalis aglycons, D 382
Strophanthus digitalis glycosides, D 382
Strophanthus divaricatus, D 402
Strophanthus eminii, D 385
Strophanthus gerrardii, D 398
Strophanthus grandiflorus, D 398
Strophanthus gratus, D 401
Strophanthus hispidus, D 398
Strophanthus hypoleucus, D 398
Strophanthus kombé, D 361, 373, 385; E 112
Strophanthus nicholsonii, D 398
Strophanthus petersianus, D 398
Strophanthus preussi, D 398
Strophanthus sarmentosus, D 398, 400
Strophanthus species, D 373
Strospeside, D 362, **391**
Structure determination, natural products, C 477
—, nuclear magnetic resonance, C 477
Strychnos species, B 214
Stryphnodendron sapogenin B, C 441
Stryphnodendron sapogenin F, C 441
Styrenes, reaction with carbenes, A 36
Styrene, A 76, 92
—, reaction with diazoacetic ester, A 53
Suberanecarboxylic acid, B 356
Suberic acid, B 354; D 413
—, pyrolysis of salts, B 407
Suberic dinitrile, B 354
Suberites compacta, D 157
Suberoin, B 412
Suberone, B **353**, 407
Suberylarginine, D 413
Substances A, C, D and V, D 290
Substance 20-epi-K, D 292, 305, 310
Substance 11-epi-M, D 312
Substance F, D 324
—, acetate, D 326
Substances J and O, D 306, 307, 308
Substance J, O and L, D 305
Substance K, D 291, 292, 307, 309, 310
Substance L, D 305, 307
Substance M, D 312, 329
Substance P, D 292, 307, 308, 310
Substances P, K, 20-epi-K and S, D **305**
Substances P, K, R and N, D 291
Substance Q, D 300
Substance S, D 296, 299, 310, 312
—, acetate, D 310, 311, 312
Substance T, D 296
Substance U, D 324, 325
—, diacetate, D 325
Succindialdehyde, A 194
Succinic acids, alkylated, B 154
Succinic acid, B 416; C 259, 282, 306
Succinic ester, A 194; C 113
—, condensation with γ-keto-esters, A 169
—, reaction with cyclohexanone, B 150
Succinic methyl ester, D 221
3α-Succinoyloxy-12-oxo-5β-cholanoic acid, D 182
Sugasterol, C 410

Sugiol, C 392
Sulcatoxanthin, B 331
Sulphanilic acid, D 199
Sulphides, cyclohexanes, B 76
Sumaresinolic acid, C 440
Suprasterols, D **148**
Suprasterol, D 71, 138, 140
Swietenine, C 431
Swietenolide, C 431
Sylvestrene, B 183, **184**
Sylvestrene dibromide, B 184
Sylveterpineol, B 184
Sylveterpinolene, B 184

Tachysterol, D 138, 139, 140, 143, 144, 148
Tagetenones, B 168
Tagetes glandulifera, B 168
Tagetes patula, B 304
Tagetone, B **168**
Tall oil, D 154
L-Talomethylose, D 361, 365, 399, 400
DL-Talomucic acid, B 61
Tamus edulis, E 26
Tamusgenin, B 26
Tanacetonedicarboxylic acid, A 171
α-Tanacetonic acid, A 58
Tanacetophorone, A 171; C 145, 146
Tanacetum vulgare, C 145
Tangeraxanthin, B 332
Tanghiferigenin, D **406**
Tanghiferin, D **406**
Tanghinia venenifera, D 406
Tanghinigenin, D **406**
Tanghinin, D **406**
Tannins, fern, B 113
Tansy oil C 145
Tarachrome B 332
Taraxanthin, B 237, 332
ψ-Taraxastandiol, C 449
Taraxastane group, triterpenoids, C 449
Taraxasterol, C 413, 415, 416
ψ-Taraxasterol, C 415, 416
Taraxerane group, triterpenoids, C 445
Taraxerene, C 445
Taraxerol, C 413, 414, 415, 445
Taraxerol acetate, C 445
Taraxerone, C 445
Tasmanone, B **130**
Tauremisin A, C 347
Taurine, D 169
Tauro-acids, D 169, 170
Taurochenodeoxycholic acid, D 177
Taurocholanic acid, D 177
Taurocholic acid, D 177
Taurodehydrocholic acid, D 177
Taurodeoxycholic acid, D 177
Taurolithocholic, D 177
Tauroursodeoxycholic acid, D 177
Taxarasterol, C 449
ψ-Taxarasterol, C 449
Taxicins, C 381
Taxicin-I, C 374
Taxines, E 87, 91
Taxinin, C 402, 478
Taxus baccata, B 313, 314; E 114
Tecomanine, B 212
Tecoma stans, B 213
Telekia speciosa, C 339
Telekin, C **339**, 340, 366
Telocinobufagin, D 13, 413, 416
—, *nomenclature*, D *432*
Telocinobufogenin, D 414, **417**
Tenulin, C **358**, 359, 368
Tercyclohexyls, C 15
Tercyclopropyl, C 8
Terebic acid, A 61, 66; B 206; C 152, 169
Terephthalic acid, C 267
—, reduction, B 148
Terephthalic dimethyl ester, hydrogenation, B 145
Teresantalal, C 192
Teresantalic acid, C 187, 190, 192, 264
Teresantalol, C 192, 264
Teresantol, E 69
Terferrocenyls, A 135
Terminolic acid, C 441
Terpenes, B 347
—, acyclic, A 51
—, biogenesis, E 54 ff
—, biological aromatization, E 77
—, degradation products, A 166
Terpene alcohols, B 44
Terpenoids, aromatic, E 77
—, biosynthesis, C 460 ff; E 136
—, polycyclic, C 478
—, synthesis in seedlings, E 135
Terpenoid aldehydes, B 195
Terpenoid cyclic ethers, B 204
Terpenoid diols, B 193
Terpenoid epoxyketones, B 202
Terpenoid hydroxyketones, B 202
Terpenoid ketones, B 196
Terpenoid nitrosochlorides, B 179
Terpenoid quinones, E 134
Terpenylic acid, B 188, 206; C 152, 167
Terphenyl pyrophsophates, B 335
1,4-Terpin, B **194**, 195, 206, 207
1,8-Terpin, B 188, 193, **194**, 195, 205

α-Terpinene, B 25, 180, **181**, 182, 207; C 161, 171
β-Terpinene, B **181**
γ-Terpinene, B 25, **181**, 182
Terpinene dihydrochloride, B 206
Terpineols, B **188**
Terpineol, commercial, B 190
α-Terpineol, B 163, 164, 178, 180, **188**, 189; C 145, 163, 169, 171, 250, 251; E 68
β-Terpinol, B **188**, 189, 220
γ-Terpinol, B **189**
α-Terpineol dibromide, B 206
Terpin hydrate, C 163
Terpinolene B **180**, 181, 189; E 74
Terrein, A 176
Testosterone, D 2, 7, 61, 185, 197, 213, 241, 254, 256, **258**, 259, 261, 262, 264, 277, 282, 283, 338, 340, 341, 342, 343, 344
—, anabolic properties, D 272
—, assay, D 270, 271
—, isomerides, D 269
—, *nomenclature*, D *439*, *440*, *442*
—, physiological properties, D 271
—, reactions,D 260
—, total synthesis, D 102
Δ^1-Testosterone, D 271
$\Delta^{9(11)}$-Testosterone, D 302
Testosterone acetate, D 261, 262, 264, 341
Testosterone benzoate, D 258
Testosterone esters, thio-acids, D 271
Testosterone propionate, D 261, 274, 276, 340
Tetra-acetoxydihydroxyetianic acid, D 402
2,3,4,6-Tetra-*O*-acetyl-α-L-mannosyl bromide, D 383
Tetra-alkyldiboranes, A 41
Tetra-aminocyclobutane, A 91
Tetrabromoadamantane, C 130
Tetrabromocholestan-3-ones, D 120, 122
Tetrabromocyclobutane, A 88; B 393
Tetrabromocyclohexane, B 27, 28, 36
Tetrabromocyclopent-3-ene-1,2-dione, A 184
Tetrabromodimedone, B 112
Tetrabutylammonium chloride, D 50
Tetracarboxycobalticinium ion, A 139
Tetrachloro-1,2-benzoquinone, B 4
Tetrachloro-1,4-benzoquinone, B 265
Tetrachlorocyclohexane-1,2-diol, B 65
Tetrachlorocyclohexanone, B 49, 102, 104
Tetrachlorocyclohexene, B 65
1,2,5,6-Tetrachlorocyclo-octa-1,5-diene, B 383
Tetrachlorocyclopentadiene, C 13
Tetrachlorocyclopentadienone, diethyl acetal, A 152
Tetrachlorocyclopropane, A 40
Tetrachlorocyclopropene, A 38
Tetrachloro-2′,3′-diphenyltripentafulvalene, C 13
Tetrachlorohexafluorocyclopentane, A 150, 151
Tetrachlorohexahydro-1,4:5,8-dimethanonaphthalene, C 105
Tetrachlorotetrafluorocyclopent-1-ene, A 150, 151
Tetracyanocyclopropanes, A 62
Tetracyanocyclopropane-3,3-dicarboxylic ester, A 63
Tetra(cyanoethyl)cyclohexanone, B 100
Tetracyanoethylene, A 77, 97, 173; B 29; C 17, 90, 123
Tetracyano-3-methoxycyclobutane, A 77
Tetracyano-7-methylenebicyclo-[2.2.2]-oct-5-ene, B 29
Tetracyclic diterpenoids, C 379
Tetracyclodecane, C 45
Tetracyclododec-3-ene, C 104
Tetracyclodec-9-ene, C 45
Tetracyclohexane, C 45
Tetracyclone, A 173
Tetracyclopentadienes, C **103**
Tetradecane-2,13-dione, B 74
n-Tetradecyclcyclohexane, B 14
1-*n*-Tetradecylcyclopentan-1-ol, A 157
Tetradehydro-[24]-annulenes, B 398
Tetraethoxycarbonylethene, B 151
Tetraethylammonium acetate, C 53
Tetraethylammonium cyclononatetraenide, B 399
Tetrafluoro-1,3-bis-(trifluoromethyl)-bicyclo[1.1.0]butane, C 34
Tetrafluorocyclobutane, A 88
Tetrafluorocyclobutene, A 88, 89
Tetrafluorocycloheptadienes, B 368
Tetrafluoro-6,8-dimethylenespiro-[3.5]-nonane, B 25
Tetrafluoroethylene, A 77, 87; B 25, 28; C 9, 23
Tetrafluoro-6-methylenespiro-[3.5]-non-7-ene, B 28
Tetrafluorospiro[2.3]hexane, C 23
Tetrahalogenomethylcyclobutanes, A 84

Tetraheptylammonium chloride, D 170
Tetrahydroalantolactone, C 302, 337, 338, 339
Tetrahydroambrosin, C 357
Tetrahydroanhydrodigitoxigenone, D 395
Tetrahydroanhydrouzarigenins, D 407
Tetrahydroartabsins, C 352, 354
Tetrahydroartemisia ketone, B 165
Tetrahydroartemisin, C 346
Tetrahydrobalduilin, C 359
Tetrahydrobenzoic acids, B 136
Tetrahydrobisabolene, C 259
Tetrahydrobufotalins, D 417
20,22-Tetrahydrobufotalinin, D 418
Tetrahydrocadinene, C 262
Tetrahydrocarbazole, B 100
Tetrahydrocarvone, C 251
Tetrahydrodicyclopentadiene, C 103, 130
Tetrahydroelemol, C 295
Tetrahydroestafiatone, C 354
Tetrahydroeucarvone, C 155, 156
Tetrahydrofurans, thermal ring contraction, A 23
Tetrahydrofuran, A 109
Tetrahydro-2-furoic ester, A 23
Tetrahydrohelenalin, C 359
Tetrahydrohumulene, C 283
Tetrahydro-1β-hydroxy-8-methylidan-5-one, D 205
Tetrahydroionones, B 176
Tetrahydroirone, B 177
Tetrahydroivalin, C 338
Tetrahydrolycopene, B 276, 334; E 128, 129
20,22-Tetrahydromarinobufagin, D 419
Tetrahydro-6-methylionone, B 177
1,4,5,8-Tetrahydronaphthalene, C 71
Tetrahydro-β-naphthoic acid, C 67
5,6,7,8-Tetrahydro-2-naphthol, C 30
Tetrahydroparthenin, C 356, 357
Tetrahydropartheniol, C 316, 317
Tetrahydrophthalic acids, B 145, 146
Tetrahydropinnatifidin, C 339
17β-Tetrahydropyranyl ethers, D 273
Tetrahydropyrethrin, A 183
Tetrahydropyrethrolone, A 179
Tetrahydropyrethrone, A 180
Tetrahydropyrethrosin, C 363
Tetrahydropyridazines, A 74
Tetrahydrosantonins, C 343
Tetrahydrosaussurea lactone, C 295, 337
Tetrahydrosmilagenin, D 60
Tetrahydrotelekin, C 339
Tetrahydrotigogenin, D 60
Tetrahydroxy-5β-bishomochol-23-ene, D 195
Tetrahydroxycholestanes, D 194, 195
3β,5β,14β,19-Tetrahydroxy-14-isoetianic acid, D 374
Tetrahydroxyisosterocholanic acid, D 193
Tetrahydroxyquinone, B 61
Tetrahydroxysterocholanic acid, D 193
Tetrahydrozerumbone, C 330
Tetrahydrozingiberene, C 262
Tetrahymanol, C 417, 454, 475; E 119
Tetrahymena pyriformis, E 119
Tetrakiscyclopentadienylmetal complexes, A 125
Tetralin, A 143
—, reduction, C 70
Tetralols, hydrogenation, C 63
Tetramethylacetonedicarboxylic acid, B 131
Tetramethyladipic acid, A 105
1,2,3,4-Tetramethylbenzene, oxidation, B 121
Tetramethylbicyclo[1.1.0]butane-1-carboxylic ester, C 33
Tetramethylcycloalkane-2,3-diones, A 16
Tetramethylcyclobutadiene, A 86, 87
Tetramethylcyclobutane-1,3-diol, A 90
Tetramethylcyclobutane-1,3-dione, A 48, 90, 93; B 113
Tetramethylcyclobutene, A 81
Tetramethylcyclohepta-1,3,5-triene, B 350
Tetramethylcyclohexa-1,3-diene, B 26
Tetramethylcyclohexadienecarbaldehydes, B 88
Tetramethylcyclohexa-2,4-dien-1-one, B 121
Tetramethylcyclohexanes, B 14
Tetramethylcyclohexanecarbaldehyde, B 88
Tetramethylcyclohexane-1,3,5-trione, B 113
Tetramethylcyclohexan-1-ol, B 47
Tetramethylcyclohexanone, B 95
Tetramethylcyclohex-3-enecarbaldehyde, B 88
Tetramethylcyclo-octatetraene, B 392
Tetramethylcyclopentanes, A 113
Tetramethylcyclopentan-1-one, A 167
Tetramethylcyclopent-1-ene, A 116
Tetramethylcyclopropanol, A 48
Tetramethylcyclopropanone, A 48
Tetramethylcyclopropene, A 33

Tetramethyldicyano-barrelene, C 110
2α,3β-Tetramethylene-5α-androstane, D *451*
Tetramethylenecyclobutane, A 84
Tetramethylene dibromide, A 80
Tetramethylmethane, C 21
Tetramethyl-2-methylenebicyclo-(3,1,0)hex-3-ene, A 163
Tetramethylphloroglucinol, B 131
3,4:3′,4′-Tetraoxo-β-carotene, B 320
Tetraphenylbutatriene, A 84
Tetraphenylcyclobutadiene, A 87; B 393; C 44
—, cyclopentadienylmetal complexes, A 143
Tetraphenylcyclobutadienecyclopentadienylcobalt, A 139
Tetraphenylcyclobutenium difluoroborate, A 85
Tetraphenylcyclopentadiene, A 109
Tetraphenylcyclopentadienol, A 156
Tetraphenylcyclopentadienone, A 156, 157, 173
Tetraphenylcyclopentadienyllithium, C 13
Tetraphenylcyclopent-2-en-1-one, A 156
Tetra-poly-ynes, B 388
Tetraterpenoids, E 120 ff
—, biosynthesis, E 120
Tetratriacontane-1,34-dioic ester, B 423
Tetraynes, B 388
Thapsoin, B 412
Thelepogine, C 385, 387, 481
Thesine, A 100
Thesium minkwitzianum, A 100
Thevebioside, D 362, **386**
Thevetia neriifolia, D 386
Thevetin, D 362, **386**
Thevetobiase, D 386
Thevetose, D 363, 367
L-Thevetose, D 361, 406
2-Thia-adamantane, C 124
Thiazoles, B 380
Thilobilianic acids, D 21, **170**
Thin-layer chromatography, D 369
—, bile acids, D 178
—, steroids, D 72
Thiocholesterol, D **110**
2-Thiocyanatocyclopentylmethane sulphonate, A 161
Thiols, cyclohexanes, B 76
Thiorhodeceae, B 238
Thiospirillum jenense, B 307
Thiosterols, D **110**
Thiotropones, B 377
Thorpe reaction, A 106; C 60; D 220
α-Thujadicarboxylic acid, C 138, 140, 143, 145
β-Thujadicarboxylic acid, C 145
Thujaketone, C 145
α-Thujaketonic acid, A 58; C 138, 145, 146
β-Thujaketonic acid, C 145
α-Thujamenthol, C **146**
β-Thujamenthol, C 146
Thujamenthone, B 208
α-Thujamenthone, C 146
β-Thujamenthone, C 146
Thujane, C 36, 136, **137**, 138, 139; E 69
Thujane alcohols, C 141
Thujane group, C 137
Thujane ketones, C **145**
Thujan-2-ol, C 148
Thujan-3-ol, C 138, **141**
Thujan-4-ol, C **144**
Thujan-2-one, C 148
Thujan-3-one, C 140, 143, **145**, 146, 148
Thuja occidentalis, C 234, 306
Thuja oil, C 234
Thuja plicata, B 215, 357, 365
Thujaplicins, B 365, 374
β-Thujaplicin, B 372, 375
γ-Thujaplicin, B 372, 374; E 78
Thujaplicinol, B **372**
Thujarhodin, B 313
Thuja standishii, C 234
2-Thujene, C 137, **138**, 139
3-Thujene, C 137, **138**, 139, 140
4(10)-Thujene, C 137, **139**
α-Thujene, C 137, **138**, 144; E 79
β-Thujene, C 137, **138**
Thuj-3-en-10-ol, C 141
4(10)-Thujen-3-ol, C **142**
3-Thujene-2-one, C **147**
Thujic acid, B **215**, 216, **357**; C 157
Thujone, A 58; C 141, **145**, 147, 252, 253, 254; E 71, 73
α-Thujone, B 197
Thujopsene, C **275**, 277, 288, 327
Thujopsis dolobrata, B 365; C 268, 275, **326**
Thujyl alcohols, properties, C 142
Thujyl alcohol, C 138, **141**, 142, 143
Thundergene, E 91
Thurberogenin, C 436
Thymohydroquinone, E 77
Thymol, B 198; C 149, 150; E 75
—, hydrogenation, B 184

Thymoquinol, B **221**
Thymoquinone, E 77
Tiglic acid, E 61
Tigogenin, D 2, 251; E 2, 7, 11, 12, 14, 21, 22, 23, 25, **30**, 32, 33, 44, 46, 52, 112
—, *nomenclature*, D *435*
ψ-Tigogenin, D 251, 253
Tigogenin acetate, E 7, 8, 14
Tigonin, E **2**, 5
Tinophyllone, C 385, 387
5α-Tirucallane, D 429
Tirucallols, E 115, 116
Tirucallol, C 410, 411, 426
Tlatlancuaya, C 348
Toad poisons, D 360 ff, **412**
Tobacco leaf, D 154
Tocopherols, E 134
α-Tocopherol, E 87, 134, 136
Tocopherolquinones, E 87, 134, 135, 137
α-Tocopherylquinone, E 136
Tocotrienols, E 87
Toddalolactone, E 63
Todomatuic acid, C **293**, 334
Tohogeninol, C 458
Tohogenol, C 458
Tokorogenic acid, E 38
Tokorogenin, E 3, 27, 30, **38**, 41
Tokorogenin triacetate, E 40, 41
Tokoronin, E 3
Tollens' test, D 371
Toluene, from carotenoids, B 258
2-*p*-Toluenesulphonamidoisoindoline, B 29
p-Toluenesulphonic acid, D 221
2-*p*-Toluenesulphonylcyclohexan-1-ol, B 18
1-*p*-Toluenesulphonylcyclohex-1-ene, B 18
Toluic acids, hydrogenation, B 134
m-Toluic acid, B 374
o-Toluidine, hydrogenation, B 81
3-*p*-Tolylthiotropolones, B 373
Tomatanine, D *449*, *450*
Tomatidane, D **2**
Tomatidine, D **2**
Tomatine, D 40
Tomentosic acid, C 441
Tomentosolic acid, C 450
Topological isomerism, A 18
Torreyal, C **290**, **334**
Torreya nucifera, C 290, 293, 297
Torreyol, C **290**, 297, 334
Tortelli-Jaffé test, D 76, 150
Torularhodin, B 247, **325**, 326, 345; E **120**, 121, 128
Torula rubra, B **280**, 325
Torularhodin, E 119
Torulene, B **280**, 325, 326, 345
Torulosal, C 383
Torulosol, C 382
3-*O*-Tosylinositol, B 61
6-*O*-Tosylinositol, B 61
7β-Tosyloxy-5α-cholestan-4-one, D 81
4-Tosyloxycyclohexanone, C 37
3-Tosyloxymethylcyclohexanone, C 39
3-Tosyloxy-2,2,4,4-tetramethylcyclobutane-1-carboxylic ester, C 33
Totarols, diterpenoids, physical properties, C 393
Totarol, C 379, 393
18-Totaroloic acid, C 393
Trachylobane, C 379
Trachylobane diterpenoids, C 400
Trachylobane group, C 381
Trachylobanic acid, C 400
Trachylobanol, C 379, 380, 400
Trametenolic acid, C 407, 408, 421
Tranetes species, C 407
Transannular reactions, bicyclo-[3.3.1]nonanes, C **127**
α-Trans-π-oxocamphor, C 191
Transvaalin, D 366, **412**
Transvaal slangkop, D 412
Treleasagenic acid, C 441
Triacetate B, C 429
Triafulvenes, A 37
1,3,3-Trialkoxy-2-chlorotetrafluorocyclopent-1-ene, A 150
Trialkylaluminiums, C 117
2,2,6-Trialkylcyclohexa-3,5-dien-1-ones, B **120**
Trialkyl phosphites, B **126**
Triamantane, C **135**
1,2,3-Triaminocyclopropane, A 46
Trianhydrohydroxylactone, D 401
Trianhydrostrophanthidin, D 375, **378**
Triapentafulvalene, C 13
Triazoles, B 379
1,3,5-Tribromoadamantane, C 130
1,2,4-Tribromobenzene, B 37
Tribromobiscyclopentadienylniobium, A **126**
Tribromocamphors, physical properties, C **211**
Tribromocholestan-3-ones, D **120**, **122**, **123**, **130**
2,2,4-Tribromocholest-4-en-3-one, D 120
1,1,2-Tribromocyclobutane, A 87, 88

1,5,9-Tribromocyclododeca-1,5,9-triene, B 397, 402
2,4,5-Tribromocyclopentane-1,3-dione, A 186
3,3,5-Tribromocyclopentane-1,2,4-trione, A 188
Tribromocyclopentanone, A 168
3,4,5-Tribromocyclopent-1-ene, A 149
2,2,4-Tribromocyclopent-4-ene-1,3-dione, A 186
Tribromodimedone, B 112
2,4,7-Tribromo-3-hydroxytropone, B 355
Tribromo-6-methylcyclohexanone, B 104
6-Tribromomethyl-4-en-3-ones, D 95
Tribromoneopentanes, A 72
1,1,1-Tribromopropane, B 126
Tribromosantonin, C 345
3,5,7-Tribromotropolone, B 372, 373
Tri-*tert*-butoxyaluminium hydride, D **207**
Tri-*n*-butylamine, D 353
1,2,4-Tri-*tert*-butylbenzene, C 45
1,2,5-Tri-*tert*-butylbenzene, C 42, 45
1,3,5-Tri-*tert*-butylbenzene, C 45
Tributyltin hydride, A 64
Tricarballylic ester, A 193
Tri-μ-carbonyl-bis(tricarbonyliron), A 85
Tricarbonyldihydropentalenylmanganese, C **52**
Tricarbonyliron, A 86
Tricarbonylnorbornadienyliron, C 91
Tricarbonyltetraphenylbutadieneiron, A 85
Tricarbonyltetraphenylcyclobutadieneiron, A 86
Trichloroacetic acids, D 344
Trichloroacetic ester, A 25
Trichlorobenzenes, B 37
Trichlorocholestan-3-ones, D 122
Trichlorocyclopentadienyl titanium, A 126
3,3,5-Trichlorocyclopentane-1,2,4-trione, A 188
2,2,3-Trichlorocyclopentanone, A 184
1,1,2-Trichlorocyclopropane, A 40
1,4,4-Trichloro-3,3-difluorocyclobutene, A 88
2,2,2-Trichloroethylidenemalonic ester, C 55
4-Trichloromethyl-3,4-dimethylcyclohexa-2,5-dien-1-one, B 126
Trichloropentafluorocyclopentene, A 150, 151
1,3,3-Trichloro-2-phenylcyclopropene, A 38
Trichothecin, E 82
Trichothecium roseum, E 82, 88
1,2,3-Tricyanocyclopropane-1,2,3-tricarboxylic ester, A 63
Tricyclene, C 160, 188, 192, 193, 198, **206**, **229**
Tricyclic diterpenoids, C 377
Tricyclic hydrocarbons, C **115**
Tricyclobutonium ions, A 24
Tricyclo[4.2.2.0]deca-3,7-dienedicarboxylic acid anhydride, B 396
Tricyclo[4.3.0.1^{2,5}]deca-3,7 diene, C 5, 99
Tricyclo[3.3.1.1^{3,7}]decane, C 129
Tricyclo[4.3.0.1^{2,5}]decane, C 100
Tricyclo[4.4.0.0^{3,8}]decane, C 133
Tricyclo[5.3.0.0^{2,6}]decane, A 171
Tricyclodecane-7,9-dicarboxylic acid, A 103
Tricyclo[4.3.0.1^{2,5}]dec-7-ene, C 100
Tricyclo[5.3.2.0^{4,9}]dodecane, C 5
Tricycloekasantalal, C 263, 264, 293
Tricycloekasantalic acid, C 264
Tricyclo[4.1.0.0^{1,3}]heptane, C 22
Tricyclo[4.1.0.0^{2,7}]heptane, C 34
Tricyclo[3.1.0.0^{2,6}]hexane-1-*N*-*tert*-butylcarboxamide, C 34
Tricyclo[4.2.0.0^{2,5}]octa-3,7-diene, C 43
Tricyclo-octane, C 43, 55, 84, 109
Tricyclopentadiene, C 102
Tricyclopentadienide complexes, A 125
Tricyclopropylborane, A 42
Tricyclopropylcarbinol, C 9
Tricyclopropylmethane, C 9
Tricyclo[4.3.1.1^{4,8}]undecane, C 133
Tricyclo[5.4.0.0^{2,9}]undecane, C 5
Tricyclovetivene, C 278, **279**, 289, 323
Tricyclovetivenol, C **323**, 336
Tridehydro-[12]-annulene, B 397
Tridehydro-[18]-annulene, B 398
Tridehydro-[26]-annulene, B 398
Trienic gestogens, D **247**
Triethylaluminium, B **2**
Triethylamine, C 46
Triethylcyclobutanone, A 92
Triethyloxonium fluoroborate, B 363
Triethyl phosphite, B 77
Trifluoroacetic acid, C 120
Trifluoroacetic anhydride, B 116, 121
Trifluoroiodomethane, B 42
Trihalogenocyclohexanes, **34**
Trihexylborane, A 112
Trihydroxy-5β-bishomochol-24-ene, D 195

3β,5,14-Trihydroxy-5β,14β-bufa-20,22-dienolide, D 432
3β,11α,14β-Trihydroxy-5β-bufa-20,22-dienolide, D 420
Trihydroxybufosterocholenic acid, D 193
3α,11,12-Trihydroxycholanic acids, D 315
3β,5,6β-Trihydroxy-5α-cholanoate 3,6-diacetate, D 49
Trihydroxy-5α-cholanoic acids, D 172, 173
Trihydroxy-5β-cholanoic acids, D 49, 172, 173, 180
3α,7α,12α-Trihydroxy-5β-cholan-24-oic acid, D *439*
Trihydroxy-5β,25α-cholestan-6(or 27)-oic acid, D 192
Trihydroxycoprostanic acids, D 2, 192
Trihydroxycoprost-22-ene-24-carboxylic acid, D 193
Trihydroxycoprostenic acid, D 193
Trihydroxycyclohexanones, B 56
3,4,5-Trihydroxycyclohex-1-ene-1-carboxylic acid, B 140
3β,5β,16β-Trihydroxy-14β,15β-epoxy-5β-bufa-20,22-dienolide 16-monoacetate, D 420
3α,5,8α-Trihydroxy-5α-ergost-22-ene, D 136
3,16α,17-Trihydroxyestra-1,3,5(10)-triene, D 201
3β,5β,19-Trihydroxyetianic acid, D 374
Trihydroxyglutaraldehyde, B 80, 85
Trihydroxyhomochol-24-ene, D 195
3,12,14β-Trihydroxy-14-isocholanic acid, D 411
3β,5β,14β-Trihydroxy-14-isoetianic acid, D 417
3β,5β,14β-Trihydroxy-14-iso-21-norpregnane-19,20-dicarboxylic acid, D 412
Trihydroxyisosterocholenic acid, D 193
Trihydroxykaurenolide, C 396
Trihydroxy-24-methylcoprostanic acid, D 193
Trihydroxy-24-methylcoprost-22-enic acid, D 193
6,7,8-Trihydroxy-1-naphthoic acid, B 375
11β,17α,21-Trihydroxy-A-norpregn-3(5)-ene-2,20-dione, D 283
11β,17α,21-Trihydroxy-A-norpregn-3(5)-en-1,2,20-trione, D 281
3β,11α,14β-Trihydroxy-12-oxo-5β-bufa-20,22-dienolide, D 421
3β,5,14-Trihydroxy-19-oxo-5β-card-20(22)-enolide, D 373
3β,5,14-Trihydroxy-19-oxo-5β,14β-card-20(22)-enolide, D 431
3β,12,14β-Trihydroxy-19-oxo-5α-card-20(22)-enolide, D 406
8,9,15-Trihydroxypentadecane-1-carboxylic acid, B 415
3α,11α,17α-Trihydroxy-5β-pregnan-20-one, D 328
3α,11β,17α-Trihydroxy-5β-pregnan-20-one, D 328
11β,17β,21-Trihydroxy-4-pregnene-3,20-dione, D *439*
Trihydroxyspirostan, E 50
2,4,6-Tri-iodophenol, B 375
Tri-iron dodecarbonyl, B 363
Triisopropylidenecyclopropane, A 34
Triketones, ring, A 187
Trillarin, E 2
Trillin, E 3
Trillioside, E 3
Trillium erectum, E 2, 3, 43
Trillium grandicorum, E 3
Trillium species, E 25
Trimesic methyl ester, B 151
1,2,4-Trimethoxybenzene, B 367
4-Trimethoxyestra-1,3,5(10)-trien-17-ols, D 216
Trimethylacetic acid, D 333
ααβ-Trimethyladipic acid, B 177
5-Trimethylaminopentan-2-one bromide, A 21
2,3,4-Trimethylbenzaldehyde, B 277
2,3,6-Trimethylbenzaldehyde, B 277
Trimethylbenzenes, B 25
2,3,6-Trimethylbenzyl bromide, B 279
2,6,6-Trimethylbicyclo[3.1.1]heptane, C 137
1,4,4-Trimethylbicyclo[3.2.0]hept-6-en-2-one, C 46, 157
4,4,6-Trimethylbicyclo[3.2.0]hept-6-en-2-one, C 157
1,5,5-Trimethylbicyclo[2.1.1]hexane, C 84
1,6,6-Trimethylbicyclo[2.1.1]hexanecarboxaldehyde, B 167
1,6,6-Trimethylbicyclo[2.1.1]hexane-5-carboxylic acid, C 233
1,6,6-Trimethylbicyclo[2.1.1]hexane-5-*exo*-carboxylic acid, C 85
Trimethylbicyclo[2.1.1]hexan-5-ol, C 86
2,2,7-Trimethylbicyclo[4.3.0]nonan-7-ol-5-one, B 120

4,4,14α-Trimethyl-5α-cholesta-7,9(11)-dien-3β-ol, B 75
4,4,14α-Trimethyl-5α-cholesta-8(9),24-dien-3β-ol, D 75
4,4,14α-Trimethyl-5α-cholestane, D 3, 4
—, *nomenclature*, D *429*
4,4,14α-Trimethyl-5α-cholestan-3β-ol, D 75
4,4,14α-Trimethyl-5α-cholesta-7,9(11),24-trien-3β-ol, D 75
4,4,14α-Trimethyl-5α-cholest-8-en-3β-ol, D 75
6,7,7-Trimethylcholesterol, D 103
Trimethylchromium, A 109
2,6,6-Trimethylcyclohepta-2,4-dienone, B 355
4,6,6-Trimethyl-3,4-cycloheptadien-1-one, C 157
1,7,7-Trimethylcyclo[2.2.1]heptane, C 137
Trimethylcycloheptanones, B 353
3,7,7-Trimethylcyclohepta-1,3,5-triene, B 350
Trimethylcycloheptenecarbaldehydes, B 353
Trimethylcyclohexadienes, B 26
Trimethylcyclohexadienecarbaldehydes, B 88
Trimethylcyclohexanes, B 14, **16**, 172
1,1,3-Trimethylcyclohexane, B 158
Trimethylcyclohexanecarbaldehydes, B 88
4,4,5-Trimethylcyclohexane-1,2-dione, B 109
4,4,5-Trimethylcyclohexane-1,3-dione, B 109
Trimethylcyclohexan-1-ols, B 47
Trimethylcyclohexanones, B 95
2,2,6-Trimethylcyclohexanone, B 286
3,3,4-Trimethylcyclohexanone, B 177
Trimethylcyclohex-1-enes, B 21
Trimethylcyclohexenecarbaldehydes, B 88
2,2,4-Trimethylcyclohex-3-ene-1-carboxylic acid, C 185
Trimethylcyclohex-2-enone, B 117
1,3,3-Trimethylcyclohex-1-yl hydroperoxide, B 75
Trimethylcyclopentanes, A 113
1,1,4-Trimethylcyclopentane-2,4-dicarboxylic acid, C 241
Trimethylcyclopentan-1-ols, A 155, 157
Trimethylcyclopentanone, A 166, 167
Trimethylcyclopent-1-ene, A 116
1,1,2-Trimethylcyclopropane, A 32
1,2,3-Trimethylcyclopropane, A 33
1,3,3-Trimethylcyclopropene, A 33
Trimethylene, A 32
Trimethylene bromide, A 95
Trimethylenecyclobutane, A 84
Trimethylenecyclohexanes, B 25
Trimethylenecyclopropane, A 34
Trimethylferrocenylmethylammonium iodide, A 133
ααβ-Trimethylglutaric acid, B 177
1,2,2-Trimethyl-4-hydroxycyclopentane-1-carboxylic acid, B 323
Trimethylmethylbenzylammonium hydroxide, B 29, 30
Trimethyl-1-methylcyclopentylammonium hydroxide, A 117
1,6,6-Trimethyl-5-methylenebicyclo[2.1.1]hexane, C 86
1,2,5-Trimethylnaphthalene, C 311
1,3,3-Trimethylnorbornane, C 243
1,7,7 Trimethylnorbornane, C 137, 233
2,2,3-Trimethylnorbornane, C 137, 197
2,7,7-Trimethylnorbornane, C 245
1,3,3-Trimethylnorbornan-2-ol, C 238
1,5,5-Trimethylnorbornan-2-ols, C 240
1,5,5-Trimethylnorbornanone-2, C 240
Trimethylnorborn-2-enes, C 234
1,5,5-Trimethylnorborn-2-en-7-one, B 215; C 157
1,3,3-Trimethylnorborn-2-one, C 234
3,7,7-Trimethylnorcarane, C 137
2,6,6-Trimethylnorpinane, C 137
1,3,3-Trimethyl-2-oxabicyclo[2.2.2]-octane, B 205
2,2,3-Trimethyl-4-oxocyclohexanecarboxylic acid, C 220
2,2,4-Trimethylpent-3-en-1-al, A 90
2,4,5-Trimethylphenol, B 121
Trimethylphosphite, D 107
ββγ-Trimethylpimelic acid, B 177
Trimethylplatinum iodide, A 129
Trimethylsilyl ethers, D 32
Trimethylsteroids, D 4
—, *nomenclature*, D *429*
Trimethylsuccinic acid, C 202
Trimethylsulphonium iodide, D 268
Trimethylsulphoxonium chloride, A 51
2,2,6-Trimethyl-1-*m*-tolylcyclohexane, B 258
1,3,5-Trimethyltricyclo[$2.2.1.0^{3,5}$]-heptane, C 242
1,7,7-Trimethyltricycloheptane, C 192
1,3,3-Trimethyltricyclo[$2.2.1.0^{2,6}$]-heptane, C 247

1,3,5-Trimethyltricyclo[2.2.1.$0^{3,5}$]-heptan-2-one, C 242
2,6,6-Trimethylvinylacetylene, B 32
Trimethylvinylsilane, A 42
1,3,5-Trinitrobenzene, B 216
Triothecin, E 88
Trioxocholanoic acids, D 189, 190
Trioxo-5β-cholanoic acid, D 20, 180, 190
3,16,22-Trioxo-5α-cholestan-26-al, E 23
Triphenylcyclobutanone, A 92
Triphenylcyclopentadiene, A 110; C 52
1,2,3-Triphenylcyclopentadienone, D 173
2,4,5-Triphenylcyclopentane-1,3-dione, A 187
1,2,3-Triphenyl-1*H*-cyclopenta(1)-phenanthrene, A 157
2,4,5-Triphenylcyclopent-4-ene-1,3-dione, A 186
1,2,3-Triphenylcyclopropene, A 33
Triphenylcyclopropenyl bromide, A 38; C 8
1,2,3-Triphenylcyclopropenylium cation, A 37
Triphenylcyclopropenylium fluoroborate, A 37
3,4,6-Triphenylgentisic ester, A 187
Triphenylmethyl acetate, A 136
Triphenylmethyl chloride, C 213
Triphenylmethyl fluoroborate, C 73
Triphenylmethylpotassium, B 357
Triphenylmethyl salts, B 359
Triphenylmethylsodium, A 52; B 417; C 32
Triphenylphosphine, A 108, 164; B 77, 363
Triphenylphosphinemethylene, D 157
Triphenylphosphite methiodide, D 42
Triphenylphosphoniumcyclopentadienylide, A 122
Triphenylphosphonium ethylidene, D 153
1,2,3-Triphenylpropenone, C 52
1,2,3-Triphenyl-3-vinyl-cyclopropene, A 110
Tripropylaluminium, A 32
Tripropylboron, A 32
1,2,3-Tripropylcyclopropene, A 33
1,2,5-Tris-*tert*-butylbicyclo[2.2.0]-hexa-2,5-diene, C 42
Triscycloheptenobenzene, C 39
Trisdehydro-[18]-annulene, B 398
Tris-(diphenylcyclopropenone) nickel (0)carbonyl, A 49
Trishomocyclopropenyl cations, C 37
Trisporic acid, E 86
Tristrimethylammoniomethyl-cyclopropane hydroxide, A 34
Tris-(triphenylphosphine)chlororhodium catalyst, D 94
Triterpenes, E 93 ff
—, biosynthesis, E 116
—, degraded, E 120
—, *nomenclature*, D *423*
—, pentacyclic biosynthesis, E 116
Triterpenoids, C 369 ff, 406, D 4
—, adianane group, C 456
—, allobetulin group, C 437
—, andirobin group, C 431
—, arborinane group, C 457
—, biosynthesis, C 371, 464; E 94
—, cedrelane group, C 428
—, cedronine group, C 434
—, cucurbitacin group, C 424
—, cycloartane groups, C 423
—, dammarane group, C 425
—, euphane-tirucallane group, C 426
—, fernane group, C 455
—, filicane group, C 456
—, friedelane group, C 447
—, fusidane group, C 420
—, gedunin group, C 429
—, glutinane group, C 446
—, hopane group, C 452
—, lanostane group, C 421
—, lupane group, C 435
—, mexicanolide group, C 431
—, nimbin group, C 432
—, *nomenclature*, D *429*
—, oleanane group, C 438
—, quassin group, C 433
—, synthesis, C 471
—, taraxastane group, C 449
—, tetracyclic biosynthesis, E 94
—, ursane group, C 450
Triterphenols, taraxerane group, C 445
Tri-(trimethylene)benzene, A 165
Tritylcamphor-3-nitronate, C 213
1,3,5-Trivinylbenzene, B 387
Trixis pipitzahuae, C 292
Trollichrome, B 332
Trolliflavin, B 332
Trolliflor, B 332
Trollius europaeus, B 332
Trollixanthin, B 332
Trop-6-en-3-one, B 362
Trophanthidinic acid, D 362
Tropidine, B 354
Tropilidene, C 46
Tropine, B 354

Tropinone, B 362
Tropolones, B 347, 356, **365**, 367, 377
—, biosynthesis, B 366
—, chelate complexes, B 369
—, conversion to tropones, B 362
—, irradiation with ultraviolet light, C 46
—, nomenclature, B 365
—, nuclear substitution, B 371
—, physical constants, B 369
—, rearrangement reactions, B 374
Tropolone, B 359, 362, 365, 368, 369, 370, 374, 375
—, adducts, B 370
—, nuclear substitution products, B 373
α-Tropolone, A 110
Tropolonecarboxylic acids, B 373
Tropolone ethers, B 362, 375, 377
Tropolone methyl ether, B 376; C 46
Tropolone-5-sulphonic acid, B 373
Tropolone-γ-sulphonic acid, B 372
Tropones, A 49; B 374
—, conversion to tropolones, B 368
—, properties, B 363
—, substituted, B 362
—, synthesis, B 361
Tropone, B 350, 355, **359**, 360, **361**, **363**, 364, 375; C 46
—, reaction with maleic anhydride, C 124
Tropone-4-carboxylic acid, B 374
Troponeimes, B 379
Tropone-iron carbonyl complex, B 363
Tropylidene, B 349
Tropylium bromide, B 351, 359, **360**
Tropyliumcyclopentadienylmetal cations, A 144
Tropylium oxide, B 361
Tropylium salts, B 350, 360, 394
Truxillic acids, A 1, 56, 72, 76, 98, 101
—, configurations, A 99
α-Truxillic acid, A 72
α-Truxillic acid *NN'*-diamide, A 101
Truxillic ketone, A 94
Truxinic acids, A 56, 72, 76, 98, 102
—, configurations, A 101
Truxinic ketone, A 94
Truxone, A 101
Truxonic acids, A 100
Tryptamine, D 412
Tryptophan, B 140
Tschugaev reaction, B 17
Tubocurarine, D 359
Tumulosic acid, C 407, 421
Tungsten hexacarbonyl A 130; C 91
Turbincorytin, C 396, 397
Turpentine oil, B 172, 193, 194; C 167, 199, 227, 229, 281
Turreanthin, C 426
Tutin, E 83
Twistane, C **133**
Tyrosine, B 140

Ubiquinone, E 135, 136
Ubiquinone-50, E 136, 137
Ultraviolet absorption chromophores, D 68
Ultraviolet absorption spectrometry, steroids, D 67
Ultraviolet spectroscopy, structure determination, C 479
Umbelliprenin, E 81, 84
Umbellularia californica, C 148
Umbellularic acid, A 61; C 146, 148, 150, 254
Umbellulone, A 61; C **147**, 148, 149, 150, 254, 255; E 69
ψ-Umbellulone, C **150**
Umbellulonic acid, C 148, 150
Undecadeuterocyclohexane, B 7
Undecafluorocyclohexane, B 37, 38, 39
Undecane-1,11-dioic acid, B 417
Undephanthontrioic acid dimethyl ester, D 380
Urane, D 281
Uranediol, D 280
Uranetriol, D 280
Uranolone, D 280
Urezigenin, D 364, 398
Urezin, D 364, 369, **398**
Urguinea burkei, D 412
Urginea maritima, D 411
Ursanes, E 116
Ursane group, triterpenoids, C 414, 450
Ursodeoxycholic acid, D 51, 171, 180, 182, 185
Ursolic acid, C 415, 416, 450
Ursonic acid, C 450
Uscharidin, D **406**
Uscharin, D **406**
Uvaol, C 450
Uzarigenin, D 364, 369, 370, 384, **396**, 397; E 112
Uzarigenin acetate, D 397
Uzarigenin diginoside, D 396
Uzarigenin-[glucoside]$_2$, D 396
Uzarigenin-[glucoside]$_3$, D 396
Uzarigenol, D 364, 398
Uzarin, D 364, 368, 370, 396
Uzaroside, D 364, **396**

Vaccenic acid, A 69
Valdiviolide, C **349**, 367
Valenciaxanthin, B 333
Valeranone, C **309**, 310, 335
Valerenic acid, C **332**, 333, 336
Valerenolic acid, C 332
Valeriana officinalis, C 180, 204, 309, 317 332
Valeriana wallichi, C 266, 310
Valeric acid, A 95; B 74, 98
Valine, E 61
Vallaris solanacea, D 386, 391
Vallarose, D 387
Vallaroside, D **386**
Vallarosolanoside, D **387**, **391**
Vanadium biscyclopentadienyls, A 131
Vanguerolic acid, C 415, 416, 450
Varanic acid, D 193
Veatchine, C 471
Veneniferin, D 386, 406
Veprisone, C 430
Veratranine, D *449*, *450*
Verbanols, physical constants, C 184
Verbanol, C **181** 182, 183
Verbanol methyl xanthate, C 176
Verbanone, C 170, 171, 177, **181** 182, 183, 184
Verbenalin, B **213**; E 72, 75
Verbena officinalis, B 213; E 72
Verbena triphylla, C 181
Verbenol, C **181** 182, 184; E 76
Verbenone, C 181, 182, 184, 185; E 76
Verbenyl acetate, C 183
Verbesina virginica, C 300
Verbesinol, C **300** 334
Verrucarin A, C 333
Verrucarol, C 333, 336
Verticillol, C 374, 381, 402
Veticadinene, C 297
Veticadinol, C **296**, 334
Vetivalene, C **304**
Vetivazulene, C 322, 323
Vetiver zizanoides, C 259, 269, 297
Vetivone, B 347; C **323**, 336
β-Vetivone, E 80
Viboquercitol, B 58
Vilsmeier reagent, D 95
Vinaconic acid, A 58
Vinca alkaloids, B 75
Vinhaticoic acid, C 378, 391
Vindoline, E 75, 76
Vinyl acetate, A 42; C 91, 92, 97, 105, 111
17α-Vinylandrostane-3β,17β-diol diacetate, D 309
Vinylcyclobutane, C **36**
1-Vinylcycloheptanol, B 351
1-Vinylcyclohept-1-ene, B 348
Vinylcyclohexane, B **24**
1-Vinylcyclohexanols, B 28, 66
Vinylcyclohexene, B 21, 24, 28, 71; C 68, 108, 117
3-Vinylcyclohex-2-enone, B 117
Vinylcyclopropanes, A 22, 35, C 8
Vinylcyclopropane, A 26, 35, 110
Vinylcyclopropanecarboxylic acids, A 53
2-Vinylcyclopropanecarboxylic ester, A 53
2-Vinylcyclopropane-1,1-dicarboxylic ester, A 27, 58
2-Vinylcyclopropyl isocyanate, A 53
Vinylene carbonate, B 63, 64
Vinyl ethers, reaction with ketens, A 76
7-Vinyl-7-hydronorbornene, C 122
Vinylidene carbenes, A 34
Vinylidene fluoride, A 75
Vinyl ketones, A 50
7-Vinylnorbornenes, C 121
4-Vinylpyridine, A 76
Vinylsulphides, A 45
17α-Vinyltestosterone, D 311
Viola tricolor, B 300
Violaxanthin, B 237, 238, 253, 294, 295, 299, **300**, 332, 344, 345; E 127
Viperidinone, D 152, 164
Viperidone, D 69, **152** 164
Viridiflorol, C **320** 321, 336
Visamminol, E 64
Visual process, B 289
Visual purple, B 289
Vitamin A, A 51; B 32, 65, 231, 243, 249, 254, 257, **282**; E 86, 132
—, activity, carotenoids, B 244
—, aldehyde, B 251, 284
—, conversion to β-carotene, B 271
—, from carotenoids, B 283
—, manufacture, B 285
—, methyl ester, B 286
—, methyl ether, B 285
—, precursors, B 239
—, structure, B 233, 234
—, synthesis, B 167, 246, 284, 285, 286
all trans-Vitamin A, B 236
Vitamin A_1, B 289, 290, **291**
—, Diels-Alder reaction with maleic anhydride, B 292
—, dimer, B 290
—, esters, B 292

Vitamin A_1, see also Vitamin A, B 282
Vitamin A_2, B **287**, **288**, 290, 291, 346
—, aldehyde, B 289
—, from Vitamin A_1, B 288
trans-Vitamin A_1 acetate, B 292
Vitamin A acid, B 282, 286, 287
—, isomers, B 287
Vitamin A_1 acid, B 287
Vitamin A_2 acid, B 288
Vitamins D, D 69, **143**
Vitamin D_1, D 143
Vitamin D_2, D 12, 138, 139, 140, 142, 143, 144, 145, 146, 147
Vitamin D_3, D 107, **145**, 146, 147
Vitamin E, E 134
Vitamin K_1, E 87, 134, 135, 136
Vitex megapotamica, D 161
Viverra zibetha, B 414
Von Auwers rule, B 44, 225
Voruscharin, D **406**
Vouacapenic acid, C 391
Vulgarobufogenin, D 413, **417**
Vulgarin, C **347**, 367

Wagner-Meerwein rearrangement, C 139, 159, 165, 166, 210, 216, 238, 244, 327, 377, 378; D 61
Warmingone, B **332**
Water fennel oil B 183, 195
Wessely oxidation, B 120
Westphalen rearrangement, D 82
Westphalen's diacetate, D 49
Wheat germ oil, D 154, 156, 165
Widdrene, C 275
Widdrenic acid, C 275
Widdringtonia dracomontana, C 301
Widdringtonia species, C 275, **326**
Widdrol, C **326**, 327, 336
Wieland-Barbier degradation, D 370
Wilcoxia viperiana, D 152
Wild bergamot oil, E 77
Willagenin, E 26, **35**
Winterin, C **349**, 367
Wittig reaction, A 109; B 248, 274; C 15, 73, 117, 286, 321; D 146, 152, 153, 158
Wittig reagents, B 247
Wolff-Kishner reduction, A 31; C 8, 9, 14, 29, 30, 31, 45, 59, 68, 89, 108, 117, 124, 127, 192, 206, 290, 293, 298; D 56, 77, 99, 115, 126, 153, 157, 184, 185, 217, 277, 304, 313, 315, 319, 374, 379, 398, 403, 418; E 30, 33, 34, 35, 36
Wolff rearrangement, B 118
Woodward synthesis, cholesterol, D 101
Wool fat, D 96
Wool wax, D 75, 109
Wormseed oil, B 204
Wormwood oil, C 141; E 86
Wurtz reaction, C 6, 9, 14, 18

Xanthates, pyrolysis, B 18
Xanthogallol, A 152, 184, 188
Xanthogallolic acid, A 188
Xanthoperol, C 392
Xanthophyceae, B 238
Xanthophylls, B 237, 238, 239, 240, 242, 250, 295, 296, 344; E 120, 126
—, biosynthesis, B 344
—, epoxidation, E 127
—, separation from carotenes, B 242
Xanthophyll epoxide, B 296
Xanthophyll esters, B 234, 241
Xanthostemone, B **130**
Xanthostemon opporitofolius, B 130
X-ray crystallography, steroids, D 71
—, structure determination, C 480
Xylenes, hydrogenation, B 15
m-Xylene, B 28
—, from carotenoids, B 258
p-Xylene, C 306
—, pyrolysis, B 30
D-Xylose, E 2, **3**
Xysmalobrium undulatum, E 110

Yamogenin, E 11, 23, 24, 25, 34
Yeast fat, D **150**
Yeast sterols, D **150**
Ylangenes, D **271**, **272**, 288
Ylang-ylang oil, C 258
Yonogenin, E 3, 25, **31**
Yononin, E 3, 5
Yucca filifera, E 26
Yuccagenin, E 25, **33**, 35, 40
Yuccagenin diacetate, E 40
Yucca shottii, E 3, 25, 27
Yucca species, E 25, 27

Zanzibaric acid, C 383
Zeacarotenes, E 126
α-Zeacarotene, B **277**; E 125
β-Zeacarotene, B **277**, 338; E 125
Zeaxanthin, B 237, 238, 249, 259, 260, 294, 295, **297**, 300, 312, 313, 314, 315, 332; E 127, 128, 129, 131
—, oxidation, B 259
—, synthesis, B 298

all trans-Zeaxanthin, B 252
Zeaxanthin diacetate, B **299**
Zeaxanthin diepoxide, B 294, 299, 300
Zeaxanthin dipalmitate, B 306, 315
Zeaxanthin monoepoxide, B 299
Zeinoxanthin, B **331**
Zeorin, C 452
Zerumbone, C **330**, 331, 336
Ziegler's method, cyclanones, B 406
Zierazulene, C 319
Zieria marophylla, C 318
Zierone, C **318**, 319, 336
Zimmermann reaction, D 76, 199, 270
Zinc alkyls, A 94
Zingiberene, C 259, **260**, 261, 288
Zingiber officinale, C 260
Zingiber zerumbet, C 330, 331
Zizanin A and B, C 403, 405; E 92
Zoosterols, D 2, 3, 73
Zymostenol, D 106, 151
Zymostenone, D **130**
Zymosterol, D 2, 68, 74, 106, **107**, 115, 131, **150**, 151, 164; E 104, 108